# Aeroacoustics of Low Mach Number Flows

# Aeroacoustics of Low Mach Number Flows

## Fundamentals, Analysis, and Measurement

*Stewart Glegg and William Devenport*

ACADEMIC PRESS

An imprint of Elsevier
elsevier.com

Academic Press is an imprint of Elsevier
125 London Wall, London EC2Y 5AS, United Kingdom
525 B Street, Suite 1800, San Diego, CA 92101-4495, United States
50 Hampshire Street, 5th Floor, Cambridge, MA 02139, United States
The Boulevard, Langford Lane, Kidlington, Oxford OX5 1GB, United Kingdom

**Notices**
Knowledge and best practice in this field are constantly changing. As new research and
experience broaden our understanding, changes in research methods, professional practices, or
medical treatment may become necessary.

Practitioners and researchers must always rely on their own experience and knowledge in
evaluating and using any information, methods, compounds, or experiments described herein.
In using such information or methods they should be mindful of their own safety and the
safety of others, including parties for whom they have a professional responsibility.

To the fullest extent of the law, neither the Publisher nor the authors, contributors, or
editors, assume any liability for any injury and/or damage to persons or property as a matter
of products liability, negligence or otherwise, or from any use or operation of any methods,
products, instructions, or ideas contained in the material herein.

**Library of Congress Cataloging-in-Publication Data**
A catalog record for this book is available from the Library of Congress

**British Library Cataloguing-in-Publication Data**
A catalogue record for this book is available from the British Library

ISBN 978-0-12-809651-2

For information on all Academic Press publications visit our
website at https://www.elsevier.com/books-and-journals

Working together
to grow libraries in
developing countries

www.elsevier.com • www.bookaid.org

*Publisher:* Joe Hayton
*Acquisition Editor:* Brian Guerin
*Editorial Project Manager:* Edward Payne
*Production Project Manager:* Anusha Sambamoorthy
*Cover Designer:* Greg Harris
*Front cover image credit:* Photo courtesy of Virginia Tech/John McCormick.

Typeset by SPi Global, India

# Dedication

To our wives, Lisa and Anne

# Contents

# Preface

As its name suggests, aeroacoustics is a subject that connects the fields of aerodynamics and acoustics. It is relatively young for a mechanical science and of an age where many of us were taught by or have worked with the pioneers in this field. Part of the appeal of aeroacoustics is in the creativity and generous spirit of cooperation unleashed by this cross-disciplinary collaboration of experts, where extraordinary accomplishments have been made possible by fusion of fields and expertise. Aeroacoustics is a field where applied mathematics, science, and engineering play a pivotal role; a field where experiments in their classical role of testing hypotheses are very much part of current progress. Consider, for example, the sound radiated by an airfoil in turbulence. The science of turbulence is hard enough, and putting an airfoil into it feels likely to be an unmanageable complication. At low Mach number the sound radiated by the interaction is perhaps one millionth part of the pressure field generated. It is hard to imagine a problem that appears more badly posed. Remarkably, it has been rendered not only tractable but, with some simplification, solvable by entirely analytical methods with results that pass near, if not through, experimental measurements. It is very satisfying to be working in a field that regularly produces extraordinary achievements of this type. A number of these achievements are detailed in this book.

Stewart Glegg obtained his BSc degree in engineering science from Southampton University and went on to also obtain his Masters and PhD at Southampton studying under the direction of Mike Fisher at the ISVR. At Southampton he was fortunate to be introduced to the fields of fluid dynamics and acoustics by some of the pioneers in aeroacoustics, notably Geoff Lilley, Chris Morfey, Peter Davies, Phil Doak, and of course Mike. After graduating with his PhD he worked at Westland Helicopters under the direction of Dave Hawkings who introduced him to the field of rotor noise. In 1979 he returned to the ISVR as a faculty member, and in 1985 he moved to Florida Atlantic University in search of a warmer climate and new challenges. It was not long before he was conscripted by Feri Farassat to work on helicopter rotor noise under NASA sponsorship, and since that time he has been actively involved in aeroacoustics in the United States.

William Devenport also graduated in engineering science in the United Kingdom, at the University of Exeter, and went on to do a PhD at the University of Cambridge. His PhD research was in the experimental and computational study of turbulent separated flows under the capable guidance of Peter Sutton. While it was not his focus then, he was fortunate enough to be able to attend lectures on aeroacoustics given by Shôn Ffowcs Williams, to interact with Ann Dowling, and to share an office with some of their graduate students. His experimental work continued when he joined Virginia Tech as a postdoc, under the guidance of Roger Simpson. His interest in aeroacoustics truly began just after he was appointed as assistant professor in 1989. Ever supportive, Roger Simpson introduced him to Tom Brooks at NASA Langley. Tom had already been sponsoring Stewart to do predictions of broadband noise generated by blade wake interactions in helicopter rotors, and had decided that some experiments were needed to define the turbulence structure. Stewart visited Virginia Tech in the Spring of 1990 initiating a collaboration that continues today. Perhaps some of the longevity of our collaboration and interest derives from our different technical backgrounds and areas of expertise.

This book began its life on paper in 2005 when the authors jointly organized a course on aeroacoustics and hydroacoustics taught from Florida Atlantic University to graduate students there and at Virginia Tech. The initial core of the text (written by S.G.) was drafted as notes for that course with the intent of bringing together the fundamentals of aeroacoustics in a form and at a level that would be appropriate to graduate students, and that would give them the background needed to understand most modern papers and developments in the field. Over 11 years later this remains the goal of this book.

We have many to thank for their help in making this book possible. In terms of those who have enabled the long-term study of aeroacoustics of which this book is a part, we first thank the Department of Ocean and Mechanical Engineering at Florida Atlantic University and the Kevin T. Crofton Department of Aerospace and Ocean Engineering at Virginia Tech for providing us with secure and supportive research and teaching environments, over many years. Without their support it would not have been possible to develop the material for this text or to write it up. We also acknowledge the invaluable support and encouragement of our colleagues at these institutions including the members of the Center for Acoustics and Vibration at FAU, and Eric Paterson, Aurélien Borgoltz, and the members of ATFRG and CREATe at VT. We owe a debt of gratitude to the many graduate students we have had the privilege to work with. Some (mentioned below) directly contributed to this book, but all have contributed indirectly through the hard work and inspiration they dedicated to the mutual advance of understanding that underpins the relationship between student and advisor. Our research sponsors have in many ways enabled this book, and we are grateful to all. Among them are Tom Brooks, who along with his colleagues at NASA Mike Marcolini and Casey Burley, found a way to sponsor us continuously for the first 10 years of our collaboration; sponsorship that was fundamental in initiating, building, and linking our research programs. We have been fortunate enough also to receive long-term support from the Office of Naval Research, in particular through the programs in hydroacoustics and turbulence and wakes managed by Ki-Han Kim and Ron Joslin. We are very grateful not only for the sustained funding, but also for the motivation they have provided to move into new and exciting technical areas, and to take on challenges outside our comfort zone. We have also benefited greatly from collaborations and support associated with the Virginia Tech Stability Tunnel and its evolution as an aeroacoustic facility. In particular, we owe debts here to Ricardo Burdisso who played a central role in its transformation and who, along with his company AVEC Inc., have been inspirational partners in its further development and advancement. We also thank Wing Ng, and his company Techsburg Inc., without whom this upgrade would never have been possible. We owe a debt of gratitude to ONR, and particularly Ron Joslin, for his courage to be the first to support this venture when it must have appeared both unconventional and risky, and to ONR and General Electric for their encouragement and support in maturing many of its state of the art capabilities.

In terms of those whose efforts have directly contributed to this book we would first like to thank those who encouraged or inspired the book, in particular, Phil Joseph, Chris Morfey, Bill Blake, and Nigel Peake. We are particularly grateful to Chris who also carefully reviewed and commented on a number of chapters, and whose comments often challenged us to maintain the technical rigor of the book. A number of people provided material for or read and commented on specific sections or components of the book and we thank them all. They include Nathan Alexander, Jason Anderson, Manuj Awasthi, Neehar Balantrapu, Andreas Bergmann, Ken Brown, Ricardo Burdisso, Lou Cattafesta, Ian Clark, Dan Cadel, Alexandra Devenport, Mitchell Devenport, Mike Doty, Marty Gerold, Christopher Hickling, Florence Hutcheson, Remy Johnson, Liselle Joseph, Phil Joseph, Emilia Kawashima, Jon Larssen, Todd Lowe, Lin Ma, Henry Murray, Mike Marcolini, Patricio Ravetta, Michel Roger, David Stephens, Ian Smith, and Hiroki Ura.

Last but not least, above all we thank our spouses Inger Hansen (known to family and friends as Lisa) and Anne Devenport for their constant support of us in our careers, in this endeavor, and for their dedicated efforts in reviewing and correcting large fractions of the book.

**Stewart Glegg**
Florida Atlantic University, Boca Raton, FL, United States

**William Devenport**
Virginia Tech, Blacksburg, VA, United States

# Part 1

# Fundamentals

# Introduction

**1**

## 1.1 Aeroacoustics of low Mach number flows

Sound is a fundamental part of human life. We use it for social interaction, for art, for communication and education, and for understanding each other. Sound is also a byproduct of human activities. Unwanted sound, or noise, can adversely affect our productivity, health, security, and quality of life. Flow noise generated by fans, vehicles, wind turbines, and propulsion systems are major contributors to this unwanted sound. Aeroacoustics is the study of noise generation by air flows, and the way in which aerodynamic systems can be designed to minimize noise.

Much of our motivation for understanding flow generated noise originates from the aircraft industry. Regulations require noise certification for all new aircraft and so new products must meet increasingly high standards for noise emissions. Historically, most key developments can be traced to the advent of the jet engine at the end of World War II and its role in ushering in a new era in commercial air transportation. The original engines were unacceptably loud and it was clear that for the jet engine to be viable the noise had to be reduced. At that time there was no understanding of how sound waves could be generated by a turbulent flow, although theories did exist for sound radiation from vibrating surfaces, and even propellers. In 1952 Sir James Lighthill [1] published his theory of aerodynamic sound and the subject of aeroacoustics was born. This theory, which is known as Lighthill's Acoustic Analogy, provides the basis for our understanding of sound generation by flow. It is an exact re-arrangement of the equations of fluid motion, but has certain limitations, which must be carefully understood if it is to be applied correctly. Lighthill's theory has been frequently challenged, but remains the most important and effective analytical tool for the understanding and reduction of flow noise.

The purpose of this book is to provide an introduction to the basic concepts that describe the sound radiation from low Mach number flows, in particular flows over moving surfaces, which are the most widespread cause of flow noise in engineering systems. This includes fan noise, rotor noise, wind turbine noise, boundary layer noise, airframe noise, and aircraft noise, with the exception of jet and shock associated noise. The basic principles are also applicable to hydroacoustics, the study of flow generated noise in underwater applications. The primary difference between aero and hydroacoustics is that water is an almost incompressible fluid and thus usually involves Mach numbers that are an order of magnitude smaller than in aerodynamic applications.

This book is intended to be an introductory text on low Mach number aeroacoustics for graduate students who do not have a background in acoustics but who are familiar with the mathematical foundations of fluid dynamics and thermodynamics. It is written in four parts. Part 1 introduces the fundamentals, including the basic equations of unsteady fluid flow, a review of fluid dynamics concepts, and an introduction to linear

Aeroacoustics of Low Mach Number Flows. http://dx.doi.org/10.1016/B978-0-12-809651-2.00001-1

acoustics. Lighthill's acoustic analogy is then introduced, leading to Curle's theorem, the Ffowcs Williams and Hawkings equation, the linearized Euler equations, vortex sound, and a discussion of turbulence and turbulent flows. This is followed by Part 2 that details experimental measurement techniques, including the use of microphones and flow measurement devices, signal processing and phased arrays, and wind tunnel testing methods. Part 3 of the text deals with edge and boundary layer noise including leading edge noise, trailing edge noise, and the flow over rough surfaces. Finally, Part 4 considers rotating sources such as fans, propellers, and rotors. It includes chapters on open rotor noise, duct acoustics, and fan noise.

## 1.2　Sound waves and turbulence

Sound waves cause small perturbations in pressure that propagate through a fluid medium. The speed of propagation, $c_o$, depends on the local properties of the fluid, but is typically about 343 m/s in air and 1500 m/s in water. The simplest form of a sound wave is a harmonic wave that has a sinusoidal variation in space with a wavelength $\lambda$ and a frequency $f = c_o/\lambda$, Fig. 1.1. The human ear responds to frequencies between 20 Hz and 20 kHz, so typical wavelengths of interest are from 17 m to 17 mm in air.

**Fig. 1.1** A turbulent flow incident on a structure that radiates sound waves.

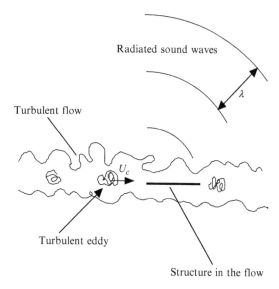

In contrast, turbulence is caused by instabilities in viscous shear flows breaking down into random chaotic motion. Common examples are boundary layers on wings and aircraft fuselages, and the wakes behind moving vehicles. At higher speeds (high Reynolds number) the loss of energy due to turbulent mixing far exceeds the loss that occurs directly from viscous action. Turbulence is sometimes conceptualized as if it consisted of eddies convected at constant velocity through the medium without evolving in the convected frame of reference. This simplification is known as Taylor's hypothesis and the average convection speed $U_c$ is typically taken to be about 60–80% of the free stream velocity. The size of the eddies $L$ is usually the same order of magnitude as the smallest dimension of the mean flow, for example the boundary layer, shear layer, or wake thickness.

When convected turbulence encounters a solid body, it generates rapid changes of pressure on the surface of the body that radiate as sound waves through the fluid. The frequencies of the fluctuations which result from this type of interaction are determined by the eddy size and its convection velocity, and can be estimated as $f \approx U_c/L$. The sound waves generated at this frequency will therefore have a wavelength $\lambda \approx Lc_o/U_c$. In low Mach number applications, the convection speed is always much less than the speed of sound ($U_c \ll c_o$), so the acoustic wavelength is always much larger than the size of the eddies generating the sound. The disparity in scales between the turbulent eddy dimensions and the acoustic wavelength is one of the most important features of aeroacoustics and hydroacoustics and one of the reasons that the subject is so challenging.

## 1.3   Quantifying sound levels and annoyance

The ear responds to the pressure fluctuations of sound waves to cause the sensation of hearing. When carrying out calculations of noise levels and designing for quiet systems it is always important to keep in mind the end goal. This is most often the level of annoyance experienced by a human listener.

The human ear has a remarkable dynamic range and can hear sound waves with amplitudes as low as 20 μPa and as high as 200 Pa before encountering the threshold of pain. The ear's sensitivity is logarithmic and so sound is measured using a decibel scale, referred to as the sound pressure level (SPL). This is given in terms of the root mean square of the fluctuating pressure time history $p_{rms}$ and a reference pressure $p_{ref}$ as

$$\text{SPL} = 20\log_{10}\left(p_{rms}/p_{ref}\right) \tag{1.3.1}$$

where the units are stated as dB(re $p_{ref}$). It is important to specify the reference sound pressure $p_{ref}$ when quoting a result in decibels because different units are used in different applications. For almost all airborne applications the standard is $p_{ref} = 20$ μPa, but results in underwater applications are not as well standardized. The most common reference pressure used in underwater acoustics is $p_{ref} = 1$ μPa, but historical data is also given in other units such as $p_{ref} = 1$ μbar.

The root mean square pressure $p_{rms}$ is the time average of the square of the fluctuating pressure. At any instant the pressure at a point is given as the sum of the mean background pressure $p_o$ and a time varying perturbation $p'(t)$. If $p(t)$ is the pressure at a point in the fluid then the pressure perturbation is $p'(t) = p(t) - p_o$. The root mean square or "rms" pressure is then defined as

$$p_{rms} = \sqrt{\frac{1}{2T}\int_{-T}^{T}(p(t) - p_o)^2 dt} \tag{1.3.2}$$

where the averaging time must be large enough to include many cycles of the lowest frequency contained in the signal.

The human ear also discriminates the frequency of sound in a logarithmic manner. The frequency content of noise is therefore often characterized by summing up the sound into one-third octave bands that appear as equal intervals on a logarithmic scale. In acoustics, an octave is a doubling of the frequency and thus a third-octave band integrates the sound over a band for which the upper limit is $2^{1/3}$ times its lower limit. The mid-band frequencies in Hertz are given by the relation

$$f_n = 1000 \times 2^{(n-30)/3} \qquad\qquad (1.3.3)$$

where $n$ is known as the band number [2]. Thus the band extends from $f_n \times 2^{-1/6}$ to $f_n \times 2^{1/6}$.

The typical response of the ear at different frequencies is plotted in Fig. 1.2 and we see that low and high frequencies are significantly less important than the mid-frequency range around 1 kHz. Note especially how the ear does not respond well to frequencies below 100 Hz, although sometimes low-frequency sounds are identified as being most irritating. To obtain a measure of the perceived loudness of a sound the most commonly used metric is the dB(A) level. This applies a weighting to each one-third octave band level and then each band level is summed on an energy basis. Since the dB(A) level corrects for the sensitivity of the ear, and is easy to measure directly using a sound level meter, it is used extensively in noise control applications. It has also been found to be a good measure of annoyance and so most noise ordinances are defined using this unit with corrections for duration, pure tones, and day/night levels. Other measures of annoyance such as perceived noise level (PNL) and effective perceived noise level (EPNL) are used in assessing aircraft noise.

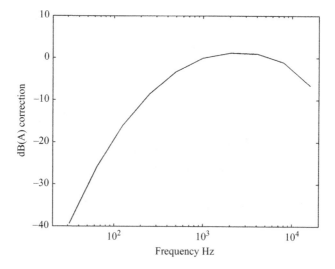

**Fig. 1.2** The dB(A) weighting scale which represents the typical sensitivity of the human ear.

It is important to appreciate that simple physical scaling of sound levels from model scale tests to full scale applications must also incorporate a correction that accounts for annoyance. For example, slowing the rpm of a propeller will not only reduce its source level, but will generate lower frequencies which receive a larger negative A-weighting correction. The dB(A) level will therefore scale quite differently from the SPL, and significant additional noise benefits can be achieved by changing the frequency content of a sound.

In this text we will not attempt to rate noise source levels in subjective units such as dB(A), PNL, or EPNL, but rather the focus will be on the fundamental physics behind the sound generation process. However, it is important to keep in mind that based on the sensitivity of the ear (Fig. 1.2), the frequencies of most concern to humans usually lie between about 350 and 10,000 Hz. This roughly corresponds to wavelengths between 1 m and 3 cm in air.

## 1.4    Symbol and analysis conventions used in this book

While there is no universally agreed upon nomenclature for aeroacoustics, we use symbol conventions that will be familiar to many aeroacousticians. In this way analysis methods or results described in this book will be, as far as possible, recognizable when seen elsewhere in the aeroacoustics literature. We have also strived to be as consistent as possible and avoid using the same symbol for quantities that may be confused. Recognizing that these desires often conflict with that of familiarity, we have included in Appendix A both a list of symbols for basic quantities that appear repeatedly throughout the book, as well as a chapter-by-chapter nomenclature for more specialized symbols.

Our conventions for symbol modifiers are also listed in Appendix A and the most important ones are summarized here. Vector quantities are represented using bold symbols, such as $\mathbf{x}$ for position, or using subscript notation to refer to their Cartesian components, e.g., $x_1$, $x_2$, and $x_3$. Except in special cases, the mean part of a variable will be denoted by subscript "$o$" or an overbar, the latter being used to indicate an averaging or expected value operation. The fluctuating part (such as $p'(t)$ discussed above) will usually be denoted using a prime. A tilde accent, such as $\tilde{p}(\omega)$ is used to denote the Fourier transform of the variable in question with respect to time, whereas a caret accent, such as $\hat{p}$ denotes the complex amplitude of a signal with harmonic time dependence.

Fourier transforms are perhaps the most important mathematical tool of aeroacoustic analysis. Except where otherwise noted we define the Fourier transform of a time history as

$$\tilde{p}(\omega) = \frac{1}{2\pi} \int_{-T}^{T} p'(t) e^{i\omega t} dt \tag{1.4.1}$$

where $T$ tends to infinity, and the inverse Fourier transform as

$$p'(t) = \int_{-\infty}^{\infty} \tilde{p}(\omega) e^{-i\omega t} d\omega \tag{1.4.2}$$

where $\omega$ is angular frequency and we are using the symbol $i$ to represent the square root of $-1$. We define the one-dimensional Fourier transform of a variation over distance as

$$\tilde{\tilde{f}}(k_1) = \frac{1}{2\pi} \int_{-R_\infty}^{R_\infty} f(x_1) e^{-ik_1 x_1} dx_1 \tag{1.4.3}$$

where $R_\infty$ tends to infinity, and the inverse transform as

$$f(x_1) = \int_{-\infty}^{\infty} \tilde{\tilde{f}}(k_1) e^{ik_1 x_1} dk_1 \tag{1.4.4}$$

with two- and three-dimensional forms that are the result of repeated application of Eqs. (1.4.3) or (1.4.4). Here $k_1$ is the wavenumber. Note that in the forward time transform the exponent is positive, and the transformed variable is identified with a tilde, whereas it is negative in the forward spatial transform, and the transformed variable is identified with a double tilde.

It is important to realize that the above Fourier transform definitions are not universal to aeroacoustics analysis since there is no common standard. Our definitions are consistent with the definitions used in the previous aeroacoustics texts by Goldstein [3], Howe [4], and Blake [5]. Other texts on aeroacoustics, such as Dowling and Ffowcs Williams [6], and books focused on related topics, such as signal analysis [7], use different conventions, and it is obviously important to be aware of these distinctions when applying or comparing results.

# References

[1] M.J. Lighthill, On sound generated aerodynamically. I. General theory, Proc. R. Soc. A Math. Phys. Sci. 211 (1952) 564–587.
[2] Acoustical Society of America, Specification for Octave-Band and Fractional-Octave-Band Analog and Digital Filters, ANSI S1.11-2004, Acoustical Society of America, Melville, NY, 2004.
[3] M.E. Goldstein, Aeroacoustics, McGraw-Hill International Book Company, New York, NY, 1976.
[4] M.S. Howe, Acoustics of Fluid–Structure Interactions, Cambridge University Press, Cambridge, 1998.
[5] W.K. Blake, Mechanics of Flow Induced Sound and Vibration, Academic Press, Orlando, FL, 1986.
[6] A.P. Dowling, J.E. Ffowcs Williams, Sound and Sources of Sound, Ellis Horwood, Chichester, 1983.
[7] J.S. Bendat, A.G. Piersol, Random Data: Analysis and Measurement Procedures, fourth ed., Wiley, New York, NY, 2011.

# The equations of fluid motion

**2**

This chapter is a review of the basic equations and concepts of fluid dynamics. These also form the foundations of aeroacoustics. We will start by considering the equations of fluid motion, the thermodynamics of small perturbations, and the role of vorticity. We will then evaluate the rate of change of energy in the fluid, including the energy associated with sound waves in a moving medium. Finally, we will review some basic concepts of fluid dynamics and summarize some results and methods of ideal flow theory that are most relevant to low Mach number aeroacoustics.

## 2.1 Tensor notation

Cartesian tensor notation is useful in aeroacoustics because it provides relatively simple expressions for tensor products. In this section we will give a brief overview of the notation to be used in the following sections.

We are typically concerned with position vectors such as $\mathbf{x}$ and $\mathbf{y}$ which describe the locations of observers and sources, and flow variables, such as the velocity vector $\mathbf{v}$, which defines the velocity of a fluid particle at a fixed location. In Cartesian coordinates these vectors have three components and if we use tensor notation, each component of the vector is defined by a subscript, say $i$, which has the values 1, 2, or 3. Hence we define $\mathbf{x} = (x_1, x_2, x_3)$ giving the three components of the position vector $\mathbf{x}$. To simplify the notation, we replace $\mathbf{x}$ by $x_i$ where the subscript $i = 1, 2, 3$ defines each component. Using this approach, the definition of a dot product between the vectors $\mathbf{q}$ and $\mathbf{v}$ is

$$\mathbf{q} \cdot \mathbf{v} = \sum_{i=1}^{3} q_i v_i = q_1 v_1 + q_2 v_2 + q_3 v_3 \tag{2.1.1}$$

In general, the summation signs in the above definition are found to be cumbersome and so we introduce the convention that whenever repeated indices occur in a product there is an implied summation over all the components. Hence we have

$$q_i v_i \equiv \sum_{i=1}^{3} q_i v_i \tag{2.1.2}$$

For example, we can define the magnitude squared of a vector as

$$|\mathbf{x}|^2 = \mathbf{x} \cdot \mathbf{x} = x_i x_i = x_i^2 = \sum_{i=1}^{3} x_i^2 = x_1^2 + x_2^2 + x_3^2 \tag{2.1.3}$$

Aeroacoustics of Low Mach Number Flows. http://dx.doi.org/10.1016/B978-0-12-809651-2.00002-3

This notation is particularly useful in the definition of the gradient of a scalar $\phi$

$$\nabla \phi = \frac{\partial \phi}{\partial x_i} \equiv \frac{\partial \phi}{\partial x_1} \mathbf{i} + \frac{\partial \phi}{\partial x_2} \mathbf{j} + \frac{\partial \phi}{\partial x_3} \mathbf{k} \qquad (2.1.4)$$

where $\mathbf{i}$, $\mathbf{j}$, and $\mathbf{k}$ are unit vectors in the $i = 1, 2$, and 3 directions. Similarly, the divergence of a velocity $\mathbf{v}$ is

$$\nabla \cdot \mathbf{v} = \frac{\partial v_i}{\partial x_i} = \frac{\partial v_1}{\partial x_1} + \frac{\partial v_2}{\partial x_2} + \frac{\partial v_3}{\partial x_3} \qquad (2.1.5)$$

This approach is most valuable when dealing with tensors. For example, the product $S_{ij} = v_i v_j$ is a tensor which represents a matrix with nine elements corresponding to $i = 1, 2, 3$ and $j = 1, 2, 3$.

$$S_{ij} = v_i v_j = \begin{bmatrix} v_1^2 & v_1 v_2 & v_1 v_3 \\ v_2 v_1 & v_2^2 & v_2 v_3 \\ v_3 v_1 & v_3 v_2 & v_3^2 \end{bmatrix} \qquad (2.1.6)$$

A common expression found in the equations of motion is the velocity gradient tensor $\partial v_i / \partial x_j$, which expands as

$$\frac{\partial v_i}{\partial x_j} = \begin{bmatrix} \dfrac{\partial v_1}{\partial x_1} & \dfrac{\partial v_1}{\partial x_2} & \dfrac{\partial v_1}{\partial x_3} \\ \dfrac{\partial v_2}{\partial x_1} & \dfrac{\partial v_2}{\partial x_2} & \dfrac{\partial v_2}{\partial x_3} \\ \dfrac{\partial v_3}{\partial x_1} & \dfrac{\partial v_3}{\partial x_2} & \dfrac{\partial v_3}{\partial x_3} \end{bmatrix} \qquad (2.1.7)$$

Care needs to be exercised when we consider the expression $S_{ii}$ since this has repeated indices and so, by the rules defined above

$$S_{ii} = \sum_{i=1}^{3} S_{ii} = S_{11} + S_{22} + S_{33} \qquad (2.1.8)$$

Hence if we want to isolate only one of the diagonal terms of the tensor we write $S_{ii}$ (*no summation implied*).

**Example** Consider the tensor defined by the Kronecker delta function $\delta_{ij}$ (which is defined as zero when $i \neq j$ and one when $i = j$) and evaluate (a) $p\delta_{ij}$, (b) $\delta_{kj}\delta_{ik}$, and (c) $S_{ij}\delta_{ij}$.

Using the summation rule we obtain

$$p\delta_{ij} = \begin{bmatrix} p & 0 & 0 \\ 0 & p & 0 \\ 0 & 0 & p \end{bmatrix} \quad \delta_{ik}\delta_{kj} = \sum_{k=1}^{3} \delta_{ik}\delta_{kj} = \begin{bmatrix} 1 & 0 & 0 \\ 0 & 1 & 0 \\ 0 & 0 & 1 \end{bmatrix} \quad S_{ij}\delta_{ij} = \sum_{i=1}^{3}\sum_{j=1}^{3} S_{ij}\delta_{ij} = S_{ii}$$

$$(2.1.9)$$

Throughout this text the velocity vector is denoted as $\mathbf{v}$ or $v_i$ and is often considered as the sum of the mean, time invariant, velocity $\mathbf{U}$ and the velocity fluctuation about the mean $\mathbf{u}$, so $\mathbf{v} = \mathbf{U} + \mathbf{u}$. Coordinates are defined using the vectors $\mathbf{x}$, $\mathbf{y}$, or $\mathbf{z}$ with the coordinates $\mathbf{x}$ and $\mathbf{y}$ used to denote observer and source location, respectively, where relevant. A volume is denoted as $V$ and a surface area by $S$.

The thermodynamic variables of pressure, density, temperature are given their usual symbols $p$, $\rho$, $T_e$, and the internal energy, enthalpy, and entropy are expressed, per unit mass, using the variables $e$, $h$, and $s$, respectively. Stagnation enthalpy and specific total energy are $H$ and $e_T$, respectively. A subscript "$o$" is used to denote the mean, time invariant, values of the thermodynamic variables at a fixed location, whereas a prime is used to indicate the fluctuating part. The mean speed of sound is given by the symbol $c_o$, and if this is constant throughout the fluid the symbol $c_\infty$ is used.

## 2.2  The equation of continuity

The concept of conservation of mass requires that mass is neither created nor destroyed in any fluid element. In Fig. 2.1 we show a region of the fluid, which is of volume $V$, and is bounded by the surface $S$. We will refer to this as a control volume. The outward pointing normal to the surface is $\mathbf{n}^{(o)}$ and the velocity of the fluid is $\mathbf{v}$. If the volume is fixed in space, and the density of the fluid is written as a function of space and time $t$ as $\rho(\mathbf{x},t)$, then the mass of the fluid contained in $V$ is

$$\int_V \rho dV \tag{2.2.1}$$

The flow transports fluid in and out of the control volume and the mass flux out of $V$, across the surface element $dS$, per unit time, is given by

$$\rho \mathbf{v} \cdot \mathbf{n}^{(o)} dS \tag{2.2.2}$$

The rate at which the control volume loses mass must equal the net outward flux of mass, which is given by the integral of Eq. (2.2.2) over the bounding surface $S$. Hence, for a fixed stationary surface

$$\int_S \rho \mathbf{v} \cdot \mathbf{n}^{(o)} dS = -\frac{d}{dt} \int_V \rho dV \quad \text{and} \quad \frac{d}{dt} \int_V \rho dV = \int_V \frac{\partial \rho}{\partial t} dV \tag{2.2.3}$$

We can simplify this equation by using the divergence theorem to turn the surface integral into a volume integral. The divergence theorem states that the integral of the divergence of a vector over a volume is related to the component of the vector normal to the surface enclosing the volume by

$$\int_V \nabla \cdot \mathbf{v} dV = \int_S \mathbf{v} \cdot \mathbf{n}^{(o)} dS$$

Using this relationship gives

$$\int_V \left[ \frac{\partial \rho}{\partial t} + \nabla \cdot (\rho \mathbf{v}) \right] dV = 0 \qquad (2.2.4)$$

This result is independent of the volume $V$ and so the integral can only be identically zero if the integrand is zero. Hence we obtain the continuity equation in differential form as

$$\frac{\partial \rho}{\partial t} + \nabla \cdot (\rho \mathbf{v}) = 0 \qquad (2.2.5)$$

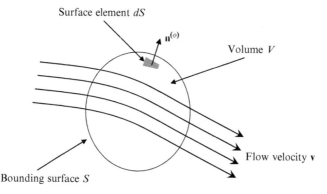

**Fig. 2.1** A control volume of size $V$ bounded by the surface $S$ with fluid moving through the volume at velocity $\mathbf{v}$.

In tensor notation this is given as

$$\frac{\partial \rho}{\partial t} + \frac{\partial (\rho v_i)}{\partial x_i} = 0 \qquad (2.2.6)$$

This is one of the most important equations in fluid dynamics and defines how mass is conserved in a fixed volume. It shows that the rate of change of density with time added to the divergence of the mass flux is zero at a fixed point.

It is also important to consider how mass is conserved in a frame of reference that moves with a differentially small piece of the fluid material, defined as a *fluid particle*. To consider this, we introduce the substantial or material derivative $Df/Dt$, which defines the rate of change of the function $f$ in a frame of reference that moves with the particles rather than in the fixed frame which was used above. To define the moving frame of reference, we specify the position of the fluid particles at time $t_1$ by the vector $\eta_i$. Since the particles move with velocity $v_i$ their location at the time $t > t_1$ will be

$$z_i(t) = \eta_i + \int_{t_1}^{t} v_i dt \tag{2.2.7}$$

The rate of change in the convected frame of reference is the rate of change of $f(z_i(t),t)$ when $\eta_i$ is fixed. Hence

$$\frac{Df}{Dt} = \left[\frac{\partial f}{\partial t}\right]_{\eta_i = const} = \frac{\partial f(z_i(t),t)}{\partial t} \tag{2.2.8}$$

Then by using Eq. (2.2.7) and evaluating the partial derivatives we find the substantial derivative for the location $x_i = z_i(t)$ in fixed coordinates is

$$\frac{Df}{Dt} = \frac{\partial f(x_i,t)}{\partial t} + \frac{\partial z_i(t)}{\partial t}\frac{\partial f(x_i,t)}{\partial x_i} = \frac{\partial f(x_i,t)}{\partial t} + v_i \frac{\partial f(x_i,t)}{\partial x_i} = \frac{\partial f}{\partial t} + \mathbf{v} \cdot \nabla f \tag{2.2.9}$$

The substantial derivative can now be used to rewrite the continuity equation: by using the vector identity $\mathbf{v} \cdot \nabla \rho + \rho \nabla \cdot \mathbf{v} \equiv \nabla \cdot (\rho \mathbf{v})$, we expand Eq. (2.2.5) as,

$$\frac{\partial \rho}{\partial t} + \mathbf{v} \cdot \nabla \rho + \rho \nabla \cdot \mathbf{v} = \frac{D\rho}{Dt} + \rho \nabla \cdot \mathbf{v} = 0 \tag{2.2.10}$$

This shows that if the fluid density is constant in the frame of reference moving with the fluid particles, so $D\rho/Dt = 0$, then the divergence of the flow velocity is zero. In an incompressible fluid such as water, the density is almost constant and so we can approximate the requirement for conservation of mass as $\nabla \cdot \mathbf{v} = 0$. However acoustic waves are, by definition, compressible and so this approximation cannot be used in the analysis of sound.

In the following sections we will consider thermodynamic quantities such as entropy or enthalpy, which are defined per unit mass rather than per unit volume. We will therefore need to consider the volume per unit mass, or specific volume, equivalent to the inverse of the density. We then have

$$\frac{D(1/\rho)}{Dt} = -\frac{1}{\rho^2}\frac{D\rho}{Dt} \tag{2.2.11}$$

and so Eq. (2.2.10) gives

$$\frac{D(1/\rho)}{Dt} = \frac{1}{\rho}\nabla \cdot \mathbf{v} \tag{2.2.12}$$

This is an alternative form of the continuity equation, which is useful in applications that involve compressible flow.

## 2.3  The momentum equation

### 2.3.1  General considerations

To determine the momentum balance in a fluid we recall that the time rate of change of momentum of a body of fluid is equal to the net force exerted on it. We now apply this principle to the control volume illustrated in Fig. 2.1. The net momentum of the fluid in the control volume is given by

$$\int_V \rho \mathbf{v} dV \tag{2.3.1}$$

According to the conservation of momentum, the rate of change of this quantity equals the force $\mathbf{F}$ applied to the control volume less the net rate at which momentum leaves the volume due to the movement of fluid across its surface. We saw above that the mass flow rate of fluid across a single element of the control surface $dS$ is $\rho \mathbf{v} \cdot \mathbf{n}^{(o)} dS$. The rate at which momentum is lost due to this motion is therefore

$$\left( \rho \mathbf{v} \cdot \mathbf{n}^{(o)} dS \right) \mathbf{v} \tag{2.3.2}$$

Hence the rate of change of momentum in the fixed stationary control volume is

$$\frac{d}{dt} \int_V \rho \mathbf{v} dV = \mathbf{F} - \int_S \left( \rho \mathbf{v} \cdot \mathbf{n}^{(o)} \right) \mathbf{v} dS \quad \text{and} \quad \frac{d}{dt} \int_V \rho \mathbf{v} dV = \int_V \frac{\partial (\rho \mathbf{v})}{\partial t} dV \tag{2.3.3}$$

The forces which are applied to the control volume are of three different types: (i) body forces, such as gravity which are almost never important in aeroacoustic applications and so will be ignored, (ii) pressure forces which apply a net force $-p\mathbf{n}^{(o)} dS$ to each surface element shown in Fig. 2.1, and (iii) viscous shear stresses that introduce a net shear force on the surface. Viscous forces are rarely important in sound waves but are often important to the flows that produce them. They are most conveniently expressed using a viscous stress tensor $\sigma_{ij}$ which gives the force per unit area in the $j$ direction applied to a surface whose outward normal lies in the $i$ direction. The stress tensor is symmetric so $\sigma_{ij} = \sigma_{ji}$ and the indices are interchanged without consequence. We can define the viscous shear stress applied to the surface of the control volume as

$$\sigma_{ji} n_j^{(o)} dS = \sigma_{ij} n_j^{(o)} dS \tag{2.3.4}$$

where $n_j^{(o)}$ is the tensor notation for the outward pointing normal to the surface. It is convenient to combine the viscous stress and pressure force into a single tensor $p_{ij}$ defined as

$$p_{ij} = p\delta_{ij} - \sigma_{ij} \tag{2.3.5}$$

This tensor is called the compressive stress tensor and is often used in aeroacoustics to replace the tensor $\tau_{ij} = -p_{ij}$ which is more commonly used in texts on fluid dynamics.

These conventions allow the force on the fluid **F** in the control volume to be written as

$$F_i = -\int_S p_{ij} n_j^{(o)} dS \tag{2.3.6}$$

Combining this with Eq. (2.3.3) yields, in tensor notation,

$$\int_V \frac{\partial(\rho v_i)}{\partial t} dV = -\int_S (\rho v_i v_j + p_{ij}) n_j^{(o)} dS \tag{2.3.7}$$

and applying the divergence theorem then gives

$$\int_V \left[ \frac{\partial(\rho v_i)}{\partial t} + \frac{\partial(\rho v_i v_j + p_{ij})}{\partial x_j} \right] dV = 0 \tag{2.3.8}$$

As with the continuity equation this integral can only be zero if the integrand is zero so we obtain the momentum equation in the absence of body forces as

$$\frac{\partial(\rho v_i)}{\partial t} + \frac{\partial(\rho v_i v_j + p_{ij})}{\partial x_j} = 0 \tag{2.3.9}$$

This is the conservation form of the momentum equation, which shows that the rate of change of momentum of a fixed volume of fluid is balanced by the flux of momentum out of the volume and the stresses applied to its surface.

We can also write the momentum equation in a nonconservation form which relates the forces applied to a fluid particle to its acceleration $Dv_i/Dt$. Newton's Second Law of motion then requires that the mass per unit volume $\rho$ times the acceleration equals the force applied to the particle per unit volume, which from Eq. (2.3.6) and the divergence theorem is $-\partial p_{ij}/\partial x_j$, so

$$\rho \frac{Dv_i}{Dt} + \frac{\partial p_{ij}}{\partial x_j} = 0 \tag{2.3.10}$$

It is a relatively simple exercise to show that Eqs. (2.3.9), (2.3.10) are the same by expanding the derivatives in Eq. (2.3.10) and using the continuity equation (2.2.6). Eq. (2.3.10) is the well-known Navier Stokes equation.

## 2.3.2 Viscous stresses

Viscous stresses are caused by molecular diffusion across the boundary enclosing the control volume. If the molecular diffusion causes fluid molecules to move into a region of fluid with a different velocity, then momentum is transferred and a viscous stress exists. The viscous stress causes the velocity parallel to the boundary, say $v_s$, to be sheared in the direction normal to the boundary, so $n_j(\partial v_s/\partial x_j) \neq 0$ or the velocity normal to the boundary, say $v_n$, to be sheared in the direction parallel to the boundary, so $(\partial v_n/\partial x_j) \neq 0$. Detailed consideration of the viscous shear stress for a compressible fluid is discussed in texts on fluid dynamics [1,2], and requires the definition of a coefficient of bulk viscosity in addition to the shear viscosity $\mu$. The bulk viscosity is usually assumed to be zero (Stokes hypothesis) in which case the viscous stress term is

$$\sigma_{ij} = \mu \left( \frac{\partial v_i}{\partial x_j} + \frac{\partial v_j}{\partial x_i} - \frac{2}{3} \frac{\partial v_k}{\partial x_k} \delta_{ij} \right) \qquad (2.3.11)$$

where $\mu$ is the coefficient of viscosity (for a detailed derivation of this equation see Batchelor [1] or Kundu [2]).

The viscous stress term in the momentum equation can be simplified for an incompressible flow. From Eq. (2.3.10) we note that the contribution of the viscous stresses will depend on $\partial \sigma_{ij}/\partial x_j$ and that for an incompressible flow $\partial v_j/\partial x_j = 0$ so we obtain from Eq. (2.3.11)

$$\frac{\partial \sigma_{ij}}{\partial x_j} = \mu \left( \frac{\partial^2 v_i}{\partial x_j^2} \right) = \mu \nabla^2 v_i \qquad (2.3.12)$$

assuming constant viscosity.

The importance of the viscous term can be assessed by writing the Navier Stokes equations in terms of dimensionless variables. We define the dimensionless velocity, distance, time, density, and pressure $v_i^\# = v_i/U$, $x_i^\# = x_i/L$, $t^\# = Ut/L$, $\rho^\# = \rho/\rho_\infty$, and $p^\# = p/\rho_\infty U^2$ where $U$, $L$, and $\rho_\infty$ are constant reference values with $L$ representing the overall scale of the flow. Substituting these definitions into Eq. (2.3.10), expanded using Eqs. (2.3.5), (2.3.11), gives

$$\frac{\rho_\infty U^2}{L} \rho^\# \frac{Dv_i^\#}{Dt^\#} + \frac{\rho_\infty U^2}{L} \frac{\partial p^\#}{\partial x_i^\#} - \frac{\partial}{\partial x_j^\#} \left[ \frac{\mu U}{L^2} \left( \frac{\partial v_i^\#}{\partial x_j^\#} + \frac{\partial v_j^\#}{\partial x_i^\#} - \frac{2}{3} \frac{\partial v_k^\#}{\partial x_k^\#} \delta_{ij} \right) \right] = 0$$

Dividing throughout by $\rho_\infty U^2/L$ we obtain

$$\rho^\# \frac{Dv_i^\#}{Dt^\#} + \frac{\partial p^\#}{\partial x_i^\#} - \frac{\partial}{\partial x_j^\#} \left( \frac{1}{Re} \left[ \frac{\partial v_i^\#}{\partial x_j^\#} + \frac{\partial v_j^\#}{\partial x_i^\#} - \frac{2}{3} \frac{\partial v_k^\#}{\partial x_k^\#} \delta_{ij} \right] \right) = 0 \quad \text{where } Re \equiv \frac{\rho_\infty UL}{\mu}$$

We see that in normalized form the viscous term is divided by the ratio of the scale of the inertial forces $\rho_\infty U^2/L$ to the scale of the viscous forces $\mu U/L^2$. This ratio is defined as the Reynolds number Re. In high speed and/or large scale flows the Reynolds number is high and the viscous term small, indicating that the effects of viscosity can often be ignored.

## 2.4 Thermodynamic quantities

Acoustic waves are a result of compressible effects in the fluid that cause small per-turbations in the local pressure. It is therefore important that we understand how pres-sure changes are associated with changes in density and temperature. We make the assumption that an acoustic wave is a thermodynamic process which does not involve any exchange of heat or dissipative processes. In this case, the pressure perturbation $p'$ about the mean pressure $p_o$ is directly proportional to the density perturbation $\rho'$ about the mean density $\rho_o$ with the constant of proportionality given by the isentropic bulk modulus $(dp/d\rho)_s$.

$$p' = \rho' c_o^2 \qquad c_o^2 \equiv \left(\frac{\partial p}{\partial \rho}\right)_s \tag{2.4.1}$$

We will show later that $c_o$ is the speed at which sound waves propagate through the medium.

We cannot ignore the role of dissipative processes or heating on the generation of sound or on the flows that generate it. For this reason, we need to discuss the thermo-dynamic properties of gases in some detail and define the role of quantities such as the *enthalpy* and *entropy*, which along with *pressure, density, internal energy, kinetic energy,* and *temperature,* define the "state" of the gas.

The First Law of Thermodynamics requires that the energy of a system can only be changed by the addition of heat or by work done on the system. In this case our "system" is a fluid particle that (by definition) moves with the flow and has constant mass. The change of internal energy per unit mass of the particle $de$ is given by the sum of the heat added per unit mass $\delta q$ and the work done on the system per unit mass, $\delta w$, so

$$de = \delta q + \delta w \tag{2.4.2}$$

Note here that the internal energy represents the state of the particle and is independent of how the energy got there, whereas the heat and the work represent path functions and are dependent on the process taking place.

First consider what happens when the molecules in a particle expand to fill a larger volume. When the particle expands from a volume $V$ to a volume $V+dV$, the work done on the surrounding fluid is given by the force exerted on the surrounding medium times the distance it moves during the expansion. Writing this as a surface integral, the force on each surface element is $pdS$, and the distance it moves is $\Delta x$, giving the total work done *by* the particle as

$$\int_S p\Delta x dS$$

where the surface $S$ encloses the volume $V$. Since the particle is very small the pressure may be considered as constant over the surface and, since

$$dV = \int_S \Delta x dS$$

the work done by the system on its surroundings is $pdV$. This represents a loss of energy to the system and so Eq. (2.4.2) becomes

$$de = \delta q - pdv \tag{2.4.3}$$

In this equation $dv$ represents a change in volume per unit mass and can be related to the density of the fluid in the particle using $v = 1/\rho$. This example considers a volumetric change which occurs at constant pressure. In many cases we need to consider changes in pressure which occur without a change in volume. In this case the particle increases its "capacity to do work." We can define this capacity using a state variable called the *enthalpy* which is related to the internal energy as $h = e + pv$. We note that $dh = de + vdp + pdv$ which allows the first law (2.4.3) to be written for a change at constant volume as

$$dh = \delta q + vdp = \delta q + \frac{dp}{\rho} \tag{2.4.4}$$

If heating takes place at either constant volume or constant pressure, then the temperature $T_e$ of the system is increased. The sensitivity coefficients are defined as specific heats for constant volume and constant pressure as

$$c_v = \left(\frac{\partial q}{\partial T_e}\right)_V \qquad c_p = \left(\frac{\partial q}{\partial T_e}\right)_p \tag{2.4.5}$$

Hence if heating takes place at constant volume then the change in internal energy can be related to the heat input by Eq. (2.4.3). Specifically,

$$\left(\frac{\partial e}{\partial T_e}\right)_V = \left(\frac{\partial q}{\partial T_e}\right)_V = c_v$$

In contrast if heating takes place at constant pressure then we can use Eq. (2.4.4) to give the change in enthalpy as,

$$\left(\frac{\partial h}{\partial T_e}\right)_p = \left(\frac{\partial q}{\partial T_e}\right)_p = c_p$$

If $e$ and $h$ are only functions of temperature (the assumption of a perfect gas), then these expressions reduce to $de = c_v dT_e$ and $dh = c_p dT_e$. Subtracting Eq. (2.4.3) from Eq. (2.4.4) and using these relationships give

$$dh - de = pdv + vdp$$

or

$$d(pv) = (c_p - c_v)dT_e$$

which integrates to the perfect gas law $p = \rho R T_e$ with $R = (c_p - c_v)$.

For an ideal gas the specific heats are constant and allow us to relate the internal energy and the enthalpy to the temperature as

$$e = c_v T_e \qquad h = c_p T_e \tag{2.4.6}$$

To proceed further we introduce the concept of the *entropy* of a system, which describes its state of disorder. Entropy is a variable like temperature or pressure that gives the state of a gas. It is increased by the addition of heat and by any irreversible process, like heat conduction or molecular diffusion. The Second Law of Thermodynamics gives the entropy change in a process as

$$ds = \frac{\delta q_{rev}}{T_e} \tag{2.4.7}$$

where $\delta q_{rev}$ is called the *reversible heat addition*. This is an imaginary quantity—heat cannot be added reversibly as the addition will always be accompanied by some dissipative process that would further increase disorder. So $\delta q_{rev}$ is the amount of heat that would have to be added, in the absence of any accompanying dissipative process, to account for the thermodynamic changes that occur due to irreversible processes and any actual addition of heat. We can use Eq. (2.4.7) and the first law to obtain a relationship for entropy in terms of the other state variables. Imagine a process containing only reversible heat addition. In this case we would have $\delta q = \delta q_{rev}$ and we could substitute $T_e ds$ in Eq. (2.4.7) for $\delta q$ in Eqs. (2.4.3), (2.4.4). This would give

$$\begin{aligned} de &= T_e ds - pd(1/\rho) \\ dh &= T_e ds + dp/\rho \end{aligned} \tag{2.4.8}$$

Notice that these expressions give the entropy change only in terms of the other state variables $p, \rho, T_e, e$, and $h$. We can therefore infer that Eq. (2.4.8) must hold in general (not just for our reversible example), unless entropy defines a property of the gas independent of these other variables. This is not the case. For example, for a thermally perfect gas the Kinetic Theory of Gases tells us that any of the state variables can be expressed as functions of two others.

So, by rearranging these equations using Eq. (2.4.6),

$$\frac{T_e ds - pd(1/\rho)}{c_v} = \frac{T_e ds + dp/\rho}{c_p}$$

then using the perfect gas law, we find

$$ds = c_v \frac{dp}{p} - c_p \frac{d\rho}{\rho} \tag{2.4.9}$$

from which it follows that the specific entropy is

$$s = c_v \ln\left(Cp/\rho^{\gamma}\right)$$

where $C$ is the constant of integration and $\gamma = c_p/c_v$ is the ratio of specific heats. We see then that for $s = const$ we have

$$\frac{p}{\rho^{\gamma}} = const \tag{2.4.10}$$

Differentiating this equation, we obtain the isentropic bulk modulus,

$$\left(\frac{\partial p}{\partial \rho}\right)_s = \frac{\gamma p}{\rho}$$

This expression is of course, equal to the sound speed squared (Eq. 2.4.1). However, for sound waves the pressure fluctuations $p'$ and density fluctuations $\rho'$ are very much less than their mean absolute values, so in this case we can replace the pressure and density with their mean values $p_o$ and $\rho_o$ to give

$$c_o^2 = \frac{\gamma p_o}{\rho_o} \tag{2.4.11}$$

We can also reinterpret Eq. (2.4.9) as an expression for the small changes in a sound wave. For fluid flows where the entropy is the same everywhere (i.e., only isentropic processes occur, and the entropy at all points has the same initial value) the flow is defined as *homentropic*, so $ds = 0$, and

$$0 = c_v \frac{p'}{p_o} - c_p \frac{\rho'}{\rho_o}$$

giving $p' = \rho' c_o^2$, which is the expression given by Eq. (2.4.1).

A more general specification is *isentropic flow*, in which the specific entropy of each moving particle is constant so $Ds/Dt = 0$. This leads to a generalized relationship between pressure and density which is obtained by taking the substantial derivative of Eq. (2.4.8), and using Eq. (2.4.6), to obtain

$$\frac{De}{Dt} = \frac{p}{\rho^2}\frac{D\rho}{Dt} = c_v \frac{DT_e}{Dt}$$
$$\frac{Dh}{Dt} = \frac{1}{\rho}\frac{Dp}{Dt} = c_p \frac{DT_e}{Dt} \tag{2.4.12}$$

We solve these equations to obtain a relationship between pressure and density fluctuations for an isentropic flow as

$$\frac{Dp}{Dt} = \frac{\gamma p}{\rho}\frac{D\rho}{Dt} \tag{2.4.13}$$

where $\gamma p/\rho$ is the local instantaneous sound speed squared, or,

$$\frac{Dp}{Dt} = c_o^2\frac{D\rho}{Dt} \tag{2.4.14}$$

if the flow generated fluctuating pressure and density are small compared to their absolute values, as is usually the case in low Mach number flows. Comparing these results with Eq. (2.4.11) shows that for an isentropic flow the relationship between pressure and density depends on their material derivatives, as distinct from their perturbations from the mean, but the constant of proportionality is the same when the mean pressure and density are used. In the following section we will consider the entropy of the fluid in more detail and show when we can make the assumptions of homentropic and isentropic flow.

## 2.5   The role of vorticity

### 2.5.1   Crocco's equation

Irrotational flows are important in fluid dynamics and to put their features in perspective we need to consider the role of vorticity. We start by rearranging the momentum equation into a form which was originally derived by Crocco. This relates the rate of change of velocity to terms such as the vorticity. To obtain Crocco's equation we expand the momentum Eq. (2.3.10) term by term and divide by $\rho$ so

$$\frac{\partial \mathbf{v}}{\partial t} + \mathbf{v}\cdot\nabla\mathbf{v} + \frac{1}{\rho}\nabla p - \mathbf{e} = 0 \tag{2.5.1}$$

where $\mathbf{e}$ is a vector specifying the viscous force per unit mass defined in tensor notation as $e_i = (1/\rho)\partial\sigma_{ij}/\partial x_j$ and the kinematic viscosity $\nu = \mu/\rho$. We can rearrange this expression by making use of the vector identity

$$\nabla(\mathbf{u}\cdot\mathbf{v}) = (\mathbf{u}\cdot\nabla)\mathbf{v} + (\mathbf{v}\cdot\nabla)\mathbf{u} + \mathbf{u}\times(\nabla\times\mathbf{v}) + \mathbf{v}\times(\nabla\times\mathbf{u}) \tag{2.5.2}$$

with $\mathbf{u} = \mathbf{v}$ to obtain

$$(\mathbf{v}\cdot\nabla)\mathbf{v} = \frac{1}{2}\nabla(\mathbf{v}\cdot\mathbf{v}) - \mathbf{v}\times(\nabla\times\mathbf{v})$$

and so

$$\frac{\partial \mathbf{v}}{\partial t} + \frac{1}{2}\nabla(\mathbf{v}\cdot\mathbf{v}) + (\boldsymbol{\omega}\times\mathbf{v}) + \frac{1}{\rho}\nabla p - \mathbf{e} = 0 \tag{2.5.3}$$

where we have introduced the vorticity, defined as $\boldsymbol{\omega} = \nabla\times\mathbf{v}$. The term $\frac{1}{2}\nabla(\mathbf{v}\cdot\mathbf{v}) = \frac{1}{2}\nabla v_i^2$ is replaced by introducing the stagnation enthalpy, defined as $H = h + \frac{1}{2}v_i^2$. We then rewrite Eq. (2.4.8) in terms of gradients as

$$\nabla h = T_e\nabla s + \frac{1}{\rho}\nabla p$$

This assumes that at some upstream point the fluid properties are initially uniform in space, so that their values at adjacent points can be thought of as being connected by a single thermodynamic process. Substituting from the definition of the stagnation enthalpy we obtain,

$$\nabla H = T_e\nabla s + \frac{1}{\rho}\nabla p + \frac{1}{2}\nabla v_i^2$$

This is substituted into Eq. (2.5.3) to give

$$\frac{\partial \mathbf{v}}{\partial t} + \nabla H - T_e\nabla s + (\boldsymbol{\omega}\times\mathbf{v}) - \mathbf{e} = 0 \tag{2.5.4}$$

This is Crocco's form of the momentum equation, which highlights some of the most important features of inviscid, irrotational flow.

If the flow is irrotational ($\nabla\times\mathbf{v} = \boldsymbol{\omega} = 0$), then the velocity can be expressed as the gradient of a scalar field, $\mathbf{v} = \nabla\phi$ since, by definition the curl of the gradient of any scalar field is zero. We refer to $\phi$ as the *velocity potential*. If the flow is also homentropic ($\nabla s = 0$) and inviscid ($\mathbf{e} = 0$) Eq. (2.5.4) reduces to $\nabla(\partial\phi/\partial t + H) = 0$, from which Bernoulli's equation is obtained by integration to give

$$\frac{\partial \phi}{\partial t} + H = f(t) \tag{2.5.5}$$

where $f(t)$ is constant across all space. This is important because it shows that for an irrotational, inviscid, and steady flow, $H = const \equiv H_o$ throughout the flow. This is the basis for potential flow calculations in fluid dynamics, which provide great insight into simple flows around streamlined bodies. If the flow is unsteady, but the unsteadiness occurs over a limited region of space embedded in an ambient steady flow, then Eq. (2.5.5) applies but $f(t)$ must be a constant equal to the ambient stagnation enthalpy. Thus we write

$$H - H_o = -\partial\phi/\partial t \equiv H'$$

where $H'$ is the unsteady stagnation enthalpy.

## 2.5.2 The vorticity equation

We can now derive an equation for the vorticity from the curl of Eq. (2.5.4) as

$$\frac{\partial \boldsymbol{\omega}}{\partial t} + \nabla \times (\boldsymbol{\omega} \times \mathbf{v}) - \nabla T_e \times \nabla s - \nabla \times \mathbf{e} = 0 \qquad (2.5.6)$$

This may be written in nonconservative form by using the vector identity

$$\nabla \times (\boldsymbol{\omega} \times \mathbf{v}) = \boldsymbol{\omega}(\nabla \cdot \mathbf{v}) - \mathbf{v}(\nabla \cdot \boldsymbol{\omega}) + (\mathbf{v} \cdot \nabla)\boldsymbol{\omega} - (\boldsymbol{\omega} \cdot \nabla)\mathbf{v} \qquad (2.5.7)$$

We note that, since the divergence of any curl field is zero, $\nabla \cdot \boldsymbol{\omega} = 0$. We can then make use of Eq. (2.2.12) to rewrite Eq. (2.5.6) as

$$\frac{\partial \boldsymbol{\omega}}{\partial t} + (\mathbf{v} \cdot \nabla)\boldsymbol{\omega} + \rho \boldsymbol{\omega} \frac{D(1/\rho)}{Dt} - (\boldsymbol{\omega} \cdot \nabla)\mathbf{v} - \nabla T_e \times \nabla s - \nabla \times \mathbf{e} = 0$$

We then use Eq. (2.3.11) and the definition $e_i = (1/\rho)\partial \sigma_{ij}/\partial x_j$ to evaluate $\nabla \times \mathbf{e}$. Since the curl of a gradient is zero we obtain $\nabla \times \mathbf{e} = \nu \nabla^2 \boldsymbol{\omega}$, assuming constant viscosity. Finally, dividing by $\rho$ and combining terms give the compressible vorticity equation

$$\frac{D(\boldsymbol{\omega}/\rho)}{Dt} - \left(\frac{\boldsymbol{\omega}}{\rho} \cdot \nabla\right)\mathbf{v} - \frac{1}{\rho}\nabla T_e \times \nabla s - \frac{\mu}{\rho}\nabla^2 \boldsymbol{\omega} = 0 \qquad (2.5.8)$$

This equation defines the generation and modification of vorticity in any fluid. When viscous and heating effects are absent the evolution of vorticity within the flow is determined by the second term in the equation which represents the distortion of the fluid particle by the velocity gradients. In two-dimensional flow the vorticity vector is normal to the direction of the flow gradient and so this term is zero. We also note that Eq. (2.5.8) is a set of nonlinear homogeneous differential equations for the vorticity vector, whose terms describe the transport, deformation, generation, and diffusion of vorticity. We will discuss the fluid dynamics of vorticity and its implications, in more detail in Section 2.7.

## 2.5.3 The speed of sound in ideal flow

Before concluding this section, it is worthwhile considering the definition of the speed of sound in a little more detail and showing how it depends on the local flow conditions for an ideal gas in steady flow. We have shown that for a homentropic (inviscid), steady, irrotational flow originating in a uniform free stream, the stagnation enthalpy is constant. If we define upstream or reference conditions using the subscript $\infty$ then we use Eq. (2.4.10) and $H_o = const$ to give

$$\left(p_o/\rho_o^{\gamma}\right) = \left(p_\infty/\rho_\infty^{\gamma}\right) \qquad h_o = h_\infty + 1/2\left(U_\infty^2 - U^2\right) \qquad (2.5.9)$$

where $U$ is the mean flow speed and the subscript $o$ represents mean quantities at the downstream location, while the subscript $\infty$ refers to upstream conditions. To obtain the speed of sound we note that $h = e + p/\rho$ and that Eq. (2.4.6) allows us to write $h = \gamma e$. Combining these relationships gives $h = \gamma p/\rho(\gamma - 1)$, and using Eq. (2.4.11) we obtain from the second part of Eq. (2.5.9)

$$c_o^2 = c_\infty^2 + \frac{(\gamma - 1)}{2}\left(U_\infty^2 - U^2\right) \tag{2.5.10}$$

where $c_o$ is the local speed of sound and $c_\infty$ is the speed of sound at the upstream location. Furthermore, using the first equation of Eq. (2.5.9) gives relationships for the local mean pressure and density as

$$\left(\frac{\rho_o}{\rho_\infty}\right)^{\gamma - 1} = \frac{c_o^2}{c_\infty^2} = 1 + \frac{(\gamma - 1)}{2c_\infty^2}\left(U_\infty^2 - U^2\right) \qquad p_o = \frac{\rho_\infty c_\infty^2}{\gamma}\left(\frac{\rho_o}{\rho_\infty}\right)^\gamma \tag{2.5.11}$$

Hence given a set of upstream conditions and a homentropic mean flow, we can determine the variation of speed of sound, mean density, and mean pressure from the local flow velocity alone. We conclude that the speed of sound and the density can be taken as constant in applications where $\left(U_\infty^2 - U^2\right) \ll c_\infty^2$, which is usually the case in low Mach number flows.

## 2.6 Energy and acoustic intensity

### 2.6.1 The energy equation

The equations of state which define the speed of sound propagation have been shown to be dependent on the distribution of the mean thermodynamic properties of the gas and can be simplified if it can be assumed that the flow is either homentropic or isentropic. We therefore need to define an equation that determines the conditions that allow these assumptions to be made. This is achieved by considering the rate of change of energy of a fluid particle as it moves through space. The total energy per unit mass of the particle $e_T$ is given by the sum of its specific internal energy $e$ and its kinetic energy per unit mass $\frac{1}{2}v_i^2$. The rate of change of energy for the fluid particle is then

$$M\frac{De_T}{Dt} \tag{2.6.1}$$

where $M$ is the mass of the particle, and the material derivative is required because we are considering a constant mass of fluid which is moving through the medium. From the First Law of Thermodynamics the energy of the particle only changes if the particle does work on the surrounding fluid or it is the recipient of heat. If the particle occupies a small volume $V(t)$ at time $t$ then the rate of work done by the particle is determined by the stresses on the surface $S(t)$ enclosing the particle, and the velocity of the surface. The rate of work done on the particle by the surrounding fluid is the negative of this,

$$-\int_{S(t)} v_j p_{ij} n_i^{(o)} dS$$

where $n_i^{(o)}$ is the outward pointing unit normal to the surface. If heat flux per unit area through the fluid material is given by the vector field $\mathbf{Q}$, then the flow rate of heat energy into the particle is,

$$-\int_{S(t)} Q_i n_i^{(o)} dS$$

Using the divergence theorem, we can change the surface integrals in the above two equations into volume integrals and write the rate of *gain* of energy of the particle as

$$-\int_{V(t)} \frac{\partial}{\partial x_i}\left(v_j p_{ij} + Q_i\right) dV \tag{2.6.2}$$

We can then equate (2.6.2) to the net *increase* in energy given by Eq. (2.6.1), as

$$M\frac{De_T}{Dt} = -\int_{V(t)} \frac{\partial}{\partial x_i}\left(v_j p_{ij} + Q_i\right) dV \tag{2.6.3}$$

Since the particle is very small we can assume the integrand is constant throughout the volume and define the density as $\rho = M/V(t)$ to obtain the energy equation

$$\rho\frac{De_T}{Dt} + \frac{\partial}{\partial x_i}\left(v_j p_{ij} + Q_i\right) = 0 \tag{2.6.4}$$

This equation relates the rate of change in total energy of a fluid particle to the rate of work done by surface stresses and the rate of heat addition. We have neglected body forces such as gravity in this derivation since these effects are rarely important in acoustics.

We can also derive the rate of change of energy directly from the relationships given in Section 2.4. First we rewrite Eq. (2.4.8) in terms of substantial derivatives,

$$\frac{De}{Dt} = T_e\frac{Ds}{Dt} - p\frac{D}{Dt}\left(\frac{1}{\rho}\right)$$

This is permissible because the rates of change in the thermodynamic variables experienced by a specific fluid particle over time constitute a thermodynamic process, regardless of the particle motion, and viscous action that may cause heating is considered negligible. Incorporating the specific total energy and multiplying by density gives,

$$\rho\frac{De_T}{Dt} = \rho T_e\frac{Ds}{Dt} - \rho p\frac{D}{Dt}\left(\frac{1}{\rho}\right) + \rho\frac{D}{Dt}\left(\frac{v_i^2}{2}\right) \tag{2.6.5}$$

The second term on the right of this equation can be simplified using Eq. (2.2.12) and the third term may be reduced using the momentum equation (2.3.10) so

$$\rho\frac{De_T}{Dt} = \rho T_e\frac{Ds}{Dt} - p\frac{\partial v_i}{\partial x_i} - v_i\frac{\partial p_{ij}}{\partial x_j} \tag{2.6.6}$$

This gives the energy equation in terms of the entropy rather than the heat flux vector, which may be useful in some applications. An important result is obtained if we subtract Eq. (2.6.6) from Eq. (2.6.4) to give

$$\rho T_e\frac{Ds}{Dt} = \sigma_{ij}\frac{\partial v_i}{\partial x_j} - \frac{\partial Q_i}{\partial x_i} \tag{2.6.7}$$

This shows that if we can ignore viscous effects and heat conduction then the flow can be considered as *isentropic*, and relationships such as Eq. (2.4.14) can be used to relate the pressure and density fluctuations in the fluid. This is an important result because direct viscous effects are negligible over large parts of most engineering flows. Similarly, in acoustic waves the momentum flux is almost completely balanced by pressure perturbations and viscous effects on acoustic propagation are found to be very small. Further, in most fluids flows the temperature is uniform or varies only slowly on the scale of wave propagation, so heat conduction effects are not significant. Hence the assumptions required of an isentropic flow apply to most of our applications. Entropy fluctuations are caused by a burst of heat such as a combustion event in the flow, and entropy is a convected quantity in the absence of viscous effects or heat generation. Therefore, if $s = 0$ at an upstream location in an inviscid unheated flow, then it remains zero throughout the flow.

In Eq. (2.6.4) the energy equation is given for a material element, which is convected through the fluid. We can also express this for a stationary point in space by using the continuity equation to provide the expansion

$$\rho\frac{De_T}{Dt} = \frac{D\rho e_T}{Dt} - e_T\frac{D\rho}{Dt} = \frac{D\rho e_T}{Dt} + e_T\rho\nabla\cdot\mathbf{v}$$

Writing in tensor notation we have,

$$\begin{aligned}\rho\frac{De_T}{Dt} &= \frac{\partial\rho e_T}{\partial t} + v_i\frac{\partial\rho e_T}{\partial x_i} + \rho e_T\frac{\partial v_i}{\partial x_i}\\ &= \frac{\partial\rho e_T}{\partial t} + \frac{\partial\rho e_T v_i}{\partial x_i}\end{aligned} \tag{2.6.8}$$

Using Eqs. (2.6.8), (2.6.4) then gives,

$$\frac{\partial(\rho e_T)}{\partial t} + \frac{\partial}{\partial x_i}\left(\rho e_T v_i + v_j p_{ij} + Q_i\right) = 0 \qquad (2.6.9)$$

Further simplification is possible by introducing the stagnation enthalpy which is equal to $H = e_T + p/\rho$, so

$$\frac{\partial}{\partial t}(\rho e_T) + \frac{\partial\left(\rho v_i H - v_j \sigma_{ij} + Q_i\right)}{\partial x_i} = 0 \qquad (2.6.10)$$

This result gives the rate of change of total energy at a point in terms of the flux of enthalpy, the viscous stresses, and the flux of heat. It is valuable because it can be integrated over a fixed control volume of any size, to give the rate of change of total energy of a system. For example, if we consider a jet engine and draw a control volume as a spherical surface of very large diameter enclosing the engine and its exhaust, the volume integral of $\partial(\rho e_T)/\partial t$ inside the control volume gives the net rate that energy is added to the fluid. If energy is being lost from the control volume, then it must be replaced at the same rate by the engine to maintain a steady outflow. The rate at which energy leaves the control volume is equal to the instantaneous power being generated by the engine $W_T$. If the control surface is far enough from the engine, so that viscous effects and heat conduction are zero on the control surface, then Eq. (2.6.10) shows that the total power generated by the engine can be obtained from

$$W_T = \int_V \nabla \cdot (\rho H \mathbf{v}) dV = \int_S \rho H \mathbf{v} \cdot \mathbf{n}^{(o)} dS \qquad (2.6.11)$$

This result is valuable in the experimental evaluation of engineering systems because it allows for the mechanical input of power to the fluid to be determined directly from measurements on a surface surrounding the power source. It is also useful in acoustics because it leads to the concept of sound power generation, and allows us to define acoustic source strength from measurements at great distances from the source of sound.

### 2.6.2 Sound power

As noted above the power generated by a system is an instantaneous quantity, which means that $W_T$ will vary with time. Furthermore Eq. (2.6.11) includes power needed to drive the system as well as the power required to maintain the unsteady flow and the acoustic waves, which may be a small fraction of the total. In acoustic applications we are interested in making this distinction and separating the mechanical power generated by the system from the "power" required to feed the acoustic motion of the fluid particles at large distances from the source. To make this distinction let us consider the average power generated by the system over a period of time. We will define this as the "expected value" of the instantaneous power and split it into contributions from the steady $W_s$ and unsteady part (time varying) $W_a$, so

$$E[W_T] = W_s + W_a \tag{2.6.12}$$

where $E[\,]$ represents an expected value. The steady components of the flow quantities are defined using the subscript $o$ and unsteady parts with primes so $H = H_o + H'$ etc. The mass flux per unit area is then $\rho v = E[\rho v] + (\rho v)'$. The unsteady components will have a zero mean value, but their correlations can be nonzero. Hence terms like $E[(\rho v)' H_o]$ are zero, but terms like $E[(\rho v)' H']$ are nonzero and contribute to the unsteady power budget. As discussed above we choose to split the power budget as

$$W_s = \int_S E[\rho v] H_o \mathbf{n} dS \tag{2.6.13}$$

and

$$W_a = \int_S E[(\rho v) H'] \cdot \mathbf{n} dS = \int_S E\left[(\rho v)' H'\right] \cdot \mathbf{n} dS \tag{2.6.14}$$

The reason for this choice is that in a homentropic potential flow we can show that $W_s$ is zero because, according to Crocco's equation, $H_o$ is constant over space and Eq. (2.6.13) reduces to

$$\begin{aligned} W_s &= H_o \int_S E[\rho v] \mathbf{n} dS = H_o \int_V E[\nabla \cdot (\rho v)] dV \\ &= -H_o \int_V E\left[\frac{\partial \rho}{\partial t}\right] dV = 0 \end{aligned} \tag{2.6.15}$$

using the continuity equation. We conclude that a steady inviscid potential flow does not need to be driven by a power source. In contrast we would expect a real flow, which includes turbulent wakes and viscous effects, to require a source of power to maintain it in a steady state, so $W_s \neq 0$. On this basis we define $W_s$ as the mean power and $W_a$ as the power from unsteady enthalpy, which we will later relate to acoustic processes. The splitting of power in this way allows us to define a procedure for measuring the sound power output $W_a$ of an acoustic system. In general, we define

$$W_a = \int_S \mathbf{I} \cdot \mathbf{n} dS \qquad \mathbf{I} = E\left[(\rho v)' H'\right] \tag{2.6.16}$$

where $\mathbf{I}$ is the acoustic intensity in the region outside of the turbulent flow. Then, if we can measure the acoustic intensity on a surface enclosing the sources of sound, we can determine their sound power output. This has proven to be a useful concept in noise control applications provided we can accurately determine $\mathbf{I}$. For a homentropic flow we use Eq. (2.4.8) to give $H' = p'/\rho_o + \mathbf{U} \cdot \mathbf{u}$ (note that to first order accuracy $v_i^2 = U_i^2 + 2U_i u_i + \cdots$ so the unsteady part of $\frac{1}{2} v_i^2$ is $U_i u_i$) so

$$\mathbf{I} = E[(\rho_o \mathbf{u} + \rho' \mathbf{U})(p'/\rho_o + \mathbf{U} \cdot \mathbf{u})] \tag{2.6.17}$$

We conclude that in order to measure the acoustic intensity in a flow, we need to know the flow speed as well as the acoustic perturbations of density, particle velocity, and pressure. The measurement is much easier in a stationary medium for which $\mathbf{I} = E[p'\mathbf{u}]$ and instrumentation is commercially available to measure this quantity directly. However, some care needs to be exercised because Eq. (2.6.17) is not uniquely related to the acoustic processes in turbulent flow where unsteady pressure and velocity fluctuations can exist that do not propagate as acoustic waves.

## 2.7  Some relevant fluid dynamic concepts and methods

In addition to the governing equations, an appreciation of some of the physics of fluid dynamics and the methods used in its analysis is needed as a basis for aeroacoustics. The purpose of this section is to provide a short review of this topic. Readers who need or desire a more in-depth treatment are referred to dedicated texts in this field such as Karamcheti [3], Howe [4], or Katz and Plotkin [5].

### 2.7.1  Streamlines and vorticity

One of the most commonly used concepts of fluid dynamics is that of a *streamline*. A streamline is simply a line that is everywhere tangent to the velocity vector. It follows that the streamlines are given by the equation $d\mathbf{s} \times \mathbf{v} = 0$ where $d\mathbf{s}$ is change in position along the streamlines. The concept of a streamline is particularly useful in steady flow, since in this case the streamlines are also the paths taken by fluid particles. In unsteady flows where the velocity fluctuations are small compared to the mean, the streamlines of the mean flow can still be taken to represent the overall paths taken by the fluid, as well as the paths along which turbulence is convected.

Consider a curve in space that cuts across a flow as shown in Fig. 2.2. All the streamlines that pass through this curve will form a *stream surface*. By definition there can be no flow through a stream surface as it is everywhere tangent to the flow. Mathematically we can define a family of stream surfaces by representing them as contour surfaces of a scalar function $\psi$, known as the stream function. Charting all the streamlines that pass through a pair of intersecting curves (Fig. 2.2) maps out two intersecting stream surfaces of different families. The intersection of these surfaces must be a streamline and aligned with $\mathbf{v}$, and the perpendiculars of these surfaces (aligned with the gradient of their respective stream functions) must both be perpendicular to $\mathbf{v}$. We therefore have that

$$\alpha\mathbf{v} = \nabla\psi_1 \times \nabla\psi_2 \tag{2.7.1}$$

where $\alpha$ is some scaling factor. Since the divergence of the cross product of two gradient fields is identically zero, we must have that $\nabla\cdot(\alpha\mathbf{v}) = 0$. In steady flow we can choose $\alpha$ to be the flow density since then this condition is satisfied by virtue of conservation of mass, Eq. (2.2.6). When the flow is incompressible $\alpha$ can equally well be taken as 1. Stream functions and streamlines are key quantitative concepts particularly with reference to rapid distortion theory and the drift function, topics to be covered in Chapter 6.

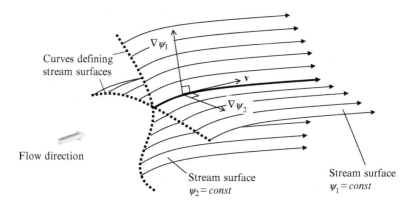

**Fig. 2.2** Relationship between stream surfaces and streamlines.

We can analogously define *vortex lines*, lines everywhere tangent to the vorticity vector that can be thought of as instantaneously stringing together the axes of rotation of adjacent fluid particles (the vorticity is twice the angular velocity of the fluid particles). Being a rotational field, the divergence of the vorticity is identically zero and thus from the divergence theorem we have that the net flux of vorticity out of any closed surface is zero

$$\int_S \boldsymbol{\omega} \cdot \mathbf{n}^{(o)} dS = \int_V \nabla \cdot \boldsymbol{\omega} dV = 0 \tag{2.7.2}$$

because the divergence of a curl operation is zero. This is valid regardless of the flow, the fluid, or the thermodynamic conditions and can be interpreted as saying that all vortex lines entering a volume must exit from it and thus, all vortex lines must continue to infinity or form loops. This is one of Helmholtz' vortex theorems.

Rotation in a fluid on a macroscopic scale is measured in terms of *circulation* $\Gamma$, the integral around a closed loop $C$ of the velocity component along the loop:

$$\Gamma \equiv \oint_C \mathbf{v} \cdot d\mathbf{s} \tag{2.7.3}$$

Circulation and vorticity are related through Stokes' theorem, which states that the circulation around a loop is equal to the net flux of vorticity through any open surface bounded by that loop:

$$\Gamma = \oint_C \mathbf{v} \cdot d\mathbf{s} = \int_S (\nabla \times \mathbf{v}) \cdot \mathbf{n}^{(o)} dS = \int_S \boldsymbol{\omega} \cdot \mathbf{n}^{(o)} dS \tag{2.7.4}$$

where $\mathbf{n}^{(o)}$ is a unit normal vector pointing out of the surface, this direction being given by the right-hand rule applied to the direction of the line integral around loop $C$. This too is a mathematical identity, but one containing subtleties. For example, Eq. (2.7.4) cannot be applied to just any loop because not all loops bound surfaces over which the vorticity is defined. A case in point is steady irrotational flow around a two-dimensional airfoil of infinite span. For any loop that contains the airfoil, no surface can be drawn bounded by that loop that does not pass inside the airfoil where the vorticity is undefined. Thus Stokes' theorem cannot be applied and we can have a circulation around the airfoil even though the flow has no vorticity. In the real flow about an airfoil the vorticity is trapped in the airfoil boundary layer and wake as we will demonstrate below. Note that in the irrotational flow, all loops that pass around the airfoil have the same circulation since two loops of different sizes form the perimeter of an annular surface on which we can apply Stokes' theorem.

The rate of change of circulation around a loop of fluid particles that convects with the flow is $D\Gamma/Dt$, which can be written, starting with Eq. (2.7.3), as:

$$\frac{D\Gamma}{Dt} = \frac{D}{Dt}\oint_C \mathbf{v}\cdot d\mathbf{s} = \oint_C \frac{D\mathbf{v}}{Dt}\cdot d\mathbf{s} + \oint_C \mathbf{v}\cdot\frac{D(d\mathbf{s})}{Dt} \tag{2.7.5}$$

Since $d\mathbf{s}$ represents the distance between adjacent fluid particles in the fluid loop, the rate of change of this seen moving with the flow $D(d\mathbf{s})/Dt$ is merely the difference in velocity between the particles $d\mathbf{v}$. Since $\mathbf{v}\cdot d\mathbf{v} = \frac{1}{2}d\left(v_i^2\right)$ and $v_i^2$ will have the same value at the beginning and end of the loop, the last integral is zero. Substituting the momentum equation (2.5.1) for $D\mathbf{v}/Dt$ we obtain,

$$\frac{D\Gamma}{Dt} = -\oint_C \frac{\nabla p}{\rho}\cdot d\mathbf{s} + \oint_C \mathbf{e}\cdot d\mathbf{s} \tag{2.7.6}$$

This is Kelvin's circulation theorem. The terms on the right-hand side are called the *pressure force torque* and *viscous force torque*, respectively. We can evaluate the pressure torque by applying Stokes' theorem to this line integral,

$$\oint_C \frac{\nabla p}{\rho}\cdot d\mathbf{s} = \int_S \nabla\times\left(\frac{\nabla p}{\rho}\right)\cdot\mathbf{n}dS = -\int_S \left(\frac{1}{\rho^2}\nabla\rho\times\nabla p\right)\cdot\mathbf{n}dS$$

If the fluid is barotropic so that the density is a unique function of pressure (as in isentropic flow), or if the density is constant, then $\nabla\rho\times\nabla p$ is always zero. In these situations, therefore, circulation can only be changed around a fluid loop as a consequence of viscous forces acting on that loop.

This has profound implications. Eq. (2.7.6) applies to a fluid loop, however small, including one wrapped around a single fluid particle. It therefore says that, in the absence of pressure and viscous force torques, the vorticity of a fluid particle will not change and if it is initially zero, then it will remain zero (this is the second of Helmholtz' theorems). Thus vorticity is convected with the flow as if it were part of the fluid material. In flows that begin with an irrotational free stream, vorticity will only appear where pressure or viscous force torques act.

A second implication is known as the *starting vortex*. Consider a stationary two-dimensional airfoil of infinite span in a stationary fluid. We define a fluid loop that encloses the airfoil (Fig. 2.3) and is large enough so as to remain outside any viscous flow generated once the airfoil starts moving. Obviously the circulation around this loop is zero and, according to Kelvin's theorem, will remain zero. The airfoil begins moving, developing a lift and therefore (as we will see below) a circulation. Vorticity is generated by viscous torques acting at the airfoil surface, and some of this is swept into the developing airfoil wake. For the circulation around the loop to remain zero, the circulation around this wake must be equal and opposite to that around the airfoil. We can apply a similar argument to any unsteadiness in the airfoil circulation once it is in motion and thus conclude that all such fluctuations must result in the shedding of vorticity of equal and opposite strength in the wake.

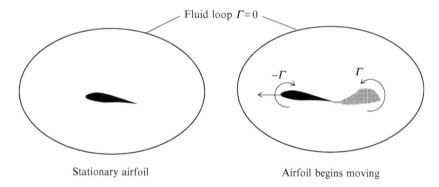

Fig. 2.3 The starting of an airfoil from rest.

In incompressible flow the viscous force per unit mass can be calculated using Eq. (2.3.12) from the Laplacian of the velocity. Expanding this using vector identities gives,

$$\mathbf{e} = \frac{\mu}{\rho}\nabla^2\mathbf{v} = \nu(\nabla(\nabla\cdot\mathbf{v}) - \nabla\times\nabla\times\mathbf{v}) = -\nu\nabla\times\boldsymbol{\omega} \tag{2.7.7}$$

where $\mu/\rho$ is the kinematic viscosity which is taken as a constant. We see that the viscous force is zero if the flow is irrotational. So, if a barotropic flow is initially irrotational then it will remain irrotational because it cannot produce viscous torques in the absence of a boundary, and without viscous torques no new vorticity can be generated. The implication is that vorticity and viscous effects can *only* originate at the flow boundaries and, in practice, this occurs at solid surfaces as a result of the *no slip condition*. To all intents and purposes, this is also true in compressible flows with low Mach number when the viscosity is constant, or nearly so, and where Eq. (2.3.12) is valid to an error of the order of the Mach number squared.

The no slip condition describes the observation that fluid immediately adjacent to a solid surface cannot move relative to it. This causes the formation of a *boundary layer*—a thin layer of fluid over the surface where the speed of flow is slowed by friction in the form of viscous forces. Boundary layers can be laminar but at high Reynolds numbers, or when disturbed, they become turbulent. Separated flows and wakes are always created from boundary layers. Vorticity generated by viscous action thus populates all these regions. When, as is usually the case, these flows are turbulent, much of the turbulence energy is contained in large organized eddies, also termed *coherent structures* or *vortices*. A classic example is the nearly regular train of eddies that can be shed into the wake of a bluff body, known as a *vortex street*. In most other flows, coherent structures are less organized than this, but are no less important in determining the dynamics of the flow and the sound it produces.

In aeroacoustics, turbulent motions are often referred to as *gusts*, commonly with the adjectives "turbulent" or "vortical." The linearity of most aeroacoustics problems means that it makes sense to decompose these motions into sinusoidal components that are then considered separately, hence the mathematically useful but physically improbable concept of a sinusoidal gust. Given Helmholtz' vortex theorems, vorticity, coherent structures, and gusts are often conceptualized as being convected by the local time-averaged flow, implying, whether true or not, that these are small disturbances and that viscous force torques act on a timescale large compared to the others controlling the flow. This latter assumption requires that the ratio of the scale of the inertial forces to that of viscous forces is large, which implies high Reynolds number flow.

### 2.7.2  Ideal flow

In general, aerodynamic surfaces are designed to operate with low drag and thus with boundary layers that remain thin compared to the overall scale of the surface. Therefore, aerodynamic performance with the exception of drag can often be modeled by ignoring the boundary layer altogether and assuming irrotational flow. If the flow is of low Mach number, homentropic, and can be considered incompressible, then the governing equations become particularly simple and we refer to the resulting motion as *ideal flow*.

The governing equations of ideal flow are obtained by first considering the condition of irrotationality $\nabla \times \mathbf{v} = 0$, which is identically solved by expressing the velocity as the gradient of a scalar function $\phi$, called the *velocity potential* (hence our previous references to potential flow in this chapter). Note that the absolute value of the velocity potential has no meaning since it is important only in that its gradient gives the velocity field.

Under the conditions of ideal flow, the momentum equation, in the form of Eq. (2.5.5), reduces to Bernoulli's equation for unsteady potential flow,

$$\frac{\partial \phi}{\partial t} + \frac{1}{2}v_i^2 + \frac{p}{\rho} = const \tag{2.7.8}$$

where the constant can be a function of time. The continuity equation (2.2.10) becomes simply $\nabla \cdot \mathbf{v} = 0$, or

$$\nabla^2 \phi = 0 \tag{2.7.9}$$

This is, of course, Laplace's equation. The challenge of ideal flow is to find solutions to Laplace's equation that satisfy the boundary conditions of the problem at hand. Since, the continuity equation is only first order in terms of velocity we can only satisfy one boundary condition and thus we choose to satisfy the nonpenetration condition $\mathbf{v} \cdot \mathbf{n} = \partial \phi / \partial n = 0$ on rigid surfaces, where $\mathbf{n}$ is a unit vector normal to the surface. The no slip condition is ignored since in any case our assumptions preclude modeling of the vortical boundary layer that it generates.

Many important aerodynamic results are obtained by considering the case of ideal flow in two dimensions. In this case we make use of the fact that any analytic function of a complex variable is a solution to Laplace's equation. As the independent variable we choose a complex coordinate $z = x_1 + ix_2$ to define positions in the flow. As dependent variables we define the complex potential $w(z)$ and complex velocity $w'(z)$, where

$$w(z) = \phi + i\psi \quad \text{and} \quad w'(z) = v_1 - iv_2 \tag{2.7.10}$$

The streamfunction $\psi$ used to complete the complex potential is obtained from Eq. (2.7.1) with $\alpha = 1$, $\psi = \psi_1$, and $\psi_2 = x_3$, the latter representing the stream surfaces coincident with the $x_1$-$x_2$ planes in which the two-dimensional flow takes place. With these values we obtain $v_1 = \partial \psi / \partial x_2$ and $v_2 = -\partial \psi / \partial x_1$ (compared to $v_1 = \partial \phi / \partial x_1$ and $v_2 = \partial \phi / \partial x_2$). Note that the complex velocity is defined with its imaginary part equal to $-v_2$ since in this case we have,

$$w' = \frac{dw}{dz} = \frac{\partial \phi}{\partial x_1} + i\frac{\partial \psi}{\partial x_1} = \frac{1}{i}\frac{\partial \phi}{\partial x_2} + \frac{\partial \psi}{\partial x_2} = v_1 - iv_2 \tag{2.7.11}$$

Since Laplace's equation is linear, the solution to complicated flow problems can be obtained through the superposition of simple solutions expressed as functions of the complex variable $z = x_1 + ix_2$. To match the nonpenetration boundary condition of a solid surface the stream function, or the imaginary part of $w(z)$, must be chosen to be a constant on that surface, and the solutions to many problems can be obtained by superimposing simple flows to meet this criterion. The complex velocity of some simple flows representing a point source, a vortex, and a doublet, are

$$w'(z) = \frac{q}{2\pi(z - z_1)} \qquad w'(z) = \frac{-i\Gamma}{2\pi(z - z_1)} \qquad w'(z) = \frac{Ae^{i\beta}}{2\pi(z - z_1)^2} \tag{2.7.12}$$
$$\quad\text{Source} \qquad\qquad\quad \text{Vortex} \qquad\qquad\quad \text{Doublet}$$

where $q$, $\Gamma$, and $A$ are real constants that denote the strength of these flows, $z_1$ denotes their positions, and angle $\beta$ denotes the orientation of the doublet. The free stream contribution to the flow is simply a constant $w'(z) = U_\infty \exp(-i\alpha)$ where $\alpha$ is the free stream angle. The flows of Eq. (2.7.12) all have singularities of infinite velocity at the point $z = z_1$ where they are not analytic. Everywhere else, however, these are valid.

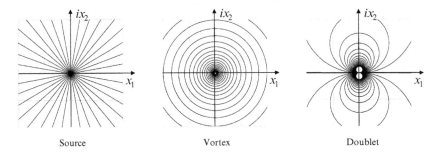

| Source | Vortex | Doublet |

**Fig. 2.4** The elementary ideal flows of Eq. (2.7.12) with $z_1 = 0$ and $\beta = 0$. For positive strength the flow is outward from the source, counter-clockwise around the vortex and exits the doublet in the $x_1$ direction.

Fig. 2.4 illustrates the flows. The source is simply flow away from the point $z = z_1$, the magnitude of the radial velocity decaying with the inverse of the radius. Despite appearances, this flow (as it must) satisfies conservation of mass everywhere except the singularity at its center, where fluid is produced at a volumetric flow rate of $q$ per unit span. By convention a source with a negative strength is referred to as a sink. The vortex produces a flow that orbits the singularity with a tangential velocity that decays inversely with radius. This is an entirely irrotational flow outside the singularity, but one that has a circulation $\Gamma$ around any loop that encloses the singularity. Consistency with Stokes' theorem is maintained in one of two ways. We can argue that Stokes' theorem doesn't apply since any surface bounded by a loop containing the singularity must pass through the singularity and therefore out of the domain where the velocity is defined. Alternatively, we can think of the point vortex as an idealization of an eddy where all the vorticity has been concentrated into a point at its center with a magnitude $\Gamma$. The vorticity is then given as $\Gamma\delta(x_1)\delta(x_2)$ and integration of the delta functions in Stokes' theorem yields the correct circulation. The doublet flow can be likened to a source and sink placed very close together. Flow is produced on one side of the doublet and is then reabsorbed on the opposite side. The orientation of the doublet, defined by the direction of flow along the single straight streamline that passes through its singularity, is set by the parameter $\beta$.

With equal validity we can think of the flow fields generated by adding these singularities as representations of steady flows or as snapshots of unsteady flows. For example, consider the flow generated by a point vortex of strength $\Gamma$ placed at a height $h$ above a plane solid surface, i.e., at $z_1 = ih$ (Fig. 2.5). To satisfy the nonpenetration condition imposed at the surface the circular vortex flow must be modified. This modification is determined using the *method of images*, which says that the influence of the wall on a singularity is identical to the effect of adding the mirror image of that singularity in the wall. In this case, where the wall is coincident with the $x_1$ axis, the image will be a counter-rotating vortex of strength $-\Gamma$ at the location $z_1 = -ih$. Essentially the addition of the mirror image flow cancels out the wall-normal velocity component at the surface, making it a streamline of the flow. The complex velocity and complex potential of this flow, illustrated in Fig. 2.5B are thus,

$$w'(z) = -\frac{i\Gamma}{2\pi(z-ih)} + \frac{i\Gamma}{2\pi(z+ih)} \tag{2.7.13}$$

$$w(z) = -\frac{i\Gamma}{2\pi}\ln(z-ih) + \frac{i\Gamma}{2\pi}\ln(z+ih) \tag{2.7.14}$$

where the complex potential is obtained by integration of the complex velocity according to Eq. (2.7.11). In these equations the second term represents the flow induced by the presence of the surface. This becomes an unsteady problem if we want the vortex singularity to represent the behavior and influence of a real flow feature, i.e., an eddy. The eddy will move as it is convected by the rest of the flow, and thus Fig. 2.5B represents only one instant of the flow history. To estimate the convection velocity in the ideal flow model we only need determine the velocity of the rest of the flow at the singularity, where "rest of the flow" includes the image representing the influence of the surface. The convection velocity $w'_c$ is therefore determined by evaluating Eq. (2.7.14) at $z = ih$, ignoring the first term since that represents the vortex itself, to give $w'_c = \Gamma/4\pi h$. Noting that this is entirely real, we conclude that the vortex convects itself parallel to the surface. The ability of a point vortex to represent actual eddies in an idealized way makes it particularly useful in aeroacoustics.

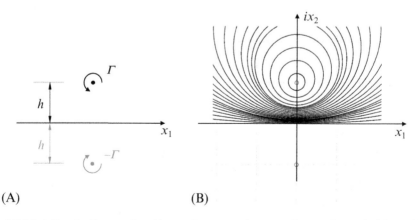

(A)                                                   (B)

**Fig. 2.5** Modeling the flow produced by a point vortex close to a plane surface coincident with the $x_1$ axis using the method of images. (A) Placement of the image vortex and (B) resulting flow.

As a second example, consider the steady flow past a circular cylinder. In its simplest form this can be simulated by placing an opposing doublet in a free stream, giving a complex velocity field of

$$w'(z) = U_\infty e^{-i\alpha} - \frac{Ae^{i\alpha}}{2\pi(z-z_1)^2} \tag{2.7.15}$$

It is left as an exercise for the reader to show that this flow, shown in Fig. 2.6A, contains a circular streamline centered at $z_1$ of radius $R = \sqrt{A/2\pi U_\infty}$. Adding a point vortex at $z_1$ does not change the circular streamline since the vortex velocity field is entirely tangential, but alters the flow to represent circulation around the cylinder (Fig. 2.6B). The complex velocity field, expressed in terms of the cylinder radius $R$ is then

$$w'(z) = U_\infty e^{-i\alpha} - \frac{U_\infty R^2 e^{i\alpha}}{(z-z_1)^2} - \frac{i\Gamma}{2\pi(z-z_1)} \tag{2.7.16}$$

Of course, neither of the ideal flows shown in Fig. 2.5 are very useful as representations of the actual flow around a cylinder, which is dominated by the shedding of a thick rotational wake. However, as will be discussed below, the circular cylinder with circulation is useful as a starting point for airfoil analysis where ideal flow does provide realistic solutions. Note that in this example we have considered the singularities to be held in place at $z_1$ and therefore representing a steady flow about a fixed cylinder.

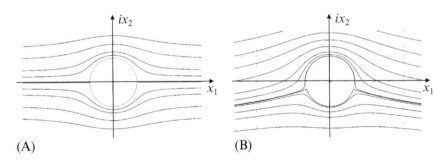

**Fig. 2.6** Ideal flow past a circular cylinder obtained from Eqs. (2.7.15), (2.7.16) with $z_1 = \alpha = 0$. (A) $\Gamma = 0$ and (B) $\Gamma < 0$.

A third example, with which we introduce the *Milne Thompson circle theorem*, combines the last two giving us the flow past a circular cylinder in the presence of a vortex. The Milne Thomson circle theorem enables us to introduce a circle of radius $R$ centered at the origin, into any ideal flow $w(z)$ as long as that flow contains no other rigid boundaries. With the circle, the complex potential becomes

$$w_1(z) = w(z) + w^*(R^2/z) \tag{2.7.17}$$

where $w^*()$ is the complex conjugate of function $w()$, obtained by conjugating all the constants in $w()$ (whether they are additive or multiplicative). This is easily proven by substituting the coordinates on the circle $z = R\exp(i\theta)$ where $\theta$ is angle measured from the real axis about the origin. We then obtain,

$$w_1(z) = w(z) + w^*(R\exp(-i\theta)) = w(z) + w^*(z^*) = w(z) + [w(z)]^* \qquad (2.7.18)$$

Thus the complex potential $w_1$ is entirely real on the surface of the cylinder, implying that the streamfunction is constant, and thus the cylinder is a streamline. To proceed with our example, consider the flow generated by an isolated point vortex located at $z_1$. From Eq. (2.7.12) we will have that

$$w(z) = \frac{-i\Gamma}{2\pi} \ln(z - z_1) \qquad (2.7.19)$$

Applying the Milne Thompson theorem to add the cylinder to this flow we obtain

$$w_1(z) = \frac{-i\Gamma}{2\pi} \ln(z - z_1) + \frac{i\Gamma}{2\pi} \ln\left(\frac{R^2}{z} - z_1^*\right) \qquad (2.7.20)$$

This flow field, shown in Fig. 2.7 can be understood by decomposing the second term in Eq. (2.7.20) which, by analogy with our plane wall example and Eq. (2.7.14), is referred to as *the image of the vortex in the circle*. Specifically, we can write

$$\frac{i\Gamma}{2\pi} \ln\left(\frac{R^2}{z} - z_1^*\right) = \frac{i\Gamma}{2\pi} \ln\left[\left(-\frac{z_1^*}{z}\right)\left(z - \frac{R^2}{z_1^*}\right)\right]$$
$$= \frac{i\Gamma}{2\pi} \ln\left(z - \frac{R^2}{z_1^*}\right) - \frac{i\Gamma}{2\pi} \ln(z) + const \qquad (2.7.21)$$

where the constant can be ignored since it has no impact on the velocity field. We see that the image consists of a point vortex at the cylinder center and another at $R^2/z_1^*$ of equal and opposite strength to the original vortex, respectively. This vortex configuration is shown in Fig. 2.7 which shows that the position $R^2/z_1^*$, referred to as the *inverse point*, lies inside the circle on the line joining the center of the circle to the position of the original vortex. Note the strength of image vortex at the center of the circle can be adjusted or set to zero so the net circulation around the cylinder is modified. This doesn't alter the circular streamline representing the cylinder since the vortex only generates tangential velocities.

As with the plane wall example, we can compute the convection velocity of the vortex by calculating the velocity produced by the rest of the flow at $z_1$. This involves differentiating Eq. (2.7.21) and substituting $z_1$ for $z$. Expressing the result in polar velocity components, which can be calculated as $v_r - iv_\theta = w_c' \exp(i\theta)$, we find that the vortex has no radial convection and moves tangentially with a velocity

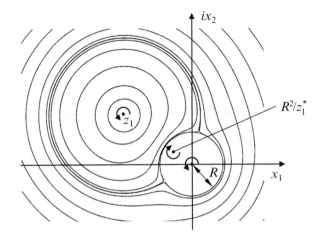

**Fig. 2.7** Ideal flow produced by a point vortex in the vicinity of a circular cylinder obtained from Eq. (2.7.20) with no adjustment to the strength of the image at the cylinder center.

$$v_\theta = -\frac{\Gamma}{2\pi r_1}\left(\frac{R^2}{r_1^2 - R^2}\right) \qquad (2.7.22)$$

where $r_1 = |z_1|$.

Returning to the uniform flow past a circular cylinder with circulation, Fig. 2.6B, it is clear that the flow passing through the cylinder is being deflected downward. To sustain this there must be an upward force on the cylinder, i.e., a lift. Lift is by definition a force perpendicular to the free stream. It is given by the *Kutta Joukowski theorem* which relates the lift force on any two-dimensional body in an otherwise undisturbed uniform flow to the circulation around it as,

$$\text{Lift per unit span} = -\rho U_\infty \Gamma \qquad (2.7.23)$$

which equals to $-F_2/b$ in terms of the span $b$ and the force $F_2$ acting on the fluid (the convention for forces in aeroacoustic analysis). Note that, the caveat "otherwise undisturbed" is important—any other disturbance to the flow around the body, say from a vortex, or a nearby surface or another body, invalidates this result. The Kutta Joukowski theorem can be derived from first principles or as a special case of the *Blasius theorem*. This gives the steady force per unit span integrated over any closed contour $C$ (not just a body surface) in any ideal flow in terms of its components in the $x_1$ and $x_2$ directions as

$$F_1 - iF_2 = -\frac{i\rho b}{2}\oint_C w(z)^2 dz \qquad (2.7.24)$$

Note that the sign of this expression has been chosen so that it gives the force on the fluid exterior to the contour rather than on the body or region inside. The force is therefore given by the residues of the function $w'(z)^2$ at singularities within $C$. For example,

Blasius theorem gives the force components on the vortex adjacent to the wall in Fig. 2.5 as $F_1 = 0$ and $F_2 = -\rho \Gamma^2 b/4\pi h$ if use is made of the identity that

$$\frac{1}{2\pi i}\oint_C \frac{dz}{z^n} = \begin{cases} 1 & n=1 \\ 0 & n>1 \end{cases}$$

### 2.7.3  Conformal mapping

The complex representation of ideal flow permits flow solutions to be modified by conformal mapping. Specifically, any complex function $z = z(\zeta)$ can be used to map one coordinate pair, say, $\zeta = \xi_1 + i\xi_2$ to another $z = x_1 + ix_2$. This enables us to take a simple flow solution in terms of $\zeta$, constructed using the methods described above, and transform it into a more sophisticated flow, in terms of $z$. Specifically, we can map a complex potential constructed in the $\zeta$ plane, $W(\zeta)$, to a new complex potential, $w(z)$, in the $z$ plane by writing

$$w(z) = W(\zeta(z)) \tag{2.7.25}$$

This transfers the value of the complex potential at the point $\zeta$ to the point $z$, and thus the streamlines of the flow are deformed in the same way that the mapping deforms the space. Since both $w$ and $W$ are functions of a complex variable, they are both solutions to Laplace's equation wherever these functions are analytic. We can get the complex velocity in the mapped domain $w'(z)$ simply by differentiating

$$w'(z) = \frac{dw}{dz} = \frac{dW}{d\zeta}\frac{d\zeta}{dz} = W'(\zeta)\frac{d\zeta}{dz} \tag{2.7.26}$$

At points where the derivative of the mapping function $dz/d\zeta = 0$, singularities can appear in the mapped flow that were not in the original flow. These are known as *critical points*. Everywhere else the mapping is referred to as *conformal*. Angles of intersection are preserved under conformal mapping, but not at critical points. Thus critical points are very useful for creating flow past a geometry with a sharp corner from one that is smooth (such as the circular cylinder). A good example and perhaps the most important example of mapping for aeroacoustics, is the *Joukowski* mapping. This is given by the function $z = \zeta + C^2/\zeta$, where $C$ is a real constant. This has critical points on the real axis at $\zeta = \pm C$ corresponding to $z = \pm 2C$. The Joukowski mapping transforms the space outside a circle of radius $C$ centered on $\zeta = 0$ in the $\zeta$ plane to the whole space by, effectively, flattening the circle on to a strip of length $4C$ on the real axis (Table 2.1). Angles of intersection are preserved everywhere except at the critical points at the end of the strip where they are doubled, for example from the 180-degree angle on the exterior of the circle to 360 degrees at the end points of the strip.

Applied to the flow past a circular cylinder placed at the origin, the Joukowski mapping can be used to produce the flow past a flat plate. This is, of course, the most elemental representation of an airfoil. For a flat plate of chordlength $c = 2a$ we choose the mapping constant and the radius of the cylinder to both be $a/2$. The flow past the cylinder is thus, from Eq. (2.7.16),

$$W'(\zeta) = U_\infty e^{-i\alpha} - \frac{U_\infty a^2 e^{i\alpha}}{4\zeta^2} - \frac{i\Gamma}{2\pi\zeta} \tag{2.7.27}$$

And thus by integration we have

$$W(\zeta) = U_\infty \zeta e^{-i\alpha} + \frac{U_\infty a^2 e^{i\alpha}}{4\zeta} - \frac{i\Gamma}{2\pi} \ln(\zeta) \tag{2.7.28}$$

The inverse of the mapping function to be substituted into the complex potential is

$$\zeta = \frac{1}{2}\left(z + \sqrt{z^2 - a^2}\right) \tag{2.7.29}$$

where the branch cut of the square root is chosen to lie along the axis between the critical points at $z = \pm 1$. The result after simplification is,

$$w(z) = U_\infty z \cos\alpha - iU_\infty\left(\sqrt{z^2 - a^2}\right)\sin\alpha - \frac{i\Gamma}{2\pi}\ln\left(\frac{z + \sqrt{z^2 - a^2}}{2}\right) \tag{2.7.30}$$

This flow field is most easily plotted by evaluating the position of streamlines in flow past the cylinder and then transforming those positions using the mapping function. The result, shown in Fig. 2.8A, for an angle of attack $\alpha$ of 5 degrees and a circulation around the cylinder $\Gamma/aU_\infty$ of $-1$, is mathematically consistent but physically unsettling. The flow is seen to pass around the sharp trailing edge of the flat plate, a behavior never observed in practice. The boundary layer present in real airfoil flows will always force the flow to detach smoothly from a sharp trailing edge, a constraint known as the *Kutta condition*. To satisfy the Kutta condition in an ideal flow the circulation around the cylinder must be chosen such that the rearward stagnation point (where the flow detaches from the cylinder) sits at the right-hand critical point that ends up forming the trailing edge of the plate. This requires that

$$\Gamma = -2\pi a U_\infty \sin\alpha \tag{2.7.31}$$

Fig. 2.8B shows the flow in the $z$ plane when the circulation is fixed at this value. Note that it can easily be shown that the circulation is unchanged by mapping. This means that the circulation around the cylinder in Eq. (2.7.31) is also the circulation around the plate, and thus the lift force per unit span on the plate is

$$-F_2/b = 2\pi a\rho U_\infty^2 \sin(\alpha) \tag{2.7.32}$$

The lift coefficient $C_l = -F_2/\frac{1}{2}\rho U_\infty^2 cb$ where $c$ is the plate chord $2a$, is thus equal to $2\pi \sin(\alpha)$, or $2\pi\alpha$ for small angles.

## Table 2.1 Conformal mappings showing their effects on space

| Mapping function | ζ plane | z plane |
|---|---|---|
| Joukowski $z = \zeta + \dfrac{C^2}{\zeta}$ | | |
| Half-plane $z = \zeta^2$ | | |
| Logarithm $z = \dfrac{h}{\pi}\ln\zeta$ | | |

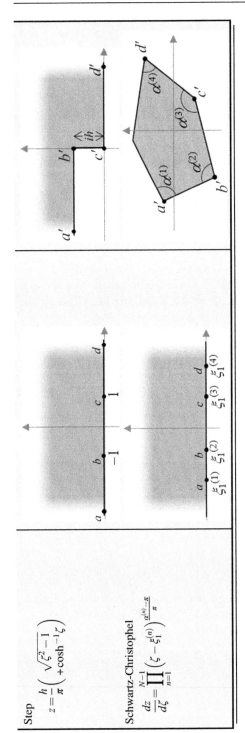

Points $a'$ through $e'$ in the mapped ($z$) plane correspond to points $a$ through $e$ in the $\zeta$ plane.

Step

$$z = \frac{h}{\pi}\left( \sqrt{\zeta^2 - 1} + \cosh^{-1}\zeta \right)$$

Schwartz-Christophel

$$\frac{dz}{d\zeta} = \prod_{n=1}^{N-1}\left( \zeta - \xi_1^{(n)} \right)^{\frac{\alpha^{(n)} - \pi}{\pi}}$$

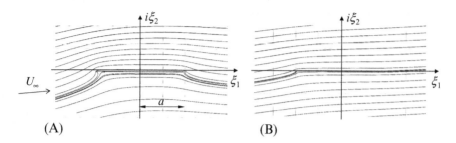

**Fig. 2.8** Uniform flow past a flat plate at an angle of attack of 5 degrees. (A) $\Gamma/aU_\infty = -1$ and (B) $\Gamma/aU_\infty = -2\pi \sin \alpha$.

The Joukowski mapping can be used to create flows past airfoils with thickness and camber as well. This is done by shifting the center of the initial circular cylinder to the left in the negative $\xi_1$ direction (adding airfoil thickness) or upwards in the direction of $i\xi_2$ (adding camber) while enlarging its radius to ensure that it still cuts the right-hand critical point at $\zeta = C$, so that a sharp trailing edge is produced (Fig. 2.9A). If we retain $a/2$ as our mapping constant the cylinder will have a radius $R > a/2$ and will produce a flowfield, in terms of the complex potential, of

$$W(\zeta) = U_\infty(\zeta - \zeta_1)e^{-i\alpha} + \frac{U_\infty R^2 e^{i\alpha}}{\zeta - \zeta_1} - \frac{i\Gamma}{2\pi} \ln(\zeta - \zeta_1) \qquad R = |\zeta_1 - C| \qquad (2.7.33)$$

where $\zeta_1$ is the position of the center of the mapping circle. Substituting the inverse of the mapping function Eq. (2.7.24) gives the airfoil flow, and the velocity field

$$\begin{aligned} w'(z) &= W'(\zeta)\left(\frac{d\zeta}{dz}\right) \\ &= \left\{ U_\infty e^{-i\alpha} - \frac{U_\infty R^2 e^{i\alpha}}{(\zeta - \zeta_1)^2} - \frac{i\Gamma}{2\pi(\zeta - \zeta_1)} \right\} \frac{1}{(1 - a^2/4\zeta^2)} \end{aligned} \qquad (2.7.34)$$

The circulation needed to satisfy the Kutta condition by placing the rearward stagnation point on the cylinder at $\zeta = C$ can be obtained from Eq. (2.7.33) as

$$\Gamma = -4\pi R U_\infty \sin(\alpha + \beta) \qquad (2.7.35)$$

so that the two-dimensional lift coefficient is

$$C_l = 8\pi \frac{R}{c} \sin(\alpha + \beta) \qquad (2.7.36)$$

As shown in Fig. 2.9A, $\beta$ is the angle between the real axis and the cylinder radius at the right-hand critical point, so that $\sin\beta = \mathrm{Im}(\zeta_1)/R$. So it is straightforward to place the center of the mapping circle so as to select the angle of attack $\alpha = -\beta$ at which the airfoil passes through zero lift. As noted above, increasing $\mathrm{Im}(\zeta_1)$ increases the camber of the airfoil, and increasing $-\mathrm{Re}(\zeta_1)$ produces a thicker foil. Sample airfoils and the flows they produce are shown in Fig. 2.9. For a symmetric airfoil ($\mathrm{Im}(\zeta_1) = 0$) the maximum thickness $t_{max}$ (which occurs near 30% chord) and the chordlength $c$ can be estimated as

$$\frac{t_{max}}{C} \approx -5.2\frac{\mathrm{Re}(\zeta_1/C)}{(R/C)^{0.8}} \quad \text{and} \quad \frac{c}{C} = 3 - 2\mathrm{Re}(\zeta_1/C) + \frac{1}{1 - 2\mathrm{Re}(\zeta_1/C)} \qquad (2.7.37)$$

where the second of these expressions is exact. These equations can be used approximately for airfoils with modest camber.

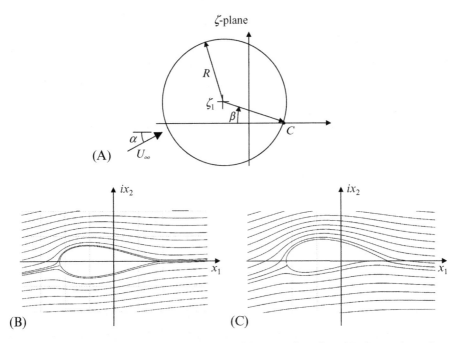

**Fig. 2.9** Joukowski mapping of an airfoil with thickness and camber. (A) Nomenclature in the $\zeta$ plane. (B) Airfoil flow generated with $\alpha = 10$ degrees and $\zeta_1 = -0.3C$ and (C) with $\zeta_1 = (-0.3 + 0.2i)C$.

Other mapping functions have been devised that produce a variety of useful results. A selection of those most relevant for aeroacoustics applications are listed in Table 2.1, along with diagrams showing their effects on the geometry of the space.

An important footnote in the application of conformal mapping to aeroacoustics problems comes when calculating the convection of a point vortex. One such example is the convection of a vortex past a lifting airfoil, which can be modeled in the $z$ plane by applying the Joukowski mapping to the flow of a free stream flow past a circular cylinder in the presence of the vortex, generated in the $\zeta$ plane using the Milne Thompson circle theorem. Superficially, it appears that we should be able to determine the convection velocity in the $z$ plane from that in the $\zeta$ plane using Eq. (2.7.26). However, Eq. (2.7.26) is not strictly valid at the singularity of the convecting vortex since the flow here is not irrotational [6]. A more careful analysis [7] shows that the convection velocity is correctly calculated as

$$w'_c = W'_c \frac{d\zeta}{dz} - \frac{i\Gamma}{4\pi} \frac{d^2\zeta}{dz^2} \frac{dz}{d\zeta} \tag{2.7.38}$$

where all terms are calculated at the location of the convecting vortex, and $\Gamma$ is the strength of that vortex. This modification is called *Routh's correction*.

### 2.7.4   Vortex filaments and the Biot Savart law

It is self-evident that all flows of practical interest are three dimensional. The two-dimensional methods we have outlined in the previous two sections are valuable because they provide tools for analyzing the fluid dynamics in regions where the flow can be assumed invariant in the third dimension. When fully three-dimensional ideal flow analysis is required conformal mapping is no longer an option, and the superposition of elementary flows becomes the main method by which solutions are found to match the boundary conditions of interest.

Viewed in three dimensions, the source, vortex, and doublet singularities of Eq. (2.7.12) extend to infinity and are known as line singularities or filaments. A source and doublet truly confined to a three-dimensional point can also be defined. The extensive engineering tools that exist to configure distributions of these flows to represent the aerodynamics of wings and other devices [5] are beyond the scope of fundamental aeroacoustics. Instead we focus here on the analysis of the vortex filament as a fundamental model of the dynamics of organized vorticity in three-dimensional flows.

In general, a three-dimensional vortex filament can trace a path of any shape through space provided that, for consistency with Helmholtz' theorems, it forms a loop or extends to infinity. The vortex strength, denoted by the circulation it generates $\Gamma$, cannot vary along the filament length since any variation would violate the requirement that Eq. (2.7.4) is independent of the surface chosen. The velocity field generated by an arbitrary filament is given by the *Biot Savart law*:

$$\mathbf{v}(\mathbf{x}) = -\frac{\Gamma}{4\pi} \int_C \frac{(\mathbf{x} - \mathbf{y}) \times d\mathbf{y}}{|\mathbf{x} - \mathbf{y}|^3} \tag{2.7.39}$$

As shown in Fig. 2.10A, this equation gives the velocity as a function of position $\mathbf{x}$ in terms of an integral with respect to distance along the filament defined by the coordinate $\mathbf{y}$. This equation can be derived [3] by solving $\boldsymbol{\omega} = \nabla \times \mathbf{v}$ for the velocity $\mathbf{v}$ in terms of the vorticity $\boldsymbol{\omega}$ and then applying this solution to the singular vorticity field of the filament.

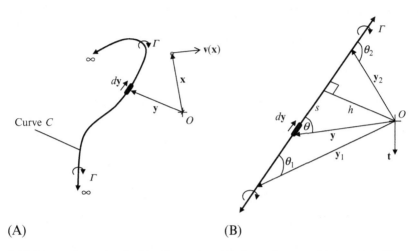

(A)                                                                      (B)

**Fig. 2.10** Nomenclature for the Biot Savart law. (A) General case and (B) straight filament.

Consider, for example, the application of the Biot Savart law to simple straight filament, Fig. 2.10B. We will use Eq. (2.7.39) to calculate the velocity induced by the filament at the origin ($\mathbf{x} = 0$). There is no loss of generality here since the origin can be placed at whatever point is of interest. We express position on the filament $\mathbf{y}$ in terms of the perpendicular distance to the filament $h$ and the angle $\theta$ between $\mathbf{y}$ and the filament. Following the geometry apparent in Fig. 2.10B we have that

$$\mathbf{v} = \frac{\Gamma}{4\pi} \int_C \frac{\mathbf{y} \times d\mathbf{y}}{|\mathbf{y}|^3} = \frac{\Gamma \mathbf{t}}{4\pi} \int_C \frac{|h d\mathbf{y}|}{|\mathbf{y}|^3} = \frac{\Gamma \mathbf{t}}{4\pi} \int_C \frac{h^2 / \sin^2 \theta}{h^3 / \sin^3 \theta} d\theta = -\frac{\Gamma \mathbf{t}}{4\pi h} [\cos(\theta)]_C \quad (2.7.40)$$

Here $\mathbf{t}$ represents the unit vector in the direction of $\mathbf{y} \times d\mathbf{y}$, i.e., perpendicular to the plane containing both the filament and the point where we obtain the velocity. For an infinite straight filament $\theta$ varies from 0 to $\pi$ and thus Eq. (2.7.40) predicts a velocity magnitude of $\Gamma / 2\pi h$. Such a flow is indistinguishable from a two-dimensional point vortex, and thus this result is identical to the velocity field implied in Eq. (2.7.12). Eq. (2.7.40) is much more generally applicable, however. We note that the Biot Savart law shows that the velocity induced by any vortex filament is equal to a linear sum of contributions from each element of the filament length. Thus we can meaningfully use Eq. (2.7.40) to extract the velocity contribution due to a finite portion of the filament, say between $\theta = \theta_1$ and $\theta_2$, as

$$\mathbf{v} = \frac{\Gamma \mathbf{t}}{4\pi h} [\cos(\theta_1) - \cos(\theta_2)] \quad (2.7.41)$$

We can now estimate the velocity field of a filament of any geometry by discretizing that geometry as a series of straight segments and using Eq. (2.7.39) to compute the velocity contribution from each segment. This calculation is sometimes more easily performed with Eq. (2.7.41) rewritten in terms of the vectors $\mathbf{y}_1$ and $\mathbf{y}_2$ that locate the ends of the segment (Fig. 2.10B):

$$\mathbf{v} = \frac{\Gamma}{4\pi} \frac{\mathbf{y}_1 \times \mathbf{y}_2}{|\mathbf{y}_1 \times \mathbf{y}_2|^2} \left[ (\mathbf{y}_1 - \mathbf{y}_2) \cdot \left( \frac{\mathbf{y}_1}{|\mathbf{y}_1|} - \frac{\mathbf{y}_2}{|\mathbf{y}_2|} \right) \right] \tag{2.7.42}$$

Just as in two dimensions we can envision the different parts of a vortex filament being convected by the velocity field of the rest of the flow at those points and thus, one would hope, mimic the behavior of a real eddy. However, there is a complication here since one part of a vortex filament can clearly convect another. A classic example is the circular vortex ring which, like a smoke ring, should convect itself along its axis. Unfortunately, the self-convection velocity turns out to be infinite at any point where a filament is curved (such as in the case of the ring) or forms a corner (such as with a discretized filament). In the latter case this is easily visualized in the case of two segments meeting at a right angle. As the corner is approached along one segment the velocity induced by the other tends to infinity since $h$ in Eq. (2.7.40) tends to zero. The solution to this problem is to recognize that a real eddy has a rotational core of finite radius and thus a scale below which its motion cannot be modeled as a vortex filament in irrotational flow. So, we introduce a "cut off" distance $d$, assumed proportional to the core radius, and exclude the section of filament within an arc-length $d$ of the point where the convection velocity from the Biot Savart law is required. An accepted value for $d$, obtained by comparison with known results for finite-core vortices due to Kelvin, is 0.642 times the core radius [8].

# References

[1] G.K. Batchelor, An Introduction to Fluid Dynamics, Cambridge University Press, Cambridge, 2000.
[2] P.K. Kundu, Fluid Mechanics, Academic Press, San Diego, CA, 1990.
[3] K. Karamcheti, Principles of Ideal-Fluid Aerodynamics, Krieger, New York, NY, 1980.
[4] M.S. Howe, Hydrodynamics and Sound, Cambridge University Press, Cambridge, 2006.
[5] J. Katz, A. Plotkin, Low-Speed Aerodynamics, second ed., Cambridge University Press, Cambridge, 2001.
[6] J. Malarkey, A.G. Davies, Use of Routh's correction in the cloud-in-cell discrete vortex method, J. Comput. Phys. 181 (2002) 753–759.
[7] R.R. Clements, An inviscid model of two-dimensional vortex shedding, J. Fluid Mech. 57 (1973) 321–336.
[8] S.C. Crow, Stability theory for a pair of trailing vortices, AIAA J. 8 (1970) 2172–2179.

# Linear acoustics

## 3

In this chapter we will review the basic concepts of linear acoustics. We will start by deriving the acoustic wave equation for small linear perturbations in a stationary fluid and then discuss the general characteristics of sound propagation, the sound radiation from volume displacement and dipole sources, the far field approximation, and acoustic scattering. We will also introduce the concept of Green's functions and solutions to the wave equation in both the time and the frequency domains. Finally, the concept of Fourier transforms and their application in acoustics will be discussed.

## 3.1 The acoustic wave equation

Consider sound generation and propagation in a medium where there is no significant flow and where the time average properties are uniform throughout the region of interest. In this situation acoustic waves are the only source of pressure and velocity fluctuations. We limit consideration to sound levels which are less than about 140 dB (re 20 μPa) in air or about 220 dB (re 1 μPa) in water. This covers almost all sound levels of practical significance, and allows us to assume that the perturbations of density caused by the sound wave are much smaller than the mean density.

To demonstrate this, consider a sound wave that propagates isentropically so that the relationship between density and pressure perturbations, discussed in Section 2.4, is given by

$$p' = \rho' c_o^2 \qquad c_o^2 = \left(\frac{\partial p}{\partial \rho}\right)_s \tag{3.1.1}$$

where $c_o$ is the sound speed, defined by the isentropic bulk modulus of the fluid. Typical values are $c_o = 343$ m/s in air and $c_o = 1500$ m/s in water. It follows that the density perturbation for a sound level of 140 dB (re 20 μPa) in air is about $1.7 \times 10^{-3}$ kg/m$^3$, which is very much less than the mean density of 1.2 kg/m$^3$. In water a sound level of 220 dB (re 1 μPa) corresponds to a density perturbation of 0.044 kg/m$^3$, which is much smaller than the mean density of 1000 kg/m$^3$. It is therefore reasonable to assume that $\rho' \ll \rho_o$ for all acoustic waves of interest, and this allows us to develop the equations of linear acoustics.

The first step in deriving the linear acoustic wave equation is to use the assumption of small density perturbations in the continuity equation given by Eq. (2.2.6).

$$\frac{\partial \rho}{\partial t} + \frac{\partial (\rho v_i)}{\partial x_i} = 0 \tag{2.2.6}$$

**Aeroacoustics of Low Mach Number Flows.** http://dx.doi.org/10.1016/B978-0-12-809651-2.00003-5

We replace $\rho$ by $\rho_o + \rho'$ and assume $\rho' \ll \rho_o$, where $\rho_o$ is independent of time and constant throughout the medium, to give

$$\frac{\partial \rho'}{\partial t} + \frac{\partial(\rho_o v_i)}{\partial x_i} + \frac{\partial(\rho' v_i)}{\partial x_i} \approx \frac{\partial \rho'}{\partial t} + \rho_o \frac{\partial v_i}{\partial x_i} \approx 0 \qquad (3.1.2)$$

Note that $v_i$ represents only velocity fluctuations due to the acoustic waves, since there is no mean flow. Next we consider the momentum equation given by Eq. (2.3.9). We neglect viscous effects and apply the assumption of small density perturbations to give

$$\frac{\partial(\rho v_i)}{\partial t} + \frac{\partial\left(\rho v_i v_j + p_{ij}\right)}{\partial x_j} \approx \rho_o \frac{\partial v_i}{\partial t} + \frac{\partial(p_o + p')}{\partial x_i} + \rho_o \frac{\partial\left(v_i v_j\right)}{\partial x_j} \approx 0$$

The first term on the right is linearly dependent on small perturbations of velocity and the second term appears to depend on both the mean pressure gradient and the gradient of the pressure perturbation. However in a stationary fluid the mean pressure gradient is zero and only matched by gravitational forces which have been ignored, so we can specify $\partial(p_o + p')/\partial x_i - \partial p'/\partial x_i$, which is also linearly dependent on small quantities. The term $\partial(v_i v_j)/\partial x_j$ is nonlinear and involves the product of two small quantities. To determine the importance of this term we assume that the perturbation has a time scale $T_s$ and lengthscale $\lambda$ so that $\lambda/T_s$ is of the order of magnitude of the speed of sound $c_o$. It then follows that $\partial\left(v_i v_j\right)/\partial x_j \sim v_i^2/\lambda$ and $\partial v_i/\partial t \sim v_i/T_s \sim v_i c_o/\lambda$. So, if the velocity perturbation associated with the wave is small compared to the speed of sound, the nonlinear term can be dropped. Using these assumptions we obtain the linearized acoustic momentum equation as

$$\rho_o \frac{\partial v_i}{\partial t} + \frac{\partial p'}{\partial x_i} \approx 0 \qquad (3.1.3)$$

If we subtract the divergence of Eq. (3.1.3) from the time derivative of Eq. (3.1.2) we obtain,

$$\frac{\partial^2 \rho}{\partial t^2} - \frac{\partial^2 p'}{\partial x_i^2} = 0$$

We then use Eq. (3.1.1) to replace the density perturbation to obtain the linear acoustic wave equation,

$$\frac{1}{c_o^2}\frac{\partial^2 p'}{\partial t^2} - \frac{\partial^2 p'}{\partial x_i^2} = 0 \quad \text{or} \quad \frac{1}{c_o^2}\frac{\partial^2 p'}{\partial t^2} - \nabla^2 p' = 0 \qquad (3.1.4)$$

Eqs. (3.1.3), (3.1.4) are the basis for all calculations in linear acoustics and their solutions can be used to address many of the problems in noise control when the acoustic medium is stationary and its mean properties are uniform throughout. Eq. (3.1.4) is a second order linear partial differential equation whose solution will depend on a set of boundary conditions and initial conditions, which, when specified, define the acoustic pressure fluctuations throughout the medium. In the following section we will consider some simple examples of sound generation and propagation in uniform media.

## 3.2 Plane waves and spherical waves

The simplest example of an acoustic wave is a one dimensional plane wave. For example, the sound propagation along a thin tube can be considered as a plane wave that is only a function of distance in the direction of propagation, and time. In this case the wave equation simplifies to

$$\frac{1}{c_o^2}\frac{\partial^2 p'}{\partial t^2} - \frac{\partial^2 p'}{\partial x_1^2} = 0 \tag{3.2.1}$$

and its solution can be obtained from the method of characteristics as

$$p'(x_1, t) = f(t - x_1/c_o) + g(t + x_1/c_o) \tag{3.2.2}$$

This is known as d'Alembert's solution of the wave equation and shows that there are two independent solutions. The first solution $f(t - x_1/c_o)$ represents a wave which propagates in the positive $x_1$ direction. It is easy to show that a pressure perturbation $f(t)$ at $x_1 = 0$ will be repeated at the location $x_1 = d$ at a time $d/c_o$ later, and so it follows that $c_o$ is indeed the speed of sound propagation. In a similar fashion, the solution $g(t + x_1/c_o)$ represents a wave which propagates in the negative $x_1$ direction so a pressure perturbation $g(t)$ at $x_1 = 0$ will be repeated at the location $x_1 = -d$ at a time $d/c_o$ later.

Of greater practical importance is the solution to the wave equation in spherical coordinates. If we limit consideration to waves that are only a function of the radial distance $r$ from the center of the coordinate system and reduce the Laplacian operator in Eq. (3.1.4) so it is only a function of the radial coordinate $r$, the wave equation becomes,

$$\frac{1}{c_o^2}\frac{\partial^2 p'}{\partial t^2} - \frac{1}{r}\frac{\partial^2}{\partial r^2}(rp') = 0 \tag{3.2.3}$$

Multiplying through by $r$ gives a one dimensional wave equation in terms of the variable $rp'$ whose solution is given by $rp' = f(t - r/c_o) + g(t + r/c_o)$. The first solution represents waves propagating outwards in the radial direction, whereas the second solution represents inwardly propagating waves that are collapsing onto the center of the spherical coordinate system. In a few applications where sound waves are focused, the inwardly propagating waves can be important, but in most cases we are only interested in outwardly propagating waves. The solution to the wave equation for outwardly propagating waves is then

$$p'(r, t) = \frac{f(t - r/c_o)}{r} \tag{3.2.4}$$

The additional feature of this result is that the pressure perturbation not only propagates outwards, but also decays in amplitude as $1/r$.

## 3.3 Harmonic time dependence

In many cases we are interested in evaluating the features of a sound field as a function of frequency rather than time. To achieve this, we can consider the time history of the signal to be at a single frequency with a harmonic time dependence. If the angular frequency is $\omega$ radians per second, then the time dependence of a harmonic wave will be $f(t) = A \cos(\omega t - \phi)$ and the solution to the wave equation for spherical waves will be

$$p'(r, t) = \frac{A \cos(\omega t - \omega r/c_o - \phi)}{r} \qquad (3.3.1)$$

where $A$ and $\phi$ are real constants which determine the amplitude of the wave and its phase, respectively. It will prove valuable to treat harmonic time series as the real part of a complex exponential with a time dependence $\exp(-i\omega t)$. The choice of the minus sign in the argument of the exponential is important and care should be exercised when comparing results with this sign convention to results in other texts which use a $\exp(i\omega t)$ time dependence. Using the $-i\omega t$ convention equation (3.3.1) can be rewritten as

$$p'(r, t) = \text{Re}\left[\hat{p}(r)e^{-i\omega t}\right] = \text{Re}\left[\frac{\hat{A}e^{-i\omega t + ikr}}{r}\right] \qquad (3.3.2)$$

where the ^ indicates a complex amplitude, defined so that $\hat{A} = A\exp(i\phi)$. Thus $\hat{p}(r)$ is the complex amplitude of the pressure as a function of position. The symbol $k$ represents the acoustic wavenumber $k = \omega/c_o$. For brevity we can often take the harmonic time dependence to be implied. For example, Eq. (3.3.2) can simply be written as

$$\hat{p}(r) = \frac{\hat{A}e^{ikr}}{r} \qquad (3.3.3)$$

Note that with this time convention a wave propagating in the $r$ direction has positive phase expressed in the term $\exp(ikr)$. The spatial dependence of the sound field is therefore also harmonic and oscillates with a length scale defined by the acoustic wavelength $\lambda$. To relate the wavelength to the acoustic wavenumber we note that the sound field starts to repeat when the phase $kr$ is incremented by $2\pi$. The distance between peaks in the wave is the wavelength and so $k\lambda = 2\pi$, and we have $k = \omega/c_o = 2\pi/\lambda$.

More generally we can write the linear acoustic wave equation (3.1.4) in terms of the complex pressure amplitude $\hat{p}(\mathbf{x})$ by substituting $p'(\mathbf{x}, t) = \text{Re}[\hat{p}(\mathbf{x}) \exp(-i\omega t)]$ to give

$$\nabla^2 \hat{p} + k^2 \hat{p} = 0 \qquad (3.3.4)$$

which is known as the Helmholtz equation.

The acoustic momentum equation for a harmonic wave is similarly obtained from Eq. (3.1.3) as

$$i\omega\rho_o\hat{\mathbf{v}} = \nabla\hat{p} \tag{3.3.5}$$

This is a result that we will use frequently in sound radiation and scattering problems.

## 3.4   Sound generation by a small sphere

To solve a sound radiation or scattering problem, we must find the appropriate solution to the wave equation and use the boundary conditions to determine the unknown constants, such as $\hat{A}$ in Eq. (3.3.3). To illustrate this procedure, consider the sound radiation from a small pulsating sphere of radius $a$ which has a normal surface velocity $u_o\exp(-i\omega t)$, as shown in Fig. 3.1.

**Fig. 3.1** A small sphere with a radial surface velocity.

The appropriate solution to the wave equation, which matches the boundary condition for this example, is given by Eq. (3.3.3). To determine the unknown constant $\hat{A}$ we match the particle velocity of the sound wave in the radial direction to the velocity of the surface, $[\hat{v}_r]_{r=a} = u_o$ and use the acoustic momentum equation (3.3.5) to give

$$\left[\frac{1}{i\omega\rho_o}\frac{\partial\hat{p}}{\partial r}\right]_{r=a} = u_o \tag{3.4.1}$$

From Eq. (3.3.3) we find

$$\frac{\partial\hat{p}(r)}{\partial r} = -\frac{\hat{A}(1-ikr)e^{ikr}}{r^2} \tag{3.4.2}$$

And so, at the sphere surface,

$$\left[\frac{\partial\hat{p}(r)}{\partial r}\right]_{r=a} = -\frac{\hat{A}(1-ika)e^{ika}}{a^2}$$

Substituting into Eq. (3.4.1) and solving for $\hat{A}$ gives

$$\hat{A} = -\frac{i\omega\rho_o a^2 u_o e^{-ika}}{(1-ika)} \tag{3.4.3}$$

The complete solution for the acoustic field is then given by Eq. (3.3.3) as

$$\hat{p} = -\frac{i\omega\rho_o a^2 u_o e^{ik(r-a)}}{(1-ika)r} \tag{3.4.4}$$

Note that if the sphere is very small so $ka \ll 1$ then $\exp(-ika) \approx 1 - ika$ and this solution approximates to

$$\hat{p} = -\frac{i\omega\rho_o a^2 u_o e^{ikr}}{r} \tag{3.4.5}$$

Eq. (3.4.4) shows how the sound field everywhere is determined by the boundary conditions. No initial conditions were required in this case because a harmonic time dependence was assumed. It is also important to note that the surface area of the sphere is $S = 4\pi a^2$ and we can define the rate of change of volume of the sphere as

$$Qe^{-i\omega t} = u_o S e^{-i\omega t}$$

so

$$\hat{p} = -\frac{i\omega\rho_o Q e^{ikr}}{4\pi r} \tag{3.4.6}$$

Consequently, the acoustic pressure is directly proportional to the rate of change of volume caused by the surface displacement, and so a radially pulsating sphere is frequently referred to as a volume displacement source. Alternatively, it is termed a simple source or an acoustic *monopole* because the sound field is only a function of distance from the center of the sphere. It is also important to note that physically $\rho_o Q \exp(-i\omega t)$ represents the "rate of change of mass" of fluid displaced by the motion of the surface.

In solving the wave equation, we specified how sound waves propagate through the medium, and by introducing the boundary conditions, we specified how the waves were initiated. However, we can only use this result for this particular boundary motion. If the surface had been an ellipsoid, then the solution to the wave equation would have to be specified in ellipsoidal coordinates, which is not as simple. Other shapes require solutions to the wave equation in other orthogonal coordinate systems of which there are only a finite number that provide analytical solutions to the wave equation. There is therefore only a limited set of problems that can be addressed using this approach, and for arbitrary shapes numerical methods have to be used. One useful example, which can be solved exactly, is for a sphere which translates back and forth in the $x_1$ direction with velocity $v_o \exp(-i\omega t)$, as shown in Fig. 3.2.

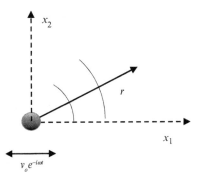

**Fig. 3.2** Sound radiation from a translating sphere.

In this case the velocity of the surface in the radial direction, normal to the surface of the sphere, provides the necessary boundary condition. The velocity tangential to the surface can be neglected because viscous effects are ignored. The surface velocity is then given by $[\hat{v}_r]_{r=a} = v_o \cos\theta$ where $\theta$ is the angle subtended by the point on the surface to the $x_1$ axis. To address this problem we need to specify a solution to the wave equation that matches this boundary condition and this is found by taking the derivative of Eq. (3.3.3) in the $x_1$ direction giving

$$\hat{p}(r) = \frac{\partial}{\partial x_1}\left(\frac{\hat{A}e^{ikr}}{r}\right) \tag{3.4.7}$$

We can readily show that this is a solution to the wave equation because any derivative of a solution to the wave equation is also a solution to the wave equation, as is easily verified from Eq. (3.1.4). Evaluating the derivative in Eq. (3.4.7) gives

$$\hat{p}(r) = \frac{\partial r}{\partial x_1}\frac{\partial}{\partial r}\left(\frac{\hat{A}e^{ikr}}{r}\right) \tag{3.4.8}$$

and since $r = \left(x_1^2 + x_2^2 + x_3^2\right)^{1/2}$ it follows that $\partial r/\partial x_1 = x_1/r = \cos\theta$. Evaluating the derivatives in Eq. (3.4.8) then gives

$$\hat{p}(r) = ik\cos\theta\left(\frac{\hat{A}e^{ikr}}{r}\right)\left(1 - \frac{1}{ikr}\right) \tag{3.4.9}$$

This solution has a $\cos\theta$ directionality associated with it, which we can use to match the boundary condition for the translating sphere. Also note how the amplitude of sound field is no longer simply a function of $1/r$ since there is now an additional factor of $1/ikr$ in the solution. This term becomes negligible at large distances from the center of the sphere when $kr \gg 1$ but will dominate in regions where $kr \ll 1$. We therefore define the region where $kr \gg 1$ as the acoustic far field, where the amplitude of the sound wave decays inversely with the distance from the source.

Evaluating the radial component of the acoustic momentum equation $\nabla \hat{p} = i\omega \rho_o \hat{\mathbf{v}}$ we obtain

$$\frac{\partial \hat{p}}{\partial r} = i\omega \rho_o \hat{\mathbf{v}} \cdot \mathbf{n}$$

where $\mathbf{n}$ is a unit vector in the radial direction on the surface of the sphere. Applying this at the sphere surface where $\hat{\mathbf{v}} \cdot \mathbf{n} = v_o \cos \theta$ we write

$$\left[\frac{1}{i\omega \rho_o}\frac{\partial \hat{p}}{\partial r}\right]_{r=a} = \frac{\cos \theta}{\rho_o c_o}\left(\frac{\hat{A}e^{ika}}{a}\right)\left(ik - \frac{2}{a} + \frac{2}{ika^2}\right) = v_o \cos \theta \tag{3.4.10}$$

For a small sphere we can assume $ka \ll 1$ and obtain the approximate solution for the sound field by only retaining the term $2/ika^2$. Then solving for $\hat{A}$ and using Eq. (3.4.9) gives

$$\hat{p}(r) = ik \cos \theta \left(\frac{i\omega \rho_o v_o a^3 e^{ikr}}{2r}\right)\left(1 - \frac{1}{ikr}\right) \tag{3.4.11}$$

The $\cos \theta$ dependence shows that the acoustic field caused by a translating sphere has a beam along the $x_1$ axis and is zero in the direction normal to the source motion. This is quite different from the field generated by uniformly pulsating sphere given (Eq. 3.4.5), which is omnidirectional. We also note that the peak value of the pressure for the translating sphere is a factor of $ka/2$ less than the level generated by the pulsating sphere at the same distance if $u_o$ and $v_o$ are taken as equal. Since $ka \ll 1$, the radiated sound levels from the translating sphere for a given surface velocity are much less than that of the pulsating sphere. The physical explanation of this observation is that the pulsating sphere displaces mass during each cycle, so the medium has nowhere to go apart from propagating away as an acoustic wave. In contrast, the translating sphere causes no net displacement of mass, and the fluid can adjust in the near field to accommodate the motion. However, some energy still escapes as sound and propagates to the acoustic far field.

## 3.5  Sound scattering by a small sphere

In the previous section we discussed the radiation of sound from a small sphere with a prescribed surface motion. In this section we will consider the scattering of sound by a small sphere. Scattering occurs because an object is placed in an acoustic field, and to satisfy the boundary conditions on the surface of the object, additional waves must be generated, which are known as the scattered field.

To illustrate this concept, we will consider scattering by a small spherical gas bubble of radius $a$ in a liquid such as water. The gas in the bubble will be assumed to be much more compressible and very much less dense than the liquid and so the bubble surface offers no resistance to the pressure perturbation of the incident wave. This boundary condition implies that the fluctuating pressure is zero on the surface of the bubble.

We will consider the incident field to be a plane wave of unit amplitude propagating in the $x_1$ direction so the incident field is defined from Eq. (3.2.2) with harmonic time dependence as

$$\hat{p}_i(r) = e^{ikx_1} = e^{ikr\cos\theta} \tag{3.5.1}$$

where $r$ is the distance from the center of the bubble and $x_1 = r\cos\theta$. If the radius of the bubble is very much smaller than the acoustic wavelength (which is always the case in underwater applications) then we can approximate the complex exponential describing the incident field on the surface $r = a$ as

$$\hat{p}_i(a) = e^{ika\cos\theta} \approx 1 + ika\cos\theta \tag{3.5.2}$$

The scattered field must match this angular dependence and be a solution to the wave equation. Both requirements can be satisfied if the scattered field is the sum of omnidirectional and directional fields of the type given in Eqs. (3.3.3), (3.4.9),

$$\hat{p}_s(r) = \frac{\hat{A}e^{ikr}}{r} + ik\cos\theta\left(\frac{\hat{B}e^{ikr}}{r}\right)\left(1 - \frac{1}{ikr}\right) \tag{3.5.3}$$

In order to satisfy the boundary condition we require that $\hat{p}_i(a) + \hat{p}_s(a) = 0$, so Eqs. (3.5.2), (3.5.3) give

$$\frac{\hat{A}e^{ika}}{a} + ik\cos\theta\left(\frac{\hat{B}e^{ika}}{a}\right)\left(1 - \frac{1}{ika}\right) = -(1 + ika\cos\theta) \tag{3.5.4}$$

Matching the $\cos\theta$ dependence we find that $\hat{A} = -ae^{-ika}$ and $\hat{B} = ika^3 e^{-ika}/(1 - ika)$, so

$$\hat{p}_s(r) = -\frac{ae^{ik(r-a)}}{r} - (ka)^2\cos\theta\left(\frac{ae^{ik(r-a)}}{r(1 - ika)}\right)\left(1 - \frac{1}{ikr}\right) \tag{3.5.5}$$

This shows that the directional term is of order $(ka)^2$ compared with the omnidirectional term and so can be neglected because $ka \ll 1$, leaving the scattered field as

$$\hat{p}_s(r) = -\frac{ae^{ik(r-a)}}{r} \tag{3.5.6}$$

It follows that bubbles in a sound field act like small sources, which generate their own waves which propagate to the acoustic far field.

## 3.6  Superposition and far field approximations

The acoustic wave equation is linear and so any number of sound fields can be added together, or superimposed. Consider a region containing a large number, say $N$, of monopole sources located at $\mathbf{y}^{(n)}$ where $n = 1,2,3,\ldots,N$, as shown in Fig. 3.3.

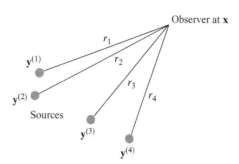

**Fig. 3.3** Sources distributed over a region, each source is located at $\mathbf{y}^{(n)}$ and the observer is at $\mathbf{x}$. The propagation distances are $r_n = |\mathbf{x} - \mathbf{y}^{(n)}|$.

The acoustic field at a point $\mathbf{x}$ is given by summing the acoustic fields of each source, given by Eq. (3.3.3), so

$$\hat{p}(\mathbf{x}) = \sum_{n=1}^{N} \frac{\hat{A}_n e^{ik|\mathbf{x}-\mathbf{y}^{(n)}|}}{|\mathbf{x} - \mathbf{y}^{(n)}|} \tag{3.6.1}$$

where $|\mathbf{x} - \mathbf{y}^{(n)}|$ is the distance from each source to the observer location $\mathbf{x}$ and $\hat{A}_n$ represents the relative contribution of each source, which, for example would be $i\omega\rho_o Q/4\pi$ if all the sources were identical pulsating spheres. At large distances from the sources this expression can be simplified by expanding $|\mathbf{x} - \mathbf{y}^{(n)}|$ as a Taylor series. We note that in general

$$|\mathbf{x} - \mathbf{y}| = r(x_i - y_i) = \left( (x_1 - y_1)^2 + (x_2 - y_2)^2 + (x_3 - y_3)^2 \right)^{1/2} \tag{3.6.2}$$

where $r = r(x_i - y_i)$ is the distance from the source to the observer explicitly defined as a function of $x_i$ and $y_i$. If we choose the coordinate origin to be located near the sources then in the far field $|\mathbf{x}| \gg |\mathbf{y}|$ we can use a Taylor series expansion to approximate the propagation distance as,

$$r(x_i - y_i) = r(x_i) - y_i \frac{\partial r(x_i)}{\partial x_i} + \frac{y_i y_j}{2} \frac{\partial^2 r(x_i)}{\partial x_i \partial x_j} + \cdots \tag{3.6.3}$$

In this expansion the term $\partial r/\partial x_i = x_i/|\mathbf{x}|$ is a direction cosine and independent of $|\mathbf{x}|$. In contrast the second term $\partial^2 r/\partial x_i \partial x_j$ is inversely proportional to $|\mathbf{x}|$ and so becomes less and less significant at large distances from the source. Retaining the first order terms gives $r(x_i - y_i) = |\mathbf{x}| - x_i y_i / |\mathbf{x}|$ so

$$\hat{p}(\mathbf{x}) \approx \sum_{n=1}^{N} \frac{\hat{A}_n e^{ik|\mathbf{x}| - ik\mathbf{y}^{(n)} \cdot \mathbf{x}/|\mathbf{x}|}}{|\mathbf{x}| - \mathbf{y}^{(n)} \cdot \mathbf{x}/|\mathbf{x}|} = e^{ik|\mathbf{x}|} \sum_{n=1}^{N} \frac{\hat{A}_n e^{-ik\mathbf{y}^{(n)} \cdot \mathbf{x}/|\mathbf{x}|}}{|\mathbf{x}| - \mathbf{y}^{(n)} \cdot \mathbf{x}/|\mathbf{x}|} \quad |\mathbf{x}| \gg |\mathbf{y}^{(n)}|$$

At large distances the term in the denominator is closely approximated by $|\mathbf{x}|$ and so to first order we obtain

$$\hat{p}(\mathbf{x}) \approx \frac{e^{ik|\mathbf{x}|}}{|\mathbf{x}|} \sum_{n=1}^{N} \hat{A}_n e^{-ik\mathbf{y}^{(n)} \cdot \mathbf{x}/|\mathbf{x}|} \left( 1 + \mathbf{y}^{(n)} \cdot \mathbf{x}/|\mathbf{x}|^2 + \cdots \right) \quad |\mathbf{x}| \gg |\mathbf{y}^{(n)}| \quad (3.6.4)$$

and the term $\mathbf{y}^{(n)} \cdot \mathbf{x}/|\mathbf{x}|^2$ in brackets can be ignored, unless, as will be shown below, the sum of the source amplitudes is zero. However, the dependence on $\mathbf{y}^{(n)}$ in the complex exponential cannot be ignored because it can dominate the result if some phase cancelation occurs in the summation. Physically this occurs when destructive interference between multiple sources leaves a residual sound field dependent upon the phase differences between those sources. The far field approximation is thus

$$\hat{p}(\mathbf{x}) \approx \frac{e^{ik|\mathbf{x}|}}{|\mathbf{x}|} \sum_{n=1}^{N} \hat{A}_n e^{-ik\mathbf{y}^{(n)} \cdot \mathbf{x}/|\mathbf{x}|} \quad |\mathbf{x}| \gg |\mathbf{y}^{(n)}| \quad (3.6.5)$$

The important feature of this result is that in the *acoustic far field* the relative position of the sources and their relative strengths determines the directionality of the field, which depends on $\mathbf{x}/|\mathbf{x}|$. The amplitude of the field decays inversely with distance and the phase increases linearly with $|\mathbf{x}|$. This is an important result that will be used extensively later in the text.

## 3.7  Monopole, dipole, and quadrupole sources

In Section 3.5 we stated that a simple volume displacement source was often referred to as a *monopole*. The acoustic field was omnidirectional and inversely proportional to the propagation distance from the center of the source. In this section we will discuss *multipoles* which are sources obtained by clustering together simple sources of equal magnitude and opposite phase.

The simplest example of a multipole source is a *dipole* which is defined as two monopole sources of equal strength and opposite phase a small distance $d$ apart, where $kd \ll 1$, as shown in Fig. 3.4.

The far field from a dipole is readily obtained from Eq. (3.6.4) by placing the positive source at $y_1 = -d/2$ and the negative source at $y_1 = d/2$, and setting $\hat{A}_1 = i\omega\rho_o Q/4\pi$, $\hat{A}_2 = -i\omega\rho_o Q/4\pi$ so that

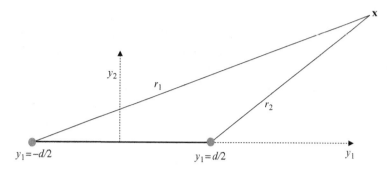

**Fig. 3.4** Two sources of opposite phase that define a dipole source.

$$\hat{p}(\mathbf{x}) \approx \frac{i\omega\rho_o Q e^{ik|\mathbf{x}|}}{4\pi|\mathbf{x}|}\left(e^{ikdx_1/2|\mathbf{x}|}\left(1 - x_1 d/2|\mathbf{x}|^2 + \cdots\right) - e^{-ikdx_1/2|\mathbf{x}|}\left(1 + x_1 d/2|\mathbf{x}|^2 + \cdots\right)\right)$$

$$(3.7.1)$$

providing that $|\mathbf{x}| \gg d$. Since we have specified $kd \ll 1$ we can simplify this result by using the expansion

$$e^{\pm ikdx_1/2|\mathbf{x}|} = 1 \pm ikdx_1/2|\mathbf{x}| - \frac{1}{2}(kdx_1/2|\mathbf{x}|)^2 + \cdots \tag{3.7.2}$$

and using this in Eq. (3.7.1) we see that the terms of order 1 cancel, leaving

$$\hat{p}(\mathbf{x}) \approx \left(\frac{i\omega\rho_o Q e^{ik|\mathbf{x}|}}{4\pi|\mathbf{x}|}\right)\left(ikd\frac{x_1}{|\mathbf{x}|} - \frac{x_1 d}{|\mathbf{x}|^2} + \cdots\right) \quad |\mathbf{x}| \gg d \tag{3.7.3}$$

which can be simplified using $\cos\theta \equiv x_1/|\mathbf{x}|$, and we find that

$$\hat{p}(\mathbf{x}) \approx ikd\cos\theta\left(\frac{i\omega\rho_o Q e^{ik|\mathbf{x}|}}{4\pi|\mathbf{x}|}\right)\left(1 - \frac{1}{ik|\mathbf{x}|} + \cdots\right) \quad |\mathbf{x}| \gg d$$

The first thing to note about an acoustic dipole is that its field is directional and depends on the cosine of the angle subtended by the observer to the line between the sources, which defines the dipole axis. Therefore, it has the same far-field characteristics as the transversely oscillating sphere, with a maximum in the direction of the dipole axis and a null or a zero at 90 degrees to the dipole axis, Fig. 3.5. Secondly we see that the dependence on distance from the source scales as $d/|\mathbf{x}|^2$ when $k|\mathbf{x}| \ll 1$ and as $kd/|\mathbf{x}|$ when $k|\mathbf{x}| \gg 1$. These represent the acoustic near field and far field approximations, respectively. The far field approximation requires that the observer is in both the geometric far field $|\mathbf{x}| \gg d$ and the acoustic far field $k|\mathbf{x}| \gg 1$, in which case only the phase shift caused by the differences in propagation distance from each source is important.

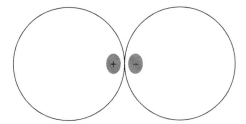

**Fig. 3.5** The cosine directionality of a dipole source.

Note also that the source has no net volume displacement because the total source strength is zero. In spite of this the acoustic field is nonzero because the two sources are displaced by the distance $d$. Finally, we note that the maximum amplitude of the acoustic field from a dipole is $ikd$ times the amplitude of the pure monopole field at the same distance, and since we assumed that $kd \ll 1$ we conclude that the dipole is an inefficient sound source.

A *quadrupole* source can be formed by placing two dipole sources back to back, with positive sources at $y_1 = -3d/2$ and $y_1 = 3d/2$, and negative sources at $y_1 = -d/2$ and $y_1 = d/2$. The net acoustic far field ($k|\mathbf{x}| \gg 1$) is

$$\hat{p}(\mathbf{x}) \approx \frac{i\omega\rho_o Q e^{ik|\mathbf{x}|}}{4\pi|\mathbf{x}|} \left( e^{3ikdx_1/2|\mathbf{x}|} - e^{ikdx_1/2|\mathbf{x}|} - e^{-ikdx_1/2|\mathbf{x}|} + e^{-3ikdx_1/2|\mathbf{x}|} \right) \quad |\mathbf{x}| \gg d \tag{3.7.4}$$

Then using Eq. (3.7.2) we find that all the terms of order 1 and all the terms of order $kd$ cancel, leaving

$$\hat{p}(\mathbf{x}) \approx -2(kd)^2 \left(\frac{x_1}{|\mathbf{x}|}\right)^2 \frac{i\omega\rho_o Q e^{ik|\mathbf{x}|}}{4\pi|\mathbf{x}|} \quad |\mathbf{x}| \gg d \;\; k|\mathbf{x}| \gg 1 \tag{3.7.5}$$

This is referred to as a longitudinal quadrupole because all the sources are in line and the net volume velocity is zero. Note how the directionality is determined by $\cos^2\theta$ (as shown on the left side of Fig. 3.6) and the source strength is now proportional to $(kd)^2$, and so the field is yet another order of magnitude less than the monopole field. We can also form quadrupoles with sources in different arrangements. For example, if we place two sources of amplitude $i\omega\rho_o Q/4\pi$ at $\mathbf{y} = (d/2, d/2)$ and $\mathbf{y} = (-d/2, -d/2)$, and two sources of strength $-i\omega\rho_o Q/4\pi$ at $\mathbf{y} = (d/2, -d/2)$ and $\mathbf{y} = (-d/2, d/2)$, we obtain in the acoustic far field

$$\hat{p}(\mathbf{x}) \approx \frac{i\omega\rho_o Q e^{ik|\mathbf{x}|}}{4\pi|\mathbf{x}|}$$

$$\left( e^{ikdx_1/2|\mathbf{x}| + ikdx_2/2|\mathbf{x}|} + e^{-ikdx_1/2|\mathbf{x}| - ikdx_2/2|\mathbf{x}|} - e^{ikdx_1/2|\mathbf{x}| - ikdx_2/2|\mathbf{x}|} - e^{-ikdx_1/2|\mathbf{x}| + ikdx_2/2|\mathbf{x}|} \right)$$

$$\tag{3.7.6}$$

Then using the expansion

$$e^{-ikdx_1/2|\mathbf{x}|\pm ikdx_2/2|\mathbf{x}|} = 1 - ikdx_1/2|\mathbf{x}| \pm ikdx_2/2|\mathbf{x}| - \frac{1}{2}(kdx_1/2|\mathbf{x}| \mp ikdx_2/2|\mathbf{x}|)^2 + \cdots$$

gives

$$\hat{p}(\mathbf{x}) \approx -(kd)^2 \frac{i\omega\rho_o Q e^{ik|\mathbf{x}|}}{4\pi|\mathbf{x}|}\left(\frac{x_1 x_2}{|\mathbf{x}|^2}\right) \quad |\mathbf{x}|\gg d \ \ k|\mathbf{x}|\gg1 \tag{3.7.7}$$

which implies a $\sin\theta\cos\theta = \frac{1}{2}\sin(2\theta)$ directionality, as shown on the right side of Fig. 3.6, and a scaling on $(kd)^2$. This type of directionality and scaling is important in aeroacoustics and we will show in Chapter 4 that turbulence generates sound fields that have the same directionality.

**Fig. 3.6** The directionality of different types of quadrupole.

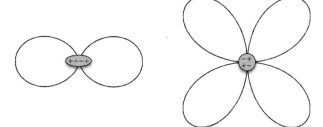

## 3.8   Acoustic intensity and sound power output

The results given above show that an arrangement of simple sources of the same level can produce a directional acoustic far field. In many applications there is a need to compare sources with different directionalities and this can be achieved by comparing their total sound power output. In Chapter 2 the acoustic sound power output of a source is defined by Eq. (2.6.16), as

$$W_a = \int_S \mathbf{I}\cdot\mathbf{n}^{(o)}dS \tag{2.6.16}$$

where $\mathbf{I}$ is the acoustic intensity, $S$ is a surface that encloses all the acoustic sources in a volume $V$ and $\mathbf{n}^{(o)}$ is a unit normal vector pointing out of the volume.

In Section 2.6 we also defined the acoustic intensity in a stationary homentropic medium as $\mathbf{I} = E[p'\mathbf{v}]$, where $E[\ ]$ represents an expected value of the terms in brackets (see Chapter 8 for a detailed discussion of expected values). For linear acoustics and harmonic time dependence, we can write this using subscript notation for the intensity vector as

$$I_j = E\left[|\hat{p}|\cos\left(\omega t + \phi_p\right)|\hat{v}_j|\cos\left(\omega t + \phi_j\right)\right]$$

where $|\hat{p}|$ and $\phi_p$ represent the amplitude and phase of the complex pressure $\hat{p}$, and $|\hat{v}_j|$ and $\phi_j$ represent these measures for the complex velocity component. Using the trigonometric expansion $\cos(\omega t + \phi_p)\cos(\omega t + \phi_j) = \frac{1}{2}(\cos(2\omega t + \phi_p + \phi_j)) + (\cos(\phi_p - \phi_j))$ gives

$$I_j = \frac{|\hat{p}||\hat{v}_j|}{2}E\left[\cos\left(2\omega t + \phi_p + \phi_j\right) + \cos\left(\phi_p - \phi_j\right)\right]$$

Since the expected value of a harmonic time history is zero only the second term remains, and we can define the acoustic intensity for a harmonic time signal as

$$I_j = \frac{|\hat{p}||\hat{v}_j|}{2}\cos\left(\phi_p - \phi_j\right) = \frac{\mathrm{Re}\left(\hat{p}\hat{v}_j^*\right)}{2} \tag{3.8.1}$$

where $\hat{v}_j^*$ represents the complex conjugate of the velocity of a harmonic wave.

We can evaluate the sound power output of a source distribution by defining a spherical surface, on $|\mathbf{x}| = const$, that encloses the sources and lies in the acoustic far field, so the outward normal vector is in the radial direction. Eq. (3.8.1) shows that the radial component of the intensity vector will depend on the particle velocity in the radial direction, which from the momentum equation (3.3.5), $i\omega\rho_o\hat{\mathbf{v}} = \nabla\hat{p}$, depends on the gradient of the acoustic pressure in that direction. We showed in deriving Eq. (3.6.5) that for all source distributions the far-field sound has a variation in the radial direction given by $e^{ikr}/r$ where $r = |\mathbf{x}|$. So, using the momentum equation (3.3.5), we obtain

$$\hat{v}_r = \frac{1}{i\omega\rho_o}\frac{\partial\hat{p}}{\partial r} = \frac{\hat{p}}{i\omega\rho_o}\left(ik - \frac{1}{r}\right) \tag{3.8.2}$$

Then, using Eq. (3.8.1), the acoustic intensity vector in the radial direction is

$$I_r = \frac{1}{2}\mathrm{Re}\left\{\frac{\hat{p}\hat{p}^*}{i\omega\rho_o}\left(ik + \frac{1}{r}\right)\right\} = \frac{1}{2}|\hat{p}|^2\frac{k}{\omega\rho_o} = \frac{|\hat{p}|^2}{2\rho_o c_o} \tag{3.8.3}$$

so we obtain the far field intensity simply in terms of the pressure amplitude. It follows that the acoustic intensity of a dipole scales as $(kd)^2$ compared to a monopole, and as $(kd)^4$ for a quadrupole compared to a monopole. Since $kd \ll 1$ we see that there is a considerable reduction in intensity for sources of higher multipole order compared to a monopole source.

To obtain the sound power output for each source type we need to evaluate the surface integral of the radial intensity as required by Eq. (2.6.16). For a monopole source defined by Eq. (3.4.6), the intensity is constant on $r = const$ and we obtain

$$W_a^{(\text{monopole})} = \frac{(\omega \rho_o Q)^2}{2 \rho_o c_o (4\pi r)^2} \int_S dS$$

Since the surface area of a sphere of radius $r$ is $4\pi r^2$ the sound power output is

$$W_a^{(\text{monopole})} = \frac{(\omega \rho_o Q)^2}{8\pi \rho_o c_o} \tag{3.8.4}$$

For a dipole source the same approach is used but the intensity has a $\cos^2 \theta$ dependence, and so the surface integral must take that into account. Using Eq. (3.7.3) to give the acoustic far field of a dipole gives, using spherical coordinates $r$, $\theta$, $\varphi$ to evaluate the surface integral

$$W_a^{(\text{dipole})} = \frac{(\omega \rho_o Q)^2 (kd)^2}{2\rho_o c_o (4\pi r)^2} \int_0^\pi \int_0^{2\pi} \cos^2 \theta \sin \theta r^2 d\varphi d\theta = \frac{(\omega \rho_o Q)^2 (kd)^2}{24\pi \rho_o c_o} \tag{3.8.5}$$

Similarly, for the quadrupole field given by Eq. (3.7.5) we obtain

$$W_a^{(\text{quadrupole})} = \frac{4(\omega \rho_o Q)^2 (kd)^4}{2\rho_o c_o (4\pi r)^2} \int_0^\pi \int_0^{2\pi} \cos^4 \theta \sin \theta r^2 d\varphi d\theta = \frac{(\omega \rho_o Q)^2 (kd)^4}{10\pi \rho_o c_o} \tag{3.8.6}$$

The important result here is that the total sound power output of a dipole source is of order $(kd)^2$ less than that of a monopole source, and that the power output of a quadrupole source is of order $(kd)^2$ less than a dipole source. Since we have taken $kd \ll 1$ it can usually be assumed that a dipole source will always be an order of magnitude more significant than a quadrupole source, and that monopole sources will dominate the sound power output from any source distribution in which they are present.

## 3.9   Solution to the wave equation using Green's functions

In this section we give a general solution to the wave equation in the presence of surface sources and scattering surfaces. This is known as the method of Green's functions and will be useful in many of the problems that will be discussed in subsequent chapters.

Consider a uniform stationary medium in which acoustic propagation is determined by the linear wave equation (3.1.4). The medium can include objects and other surfaces, which can generate and reflect sound (see Fig. 3.7). We wish to determine the acoustic pressure $p'(\mathbf{x},t)$ at observer location $\mathbf{x}$ and at time $t$ due to some sound

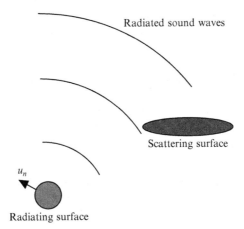

**Fig. 3.7** Radiating and scattering surfaces in a sound field.

source distributed in space $\mathbf{y}$ at time $\tau$. To do this we first write the wave equation in terms of $\mathbf{y}$ and $\tau$.

$$\frac{1}{c_o^2}\frac{\partial^2 p'(\mathbf{y},\tau)}{\partial \tau^2} - \frac{\partial^2 p'(\mathbf{y},\tau)}{\partial y_i^2} = 0 \tag{3.9.1}$$

and then introduce the Green's function $G$ defined as being a solution to the inhomogeneous wave equation

$$\frac{1}{c_o^2}\frac{\partial^2 G}{\partial \tau^2} - \frac{\partial^2 G}{\partial y_i^2} = \delta(\mathbf{x}-\mathbf{y})\delta(t-\tau) \tag{3.9.2}$$

The Green's function is distributed in space as a function of $\mathbf{y}$ and $\tau$, but also depends on the observer position $\mathbf{x}$ and time $t$, where we wish to obtain our solution and so has a functional form $G(\mathbf{x},t|\mathbf{y},\tau)$. We will show below that it gives the relationship between the acoustic field at $\mathbf{x}$ and $t$ due to an impulsive source of sound at $\mathbf{y}$ and $\tau$.

The functions on the right side of Eq. (3.9.2) represent Dirac delta functions. In one dimension a Dirac delta function has the property that

$$\delta(t-\tau) = 0 \quad \text{when } t \neq \tau, \quad \text{and} \quad \int_{-T}^{T} \delta(t-\tau)d\tau = 1 \quad -T < t < T \tag{3.9.3}$$

When the argument of the Dirac is a vector we define it so that

$$\delta(\mathbf{x}-\mathbf{y}) = 0 \quad \text{when } \mathbf{x} \neq \mathbf{y}, \quad \text{and} \quad \int_{V} \delta(\mathbf{x}-\mathbf{y})dV(\mathbf{y}) = \begin{cases} 1 & \text{when } \mathbf{x} \text{ is in } V \\ 0 & \text{when } \mathbf{x} \text{ is not in } V \end{cases}$$

$$\tag{3.9.4}$$

For Cartesian coordinates this is given as

$$\delta(\mathbf{x} - \mathbf{y}) = \delta(x_1 - y_1)\delta(x_2 - y_2)\delta(x_3 - y_3) \tag{3.9.5}$$

but for other coordinate systems the definition is more complicated. For example in spherical coordinates we have

$$\delta(\mathbf{x} - \mathbf{y}) = \lim_{\varepsilon \to 0}\left(\frac{\varepsilon^2 e^{-\varepsilon r}}{4\pi r}\right) \quad r = |\mathbf{x} - \mathbf{y}|, \quad \varepsilon > 0 \tag{3.9.6}$$

Given these special properties we can find a solution to the linear wave equation by using the following approach. We multiply the equation for the Green's function (3.9.2) by $p'(\mathbf{y}, \tau)$, and the linear wave equation (3.9.1) by $G(\mathbf{x}, t|\mathbf{y}, \tau)$, then subtract the results to give

$$\frac{1}{c_o^2}\left(p'\frac{\partial^2 G}{\partial \tau^2} - G\frac{\partial^2 p'}{\partial \tau^2}\right) - \left(p'\frac{\partial^2 G}{\partial y_i^2} - G\frac{\partial^2 p'}{\partial y_i^2}\right) = \delta(\mathbf{x} - \mathbf{y})\delta(t - \tau)p'(\mathbf{y}, \tau) \tag{3.9.7}$$

If we integrate the right side over $\tau$ and the volume $V(\mathbf{y})$ which includes $\mathbf{x}$ then we obtain $p'(\mathbf{x}, t)$, which is the result we want, so

$$\int_V\int_{-T}^T \frac{1}{c_o^2}\left(p'\frac{\partial^2 G}{\partial \tau^2} - G\frac{\partial^2 p'}{\partial \tau^2}\right) - \left(p'\frac{\partial^2 G}{\partial y_i^2} - G\frac{\partial^2 p'}{\partial y_i^2}\right)d\tau dV(\mathbf{y}) = p'(\mathbf{x}, t) \tag{3.9.8}$$

providing $\mathbf{x}$ is inside $V$ and $-T < t < T$. Considering the first term in the integrand we note that we can rearrange the differentials as

$$\left(p'\frac{\partial^2 G}{\partial \tau^2} - G\frac{\partial^2 p'}{\partial \tau^2}\right) = \frac{\partial}{\partial \tau}\left(p'\frac{\partial G}{\partial \tau} - G\frac{\partial p'}{\partial \tau}\right) \tag{3.9.9}$$

from which it follows that

$$\int_V\int_{-T}^T \frac{\partial}{\partial \tau}\left(p'\frac{\partial G}{\partial \tau} - G\frac{\partial p'}{\partial \tau}\right)d\tau dV(\mathbf{y}) = \int_V\left[p'\frac{\partial G}{\partial \tau} - G\frac{\partial p'}{\partial \tau}\right]_{\tau=-T}^{\tau=T} dV(\mathbf{y}) \tag{3.9.10}$$

If we specify initial conditions that require $p'$ and $\partial p'/\partial \tau$ to be zero at $\tau = -T$ then the integrand in Eq. (3.9.10) is zero at its lower limit. If we introduce the additional constraint on the Green's function that $\partial G(\mathbf{x}, t|\mathbf{y}, \tau)/\partial \tau$ and $G(\mathbf{x}, t|\mathbf{y}, \tau)$ be zero when $\tau \geq t$ then the upper limit of the integrand is zero if $t < T$. This constraint, called the *causality condition*, simply expresses the requirement that the sound heard at time $t$ must be produced at time $\tau < t$.

A similar expansion can be used on the second term in Eq. (3.9.8) which allows us to write

$$\int_V \left( p' \frac{\partial^2 G}{\partial y_i^2} - G \frac{\partial^2 p'}{\partial y_i^2} \right) dV(\mathbf{y}) = \int_V \frac{\partial}{\partial y_i} \left( p' \frac{\partial G}{\partial y_i} - G \frac{\partial p'}{\partial y_i} \right) dV(\mathbf{y}) \tag{3.9.11}$$

The integrand in this equation is now the divergence of a vector, so we can use the divergence theorem to turn the volume integral into a surface integral. If we choose a unit vector $n_i$, normal to the surface and pointing *into* the volume, we can write Eq. (3.9.8) as

$$p'(\mathbf{x}, t) = \int_{-T}^{T} \int_S \left( p'(\mathbf{y}, \tau) \frac{\partial G(\mathbf{x}, t | \mathbf{y}, \tau)}{\partial y_i} - G(\mathbf{x}, t | \mathbf{y}, \tau) \frac{\partial p'(\mathbf{y}, \tau)}{\partial y_i} \right) n_i dS(\mathbf{y}) d\tau \tag{3.9.12}$$

where the surface $S$ includes the surfaces of all objects in $V$ and the surface that bounds $V$ at its exterior boundary. Note the arguments of the Green's functions have been included in Eq. (3.9.12) for clarity. In free field applications the Sommerfeld radiation condition is used. This requires that there can only be outgoing waves on the exterior boundary and no sound is reflected back to the observer from the exterior boundary. This condition eliminates the need to include the exterior boundary in the surface integral.

The result given by Eq. (3.9.12) shows how any sound field, which satisfies the linear acoustic wave equation, can be evaluated from knowledge of the pressure and the pressure gradient on surfaces that bound the region of interest. To evaluate the integrals we need to know the Green's function, which must satisfy the inhomogeneous wave equation (3.9.2) and a causality condition, but is otherwise unrestricted. We can use Eq. (3.9.12) for both sound radiation problems in which the surface motion and or surface pressure is prescribed, and for sound scattering problems where passive boundary conditions relating the pressure and pressure gradient are defined on the scattering surfaces (see Fig. 3.7).

For radiating surfaces, it is useful to consider the two terms in Eq. (3.9.12) separately. The pressure gradient term can be obtained from the acoustic momentum equation (3.1.3) as

$$\frac{\partial p'}{\partial y_i} n_i = \mathbf{n} \cdot \nabla p' = -\rho_o \frac{\partial v_i}{\partial \tau} n_i \tag{3.9.13}$$

At a surface, the acoustic particle velocity normal to the surface $v_i n_i$ will be equal to the velocity normal to the surface $u_n$ imposed by the boundary condition, giving

$$p'(\mathbf{x}, t) = \int\limits_{-T}^{T}\int\limits_{S} \left( \rho_o \frac{\partial u_n(\mathbf{y}, \tau)}{\partial \tau} G(\mathbf{x}, t | \mathbf{y}, \tau) \right) dS(\mathbf{y}) d\tau + \int\limits_{-T}^{T}\int\limits_{S} \left( p'(\mathbf{y}, \tau) \frac{\partial G(\mathbf{x}, t | \mathbf{y}, \tau)}{\partial y_i} \right) n_i dS(\mathbf{y}) d\tau$$

(3.9.14)

The first term represents the sound radiated by the vibration of the surface, and the second term gives the contribution from the force per unit area $p' n_i$ exerted on the fluid by the surface. To evaluate these integrals, we need to specify a Green's function and this can be obtained most simply by comparing the result (Eq. 3.9.14) with the result given in Section 3.4 for a small pulsating sphere. In this example a small spherical surface is defined with its center at $\mathbf{y} = 0$, and the normal velocity on the surface is given by the real part of $u_o \exp(-i\omega t)$. If we allow the radius of the source to become smaller and smaller then eventually the spatial dependence of $G$ and $\partial G/\partial y_i$ will become so inconsequential that these functions can be taken outside of the surface integral, giving

$$p'(\mathbf{x}, t) = \mathrm{Re}\left[ \int\limits_{-T}^{T} [G]_{y_i=0} \int\limits_{S} \left( -i\omega\rho_o u_o e^{-i\omega\tau} \right) dS(\mathbf{y}) d\tau \right.$$
$$\left. + \int\limits_{-T}^{T} \left[ \frac{\partial G}{\partial y_i} \right]_{y_i=0} \left\{ \int\limits_{S} p' n_i dS(\mathbf{y}) \right\} d\tau \right]$$

The second integral in this equation depends on the surface integral in { } which represents the net force exerted on the fluid by the sphere, and is zero because the pressure is constant on the surface. In the first term the surface velocity is also constant on the surface and so the surface integral is simply $S = 4\pi a^2$ times the integrand, giving

$$p'(\mathbf{x}, t) = \mathrm{Re}\left[ \int\limits_{-T}^{T} \left( -i\omega\rho_o u_o S e^{-i\omega\tau} \right) [G(\mathbf{x}, t | \mathbf{y}, \tau)]_{y_i=0} d\tau \right]$$

(3.9.15)

Let us compare this result with the result we obtained in Section 3.4 when we analyzed the same problem. In that case we found (Eq. 3.4.6) that

$$p' = \mathrm{Re}\left[ -\frac{i\omega\rho_o u_o S e^{-i\omega t + ikr}}{4\pi r} \right] \qquad k = \omega/c_o \qquad r = |\mathbf{x}|$$

(3.9.16)

Comparing these two results shows that the Green's function must be

$$G(\mathbf{x}, t | 0, \tau) = \frac{\delta(t - r/c_o - \tau)}{4\pi r}$$

This is known as the free field Green's function because it defines the radiation from an infinitely small volume displacement source in an unbounded domain. Since the free field Green's function is of such significance we will give it the symbol, $G_o$. If the source is located at $\mathbf{y}$ rather than $\mathbf{y} = 0$ then we use $r = |\mathbf{x} - \mathbf{y}|$, and

$$G_o(\mathbf{x}, t | \mathbf{y}, \tau) = \frac{\delta(t - |\mathbf{x} - \mathbf{y}|/c_o - \tau)}{4\pi |\mathbf{x} - \mathbf{y}|} \tag{3.9.17}$$

The Green's function is a solution to Eq. (3.9.2) when $r = |\mathbf{x} - \mathbf{y}| > 0$ and indeterminate when $r = 0$, so we cannot use it to determine the surface pressure from Eq. (3.9.12) unless asymptotic methods are used. However, we can use it to find the pressure anywhere off the surface provided we can specify the pressure and the pressure gradient on the boundary. We also note that the free field Green's function satisfies the causality condition when $r > 0$ because $G_o = 0$ and $\partial G_o/\partial \tau = 0$ when $t < \tau < \tau + r/c_o$.

When we use the free field Green's function in Eq. (3.9.12) or (3.9.13) we need to define the gradient of $G_o$, which can be evaluated as

$$\frac{\partial G_o}{\partial y_i} = \frac{(x_i - y_i)}{r} \left( \frac{\dot{\delta}(t - r/c_o - \tau)}{4\pi r c_o} + \frac{\delta(t - r/c_o - \tau)}{4\pi r^2} \right) \qquad r = |\mathbf{x} - \mathbf{y}| \tag{3.9.18}$$

where the dot above the Dirac function represents a time derivative. Notice that the field from this term has a cosine directionality built in, and there is both a near field and far field term, as we found when we considered the translating sphere.

By using this form of the Green's function it is convenient to evaluate the time integrals in Eq. (3.9.14). Integrating the time derivative of the Dirac delta function by parts we obtain,

$$p'(\mathbf{x}, t) = \int_S \left[ \rho_o \frac{\partial u_n}{\partial \tau} \right]_{\tau = \tau^*} \frac{dS(\mathbf{y})}{4\pi |\mathbf{x} - \mathbf{y}|} + \int_S \left[ \frac{\partial p'}{\partial \tau} n_i + \frac{p' n_i c_o}{|\mathbf{x} - \mathbf{y}|} \right]_{\tau = \tau^*} \frac{(x_i - y_i) dS(\mathbf{y})}{4\pi |\mathbf{x} - \mathbf{y}|^2 c_o}$$

$$\tag{3.9.19}$$

where $\tau^* = t - |\mathbf{x} - \mathbf{y}|/c_o$.

The integrands of the surface integral are now evaluated at the *correct retarded time* $\tau^*$, also referred to as the *source time* or *emission time*. This is the observer time $t$ less the time it takes for the wave to propagate from the surface element to the observer at the speed of sound. The first term in Eq. (3.9.19) represents the sound generated by the vibration of the surface and depends on its acceleration. The second term gives the sound generated by the surface loading, and includes both near field and far field components. If we only consider the acoustic far field where $|\mathbf{x}| \gg |\mathbf{y}|$ then terms of order $1/|\mathbf{x}|^2$ can be ignored compared to terms of order $1/|\mathbf{x}|$, and so

$$p'(\mathbf{x}, t) \approx \frac{1}{4\pi |\mathbf{x}|} \int_S \left[ \rho_o \frac{\partial u_n}{\partial \tau} \right]_{\tau = \tau^*} dS(\mathbf{y}) + \frac{x_i}{4\pi |\mathbf{x}|^2 c_o} \int_S \left[ \frac{\partial p'}{\partial \tau} n_i \right]_{\tau = \tau^*} dS(\mathbf{y}) \tag{3.9.20}$$

This is an important result because it shows that in the acoustic far field the surface velocity term, often referred to as the monopole term, is inherently omnidirectional unless retarded time delays from different parts of the surface cause some kind of directionality (as was the case with the source distributions we discussed in Section 3.6). The second term has an inherent cosine directionality with a peak in the direction of the net force applied to the fluid (when retarded time effects are negligible), and so this term is often referred to as the dipole term.

In summary, we have considered sound radiation by fixed surfaces in a uniform stationary medium, and obtained a general solution to the linear wave equation using Green's functions. The acoustic field has been shown to be caused by the vibrational motion of the boundaries, and the forces exerted on the fluid by the surfaces. Surface vibration causes monopole radiation, and loading noise has an inherent dipole or cosine directionality.

## 3.10 Frequency domain solutions and Fourier transforms

In the first part of this chapter we discussed the solution to the wave equation for a harmonic wave of a single frequency, and then we discussed the general solution to the wave equation when the pressure was a function of time. The solution was found in terms of a Green's function $G(\mathbf{x},t|\mathbf{y},\tau)$ that represents the propagation from a point source at $\mathbf{y}$ to an observer at $\mathbf{x}$. The source fluctuations at time $\tau$ are received at the observer at time $t$. In many applications we are interested in the frequency content of a nonsinusoidal signal which is obtained by taking its Fourier transform with respect to time. In this section we will show how we can obtain the Fourier transform of the received signal directly from that of the source fluctuations by using a Green's function defined in the frequency domain.

There are many variations on the definition of the Fourier transform that can be found in the literature and there is no consistent convention. For this text we will be consistent with the conventions used by Morse and Ingard [1] and Goldstein [2], and define the Fourier transform of a time history as

$$\widetilde{p}(\mathbf{x}, \omega) = \frac{1}{2\pi} \int_{-T}^{T} p'(\mathbf{x}, t) e^{i\omega t} dt \tag{3.10.1}$$

and the inverse Fourier transform as

$$p'(\mathbf{x}, t) = \int_{-\infty}^{\infty} \widetilde{p}(\mathbf{x}, \omega) e^{-i\omega t} d\omega \tag{3.10.2}$$

Note that the units of the transformed pressure variable are Pascals per "radians per second." For the inverse transform relationship to be valid the time history must have finite energy during $-\infty < t < \infty$. Thus, in the forward transform, $T$ is taken to be large enough to encompass the entire (but finite) time history, so that

$$\int_{-\infty}^{\infty} |p'(\mathbf{x}, t)| dt = \int_{-T}^{T} |p'(\mathbf{x}, t)| dt < \infty$$

If we take the Fourier transform of the wave equation (3.9.1) with respect to the time variable $\tau$ then we obtain the Helmholtz equation defined as

$$\frac{\partial^2 \widetilde{p}(\mathbf{y}, \omega)}{\partial y_i^2} + k^2 \widetilde{p}(\mathbf{y}, \omega) = 0 \tag{3.10.3}$$

where $k = \omega/c_o$ is the wavenumber. This is exactly the same equation as Eq. (3.3.4). We can obtain a solution to the Helmholtz equation by using a Green's function that is the solution to the inhomogeneous equation

$$\frac{\partial^2 \widetilde{G}(\mathbf{x}|\mathbf{y})}{\partial y_i^2} + k^2 \widetilde{G}(\mathbf{x}|\mathbf{y}) = -\delta(\mathbf{x} - \mathbf{y}) \tag{3.10.4}$$

Note the similarities and differences between this equation and Eq. (3.9.2), and that there is a negative sign in front of the Dirac delta function, which is required so that the two equations are consistent.

If we now multiply Eq. (3.10.3) by $\widetilde{G}$ and Eq. (3.10.4) by $\widetilde{p}(\mathbf{y}, \omega)$, integrate over the volume and subtract we obtain

$$\widetilde{p}(\mathbf{x}, \omega) = -\int_V \left( \widetilde{p}(\mathbf{y}, \omega) \frac{\partial^2 \widetilde{G}(\mathbf{x}|\mathbf{y})}{\partial y_i^2} - \widetilde{G}(\mathbf{x}|\mathbf{y}) \frac{\partial^2 \widetilde{p}(\mathbf{y}, \omega)}{\partial y_i^2} \right) dV(\mathbf{y})$$

$$= -\int_V \frac{\partial}{\partial y_i} \left( \widetilde{p}(\mathbf{y}, \omega) \frac{\partial \widetilde{G}(\mathbf{x}|\mathbf{y})}{\partial y_i} - \widetilde{G}(\mathbf{x}|\mathbf{y}) \frac{\partial \widetilde{p}(\mathbf{y}, \omega)}{\partial y_i} \right) dV(\mathbf{y})$$

As before, the volume integral can be evaluated using the divergence theorem. If we choose a unit vector $n_i$, normal to the surface elements and pointing *into* the volume we obtain

$$\widetilde{p}(\mathbf{x}, \omega) = \int_S \left( \widetilde{p}(\mathbf{y}, \omega) \frac{\partial \widetilde{G}(\mathbf{x}|\mathbf{y})}{\partial y_i} - \widetilde{G}(\mathbf{x}|\mathbf{y}) \frac{\partial \widetilde{p}(\mathbf{y}, \omega)}{\partial y_i} \right) n_i(\mathbf{y}) dS(\mathbf{y}) \tag{3.10.5}$$

This is the frequency domain solution to the wave equation, or Helmholtz equation, and is equivalent to Eq. (3.9.12), which gives the time domain solution. The equivalence between these two results is established by taking the inverse transform of Eq. (3.10.5), and substituting Eq. (3.10.1), so

$p'(\mathbf{x}, t)$

$$= \int_{-\infty}^{\infty} \int_S \left\{ \frac{1}{2\pi} \int_{-T}^{T} \left( p'(\mathbf{y}, \tau) \left[ \frac{\partial \widetilde{G}(\mathbf{x}|\mathbf{y})}{\partial y_i} \right]_\omega - \left[ \widetilde{G}(\mathbf{x}|\mathbf{y}) \right]_\omega \frac{\partial p'(\mathbf{y}, \tau)}{\partial y_i} \right) n_i(\mathbf{y}) e^{-i\omega(t-\tau)} d\tau \right\} dS(\mathbf{y}) d\omega$$

$$(3.10.6)$$

where the frequency domain Green's function is evaluated at the frequency $\omega$.

Comparing Eq. (3.10.6) with Eq. (3.9.12) shows that the relationship between the Green's function in the time domain and the frequency domain is given by

$$G(\mathbf{x}, t|\mathbf{y}, \tau) = \frac{1}{2\pi} \int_{-\infty}^{\infty} \left[ \widetilde{G}(\mathbf{x}|\mathbf{y}) \right]_\omega e^{-i\omega(t-\tau)} d\omega \qquad (3.10.7)$$

and the frequency domain Green's function is obtained by taking the Fourier transform defined by Eq. (3.10.1) of $2\pi G(\mathbf{x}, t|\mathbf{y}, 0)$ as

$$\widetilde{G}(\mathbf{x}|\mathbf{y}) = \int_{-T}^{T} G(\mathbf{x}, t|\mathbf{y}, 0) e^{i\omega t} dt$$

Using the result given by Eq. (3.9.17) defines the free field Green's function in the frequency domain as

$$\widetilde{G}_o(\mathbf{x}|\mathbf{y}) = \frac{e^{ik|\mathbf{x}-\mathbf{y}|}}{4\pi|\mathbf{x}-\mathbf{y}|} \qquad (3.10.8)$$

In the rest of the text we will find that solutions using Green's functions in both the time domain and frequency domain are valuable, and we will use Eqs. (3.9.12), (3.10.5) to solve problems in aeroacoustics that involve solid boundaries. The Green's functions given by Eqs. (3.10.8), (3.9.17) are referred to as the free field Green's functions and we will show in the next chapters how these results can be generalized to eliminate contributions from surface source terms that are sometimes difficult to evaluate.

# References

[1] P.M. Morse, K.U. Ingard, Theoretical Acoustics, McGraw-Hill, New York, 1987.
[2] M.E. Goldstein, Aeroacoustics, McGraw-Hill International Book Company, New York, NY, 1976.

# Lighthill's acoustic analogy

**4**

The equations of motion used for linear acoustics are based on a number of assumptions that do not apply to turbulent flows. This chapter will discuss the classic theory of sound generation by turbulence, which is the basis for our understanding of aeroacoustics. Lighthill's acoustic analogy will be described and an inhomogeneous wave equation will be derived that identifies the sources of sound in an arbitrary unsteady flow. The solution to this wave equation in the presence of stationary surfaces will be given based on Curle's theorem and Green's functions. The concept of a tailored Green's function will also be introduced.

## 4.1  Lighthill's analogy

In the previous chapters we discussed sound radiation from sources in a uniform stationary medium. We showed that in ideal conditions the propagation of sound could be described by the linear acoustic wave equation with a pressure perturbation as the dependent variable. Sound was only generated by disturbances that caused a prescribed motion of a boundary, and boundary conditions were matched using the linearized momentum equation. With the invention of the jet engine it was soon realized that the theory of linear acoustics could not be used to specify the radiation of sound from this very loud source. When a jet exhausts into a stationary fluid a large region of turbulent flow is generated but there are no surfaces with prescribed motions, so some other mechanism of creating sound waves must be present. In general, turbulence occurs in high Reynolds number flows where viscous effects are small, and the motion of the fluid is dominated by nonlinear interactions. To describe the noise generated by turbulence we must be very careful about making assumptions that allow the equations of motion to be linearized, and this is the reason why the concepts of linear acoustics cannot provide the basis for sound generation by an unconfined fluid flow.

Lighthill's analogy specifically addresses the problem of sound generation by a region of high speed turbulent flow in an otherwise stationary fluid as illustrated in Fig. 4.1. The objective is to determine the equations that describe the generation of sound waves that propagate to the acoustic far field, as distinct from defining the fluid motion in the turbulent flow. The magnitude of turbulent velocity fluctuations is typically 10% of the mean flow speed, and for subsonic flows of interest in air, will probably exceed 2–3 m/s. In contrast an acoustic wave has a particle velocity $u = p'/\rho_o c_o = \rho' c_o/\rho_o$ and, for a high intensity acoustic wave in air, we showed in Chapter 3 that $\rho'/\rho_o$ is of order 1/1000. Since $c_o = 343$ m/s it follows that the acoustic wave will only be a small part of the unsteady motion and so we must be very cautious about the coupling of the turbulent flow to the acoustic waves. If, for example, we blindly assume the turbulence to be incompressible then no sound is generated because this assumption eliminates all acoustic effects. Alternatively, if we assume a weakly incompressible flow we can permit acoustic motion in a fluid flow, which

*Aeroacoustics of Low Mach Number Flows.* http://dx.doi.org/10.1016/B978-0-12-809651-2.00004-7

is dominated by incompressible turbulence, but do we get the right result? For this reason, Lighthill's analogy is based on the exact equations of fluid flow, without making any assumptions about the compressibility effects in the region of turbulence, and is valuable because it provides the correct equations for coupling the acoustic wave motion outside the turbulent region to the large velocity fluctuations, which occur inside the flow.

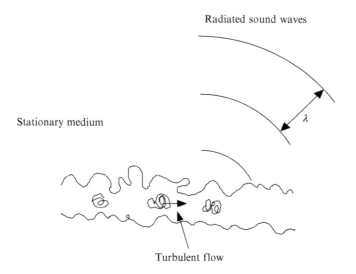

**Fig. 4.1** Sound radiation from a region of turbulent flow in an otherwise stationary medium.

To derive Lighthill's equation we take the time derivative of the continuity equation (2.2.6)

$$\frac{\partial}{\partial t}\left(\frac{\partial \rho}{\partial t} + \frac{\partial(\rho v_i)}{\partial x_i}\right) = 0 \tag{4.1.1}$$

and subtract the divergence of the momentum equation (2.3.9)

$$\frac{\partial}{\partial x_i}\left(\frac{\partial(\rho v_i)}{\partial t} + \frac{\partial(\rho v_i v_j + p_{ij})}{\partial x_j}\right) = 0 \tag{4.1.2}$$

to give

$$\frac{\partial^2 \rho}{\partial t^2} = \frac{\partial^2(\rho v_i v_j + p_{ij})}{\partial x_i \partial x_j} \tag{4.1.3}$$

In this equation the diagonal elements of the stress tensor are defined as gauge pressure, so they are relative to the surrounding ambient pressure, giving $p_{ij} = (p - p_\infty)\delta_{ij} - \sigma_{ij}$. We then define the density perturbation relative to the surrounding medium as

$\rho' = \rho - \rho_\infty$ and subtract $\partial^2(\rho' c_\infty^2)/\partial x_i^2$ from each side of Eq. (4.1.3), where $c_\infty$ is the ambient speed of sound outside the flow, which is constant, and note that $\partial \rho/\partial t = \partial \rho'/\partial t$, to obtain a wave equation in the form

$$\frac{\partial^2 \rho'}{\partial t^2} - c_\infty^2 \frac{\partial^2 \rho'}{\partial x_i^2} = \frac{\partial^2 T_{ij}}{\partial x_i \partial x_j} \qquad T_{ij} = \rho v_i v_j - \left[(p - p_\infty) - (\rho - \rho_\infty)c_\infty^2\right]\delta_{ij} - \sigma_{ij} \quad (4.1.4)$$

This is known as Lighthill's wave equation [1]. The left side specifies the propagation of an acoustic wave in a uniform medium with sound speed $c_\infty$ using density as the dependent variable. The right side is frequently referred to as a source term, which contains all the effects that generate acoustic waves. However, it would appear that the choice of the sound speed $c_\infty$ is arbitrary but that is not the case. Lighthill's equation represents an analogy for the waves that radiated from a finite volume of turbulent flow into a surrounding stationary medium. The speed of sound $c_\infty$ must be chosen as the speed of sound in the stationary medium, not the region of turbulence where it could have a very different value due to compressible flow effects, combustion, or entrained air in underwater applications. This choice is appropriate because the objective is to evaluate the acoustic waves that propagate away from the region of turbulent flow in the stationary medium, and the source term defines how the turbulence, or fluid perturbations in the turbulent region, will couple with the sound waves in the far field.

It is important to appreciate that in deriving Lighthill's equation no approximations have been made, and so Eq. (4.1.4) is an exact rearrangement of the equations of fluid flow. In linear acoustics $\rho' = \rho - \rho_\infty$ is replaced by $(p - p_\infty)/c_\infty^2$, it is assumed that nonlinear interactions are not important so $\rho v_i v_j = 0$ and viscous effects can be ignored. Consequently $T_{ij}$ is approximated as being equal to zero and so is the right side of Eq. (4.1.4).

Lighthill's equation is also referred to as *Lighthill's acoustic analogy* because it treats the turbulent flow as if it contained sound waves propagating in the same manner as in the surrounding fluid. All the issues about the modeling of the turbulence, which cannot be addressed by linear acoustics, are addressed in Lighthill's equation by the inclusion of the term on the right side. However, since this equation is all inclusive, it also hides some important details of what is happening in the flow region. It should be remembered here that Lighthill's equation, although exact, is designed to provide insight into the sound waves outside the flow, not in the region of turbulent motion where the speed of sound propagation may be distinctly different from $c_\infty$.

## 4.2 Limitations of the acoustic analogy

### 4.2.1 Nearly incompressible flow

Lighthill's equation has been used extensively in the fields of aero and hydroacoustics because of its simplicity and because it makes no approximations about the flow. However, in most situations the variable $T_{ij}$ cannot be determined and approximations have to be made to estimate the terms which couple most effectively with the sound field. The difficulty is that the very large fluctuations of the source terms may be very weakly coupled to the sound waves, and so effectively do not radiate any sound.

First consider an almost incompressible flow in which the velocity fluctuations are dominated by turbulence, and the viscous terms can be ignored. We also assume the flow is homentropic so that $(p - p_\infty) = (\rho - \rho_\infty)c_\infty^2$. Furthermore we will assume as before that $\rho' \ll \rho_\infty = const$ and $v_i^2 \ll c_\infty^2$. These approximations are perfectly legitimate in hydroacoustic applications but must be treated with care in aeroacoustic applications where the flow may be isentropic but not homentropic and flow velocities approach the speed of sound. Given these approximations Eq. (4.1.4) reduces to

$$\frac{\partial^2 \rho'}{\partial t^2} - c_\infty^2 \frac{\partial^2 \rho'}{\partial x_i^2} = \rho_\infty \frac{\partial^2 (v_i v_j)}{\partial x_i \partial x_j} \tag{4.2.1}$$

Now let us reduce the source term further by expanding it term by term

$$\frac{\partial^2 (v_i v_j)}{\partial x_i \partial x_j} = \frac{\partial}{\partial x_i}\left(v_i \frac{\partial v_j}{\partial x_j} + v_j \frac{\partial v_i}{\partial x_j}\right) = \frac{\partial}{\partial x_i}\left(v_i \frac{\partial v_j}{\partial x_j}\right) + \frac{\partial v_j}{\partial x_i}\frac{\partial v_i}{\partial x_j} + v_j \frac{\partial^2 v_i}{\partial x_i \partial x_j}$$

For an incompressible flow the divergence of the velocity is zero so the first and last term drop out giving

$$\frac{\partial^2 (v_i v_j)}{\partial x_i \partial x_j} = \frac{\partial v_j}{\partial x_i}\frac{\partial v_i}{\partial x_j} \tag{4.2.2}$$

We conclude that, given the assumptions of an almost incompressible flow, the unsteady terms in Lighthill's source term are not only nonlinear but also determined by the velocity gradients. However, as was pointed out in Chapter 1, the lengthscales of the turbulent eddies in this flow are of order $U_c/f$ where $U_c$ is the convection speed and $f$ is the frequency, and this is much less than the acoustic wavelength $\lambda = c_o/f$ at the same frequency. However, the source term Eq. (4.2.2) is nonlinear and will generate perturbations at the sum and difference frequencies of two interacting velocity fluctuations. As a result the radiated sound field will occur at a different frequency from the frequency of the velocity perturbation observed in the flow, and this can allow the acoustic and turbulence lengthscales to match. The important point to be made here is that the flow noise is generated by a nonlinear interaction, and so linear approximations will not give the correct results.

## 4.2.2  Uniform flow

In aeroacoustic applications we cannot always assume that $v_i^2 \ll c_\infty^2$ and the local speed of sound may vary significantly from its value outside the flow. The wave equation in (4.1.4) is then a poor approximation for sound waves in the flow. Specifically, there are two effects taking place, the generation of sound by turbulence at a source element, and secondly the propagation of the sound from the source element through the nonuniform flow to the acoustic far field. The latter effect causes waves to bend as they propagate, and this feature is hidden in Lighthill's source term. Separating sound propagation effects from sound generation in $T_{ij}$ has been the cause of significant debate for a number of years. To illustrate the subtleties of refraction or propagation effects, and show how Lighthill's equation includes these effects in $T_{ij}$, consider turbulence in a uniformly moving medium. Because the mean flow is uniform we can partition the velocity such that $v_i = U_i^{(\infty)} + w_i$ where $U_i^{(\infty)}$ is constant everywhere, and this gives

$$T_{ij} = \rho \left( U_i^{(\infty)} + w_i \right) \left( U_j^{(\infty)} + w_j \right) + p_{ij} - \rho' c_\infty^2 \delta_{ij} \tag{4.2.3}$$

We can expand the product of velocities term by term to give

$$\rho \left( U_i^{(\infty)} + w_i \right) \left( U_j^{(\infty)} + w_j \right) = \rho \left( U_i^{(\infty)} U_j^{(\infty)} + U_j^{(\infty)} w_i + U_i^{(\infty)} w_j + w_i w_j \right) \tag{4.2.4}$$

and so the right side of Eq. (4.1.4) becomes

$$\frac{\partial^2 T_{ij}}{\partial x_i \partial x_j} = U_i^{(\infty)} U_j^{(\infty)} \frac{\partial^2 \rho}{\partial x_i \partial x_j} + 2 U_j^{(\infty)} \frac{\partial^2 (\rho w_i)}{\partial x_i \partial x_j} + \frac{\partial^2}{\partial x_i \partial x_j} \left( \rho w_i w_j + p_{ij} - \rho' c_\infty^2 \delta_{ij} \right) \tag{4.2.5}$$

We then use the continuity equation (2.2.6) to modify the second term

$$\frac{\partial (\rho w_i)}{\partial x_i} = -\frac{\partial \rho}{\partial t} - U_i^{(\infty)} \frac{\partial \rho}{\partial x_i} \tag{4.2.6}$$

so we obtain

$$\frac{\partial^2 T_{ij}}{\partial x_i \partial x_j} = -U_i^{(\infty)} U_j^{(\infty)} \frac{\partial^2 \rho}{\partial x_i \partial x_j} - 2 U_j^{(\infty)} \frac{\partial^2 \rho}{\partial t \partial x_j} + \frac{\partial^2}{\partial x_i \partial x_j} \left( \rho w_i w_j + p_{ij} - \rho' c_\infty^2 \delta_{ij} \right) \tag{4.2.7}$$

Note the sign of the first term is changed by this last operation. Using this result in Eq. (4.1.4) then gives, since $\rho = \rho_\infty + \rho'$ where $\rho_\infty$ is constant

$$\frac{\partial^2 \rho'}{\partial t^2} + 2 U_j^{(\infty)} \frac{\partial^2 \rho'}{\partial t \partial x_j} + U_i^{(\infty)} U_j^{(\infty)} \frac{\partial^2 \rho'}{\partial x_i \partial x_j} - c_\infty^2 \frac{\partial^2 \rho'}{\partial x_i^2} = \frac{\partial^2}{\partial x_i \partial x_j} \left( \rho w_i w_j + p_{ij} - \rho' c_\infty^2 \delta_{ij} \right) \tag{4.2.8}$$

The first three terms on the left can be combined to give a convected wave equation, so

$$\frac{D_\infty^2 \rho'}{Dt^2} - c_\infty^2 \frac{\partial^2 \rho'}{\partial x_i^2} = \frac{\partial^2}{\partial x_i \partial x_j} \left( \rho w_i w_j + p_{ij} - \rho' c_\infty^2 \delta_{ij} \right)$$

$$\text{where} \quad \frac{D_\infty \rho'}{Dt} = \frac{\partial \rho'}{\partial t} + U_j^{(\infty)} \frac{\partial \rho'}{\partial x_j}$$

(4.2.9)

The important part of this exercise is that it shows how Lighthill's equation can be applied in a uniform flow where the acoustic waves satisfy a convected wave equation. It also shows that when applying Lighthill's wave equation (4.1.4) in a uniform flow there are terms associated with linear acoustic perturbations that appear in the source term. In Eq. (4.2.9) the source term depends on the velocity perturbation about the mean flow $w_i$, and the mean flow only affects unsteady terms on the left side, causing a propagation effect. Both Eqs. (4.1.4), (4.2.9) are mathematically correct, but if we wish to determine the acoustic waves in a uniformly convected flow then Eq. (4.2.9) is obviously the better choice. Note that the mean flow in this example is constant everywhere, which is never the case in a real flow, and we must resort to more detailed analyses to extract the effect of a nonuniform mean flow on sound propagation inside the flow.

## 4.3 Curle's theorem

A solution to Lighthill's wave equation can be obtained by using the method of Green's functions described in Chapter 3. Specifically, we will determine the acoustic field in a stationary uniform medium, which includes a bounded region of unsteady flow (in which the assumptions of linear acoustics need not apply). In line with Lighthill's analogy, the solution will not necessarily be correct in the turbulent flow, but should apply everywhere outside the flow where the medium is uniform and at rest. We will start by writing Lighthill's wave equation (4.1.4), in terms of $\mathbf{y}$ and $\tau$, as

$$\frac{1}{c_\infty^2} \frac{\partial^2 \left( \rho' c_\infty^2 \right)}{\partial \tau^2} - \frac{\partial^2 \left( \rho' c_\infty^2 \right)}{\partial y_i^2} = \frac{\partial^2 T_{ij}}{\partial y_i \partial y_j}$$

(4.3.1)

which has the same structure as Eq. (3.9.1) with $p'$ replaced by $\rho' c_\infty^2$ and a source term on the right side. As before we obtain a solution by multiplying this equation by a Green's function $G$ and subtracting it from $\rho' c_\infty^2$ times Eq. (3.9.2). Then, as in Eq. (3.9.7), we obtain

$$\frac{1}{c_\infty^2} \left( \rho' c_\infty^2 \frac{\partial^2 G}{\partial \tau^2} - G \frac{\partial^2 \left( \rho' c_\infty^2 \right)}{\partial \tau^2} \right) - \left( \rho' c_\infty^2 \frac{\partial^2 G}{\partial y_i^2} - G \frac{\partial^2 \left( \rho' c_\infty^2 \right)}{\partial y_i^2} \right)$$

$$= \delta(\mathbf{x} - \mathbf{y}) \delta(t - \tau) \rho'(\mathbf{y}, \tau) c_\infty^2 - G \frac{\partial^2 T_{ij}}{\partial y_i \partial y_j}$$

(4.3.2)

When this equation is integrated over time and space the left side reduces in exactly the same way as it did in the previous chapter giving a surface integral as in

Eq. (3.9.12). However, in this case we must also include the additional term on the right of Eq. (4.3.2) giving

$$
\rho'(\mathbf{x}, t)c_\infty^2 = \int\limits_{-T}^{T}\int\limits_{S} \left( \rho'(\mathbf{y}, \tau)c_\infty^2 \frac{\partial G}{\partial y_i} - G \frac{\partial\left(\rho'(\mathbf{y}, \tau)c_\infty^2\right)}{\partial y_i} \right) n_i dS(\mathbf{y}) d\tau
$$

$$
+ \int\limits_{-T}^{T}\int\limits_{V} G\left( \frac{\partial^2 T_{ij}(\mathbf{y}, \tau)}{\partial y_i \partial y_j} \right) dV(\mathbf{y}) d\tau \tag{4.3.3}
$$

This equation, while correct, is a dangerous result. The issue is that if we assume that the source region, where $T_{ij}$ is nonzero, is very small, then we can neglect the propagation times between different points $\mathbf{y}$ inside the volume and the observer at $\mathbf{x}$, then the Green's function can be taken outside the volume integral and it becomes the integral of a divergence. This can be evaluated as a surface integral, and if there are no objects in the flow, the only contributions to the integrand comes from surfaces that can be placed where $T_{ij}=0$. Consequently, for sound radiation from turbulence, in the absence of any nearby surfaces, Eq. (4.3.3) gives a result that is identically zero if the turbulent region is so small that the retarded times can be neglected in the volume integral. To use Eq. (4.3.3) we must know the value of integrand precisely at the correct source times $\tau$ and carry out an integration with no rounding error in order to get the correct result, but the danger is that numerical errors in the integration will be larger than the contribution from the source terms. Lighthill recognized this issue and proposed an important simplification of this equation that identified the leading order source terms in the volume integral, so that it could be evaluated without requiring exact retarded time calculations. This concept was later extended by both Curle [2] and Doak [3] to apply to turbulent flow in the presence of surfaces. Using Doak's approach, which describes the acoustic propagation using a Green's function, we can derive Curle's equation by using the expansion

$$
\frac{\partial}{\partial y_i}\left( G\frac{\partial T_{ij}}{\partial y_j} \right) - \frac{\partial}{\partial y_j}\left( T_{ij}\frac{\partial G}{\partial y_i} \right) = G\frac{\partial^2 T_{ij}}{\partial y_i \partial y_j} - T_{ij}\frac{\partial^2 G}{\partial y_i \partial y_j} \tag{4.3.4}
$$

The volume integral in Eq. (4.3.3) then becomes

$$
\int\limits_{-T}^{T}\int\limits_{V} G\left( \frac{\partial^2 T_{ij}(\mathbf{y}, \tau)}{\partial y_i \partial y_j} \right) dV(\mathbf{y}) d\tau = \int\limits_{-T}^{T}\int\limits_{V} \left( \frac{\partial^2 G}{\partial y_i \partial y_j} \right) T_{ij}(\mathbf{y}, \tau) dV(\mathbf{y}) d\tau
$$

$$
+ \int\limits_{-T}^{T}\int\limits_{V} \frac{\partial}{\partial y_i}\left( G\frac{\partial T_{ij}(\mathbf{y}, \tau)}{\partial y_j} \right) dV(\mathbf{y}) d\tau
$$

$$
- \int\limits_{-T}^{T}\int\limits_{V} \frac{\partial}{\partial y_j}\left( T_{ij}(\mathbf{y}, \tau)\frac{\partial G}{\partial y_i} \right) dV(\mathbf{y}) d\tau \tag{4.3.5}
$$

The volume integrals in the last two terms can be evaluated using the divergence theorem, using a unit normal $n_i$ that points into the fluid, and if we also note that $T_{ij} = T_{ji}$, we obtain

$$
\int_{-T}^{T} \int_V G\left(\frac{\partial^2 T_{ij}(\mathbf{y}, \tau)}{\partial y_i \partial y_j}\right) dV(\mathbf{y}) d\tau = \int_{-T}^{T} \int_V \left(\frac{\partial^2 G}{\partial y_i \partial y_j}\right) T_{ij}(\mathbf{y}, \tau) dV(\mathbf{y}) d\tau
$$

$$
+ \int_{-T}^{T} \int_S \left(T_{ij}(\mathbf{y}, \tau)\frac{\partial G}{\partial y_i} - G\frac{\partial T_{ij}(\mathbf{y}, \tau)}{\partial y_i}\right) n_j dS(\mathbf{y}) d\tau
$$

$$(4.3.6)$$

We are now able to replace the volume integral in Eq. (4.3.3) using Eq. (4.3.6), but we note that the integrand of the surface integral is modified giving

$$
\rho'(\mathbf{x}, t)c_\infty^2 = \int_{-T}^{T} \int_S \left((\rho' c_\infty^2 \delta_{ij} + T_{ij})\frac{\partial G}{\partial y_i} - G\frac{\partial(\rho' c_\infty^2 \delta_{ij} + T_{ij})}{\partial y_i}\right) n_j dS(\mathbf{y}) d\tau
$$

$$
+ \int_{-T}^{T} \int_V \left(\frac{\partial^2 G}{\partial y_i \partial y_j}\right) T_{ij}(\mathbf{y}, \tau) dV(\mathbf{y}) d\tau
$$

$$(4.3.7)$$

Finally, we make use of the definition of Lighthill's stress tensor to replace

$$
\rho' c_\infty^2 \delta_{ij} + T_{ij} = p_{ij} + \rho v_i v_j \tag{4.3.8}
$$

and then use the momentum equation to give

$$
\rho'(\mathbf{x}, t)c_\infty^2 = \int_{-T}^{T} \int_S \left((p_{ij} + \rho v_i v_j)\frac{\partial G}{\partial y_i} + G\frac{\partial(\rho v_j)}{\partial \tau}\right) n_j dS(\mathbf{y}) d\tau
$$

$$
+ \int_{-T}^{T} \int_V \left(\frac{\partial^2 G}{\partial y_i \partial y_j}\right) T_{ij}(\mathbf{y}, \tau) dV(\mathbf{y}) d\tau
$$

$$(4.3.9)$$

This is a very general result and we will discuss it term by term in the next section. However, before proceeding, it is valuable to compare Eq. (4.3.9) and the results for linear acoustics. In the previous chapter we derived a solution to the homogeneous wave equation (with no source terms on the right side), and showed that the acoustic field can be determined by the boundary conditions on the surfaces in the region of interest. The two important quantities that contributed to the sound generation were shown to be the surface acceleration and the force per unit area applied to the fluid by the surface (see Eq. 3.9.14). In this section a more general analysis has been carried out, which is not limited by the assumptions of linear acoustics used in Chapter 3.

The only restrictions are that the surfaces bounding the fluid are stationary and that the medium is at rest outside the region of turbulent flow. The solution to Lighthill's wave equation is then given by Eq. (4.3.9), with source terms that depend on (i) the rate of change of mass flux on the surface, (ii) the force per unit area $p_{ij}n_j$ applied to the fluid by the surface, (iii) the momentum flux across the surface, and (iv) a volume integral contribution from the distribution of Lighthill's stress tensor $T_{ij}$ throughout the fluid. The additional contributions that appear in Eq. (4.3.9) and not in Eq. (3.9.13) are the momentum flux and the contribution from the volume sources, both of which are nonlinear and so not part of the linear acoustic wave equation. These are important because they identify the sources of sound in a turbulent flow.

It is also of interest to note that the result given by Eq. (4.3.9) depends on a Green's function that only needs to be a solution to Eq. (3.9.2) and satisfy a causality condition. We can therefore choose $G$ in a fairly arbitrary manner. In the previous chapter we gave the solution for $G$ in a free field, but if there are rigid surfaces present then we can also find solutions for $G$ that satisfy the boundary conditions. This technique is useful for certain problems and will be discussed in more detail later in this chapter.

One final note on Eq. (4.3.9), we can easily show that since the free field Green's function

$$G_o(\mathbf{x},t|\mathbf{y},\tau) = \frac{\delta(t-\tau-|\mathbf{x}-\mathbf{y}|/c_\infty)}{4\pi|\mathbf{x}-\mathbf{y}|} \tag{4.3.10}$$

is only a function of $|\mathbf{x}-\mathbf{y}|$, and so has the properties

$$\frac{\partial G_o}{\partial y_i} = -\frac{\partial G_o}{\partial x_i} \qquad\qquad \frac{\partial^2 G_o}{\partial y_i \partial y_j} = \frac{\partial^2 G_o}{\partial x_i \partial x_j}$$

and so we can rewrite Eq. (4.3.9) as

$$\rho'(\mathbf{x},t)c_\infty^2 = \int_{-T}^{T}\int_S \frac{\partial(\rho v_j)}{\partial \tau}G_o n_j dS(\mathbf{y})d\tau - \frac{\partial}{\partial x_i}\int_{-T}^{T}\int_S (p_{ij}+\rho v_i v_j)G_o n_j dS(\mathbf{y})d\tau$$

$$+ \frac{\partial^2}{\partial x_i \partial x_j}\int_{-T}^{T}\int_V T_{ij}(\mathbf{y},\tau)G_o dV(\mathbf{y})d\tau \tag{4.3.11}$$

Then if we use the definition of $G_o$ in Eq. (4.3.10) and carry out the time integrals we find

$$\rho'(\mathbf{x},t)c_\infty^2 = \int_S \left[\frac{\partial(\rho v_j)}{\partial \tau}\right]_{\tau=\tau^*}\frac{n_j dS(\mathbf{y})}{4\pi|\mathbf{x}-\mathbf{y}|} - \frac{\partial}{\partial x_i}\int_S \left[p_{ij}+\rho v_i v_j\right]_{\tau=\tau^*}\frac{n_j dS(\mathbf{y})}{4\pi|\mathbf{x}-\mathbf{y}|}$$

$$+ \frac{\partial^2}{\partial x_i \partial x_j}\int_V \left[T_{ij}(\mathbf{y},\tau)\right]_{\tau=\tau^*}\frac{dV(\mathbf{y})}{4\pi|\mathbf{x}-\mathbf{y}|} \tag{4.3.12}$$

In this solution we have written the results so the source terms are evaluated at the correct retarded times $\tau=\tau^*=t-|\mathbf{x}-\mathbf{y}|/c_\infty$. The surfaces are stationary so $\mathbf{y}$ is independent of $\tau$, and fluctuations that occur in different parts of the flow at the same source time $\tau$ will contribute to the received signal at $\mathbf{x}$ at different reception times $t$. The effect of retarded time is therefore a key component of accurately predicting far field sound from flow noise sources. If the source region is sufficiently small that retarded times can be ignored, then the integrals can be evaluated approximately at a fixed source time. In contrast to Eq. (4.3.3) we now see that the volume source term is given by the integral of $T_{ij}$ and does not integrate to zero. It therefore represents the acoustic source term in a flow to leading order.

## 4.4 Monopole, dipole, and quadrupole sources

The results given in the previous section are sufficiently important that we need to spend some time fully evaluating the importance of each term. To facilitate this, we will split Eq. (4.3.12) into three parts so

$$\rho' c_\infty^2 = \left(\rho' c_\infty^2\right)_{\text{monopole}} + \left(\rho' c_\infty^2\right)_{\text{dipole}} + \left(\rho' c_\infty^2\right)_{\text{quadrupole}} \tag{4.4.1}$$

The first term of interest is given by the monopole term, which we obtain from the mass flux term of Eq. (4.3.12). For a surface on which the fluid moves with a normal velocity $v_i n_i = u_n$ the monopole term is

$$\left(\rho'(\mathbf{x}, t) c_\infty^2\right)_{\text{monopole}} = \int_S \left[\frac{\partial(\rho u_n)}{\partial \tau}\right]_{\tau=\tau^*} \frac{dS(\mathbf{y})}{4\pi|\mathbf{x} - \mathbf{y}|} \tag{4.4.2}$$

This equation shows that sound can be generated by the flux of mass across the surface, which is determined by the local density and the motion across the surface. If the surface is rigid or impenetrable (and of course stationary) then this term is zero. If the surface is sufficiently small that retarded time effects can be ignored, it is said to be acoustically compact, and in the acoustic far field, we can write Eq. (4.4.2) in terms of the position $\mathbf{x}$ (relative to a point inside or on the surface) as

$$\left(\rho'(\mathbf{x}, t) c_\infty^2\right)_{\text{monopole}} = \frac{1}{4\pi|\mathbf{x}|} \left[\int_S \frac{\partial(\rho u_n)}{\partial \tau} dS(\mathbf{y})\right]_{\tau=\tau^*}$$

This term is omnidirectional in the acoustic far field, and so is labeled the monopole term, as discussed in Chapter 3. It shows that the source term depends on the net rate of mass flux at the surface $\rho u_n$, and does not depend on the shape of the surface. Consequently the acoustic field does not distinguish between an acoustically compact spherical surface or an acoustically compact surface of any other shape. The discussion of the acoustic field from a spherical source is therefore far more general than it would appear from first sight, and the sound radiation from a vibrating surface of arbitrary shape is the same as the sound radiation from a spherical surface with the same net rate of mass flux injection.

Next consider the dipole term of Eq. (4.3.12), which is the surface integral with terms related to the surface loading

$$\left(\rho'(\mathbf{x}, t)c_\infty^2\right)_{\text{dipole}} = -\frac{\partial}{\partial x_i} \int_S \left[p_{ij} + \rho v_i v_j\right]_{\tau=\tau^*} \frac{n_j dS(\mathbf{y})}{4\pi|\mathbf{x} - \mathbf{y}|} \qquad (4.4.3)$$

To evaluate the space derivative we can use the chain rule to define

$$\frac{\partial f(\tau^*)}{\partial x_i} = \frac{\partial \tau^*}{\partial x_i}\left[\frac{\partial f(\tau)}{\partial \tau}\right]_{\tau=\tau^*}$$

but since $\tau^* = t - |\mathbf{x} - \mathbf{y}|/c_\infty$ we find that

$$\frac{\partial f(\tau^*)}{\partial x_i} = \frac{-(x_i - y_i)}{|\mathbf{x} - \mathbf{y}|c_\infty}\left[\frac{\partial f(\tau)}{\partial \tau}\right]_{\tau=\tau^*} \qquad (4.4.4)$$

An alternate form of Eq. (4.4.3) is then

$$\left(\rho'(\mathbf{x}, t)c_\infty^2\right)_{\text{dipole}} = \int_S \left[\frac{\partial\left(p_{ij} + \rho v_i v_j\right)}{\partial \tau} + \frac{\left(p_{ij} + \rho v_i v_j\right)c_\infty}{|\mathbf{x} - \mathbf{y}|}\right]_{\tau=\tau^*} \frac{(x_i - y_i)n_j dS(\mathbf{y})}{4\pi|\mathbf{x} - \mathbf{y}|^2 c_\infty} \qquad (4.4.5)$$

Note that if the surface is impermeable then $\rho v_i v_j n_j = 0$ and the only significant term is the compressive stress tensor. We also see that there is a near field and far field term in this result. The second term in square brackets will be important when the observer is close to the source, but becomes less and less important when the observer moves away from the source region, as is often the case. In most applications, we need only consider the far field approximation of Eq. (4.4.5) for a rigid stationary surface, which is

$$\left(\rho'(\mathbf{x}, t)c_\infty^2\right)_{\text{dipole}} \approx \frac{x_i}{4\pi|\mathbf{x}|^2 c_\infty} \int_S \left[\frac{\partial\left(p_{ij}n_j\right)}{\partial \tau}\right]_{\tau=\tau^*} dS(\mathbf{y}) \qquad (4.4.6)$$

We now see that for a rigid surface it is only the surface loading which causes the sound, and this depends on both the pressure and the viscous stresses, although in general the latter is assumed to be negligibly small and cannot be included without also evaluating the volume sources in the boundary layer close to the wall. Furthermore if the surface is acoustically compact then we can neglect the effects of retarded time over the surface and write Eq. (4.4.6) in the form

$$\left(\rho'(\mathbf{x}, t)c_\infty^2\right)_{dipole} \approx \frac{x_i}{4\pi|\mathbf{x}|^2c_\infty}\left[\frac{\partial F_i}{\partial \tau}\right]_{\tau=\tau^*}, \quad F_i(\tau) = \int_S p_{ij}n_j dS(\mathbf{y}) \tag{4.4.7}$$

where $F_i$ is the net force applied to the fluid by the surface, including viscous drag. It follows therefore that for an acoustically compact surface that applies a force to the fluid the sound radiation has the characteristics of a compact dipole (as discussed in Chapter 3) orientated in the direction of the force. Consequently this term is defined as the dipole term because it has an inherent cosine directionality if retarded time effects are ignored.

We can also derive a scaling law for the dipole source term because in general the force on the surface scales with the square of the mean flow velocity around the body, so $F_i \propto \rho_\infty U^2 S$, where $S$ is the surface area of the body. In addition the time scale of the fluctuations is given by a typical dimension of the turbulence, say $L$ divided by the mean flow speed $U$. Hence the time derivative of $F_i$ will scale as $\partial F_i/\partial \tau \propto \rho_\infty U^3 S/L$. It follows that the acoustic intensity in the far field, $I_r = p_{rms}^2/\rho_\infty c_\infty$, is proportional to the square of the acoustic wave amplitude in the far field (Note in Chapter 3 we considered a harmonic wave and showed that the acoustic intensity was $I_r = |\hat{p}|^2/2\rho_\infty c_\infty$. For the more general case, we can define the acoustic intensity using the root mean square pressure $p_{rms} = |\hat{p}|/\sqrt{2}$ and the speed of sound in the stationary fluid). The far field intensity of a dipole source will then scale as

$$I_r \propto \frac{\rho_\infty U^6 S^2 \cos^2\theta}{(4\pi|\mathbf{x}|)^2 c_\infty^3 L^2} \tag{4.4.8}$$

where $\theta$ is the observer angle relative to the dipole axis.

The important conclusion from Eq. (4.4.8) is that the far field sound scales on the sixth power of the mean flow velocity. This simple scaling law has proven to be invaluable in many applications where dipole sources are expected to dominate. For example in applications where we have an acoustically compact rigid surface in a turbulent flow, the flow noise is expected to scale as the sixth power of the velocity. We can conclude that noise will be significantly reduced if we lower the flow speed and in many practical applications this is the most effective way to obtain a noise level reduction.

Finally we will consider the volume integral in Eq. (4.3.12) which is referred to as the quadrupole term. Using Eq. (4.4.4) to change the space derivatives to derivatives over source time gives the quadrupole term as

$$\frac{\partial^2 [T_{ij}/r]_{\tau=\tau^*}}{\partial x_i \partial x_j} = \left[\frac{\partial^2 T_{ij}}{\partial \tau^2}\right]_{\tau=\tau^*}\frac{\partial \tau^*}{\partial x_i}\frac{\partial \tau^*}{\partial x_j}\frac{1}{r} + 2\left[\frac{\partial T_{ij}}{\partial \tau}\right]_{\tau=\tau^*}\left(\frac{\partial \tau^*}{\partial x_i}\frac{\partial}{\partial x_j}\left(\frac{1}{r}\right) + \frac{\partial}{\partial x_i}\left(\frac{1}{r}\frac{\partial \tau^*}{\partial x_j}\right)\right) \tag{4.4.9}$$

where $r = |\mathbf{x} - \mathbf{y}|$ and $\partial \tau^*/\partial x_i = -(x_i - y_i)/rc_\infty$. In the acoustic far field where Lighthill's analogy applies, this term can be simplified by dropping all the terms that are of order $1/r^2$ and $1/r^3$ so

$$\left(\rho'(\mathbf{x}, t)c_\infty^2\right)_{\text{quadrupole}} \approx \frac{x_i x_j}{4\pi c_\infty^2 r^3} \int\limits_V \left[\frac{\partial^2 T_{ij}}{\partial \tau^2}\right]_{\tau = \tau^*} dV(\mathbf{y}) \tag{4.4.10}$$

Furthermore, if the turbulence is limited to a volume that is small compared to the acoustic wavelength, then the effects of retarded time across the volume can be ignored, leading to

$$\left(\rho'(\mathbf{x}, t)c_\infty^2\right)_{\text{quadrupole}} \approx \frac{x_i x_j}{4\pi c_\infty^2 |\mathbf{x}|^3} \left[\int_V \frac{\partial^2 T_{ij}}{\partial \tau^2} dV(\mathbf{y})\right]_{\tau = \tau^*} \tag{4.4.11}$$

The inherent directionality of this source, if retarded time effects are ignored, is the same as the quadrupole sources we considered in Chapter 3. For example the $T_{11}$ term has a $(x_i/|\mathbf{x}|)^2 = (\cos\theta)^2$ directionality which is the same as a longitudinal quadrupole, Eq. (3.7.5), and $T_{12}$ has the same directionality as Eq. (3.7.7), as illustrated in Fig. 3.6.

In principle, this result gives the far field noise from a region of unbounded turbulent flow and shows that it is the second time derivative of Lighthill's stress tensor that generates the sound. We can also obtain a scaling law for the noise from turbulence using this result. The volume integral of the stress tensor $T_{ij}$ is expected to scale as $\rho_\infty U^2 V$ where $V$ is the volume of the fluid, and, as for the dipole source, we expect the time scale to be $L/U$ where $L$ is the lengthscale of the turbulence in the flow. The far field pressure is therefore proportional to $U^4$, and the far field acoustic intensity scales as

$$I_r \propto \frac{\rho_\infty U^8 V^2}{(4\pi|\mathbf{x}|)^2 c_\infty^5 L^4} \left(\frac{x_i x_j}{|\mathbf{x}|^2}\right)^2 \tag{4.4.12}$$

and shows that the noise from turbulence in the flow scales with the eighth power of the flow speed, $U^8$. This is one of the most important results of Lighthill's theory because it shows how the noise from free turbulence is very sensitive to mean flow speed. For example if the exit velocity of a jet is reduced by 30% the acoustic intensity in the far field is reduced by $80\log_{10}(0.7) = -12.4$ dB, which is a significant reduction. However, a decrease in jet exit velocity is also accompanied by a reduction in thrust, which often limits the noise reduction that can be achieved. The thrust is proportional to $\rho_\infty U^2 A$ where $A$ is the exit area of the jet, hence if we increase the jet diameter or exit area we can reduce the exit velocity for a given thrust. Since the noise is determined by the eighth power of the velocity, an increase in jet diameter for the same thrust is accompanied by a reduction in noise, and this is one of the reasons why many modern commercial aircraft have large diameter engines.

Another important conclusion from these scaling laws is to estimate the sound from a finite volume of turbulence compared to the sound from a surface dipole source. Taking the ratio of Eq. (4.4.12) to Eq. (4.4.8) gives

$$\frac{(I_r)_{\text{quadrupole}}}{(I_r)_{\text{dipole}}} < \frac{U^2 V^2}{c_\infty^2 L^2 S^2}$$

It follows that the quadrupole source strength is of order $(U/c_\infty)^2 = M^2$ times the dipole source strength where $M$ is the flow Mach number, providing the ratio $V/LS$ is of order one. For low Mach number flows $M \ll 1$ and this implies that the volume source terms of quadrupole order are negligible compared to the surface source terms that are of dipole order.

## 4.5  Tailored Green's functions

As pointed above the solution to Lighthill's wave equation given by Eq. (4.3.9) depends on a Green's function that only needs to be a solution to the inhomogeneous wave equation (3.9.2). We therefore have some latitude in choosing this function and we can also require that the Green's function satisfies boundary conditions on the surfaces within the flow. This can be important when we consider high Reynolds number flows over rigid surfaces. Eq. (4.3.9) shows that the acoustic field is given by

$$\rho'(\mathbf{x}, t)c_\infty^2 = \int\limits_{-T}^{T}\int\limits_{S} \left( (p_{ij} + \rho v_i v_j)\frac{\partial G}{\partial y_i} + G\frac{\partial(\rho v_j)}{\partial \tau} \right) n_j dS(\mathbf{y})d\tau$$

$$+ \int\limits_{-T}^{T}\int\limits_{V} \left( \frac{\partial^2 G}{\partial y_i \partial y_j} \right) T_{ij}(\mathbf{y}, \tau)dV(\mathbf{y})d\tau \qquad (4.3.9)$$

If the surface is stationary and impenetrable then $v_j n_j = 0$ and the only remaining term in the surface integral depends on $p_{ij}$. In high Reynolds number flow the viscous shear stress terms can be considered to be negligible and so the compressive stress tensor is determined by the pressure term alone, giving

$$\rho'(\mathbf{x}, t)c_\infty^2 = \int\limits_{-T}^{T}\int\limits_{S} \left( p'\frac{\partial G}{\partial y_i}n_i \right) dS(\mathbf{y})d\tau + \int\limits_{-T}^{T}\int\limits_{V} \left( \frac{\partial^2 G}{\partial y_i \partial y_j} \right) T_{ij}(\mathbf{y}, \tau)dV(\mathbf{y})d\tau \qquad (4.5.1)$$

where $p' = p - p_\infty$. The surface integral is then only dependent on the distribution of surface pressure. However, if the Green's function is chosen to satisfy the rigid wall boundary condition so that $n_i \partial G/\partial y_i = 0$ on the surface then the only remaining term in Eq. (4.5.1) is the volume integral. A Green's function that satisfies additional boundary conditions is specified as a tailored Green's function, and will be given the notation $G_T$. The acoustic field is then

$$\rho'(\mathbf{x},t)c_\infty^2 = \int\limits_{-T}^{T}\int\limits_{V}\left(\frac{\partial^2 G_T(\mathbf{x},t|\mathbf{y},\tau)}{\partial y_i \partial y_j}\right)T_{ij}(\mathbf{y},\tau)dV(\mathbf{y})d\tau \qquad (4.5.2)$$

This of course is an important simplification because it eliminates the need to calculate the surface pressure and allows the acoustic far field to be calculated directly from the volume source terms. The main difficulties with this approach are finding the tailored Green's function for bodies of arbitrary shape, and modeling the Lighthill source term with sufficient accuracy. Analytical models of the surface pressure below a turbulent boundary layer are often more readily available than models of the nonlinear terms in Lighthill's stress tensor. Furthermore, in low Mach number flows the volume source terms in Eq. (4.3.9) are of quadrupole order, as discussed in Section 4.4, and so can be neglected in comparison to the surface integral terms. For these reasons many problems can be most easily addressed by using the surface integral term in Eq. (4.3.9) with an appropriate analytical model for the surface pressure. However, when numerical calculations of the unsteady flow are available Eq. (4.5.2) gives an attractive alternative to Eq. (4.3.9) because the volume integral need only be evaluated over the volume where the flow is turbulent, and the form of the tailored Green's function may also help to window or limit that volume to the region close to discontinuities. However, for flow over streamlined bodies the wake can be an extremely important contributor to $T_{ij}$ and integrating this correctly needs to be done carefully with the correct downstream boundary conditions.

The simplest example of a tailored Green's function is a hard flat surface of infinite extent, as shown in Fig. 4.2.

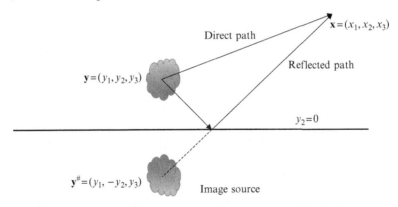

**Fig. 4.2** A source over a hard flat surface and its image source.

The acoustic waves from the source at $\mathbf{y}$ reach the observer at $\mathbf{x}$ by two different paths, the direct path and the reflected path, and the length of the reflected path is the same as the distance from an image source below the surface at the location $\mathbf{y}^{\#}$. Since the surface is rigid the boundary condition on the surface is $\partial p/\partial y_2 = 0$ so the tailored Green's function also needs to satisfy this condition. If we define

$$G_T(\mathbf{x},t|\mathbf{y},\tau) = G_o(\mathbf{x},t|\mathbf{y},\tau) + G_o(\mathbf{x},t|\mathbf{y}^\#,\tau) \tag{4.5.3}$$

then it is relatively simple to verify that $\partial G_T/\partial y_2 = 0$ for all points on the surface and $G_T$ is a solution to Eq. (3.9.2) for $y_2 > 0$. This particular Green's function is important to problems involving sources close to a large rigid surface. If the source is within a fraction of an acoustic wavelength of the surface, then we can approximate $\mathbf{y}$ and $\mathbf{y}^\#$ by their values on the surface, so $G_T = 2G_o$ where $G_o$ is evaluated on the surface at $y_2 = 0$.

Tailored Green's functions are available for simple shapes in the frequency domain, and can be computed using boundary element methods for more complex shapes. The frequency domain approach is based on taking the Fourier transform, defined by Eq. (3.10.1), of Eq. (4.5.1), using Eq. (3.10.7) and the convolution theorem, to give

$$\widetilde{\rho}(\mathbf{x},\omega)c_\infty^2 = \int_S \left( \widetilde{p}(\mathbf{y},\omega)\frac{\partial \widetilde{G}(\mathbf{x}|\mathbf{y})}{\partial y_i}n_i \right) dS(\mathbf{y}) + \int_V \left( \frac{\partial^2 \widetilde{G}(\mathbf{x}|\mathbf{y})}{\partial y_i \partial y_j} \right) \widetilde{T}_{ij}(\mathbf{y},\omega)dV(\mathbf{y})$$

$$\tag{4.5.4}$$

where $\widetilde{G}(\mathbf{x}|\mathbf{y})$ is the Green's function in the frequency domain defined by Eq. (3.10.7). We can reduce this to a volume integral by introducing a tailored Green's function that satisfies the inhomogeneous Helmholtz equation (3.10.4),

$$\frac{\partial^2 \widetilde{G}_T(\mathbf{x}|\mathbf{y})}{\partial y_i^2} + k^2 \widetilde{G}_T(\mathbf{x}|\mathbf{y}) = -\delta(\mathbf{x}-\mathbf{y}) \tag{4.5.5}$$

and matches the boundary condition on the surface. For an impenetrable surface we find a solution for which $n_i \partial \widetilde{G}_T/\partial y_i = 0$ on $S$ and so the integrand of the first integral in Eq. (4.5.4) is zero, and

$$\widetilde{\rho}(\mathbf{x},\omega)c_\infty^2 = \int_V \left( \frac{\partial^2 \widetilde{G}_T(\mathbf{x}|\mathbf{y})}{\partial y_i \partial y_j} \right) \widetilde{T}_{ij}(\mathbf{y},\omega)dV(\mathbf{y}) \tag{4.5.6}$$

The two results given by Eqs. (4.5.4), (4.5.6), or Eqs. (4.5.1), (4.5.2), give some insight into the sources of sound as they appear in the acoustic far field. From Eq. (4.5.6) it would appear that the only sound source is the turbulence in the volume of the fluid. However, that is a misconception hidden in the nature of the tailored Green's function. If we consider Eq. (4.5.4), it is seen that there are two apparent noise sources, one being the turbulence in the volume and the other being the radiation from the surface, which directly relates the outgoing waves to their "source as observed from the acoustic far field." The tailored Green's function includes not only these outgoing waves but also the waves that are reflected by the surface, and indicates that the sound radiation in the far field is "driven" by the turbulence, and so is best described as the "origin" of the sound. One must therefore be careful in choosing the terms that describe the mechanisms of sound generation and distinguish between the "origin" of the sound and the "source of sound as observed from the acoustic field."

## 4.6 Integral formulas for tailored Green's functions

To calculate the tailored Green's function for an arbitrary shape we can combine Eqs. (4.5.5), (3.10.4) which defines the free field Green's function and is rewritten as

$$\frac{\partial^2 \widetilde{G}_o(\mathbf{z}|\mathbf{y})}{\partial y_i^2} + k^2 \widetilde{G}_o(\mathbf{z}|\mathbf{y}) = -\delta(\mathbf{z} - \mathbf{y}) \tag{4.6.1}$$

Multiplying Eq. (4.6.1) by $\widetilde{G}_T(\mathbf{x}|\mathbf{y})$ and Eq. (4.5.5) by $\widetilde{G}_o(\mathbf{z}|\mathbf{y})$, subtracting the two equations, integrating over the volume $V(\mathbf{y})$, and using the divergence theorem, gives

$$\int_S \left( \widetilde{G}_o(\mathbf{z}|\mathbf{y}) \frac{\partial \widetilde{G}_T(\mathbf{x}|\mathbf{y})}{\partial y_i} - \widetilde{G}_T(\mathbf{x}|\mathbf{y}) \frac{\partial \widetilde{G}_o(\mathbf{z}|\mathbf{y})}{\partial y_i} \right) n_i dS(\mathbf{y}) = \widetilde{G}_T(\mathbf{x}|\mathbf{z}) - \widetilde{G}_o(\mathbf{z}|\mathbf{x})$$

Since $n_i \partial \widetilde{G}_T(\mathbf{x}|\mathbf{y})/\partial y_i = 0$ the first term in the surface integral is eliminated and the tailored Green's function is given by

$$\widetilde{G}_T(\mathbf{x}|\mathbf{z}) = \widetilde{G}_o(\mathbf{z}|\mathbf{x}) - \int_S \left( \widetilde{G}_T(\mathbf{x}|\mathbf{y}) \frac{\partial \widetilde{G}_o(\mathbf{z}|\mathbf{y})}{\partial y_i} \right) n_i dS(\mathbf{y}) \tag{4.6.2}$$

In principle, for a source at $\mathbf{z}$, the acoustic field at $\mathbf{x}$ can be calculated using the free field Green's function and the surface integral in Eq. (4.6.2). However, to use this result we need to know the tailored Green's function for a source on the surface and an observer at $\mathbf{x}$. This can be obtained using a boundary element method and solving Eq. (4.6.2) for a point $\mathbf{z}$ that lies on the surface. Some care is required because the surface integral becomes singular when $\mathbf{y} = \mathbf{z}$, and so the limiting form of the integral must be used. The approach is to indent the surface about the point $\mathbf{z}$ with a hemisphere of radius $\varepsilon$ and then let $\varepsilon$ tend to zero. On the hemisphere $|\mathbf{z} - \mathbf{y}| = r = \varepsilon$ where $r$ is in the radial direction pointing away from the point $\mathbf{z}$. The surface integral can be defined as

$$-\widetilde{G}_T(\mathbf{x}|\mathbf{z}) \int_{S_\varepsilon} \left[ \frac{\partial}{\partial r} \left( \frac{e^{ikr}}{4\pi r} \right) \right]_{r=\varepsilon} dS$$

where the negative sign is the result of the normal being chosen to point into the fluid in the opposite direction to the radial coordinate $r$. Evaluating the radial derivative in the limit that $\varepsilon$ tends to zero gives $1/4\pi\varepsilon^2$ and, since the integrand is constant on the surface the integral is simply the area of a hemisphere times the integrand, or $2\pi\varepsilon^2/4\pi\varepsilon^2 = 1/2$. It follows that

$$
\frac{1}{2}\widetilde{G}_T(\mathbf{x}|\mathbf{z}) = \widetilde{G}_o(\mathbf{z}|\mathbf{x})
$$

$$
- \int\limits_{S,\mathbf{z}\neq\mathbf{y}} \left( \widetilde{G}_T(\mathbf{x}|\mathbf{y})\frac{\partial\widetilde{G}_o(\mathbf{z}|\mathbf{y})}{\partial y_i} \right) n_i dS(\mathbf{y}) \quad \text{for } \mathbf{z} \text{ on } S(\mathbf{y}) \tag{4.6.3}
$$

is a suitable form for the evaluation of the tailored Green's function on the surface that can be evaluated numerically using a boundary element method. The tailored Green's function for different source and receiver positions can then be obtained from Eq. (4.6.2).

## 4.7 Wavenumber and Fourier transforms

Wavenumber methods are very important in aeroacoustics because they directly relate the source fluctuations to the propagating waves in the acoustic far field. To illustrate this we will first consider the general definition of a Fourier transform and then extend it to both time and space.

As described in Section 3.10 the Fourier transform with respect to time is defined as

$$
\widetilde{f}(\omega) = \frac{1}{2\pi} \int\limits_{-T}^{T} f(t)e^{i\omega t} dt \tag{4.7.1}
$$

The integral converges for all functions $f(t)$ provided that they are piecewise continuously differentiable and the integral of $|f(t)|$ exists over the limits $-T$ to $T$. This usually requires that $T$ is less than infinity. In the limit that $T$ becomes very large we can define the inverse transform as

$$
f(t) = \int\limits_{-\infty}^{\infty} \widetilde{f}(\omega)e^{-i\omega t} d\omega \tag{4.7.2}
$$

The frequency content of the acoustic field generated by a turbulent flow can be determined using Eq. (4.7.1). For example the far field sound generated by turbulence in the absence of scattering surfaces, is given by Eq. (4.4.10) as

$$
\left(\rho'(\mathbf{x}, t)c_\infty^2\right)_{\text{quadrupole}} \approx \frac{x_i x_j}{4\pi c_\infty^2 |\mathbf{x}|^3} \int\limits_V \left[\frac{\partial^2 T_{ij}(\mathbf{y}, \tau)}{\partial\tau^2}\right]_{\tau=\tau^*} dV(\mathbf{y}) \tag{4.7.3}
$$

By introducing the inverse Fourier transform of the Lighthill stress tensor,

$$T_{ij}(\mathbf{y}, \tau) = \int_{-\infty}^{\infty} \widetilde{T}_{ij}(\mathbf{y}, \omega) e^{-i\omega\tau} d\omega$$

and evaluating $\tau$ at the correct retarded time we find that, after rearranging the order of integration,

$$\left(\rho'(\mathbf{x}, t) c_\infty^2\right)_{\text{quadrupole}} \approx \int_{-\infty}^{\infty} \left\{ \frac{-\omega^2 x_i x_j}{4\pi c_\infty^2 |\mathbf{x}|^3} \int_V \widetilde{T}_{ij}(\mathbf{y}, \omega) e^{ikr} dV(\mathbf{y}) \right\} e^{-i\omega t} d\omega$$

Taking the Fourier transform of this with respect to time we obtain

$$\left(\widetilde{\rho}(\mathbf{x}, \omega) c_\infty^2\right)_{\text{quadrupole}} \approx \frac{-\omega^2 x_i x_j}{4\pi c_\infty^2 |\mathbf{x}|^3} \int_V \widetilde{T}_{ij}(\mathbf{y}, \omega) e^{ikr} dV(\mathbf{y}) \qquad (4.7.4)$$

This result shows that all the retarded time effects are now accounted for by the exponential factor under the integral, and we no longer require the rather complicated evaluation of emission time at the source. This can have some advantages in the evaluation of some simple problems, especially when the distribution of the sources are well defined and we can use the far field approximation that $r = |\mathbf{x}| - \mathbf{x} \cdot \mathbf{y}/|\mathbf{x}|$, so

$$\left(\widetilde{\rho}(\mathbf{x}, \omega) c_\infty^2\right)_{\text{quadrupole}} \approx \frac{-\omega^2 x_i x_j e^{ik|\mathbf{x}|}}{4\pi c_\infty^2 |\mathbf{x}|^3} \int_V \widetilde{T}_{ij}(\mathbf{y}, \omega) e^{-ik\mathbf{x} \cdot \mathbf{y}/|\mathbf{x}|} dV(\mathbf{y}) \qquad (4.7.5)$$

The concept of a Fourier transform in time can be extended to a four dimensional transform in both time and space and is defined as wavenumber transform

$$\widetilde{q}(\mathbf{k}, \omega) = \frac{1}{(2\pi)^4} \int_{-T}^{T} \int_{-R_\infty}^{R_\infty} \int_{-R_\infty}^{R_\infty} \int_{-R_\infty}^{R_\infty} q(\mathbf{y}, t) e^{i\omega t - i\mathbf{k} \cdot \mathbf{y}} dy_1 dy_2 dy_3 dt \qquad (4.7.6)$$

and its inverse

$$q(\mathbf{y}, t) = \int_{-\infty}^{\infty} \int_{-\infty}^{\infty} \int_{-\infty}^{\infty} \int_{-\infty}^{\infty} \widetilde{q}(\mathbf{k}, \omega) e^{-i\omega t + i\mathbf{k} \cdot \mathbf{y}} dk_1 dk_2 dk_3 d\omega \qquad (4.7.7)$$

An important point to note here is that the Fourier transform definition given by Eq. (4.7.6) uses an exponent $+i\omega t - i\mathbf{k} \cdot \mathbf{x}$.

Using these results the frequency dependence of the acoustic field given by Eq. (4.7.5) is defined in terms of a wavenumber transform by

$$\left(\widetilde{\rho}(\mathbf{x},\omega)c_\infty^2\right)_{\text{quadrupole}} \approx \frac{-2\pi^2 x_i x_j \omega^2 e^{ik|\mathbf{x}|} \overset{\approx}{T}_{ij}\left(\mathbf{k}^{(o)},\omega\right)}{c_\infty^2 |\mathbf{x}|^3} \qquad \mathbf{k}^{(o)} = \omega\mathbf{x}/c_\infty|\mathbf{x}| \quad (4.7.8)$$

This is an important result because it shows that the far field sound from a turbulent flow can be directly related to the wavenumber transform of the source distribution. Furthermore, it is only the wavenumbers which have a magnitude $|\mathbf{k}^{(o)}| = \omega/c_\infty$ that couple with the far field, and fluctuations at any other wavenumber are not important. In deriving this result we made use of the far field solution given by Eq. (4.7.4), and it is an interesting exercise (which will be left to the reader) to show that the same result can be obtained from Eq. (4.3.3) (in the absence of a scattering surface), in which the derivatives are applied to the Green's function.

The fact that Eq. (4.7.8) only depends on the acoustic wavenumber $k = \omega/c_\infty$ has some important implications. For example, if we consider the turbulence to be purely convected at the mean velocity $\mathbf{U}$, then $T_{ij}$ will be a function of $\mathbf{y} - \mathbf{U}t$ and the frequency associated with any wavenumber will be $\omega = \mathbf{k} \cdot \mathbf{U}$. There will be no sound radiation to the acoustic far field unless $|\mathbf{k}^{(o)}| = \omega/c_\infty = |\mathbf{k} \cdot \mathbf{U}|/c_\infty$, which requires that $|\mathbf{U}|$ is greater than or equal to the speed of sound. Eq. (4.7.8) also shows that we need only evaluate the source term for wavenumbers where $|\mathbf{k}| \leq \omega/c_\infty$, which, in principle, simplifies the modeling process. Unfortunately the measurement or numerical simulation of turbulence at these wavenumbers is very difficult, and this inhibits the application of this method.

The same approach can be used for dipole surface sources. Starting with Eq. (4.4.6) we take its Fourier transform with respect to time and apply the far field approximation $r \approx |\mathbf{x}| - \mathbf{x} \cdot \mathbf{y}/|\mathbf{x}|$ to give

$$\left(\widetilde{\rho}(\mathbf{x},\omega)c_\infty^2\right)_{\text{dipole}} \approx \frac{-i\omega x_i e^{ik|\mathbf{x}|}}{4\pi c_\infty |\mathbf{x}|^2} \int_S \widetilde{p}_{ij}(\mathbf{y},\omega) n_j e^{-ik\mathbf{x} \cdot \mathbf{y}/|\mathbf{x}|} dS(\mathbf{y}) \qquad (4.7.9)$$

For thin airfoils in an inviscid flow, which will be discussed in later chapters, the surface integral can be expressed in terms of the pressure jump $\Delta p$ across the airfoil planform where

$$[\Delta\widetilde{p}n_i d\Sigma]_{\text{planform}} = [\widetilde{p}n_i dS]_{\text{upper}} - [\widetilde{p}n_i dS]_{\text{lower}}$$

and the subscripts refer to the upper and lower surfaces of the airfoil. Choosing the planform to lie in the $y_2 = 0$ plane then gives

$$\left(\widetilde{\rho}(\mathbf{x},\omega)c_\infty^2\right)_{\text{dipole}} \approx \frac{-i\omega x_2 e^{ik|\mathbf{x}|}}{4\pi c_\infty |\mathbf{x}|^2} \int_0^c \int_{-b/2}^{b/2} \Delta\widetilde{p}(\mathbf{y},\omega) e^{-ikx_1 y_1/|\mathbf{x}| - ikx_3 y_3/|\mathbf{x}|} dy_1 dy_3$$

$$(4.7.10)$$

where the span of the blade is $b$ and the blade chord is $c$. Since the pressure jump is zero on all parts of the plane $y_2 = 0$ not on the airfoil, the limits of the integrals in

Eq. (4.7.10) can be extended to $\pm R_\infty$. The integral then represents a wavenumber transform giving

$$\left(\widetilde{p}(\mathbf{x}, \omega) c_\infty^2\right)_{\text{dipole}} \approx \frac{-i\pi\omega x_2 \Delta\widetilde{p}\left(k_1^{(o)}, k_3^{(o)}, \omega\right) e^{ik|\mathbf{x}|}}{c_\infty |\mathbf{x}|^2} \qquad (4.7.11)$$

where

$$\Delta\widetilde{p}(k_1, k_3, \omega) = \frac{1}{(2\pi)^2} \int\limits_{-R_\infty}^{R_\infty} \int\limits_{-R_\infty}^{R_\infty} \Delta\widetilde{p}(\mathbf{y}, \omega) e^{-ik_1 y_1 - ik_3 y_3} \, dy_1 \, dy_3 \qquad (4.7.12)$$

is the wavenumber transform of the surface pressure and $k_i^{(o)} = \omega x_i / |\mathbf{x}| c_\infty$ is the acoustic wavenumber in the direction of the observer.

This is an important result because it shows that the acoustic field only depends on the wavenumber transform of the surface pressure fluctuations at the acoustic wavenumbers. As with the turbulence sources discussed above, any disturbance that convects over the surface at a subsonic speed will not couple with the acoustic far field. Any attempts to measure the surface pressure fluctuations on an airfoil and relate them to the acoustic far field are therefore very difficult because the measurement needs to distinguish between hydrodynamic pressure fluctuations, convected subsonically over the surface, and acoustic fluctuations (which also may be contaminated by facility noise). A much easier approach is to infer the surface pressure fluctuations at the acoustic wavenumbers from a far field sound measurement. If the left side of Eq. (4.7.11) is measured for values of $k_1^{(o)}$ and $k_3^{(o)}$ in the range $-\omega/c_\infty < k_i^{(o)} < \omega/c_\infty$ (or for all angles at all frequencies), then we can obtain the wavenumber spectrum of the pressure jump at all the wavenumbers that are important for acoustic calculations in other environments.

Finally, we note that at low frequencies the acoustic wavenumbers tend to zero and so the wavenumber spectrum defined by Eq. (4.7.12) reduces to the net unsteady loading on the airfoil, divided by $(2\pi)^2$. In contrast at very high frequencies the wavenumber transform (4.7.12) will be dominated by edge effects because the oscillatory nature of the integrand tends to suppress the contributions from parts of the surface where the pressure varies smoothly. Because of the importance of edges on sound radiation we will consider their effect on blade noise in great detail in Part 3 of the text.

# References

[1] M.J. Lighthill, On sound generated aerodynamically. I. General theory, Proc. R. Soc. Lond. A Math. Phys. Sci. 211 (1952) 564–587.
[2] N. Curle, The influence of solid boundaries upon aerodynamic sound, Proc. R. Soc. Lond. A Math. Phys. Sci. 231 (1955) 505–514.
[3] P.E. Doak, Acoustic radiation from a turbulent fluid containing foreign bodies, Proc. R. Soc. Lond. A Math. Phys. Sci. 254 (1960) 129–146.

# The Ffowcs Williams and Hawkings equation 5

In Chapter 4 we derived the general solution to the wave equation for a medium that included stationary scattering objects. In many important applications, such as for propellers and helicopter rotor noise, the surfaces are moving and so we need to modify the analysis to take full account of surface motion. Powerful techniques to address this problem have been pioneered by Ffowcs Williams and Farassat, and the objective of this chapter is to introduce these techniques. We will start by reviewing the concept of generalized derivatives and then show how these may be used to give solutions to Lighthill's equation for a medium that includes moving surfaces and convected turbulent flow. This will be followed by a general discussion of the sound fields from moving sources and the extension of the results to sources in a moving fluid. Finally, we will show how incompressible computational fluid dynamics (CFD) codes can be used to calculate the sound radiated by stationary objects in the flow.

## 5.1 Generalized derivatives

A generalized derivative extends the concept of an ordinary derivative to discontinuous functions. For example, consider the Heaviside step function $H_s(x)$ which is defined as being zero when $x < 0$ and one when $x > 0$ as shown in Fig. 5.1.

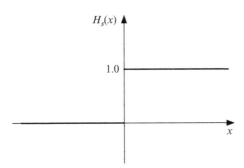

**Fig. 5.1** A Heaviside step function.

The derivative of this function is obviously zero when $x < 0$ and $x > 0$ and must be very large when $x = 0$. Also it follows that if we integrate the derivative of $H_s$ then we have

$$\int_{x_1}^{x} \frac{\partial H_s(x')}{\partial x'} dx' = \begin{cases} 1 & x > 0 \quad x_1 < 0 \\ 0 & x < 0 \quad x_1 < 0 \end{cases} \tag{5.1.1}$$

Aeroacoustics of Low Mach Number Flows. http://dx.doi.org/10.1016/B978-0-12-809651-2.00005-9

We see therefore that the derivative of the Heaviside function has exactly the same properties as the Dirac delta function so we can define,

$$\delta(x - x_o) = \frac{\partial H_s(x - x_o)}{\partial x} \tag{5.1.2}$$

This concept can be extended to surfaces which bound a region. If we define a function $f(\mathbf{x})$ which is greater than zero outside a volume enclosed by a surface $S_o$ and less than zero inside the volume, as shown in Fig. 5.2, it follows that the surface is defined by $f=0$.

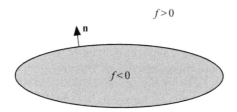

**Fig. 5.2** Function $f$ that defines the surface bounding a volume.

For example, if we want to define a spherical surface of radius $a$ we can choose $f = x_1^2 + x_2^2 + x_3^2 - a^2$. Similarly, for a cylinder of radius $a$ we have $f = x_1^2 + x_2^2 - a^2$. The unit normal to the surface, pointing out of the region as shown, is given by

$$\mathbf{n} = \nabla f / |\nabla f| \tag{5.1.3}$$

evaluated on $f=0$. This description of a surface is perfectly general and can be extended to moving surfaces by letting $f$ also be a function of time. For example, a sphere moving with velocity $\mathbf{U}$ can be defined by

$$f = |\mathbf{x} - \mathbf{U}t|^2 - a^2$$

Now consider how we can analyze a flow field in the region exterior to the surface, which may be in arbitrary accelerated motion. We want to know the flow variables in the region exterior to $S_o$ but we know nothing about the variables inside $S_o$, and so the velocity, for example, can only be defined outside $S_o$. We can however introduce a new velocity variable $\mathbf{v}H_s(f)$ which is defined everywhere. In the region outside the volume this variable is still the velocity, but inside the volume it is zero. We now have a flow variable which is specified everywhere in the presence of an arbitrary moving surface. One important property of $\mathbf{v}H_s(f)$ is its divergence

$$\nabla \cdot (\mathbf{v}H_s(f)) = H_s(f)\nabla \cdot \mathbf{v} + \mathbf{v} \cdot \nabla H_s(f) \tag{5.1.4}$$

but

$$\nabla H_s(f) = \frac{\partial H_s}{\partial f}\nabla f = \delta(f)|\nabla f|\mathbf{n} \tag{5.1.5}$$

so the divergence becomes

$$\nabla \cdot (\mathbf{v}H_s(f)) = H_s(f)\nabla \cdot \mathbf{v} + \mathbf{v} \cdot \mathbf{n}\delta(f)|\nabla f| \tag{5.1.6}$$

We can of course evaluate the divergence theorem for this new variable and integrate over an infinite region $V_\infty$ bounded only at infinity by the surface $S_\infty$, so

$$\int_{V_\infty} \nabla \cdot (\mathbf{v}H_s(f))dV = -\int_{S_\infty} (\mathbf{v}H_s(f)) \cdot \mathbf{n}dS \tag{5.1.7}$$

It follows from Eq. (5.1.6) that

$$\int_{V_\infty} (H_s(f)\nabla \cdot \mathbf{v} + \mathbf{v} \cdot \mathbf{n}\delta(f)|\nabla f|)dV = -\int_{S_\infty} \mathbf{v} \cdot \mathbf{n}dS \tag{5.1.8}$$

If we apply the divergence theorem to the first term in the integrand of this equation, then

$$\int_{V_\infty} H_s(f)(\nabla \cdot \mathbf{v})dV = \int_{V_o} (\nabla \cdot \mathbf{v})dV = -\int_{S_o + S_\infty} \mathbf{v} \cdot \mathbf{n}dS \tag{5.1.9}$$

where $V_o$ is the region outside of $S_o$ where $f > 0$, and $S_o$ is the surface of the body where $f = 0$, which may of course be a function of time. It then follows from Eq. (5.1.8) that

$$\int_{V_\infty} \mathbf{v} \cdot \mathbf{n}\delta(f)|\nabla f|dV = \int_{S_o} \mathbf{v} \cdot \mathbf{n}dS \tag{5.1.10}$$

This result is invaluable for the analysis of moving surfaces because the effect of the surface motion is completely defined by the integrand on the left of this expression.

Now let us consider the time derivatives of our new variable $\mathbf{v}H_s(f)$, which will be given by

$$\frac{\partial(\mathbf{v}H_s(f))}{\partial t} = H_s(f)\frac{\partial \mathbf{v}}{\partial t} + \mathbf{v}\delta(f)\frac{\partial f}{\partial t} \tag{5.1.11}$$

For a surface which moves with velocity $\mathbf{V}$ it follows that $f$ is a function of $\mathbf{x} - \int \mathbf{V}dt$ on the surface and so the chain rule gives

$$\left[\frac{\partial f\left(\mathbf{x} - \int \mathbf{V}dt\right)}{\partial t}\right]_{f=0} = [-\mathbf{V} \cdot \nabla f]_{f=0} \tag{5.1.12}$$

then Eq. (5.1.11) is modified to

$$\frac{\partial(\mathbf{v}H_s(f))}{\partial t} = H_s(f)\frac{\partial \mathbf{v}}{\partial t} - \mathbf{v}\delta(f)(\mathbf{V}\cdot\mathbf{n}|\nabla f|) \tag{5.1.13}$$

so that the second term on the right is specified in terms of the surface velocity.
Next consider the material derivative of $\mathbf{v}H_s$, which is given by

$$\frac{D(\mathbf{v}H_s)}{Dt} = \frac{D\mathbf{v}}{Dt}H_s + \mathbf{v}\frac{Df}{Dt}\delta(f) \tag{5.1.14}$$

Using Eq. (5.1.12) we see that

$$\frac{Df}{Dt} = \frac{\partial f}{\partial t} + (\mathbf{v}.\nabla)f = (\mathbf{v} - \mathbf{V})\cdot\mathbf{n}|\nabla f| \tag{5.1.15}$$

For an impermeable boundary the velocity of the fluid normal to the surface is equal to
the surface velocity in this direction and so $\mathbf{v}\cdot\mathbf{n} = \mathbf{V}\cdot\mathbf{n}$ and it follows that $Df/Dt - 0$.
Hence the second term in Eq. (5.1.14), which is only nonzero on the surface, can be
eliminated for impermeable surfaces.

Finally, in the next section, we need to consider integrals that include a Green's
function. For example,

$$\int_{V_\infty} G\frac{\partial}{\partial y_i}(p_{ij}n_j\delta(f)|\nabla f|)dV(\mathbf{y}) \tag{5.1.16}$$

which we can expand as

$$\int_{V_\infty} \frac{\partial}{\partial y_i}(Gp_{ij}n_j\delta(f)|\nabla f|)dV(\mathbf{y}) - \int_{V_\infty}(p_{ij}n_j\delta(f)|\nabla f|)\frac{\partial G}{\partial y_i}dV(\mathbf{y}) \tag{5.1.17}$$

The first of these two integrals can be evaluated using the divergence theorem, but
since the only boundary to $V_\infty$ is at infinity where $G$ tends to zero, its net contribution
is zero so

$$\int_{V_\infty} G\frac{\partial}{\partial y_i}(p_{ij}n_j\delta(f)|\nabla f|)dV(\mathbf{y}) = -\int_{V_\infty}(p_{ij}n_j\delta(f)|\nabla f|)\frac{\partial G}{\partial y_i}dV(\mathbf{y}) \tag{5.1.18}$$

Similarly for integrands which include time derivatives we can show that, integrating
by parts and noting that $G$ is zero when $\tau = \pm T$, then

$$\int_{-T}^{T}\int_{V_\infty} G\frac{\partial}{\partial \tau}(v_j n_j\delta(f)|\nabla f|)dV(\mathbf{y})d\tau = -\int_{-T}^{T}\int_{V_\infty}(v_j n_j\delta(f)|\nabla f|)\frac{\partial G}{\partial \tau}dV(\mathbf{y})d\tau \tag{5.1.19}$$

These results summarize some of the basic concepts of generalized derivatives. In the next section we will show how they can be used to provide a solution to Lighthill's equation in the presence of moving surfaces.

## 5.2  The Ffowcs Williams and Hawkings equation

In this section we will consider how the concept of a generalized derivative may be used to obtain the solution to Lighthill's equation in the presence of moving surfaces. We will start by evaluating the continuity and momentum equations in terms of the new variables $vH_s$, $pH_s$, and $\rho H_s$ which are defined everywhere in an unbounded infinite volume $V_\infty$. To obtain the continuity equation (2.2.6) in terms of the new variables we expand

$$\frac{\partial(\rho' H_s)}{\partial t} + \frac{\partial(\rho v_i H_s)}{\partial x_i} = H_s\left(\frac{\partial \rho'}{\partial t} + \frac{\partial(\rho v_i)}{\partial x_i}\right) + \left((\rho v_j - \rho' V_j)n_j\right)\delta(f)|\nabla f| \qquad (5.2.1)$$

where $\rho' = \rho - \rho_\infty$. The first term on the right is zero because it is the continuity equation in the region where the flow is defined, and zero outside the flow region because $H_s$ is zero. It follows that it is zero everywhere, so

$$\frac{\partial(\rho' H_s)}{\partial t} + \frac{\partial(\rho v_i H_s)}{\partial x_i} = (\rho v_j - \rho' V_j)n_j\delta(f)|\nabla f| \qquad (5.2.2)$$

A similar procedure may be applied to the momentum equation given by Eq. (2.3.9), leading to

$$\frac{\partial(\rho v_i H_s)}{\partial t} + \frac{\partial\left(\rho v_i v_j H_s + p_{ij} H_s\right)}{\partial x_j} = (\rho v_i(v_j - V_j) + p_{ij})n_j\delta(f)|\nabla f| \qquad (5.2.3)$$

(where the pressure is defined as gauge pressure as in Chapter 4). We can then obtain a wave equation for the new variable $\rho' H_s$ in exactly the same way we did for Lighthill's equation in Section 4.1. Taking the time derivative of Eq. (5.2.2), the divergence of Eq. (5.2.3) and subtracting gives an equation equivalent to Eq. (4.1.3). Then subtracting $\partial^2\left(\rho' c_\infty^2 H_s\right)/\partial x_i^2$ from both sides gives

$$\frac{\partial^2(\rho' H_s)}{\partial t^2} - c_\infty^2\frac{\partial^2(\rho' H_s)}{\partial x_i^2} = \frac{\partial^2(T_{ij} H_s)}{\partial x_i \partial x_j} - \frac{\partial}{\partial x_i}\left((\rho v_i(v_j - V_j) + p_{ij})n_j\delta(f)|\nabla f|\right)$$

$$+ \frac{\partial}{\partial t}\left((\rho v_j - \rho' V_j)n_j\delta(f)|\nabla f|\right)$$

$$(5.2.4)$$

This is the Ffowcs Williams and Hawkings equation [1], which is an inhomogeneous wave equation that includes the effects of moving surfaces on the right hand side. We can solve this equation using the method of Green's functions as was done in Section 4.3, but since the dependent variable of the wave equation is defined in an unbounded medium, the surface integrals, which appeared in Eq. (4.3.3), are not required and we obtain

$$\rho'(\mathbf{x}, t)c_\infty^2 H_s = \int_{-T}^{T} \int_{V_\infty} G\left(\frac{\partial^2 (H_s T_{ij})}{\partial y_i \partial y_j}\right) dV(\mathbf{y}) d\tau$$

$$- \int_{-T}^{T} \int_{V_\infty} G\frac{\partial}{\partial y_i}\left((\rho v_i(v_j - V_j) + p_{ij})n_j\delta(f)|\nabla f|\right) dV(\mathbf{y}) d\tau \quad (5.2.5)$$

$$+ \int_{-T}^{T} \int_{V_\infty} G\frac{\partial}{\partial \tau}\left((\rho v_j - \rho' V_j)n_j\delta(f)|\nabla f|\right) dV(\mathbf{y}) d\tau$$

Using the same procedure employed in deriving Eqs. (4.3.4)–(4.3.6) we can recast the first integral term as

$$\int_{-T}^{T} \int_{V_\infty} \frac{\partial^2 G}{\partial y_i \partial y_j}(T_{ij}H_s) dV(\mathbf{y}) d\tau$$

where the additional surface integral that appears in Eq. (4.3.6) is zero in the present case because of the unbounded domain. We can also convert the second and third terms in Eq. (5.2.5) using the identities established in Eqs. (5.1.18), (5.1.19) to give

$$\rho'(\mathbf{x}, t)c_\infty^2 H_s = \int_{-T}^{T} \int_{V_\infty} \frac{\partial^2 G}{\partial y_i \partial y_j}(T_{ij}H_s) dV(\mathbf{y}) d\tau$$

$$+ \int_{-T}^{T} \int_{V_\infty} \left((\rho v_i(v_j - V_j) + p_{ij})n_j\delta(f)|\nabla f|\right)\frac{\partial G}{\partial y_i} dV(\mathbf{y}) d\tau \quad (5.2.6)$$

$$- \int_{-T}^{T} \int_{V_\infty} \left((\rho v_j - \rho' V_j)n_j\delta(f)|\nabla f|\right)\frac{\partial G}{\partial \tau} dV(\mathbf{y}) d\tau$$

We can obtain the form preferred by Ffowcs Williams for the free field Green's function $G = G_o$, by noting that, since $\mathbf{x}$ and $\mathbf{y}$ are fixed,

$$\frac{\partial G_o}{\partial y_i} = -\frac{\partial G_o}{\partial x_i} \qquad \frac{\partial G_o}{\partial \tau} = -\frac{\partial G_o}{\partial t}$$

and thus,

$$
\rho'(\mathbf{x},t)c_\infty^2 H_s = \frac{\partial^2}{\partial x_i \partial x_j} \int_{-T}^{T} \int_{V_\infty} (T_{ij}H_s) G_o dV(\mathbf{y}) d\tau
$$

$$
- \frac{\partial}{\partial x_i} \int_{-T}^{T} \int_{V_\infty} \left( (\rho v_i (v_j - V_j) + p_{ij}) n_j \delta(f) |\nabla f| \right) G_o dV(\mathbf{y}) d\tau \quad (5.2.7)
$$

$$
+ \frac{\partial}{\partial t} \int_{-T}^{T} \int_{V_\infty} \left( (\rho v_j - \rho' V_j) n_j \delta(f) |\nabla f| \right) G_o dV(\mathbf{y}) d\tau
$$

where partial derivatives have been shifted to the observer variables, as in Eq. (4.3.11). Both these results are given because in some applications it is easier to evaluate the differentials at the source as in Eq. (5.2.6), whereas in other applications the differentials are more accurately applied at the observer location as in Eq. (5.2.7).

Evaluating the second and third integrals in Eq. (5.2.6) using Eq. (5.1.10) converts them to surface integrals

$$
\rho'(\mathbf{x},t)c_\infty^2 = \int_{-T}^{T} \int_{V_o(\tau)} \frac{\partial^2 G}{\partial y_i \partial y_j} T_{ij} dV(\mathbf{y}) d\tau
$$

$$
+ \int_{-T}^{T} \int_{S_o(\tau)} \frac{\partial G}{\partial y_i} \left( (\rho v_i (v_j - V_j) + p_{ij}) n_j \right) dS(\mathbf{y}) d\tau \quad (5.2.8)
$$

$$
- \int_{-T}^{T} \int_{S_o(\tau)} \frac{\partial G}{\partial \tau} \left( (\rho v_j - \rho' V_j) n_j \right) dS(\mathbf{y}) d\tau
$$

Notice how this result reduces to Curle's equation (4.3.9) when the surfaces are stationary so $\mathbf{V}=0$. It is important to appreciate that this result is not an obvious extension to Curle's theorem because of the change to the momentum flux term and the additional surface source $\rho' V_j$. Also note how the volume and surface integrals must now be evaluated over moving surfaces, which adds a new level of complexity because the sources are moving relative to the observer, so both $\mathbf{y}$ and the propagation distance between the source and the observer will change with emission time.

One of the key simplifications to this result is for an impenetrable surface where the flow velocity normal to the surface equals the surface normal velocity so that $v_i n_i = V_i n_i$. This eliminates the momentum flux terms in the second integral and because $\rho - \rho' = \rho_\infty$ the integrand in the third integral reduces to $\rho_\infty V_i n_i$ giving

$$
\rho'(\mathbf{x},t)c_\infty^2 = \int_{-T}^{T} \int_{V_o(\tau)} \frac{\partial^2 G}{\partial y_i \partial y_j} T_{ij} dV(\mathbf{y}) d\tau
$$

$$
+ \int_{-T}^{T} \int_{S_o(\tau)} \frac{\partial G}{\partial y_i} p_{ij} n_j dS(\mathbf{y}) d\tau \quad (5.2.9)
$$

$$
- \int_{-T}^{T} \int_{S_o(\tau)} \frac{\partial G}{\partial \tau} \rho_\infty V_j n_j dS(\mathbf{y}) d\tau
$$

The volume and surface integrals in Eqs. (5.2.8) or (5.2.9) can be evaluated by defining a coordinate system that moves with the surface. Moving coordinates will be denoted by $\mathbf{z}$ which is identical to the fixed source coordinate $\mathbf{y}$ at time $\tau_o$ so for emission times $\tau > \tau_o$ we have

$$\mathbf{y} = \mathbf{z} + \int_{\tau_o}^{\tau} \mathbf{V}(t')dt' \tag{5.2.10}$$

The surface and volume integrals can be evaluated in the moving coordinate system provided we account for the change in the size of the volume and surface elements introduced by this transformation, hence we specify the Jacobians $J$ and $K$ such that $dV(\mathbf{y}) = JdV(\mathbf{z})$ and $dS(\mathbf{y}) = KdS(\mathbf{z})$. It will be left as an exercise for the reader to show that if surfaces are moving with constant linear or angular velocity then the Jacobians are unity and $\partial G/\partial y_i = \partial G/\partial z_i$.

Now let us return to Ffowcs Williams' form of these equations given by Eq. (5.2.7). An important application of Eq. (5.2.7) is in propeller or helicopter rotor noise where the surfaces are rotating and translating in an otherwise stationary medium. In this case the free field Green's function is appropriate and we are most concerned with the acoustic far field. The Green's function is given by

$$G_o = \frac{\delta(t - \tau - r(\tau)/c_\infty)}{4\pi r(\tau)} \quad r(\tau) = |\mathbf{x} - \mathbf{z} - [\mathbf{U}_s + (\mathbf{z} \times \mathbf{\Omega})](\tau - \tau_o)| \tag{5.2.11}$$

where $\mathbf{U}_s$ is the translational velocity of the propeller and $\mathbf{\Omega}$ is its angular velocity about the origin of $\mathbf{z}$. If we convert the volume integrals to surface integrals we obtain,

$$\rho'(\mathbf{x}, t)c_\infty^2 = \frac{\partial^2}{\partial x_i \partial x_j} \int_{-T}^{T} \int_{V_\infty} T_{ij} G_o dV(\mathbf{z}) d\tau$$

$$- \frac{\partial}{\partial x_i} \int_{-T}^{T} \int_{S_o(\tau)} \left( (\rho v_i (v_j - V_j) + p_{ij}) n_j \right) G_o dS(\mathbf{z}) d\tau$$

$$+ \frac{\partial}{\partial t} \int_{-T}^{T} \int_{S_o(\tau)} (\rho v_j - \rho' V_j) G_o dS(\mathbf{z}) d\tau$$

Since the Green's function is multiplied into each integrand, we can now perform the time integrations. These are made easier by noting that, with $g = t - \tau - r(\tau)/c_\infty$,

$$\int_{-T}^{T} f(\tau)\delta(g(\tau))d\tau = \int_{g(-T)}^{g(T)} \frac{f(\tau)\delta(g)}{(\partial g/\partial \tau)} dg$$

$$= \left[ \frac{f}{|\partial g/\partial \tau|} \right]_{g=0} = \left[ \frac{f}{\left| -1 - \dfrac{1}{c_\infty} \dfrac{\partial r}{\partial \tau} \right|} \right]_{\tau=\tau^*} = \left[ \frac{f}{|1 - M_r|} \right]_{\tau=\tau^*}$$

$$\tag{5.2.12}$$

where $M_r c_\infty = -\partial r/\partial \tau$ is the velocity of the source in the direction of the observer, and the correct retarded time $\tau^*$ is the solution to $\tau^* = t - r(\tau^*)/c_\infty$. Thus the Ffowcs Williams Hawkings equation becomes,

$$\rho'(\mathbf{x}, t)c_\infty^2 H_s = \frac{\partial^2}{\partial x_i \partial x_j} \int_{V_o} \left[ \frac{T_{ij}}{4\pi r |1 - M_r|} \right]_{\tau=\tau^*} dV(\mathbf{z})$$

$$- \frac{\partial}{\partial x_i} \int_{S_o} \left[ \frac{\left( \rho v_i (v_j - V_j) + p_{ij} \right) n_j}{4\pi r |1 - M_r|} \right]_{\tau=\tau^*} dS \qquad (5.2.13)$$

$$+ \frac{\partial}{\partial t} \int_{S_o} \left[ \frac{(\rho v_j - \rho' V_j) n_j}{4\pi r |1 - M_r|} \right]_{\tau=\tau^*} dS$$

In this result the integrals are carried out over the moving surface at the correct emission time, and then the acoustic field is differentiated to obtain the final result.

In the acoustic far field, where the sound is the only disturbance and the sound waves are propagating radially away from the source, the space and time derivatives can be exchanged. This is simply a reflection of the fact that the change of the acoustic pressure in time experienced at a fixed point is the same as the change in space multiplied by the sound speed. To verify this for moving surfaces, we evaluate the gradient of the free field Green's function defined in Eq. (5.2.11) as

$$\frac{\partial G_o}{\partial x_i} = \frac{\partial}{\partial x_i} \left( \frac{\delta(g)}{4\pi r(\tau)} \right) = \frac{\partial r}{\partial x_i} \left( \frac{\partial g}{\partial r} \frac{\partial \delta(g)}{\partial g} \frac{1}{4\pi r(\tau)} - \frac{\delta(g)}{4\pi r^2(\tau)} \right)$$

$$= \frac{x_i - y_i}{r} \left( \frac{-1}{c_\infty} \frac{\partial \delta(g)}{\partial t} \frac{1}{4\pi r(\tau)} - \frac{\delta(g)}{4\pi r^2(\tau)} \right) = -\frac{x_i - y_i}{r} \left( \frac{1}{c_\infty} \frac{\partial G_o}{\partial t} + \frac{G_o}{r(\tau)} \right)$$

In the acoustic far field where $|\mathbf{x}| \gg |\mathbf{y}|$ the derivatives of the Green's function can therefore be approximated by

$$\frac{\partial G_o}{\partial x_i} \approx -\frac{x_i}{|\mathbf{x}|} \left( \frac{1}{c_\infty} \frac{\partial G_o}{\partial t} \right) \qquad (5.2.14)$$

The approximation may be applied to both the dipole and quadrupole terms in Eq. (5.2.13) giving the far field approximation for impermeable surfaces, as

$$\rho'(\mathbf{x}, t)c_\infty^2 \approx \frac{x_i x_j}{|\mathbf{x}|^2} \frac{1}{c_\infty^2} \frac{\partial^2}{\partial t^2} \int_{V_o} \left[ \frac{T_{ij}}{4\pi |\mathbf{x}||1 - M_r|} \right]_{\tau=\tau^*} dV(\mathbf{z})$$

$$+ \frac{x_i}{|\mathbf{x}|} \frac{1}{c_\infty} \frac{\partial}{\partial t} \int_{S_o} \left[ \frac{p_{ij} n_j}{4\pi |\mathbf{x}||1 - M_r|} \right]_{\tau=\tau^*} dS(\mathbf{z}) \qquad (5.2.15)$$

$$+ \frac{\partial}{\partial t} \int_{S_o} \left[ \frac{\rho_\infty V_j n_j}{4\pi |\mathbf{x}||1 - M_r|} \right]_{\tau=\tau^*} dS(\mathbf{z})$$

The result shows that for propeller or rotor noise there are three terms of importance. The first is the quadrupole term whose strength depends on $T_{ij}$ and is the sound radiated by both turbulence and flow distortions such as shock waves that are associated with the blades. The second is the dipole source term that is controlled by the surface loading $p_{ij}$. Finally, we have a term $\rho_\infty V_j n_j$ that depends only on the blade surface velocity and the density at the observer. This is a volume displacement source and is only nonzero if the observer "sees" a time varying surface velocity at emission time. This source is zero for an object in linear motion traveling toward the observer, but nonzero for a rotating blade which continuously changes direction relative to the observer. All the integrands are evaluated at emission time $\tau = \tau^* = t - r(\tau^*)/c_\infty$ and so for moving sources, the emitted time history and the observed time history are distorted relative to each other by the motion. We can shift the time differentials from the observed field onto the source terms using

$$\frac{\partial}{\partial t} = \frac{\partial \tau^*}{\partial t} \frac{\partial}{\partial \tau} = \frac{1}{1 - M_r} \frac{\partial}{\partial \tau} \tag{5.2.16}$$

which introduces an extra factor of $(1 - M_r)^{-1}$ for each time derivative in Eq. (5.2.15). A physical interpretation of this result is that the motion of the source compresses the time history as the source moves towards the observer, and this dramatically increases the radiated sound, especially when the relative Mach number $M_r$ approaches one. Also note that the integrals become singular when $M_r = 1$, and so special care needs to be exercised when evaluating these source terms for propellers or rotors with supersonic tip speeds. We will discuss this issue in more detail in Chapter 16 when we review rotor noise.

## 5.3   Moving sources

To illustrate some of the subtleties of the results given above we will evaluate the acoustic field from a moving source using Eq. (5.2.13). Consider an acoustically compact surface that is moving through a stationary fluid with constant speed $U_s$ in the $x_1$ direction. We will assume that the sound from turbulence in the fluid is negligible compared to the sound from the surface sources, and that the size of the surface is small compared to the propagation distance to the observer. If the source is acoustically compact, then the differences in retarded time from different points on the surface can be ignored and we can approximate the source to observer distance $r$ as $|\mathbf{x} - \mathbf{y}^{(c)}|$. Here $\mathbf{y}^{(c)}$ is the centroid of the surface, at the correct retarded time, in the fixed reference frame, corresponding to $\mathbf{z} = 0$ in the moving frame. Therefore, from Eq. (5.2.11),

$$r(\tau) = |\mathbf{x} - \mathbf{U}_s \tau|$$

where we have taken the time origin of the surface motion $\tau_o$ to be zero. The surface terms in Eq. (5.2.13) can then be simplified by taking the terms that depend on the propagation distance outside of the surface integrals giving

$$
\rho'(\mathbf{x}, t)c_\infty^2 \approx -\frac{\partial}{\partial x_i}\left[\frac{1}{4\pi r|1 - M_r|}\left\{\int_{S_o} p_{ij}n_j dS(\mathbf{z})\right\}\right]_{\tau=\tau^*}
$$
$$
+\frac{\partial}{\partial t}\left[\frac{1}{4\pi r|1 - M_r|}\left\{\int_{S_o} \rho_\infty V_j n_j dS(\mathbf{z})\right\}\right]_{\tau=\tau^*}
$$

Since $V_j = (U_s, 0, 0)$ is constant on the surface the last term integrates to zero and the only contribution comes from the dipole term. The surface integral of $p_{ij}n_j$ in that term is the net force applied to the fluid by the surface. For modeling purposes we take this as being a harmonic source in the frame of reference moving with the surface, with vector amplitude $\hat{F}_i$, defined as

$$
\int_{S_o} p_{ij}n_j dS(z) = \mathrm{Re}\left(\hat{F}_i e^{-i\omega_o \tau}\right)
$$

so that the acoustic field is given by

$$
\rho'(\mathbf{x}, t)c_\infty^2 \approx -\mathrm{Re}\left(\frac{\partial}{\partial x_i}\left[\frac{\hat{F}_i e^{-i\omega_o \tau}}{4\pi r|1 - M_r|}\right]_{\tau=\tau^*}\right) \tag{5.3.1}
$$

The retarded time is evaluated by solving the equation $t = \tau + r(\tau)/c_\infty$ or

$$
t - \tau = \left((x_1 - U_s\tau)^2 + x_2^2 + x_3^2\right)^{1/2}/c_\infty \tag{5.3.2}
$$

The retarded times $\tau = \tau^*$ can be obtained in terms of the observer time by squaring this equation and solving the resultant quadratic equation giving

$$
\tau^* = \frac{(c_\infty t - Mx_1) \pm \sqrt{(x_1 - U_s t)^2 + (1 - M^2)(x_2^2 + x_3^2)}}{c_\infty(1 - M^2)} \tag{5.3.3}
$$

where $M = U_s/c_\infty$. There are two possible solutions to this equation, but it is also constrained by the causality condition, which requires that $t > \tau$. For subsonic source speeds where $M < 1$ the only option is to choose the negative sign on the square root in Eq. (5.3.3) (this is readily verifiable by setting $x_2 = x_3 = 0$). However, for supersonic source speeds, two real solutions are possible if $(x_1 - U_s t)^2 > (M^2 - 1)(x_2^2 + x_3^2)$. To illustrate this Fig. 5.3A shows a series of wave fronts emitted from a source that is moving subsonically, and Fig. 5.3B shows the wave fronts for a source moving supersonically.

These figures highlight the physics of the problem. For the subsonic case (Fig. 5.3A), the wave fronts spread out from the source, but as the source moves

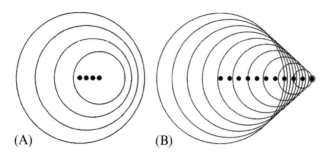

**Fig. 5.3** Wave fronts from a source moving from left to right. In (A) the source is moving with a Mach number of $M=0.5$ and in (B) the Mach number is $M=1.5$.

the center of the wavefronts changes. This causes an apparent contraction of the acoustic wavelength for an observer ahead of the source, and an apparent increase in frequency of the signal. For an observer behind the source, the apparent wavelength is increased and the observed frequency of the signal is decreased. In the far field the retarded time equation can be approximated using the far field approximation discussed in Section 3.6 with $y_1=U_s\tau$, so

$$r(\tau) \approx |\mathbf{x}| - x_1 U_s\tau/|\mathbf{x}| + \cdots$$

The velocity of the source in the direction of the observer is $U_r=U_s x_1/|\mathbf{x}|$ and the source Mach number in the direction of the observer is $M_r=U_r/c_\infty$. The far field approximation applied to Eq. (5.3.2) then gives

$$t-\tau \approx |\mathbf{x}|/c_\infty - M_r\tau \quad \text{or} \quad \tau \approx (t - |\mathbf{x}|/c_\infty)/(1-M_r) \tag{5.3.4}$$

leading to the approximation of Eq. (5.3.1) as

$$
\begin{aligned}
\rho'(\mathbf{x},t)c_\infty^2 &\approx -\mathrm{Re}\left(\frac{\partial}{\partial x_i}\left[\frac{\hat{F}_i e^{-i\omega_o(t-|\mathbf{x}|/c_\infty)/(1-M_r)}}{4\pi|\mathbf{x}||1-M_r|}\right]\right) \\
&\approx \mathrm{Re}\left(\frac{-i\omega_o x_i \hat{F}_i e^{-i\omega_o(t-|\mathbf{x}|/c_\infty)/(1-M_r)}}{4\pi c_\infty |\mathbf{x}|^2 (1-M_r)^2} + O\left(|\mathbf{x}|^{-2}\right)\right)
\end{aligned}
\tag{5.3.5}
$$

where near field terms of order $|\mathbf{x}|^{-2}$ have been ignored. This shows that the apparent frequency at the observer is multiplied by a factor $(1-M_r)^{-1}$ giving an increase of frequency when the source approaches the observer, and $M_r$ is positive, and a reduction of frequency as the source moves away from the observer, and $M_r$ is negative. This is called the Doppler frequency shift and is well known in many applications. We also note that the source has a dipole directivity given by $x_i/|\mathbf{x}|$ as expected, but there is also a directivity caused by the source motion. In this example the source is said to have two powers of Doppler amplification because there is a factor of $(1-M_r)^2$ on the bottom line of Eq. (5.3.5). This can significantly increase the levels as the source approaches the observer and reduce the levels as the source moves away.

When the source is moving supersonically as illustrated in Fig. 5.3B the situation is quite different. The source is now moving faster than the acoustic waves it generates and so it cannot be heard as it approaches the observer. This is to be expected from the solution to the retarded time equation (5.3.3) because the argument of the square root is negative if $(x_1 - U_s t)^2 > (M^2 - 1)(x_2^2 + x_3^2)$. To expand on this, we note that the term $(x_1 - U_s t)$ is the source position at *reception* time $t$ (see Fig. 5.4) and for a real solution to Eq. (5.3.3) we require

$$\frac{|x_1 - U_s t|}{\sqrt{x_2^2 + x_3^3}} = \cot \mu > \sqrt{M^2 - 1}$$

where $\mu$ is the angle of the observer to the path of the source evaluated at reception time, as shown in Fig. 5.4. We can rearrange this inequality so that it reads

$$\mu < \sin^{-1}\left(\frac{1}{M}\right) \tag{5.3.6}$$

where $\sin^{-1}(1/M)$ defines the angle of the Mach cone caused by the source motion, which is illustrated by the leading boundary of the waves in Fig. 5.3B. Inside the Mach cone we see that two wave fronts reach the same point at the same time, and this corresponds to the two possible solutions to the retarded time equation (5.3.3) when the argument of the square root is positive. An interesting point is that the propagation time for the wave front that originated from the source after it has passed the observer is shorter than the propagation time for the wave front originating from the source before it passed the observer, and hence the signal emitted by the source is both Doppler shifted and reversed in time!

The characteristics of sources moving at speed are important when we consider rotating sources such as propellers, helicopter rotors, and wind turbines, which will be considered in Part 4 of the text. We will leave further discussion of this topic until that time.

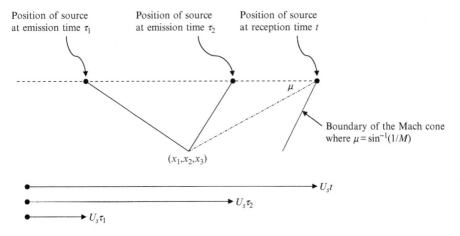

**Fig. 5.4** The source positions at emission times and reception times for supersonically moving sources.

## 5.4   Sources in a free stream

In many applications we are interested in the sound radiation from a uniform flow over a stationary object. Examples include model testing in wind tunnels, the flow in ducts, flow over large surfaces such as aircraft fuselages, and perhaps most importantly CFD calculations in body fixed coordinates. In each case there is a uniform steady flow at large distances from the region of turbulence that causes the sound and this contradicts the assumption, made in both Lighthill's acoustic analogy and the Ffowcs Williams and Hawkings equation that the medium is at rest at infinity. We showed in Section 4.2 that, in a uniform flow, the Lighthill stress tensor needed to be modified to only include the unsteady perturbations relative to the uniform flow, and that the acoustic propagation was determined by the convected wave equation. In principle, we can solve this modified equation using the techniques described earlier in this chapter using solutions to the convected wave equation. However, a simpler approach is often possible if we work in a frame of reference moving with the uniform flow, in which the medium appears stationary and the source and the observer are seen to be moving upstream at the flow speed. This approach is accommodated by modifying the results of this chapter to include a moving source and observer and by referencing flow velocity perturbations to the free stream.

An important application of this theory is to ducted fan noise where the sources are stationary or rotating next to stationary surfaces. The acoustic propagation is most readily addressed in this case by using the convected wave equation. It is possible to model this propagation by using a Green's function that satisfies the boundary conditions on the wall of the duct. We therefore need a formulation that will correctly allow for moving sources in a uniform flow that is bounded by stationary surfaces.

First consider the solution to Lighthill's wave equation in a frame of reference moving with the uniform flow velocity $\mathbf{U}^{(\infty)} = (U_\infty, 0, 0)$. The velocity perturbations relative to the moving frame of reference are given by $\mathbf{w}$, so the velocity in the fixed frame is $\mathbf{v} = \mathbf{U}^{(\infty)} + \mathbf{w}$. Lighthill's stress tensor is defined in the moving frame as

$$T'_{ij} = \rho w_i w_j + (p - p_\infty)\delta_{ij} - (\rho - \rho_\infty)c_\infty^2 \delta_{ij} - \sigma_{ij}$$

Also the velocity of the surface relative to the fixed frame is $V_j - U_j^{(\infty)}$. Eq. (5.2.8) can then be written, for the positions $\mathbf{x}'$ and $\mathbf{y}'$ in the moving frame as

$$\rho'(\mathbf{x}', t)c_\infty^2 = \int_{-T}^{T}\int_{V_o(\tau)} \frac{\partial^2 G}{\partial y'_i \partial y'_j} T'_{ij} dV(\mathbf{y}') d\tau$$

$$+ \int_{-T}^{T}\int_{S_o(\tau)} \frac{\partial G}{\partial y'_i}\left(\left(\rho w_i\left(w_j - V_j + U_j^{(\infty)}\right) + p_{ij}\right)n_j\right) dS(\mathbf{y}') d\tau \quad (5.4.1)$$

$$- \int_{-T}^{T}\int_{S_o(\tau)} \frac{\partial G}{\partial \tau}\left(\left(\rho w_j - \rho'\left(V_j - U_j^{(\infty)}\right)\right)n_j\right) dS(\mathbf{y}') d\tau$$

We wish to change this result to give a solution at the points $\mathbf{x}$ and $\mathbf{y}$, which are the observer and source locations in the fixed frame of reference. As part of this we need to replace the Green's function. In the moving frame where no free stream is observed, the Green's function is a solution to the inhomogeneous wave equation,

$$\frac{1}{c_\infty^2}\frac{\partial^2 G}{\partial \tau^2} - \frac{\partial^2 G}{\partial y_i'^2} = \delta(\mathbf{x}' - \mathbf{y}')\delta(t - \tau) \tag{3.9.2}$$

The Green's function in the fixed frame, $G_e(\mathbf{x},t|\mathbf{y},\tau)$, will be related to $G$ by straightforward translation of the coordinates,

$$G(\mathbf{x}',t|\mathbf{y}',\tau) = G_e\left(\mathbf{x} + \mathbf{U}^{(\infty)}t, t|\mathbf{y} + \mathbf{U}^{(\infty)}\tau, \tau\right)$$

and thus

$$\frac{\partial G}{\partial y_i'} = \frac{\partial G_e}{\partial y_i} \quad \text{and} \quad \frac{\partial G}{\partial \tau} = \frac{\partial G_e}{\partial \tau} + U_i^{(\infty)}\frac{\partial G_e}{\partial y_i} = \frac{D_\infty G_e}{D\tau}$$

where $D_\infty/D\tau$ is the free stream convective derivative, introduced in Eq. (4.2.9). Applying these conversions to Eq. (3.9.2) we see that the fixed frame Green's function must be a solution to the equation,

$$\frac{1}{c_\infty^2}\frac{D_\infty^2 G_e}{D\tau} - \nabla^2 G_e = \delta(t - \tau)\delta(\mathbf{x} - \mathbf{y}) \tag{5.4.2}$$

and so we obtain

$$\rho'(\mathbf{x},t)c_\infty^2 = \int_{-T}^{T}\int_{V_o(\tau)}\frac{\partial^2 G_e}{\partial y_i \partial y_j}T_{ij}'dV(\mathbf{y})d\tau$$

$$+ \int_{-T}^{T}\int_{S_o(\tau)}\frac{\partial G_e}{\partial y_i}\left(\left(\rho w_i\left(w_j - V_j + U_j^{(\infty)}\right) + p_{ij}\right)n_j\right)dS(\mathbf{y})d\tau \tag{5.4.3}$$

$$- \int_{-T}^{T}\int_{S_o(\tau)}\frac{D_\infty G_e}{D\tau}\left(\left(\rho w_j - \rho'\left(V_j - U_j^{(\infty)}\right)\right)n_j\right)dS(\mathbf{y})d\tau$$

An important application of Eq. (5.4.3) is in CFD calculations or surface flows in which the surfaces are stationary and the medium is in uniform motion at infinity. In this case $\mathbf{V} = 0$ and the nonpenetration boundary condition on the surfaces require that $(\mathbf{U}^{(\infty)} + \mathbf{w})\cdot\mathbf{n} = 0$ on all the surfaces. The integrand of the last integral in Eq. (5.4.3) then reduces to $\rho_\infty U_j^{(\infty)}n_j D_\infty G_e/D\tau$ which does not vary with time and so makes no contribution to the acoustic field. A similar reduction can be made to the second integral in Eq. (5.4.3) and, neglecting viscous stresses as we did in Section 4.5, gives

$$\rho'(\mathbf{x},t)c_\infty^2 = \int_{-T}^{T}\int_{S_o}\frac{\partial G_e}{\partial y_i}p'n_i dS(\mathbf{y})d\tau + \int_{-T}^{T}\int_{V_o}\frac{\partial^2 G_e}{\partial y_i \partial y_j}T'_{ij}dV(\mathbf{y})d\tau \qquad (5.4.4)$$

where $p' = p - p_\infty$. This result is identical to Eq. (4.5.1), which was the result for a stationary medium, the differences being that the Green's function must satisfy the convected wave equation (5.4.2), and that the Lighthill stress tensor depends on $\rho w_i w_j$.

## 5.5  Ffowcs Williams and Hawkings surfaces

One of the most useful applications of the formulas given above is in the calculation of the acoustic far field from detailed numerical simulations of a flow within a limited region. Recent advances in computational methods have enabled the accurate calculation of many time varying flows of practical interest. Provided the Mach number is not too high, these solutions provide both the unsteady flow and the acoustics inside a computational grid, but the computational domain is limited by the size of the computer, and usually cannot be extended to the acoustic far field. Furthermore, increasing the size of the computational domain soon becomes a wasteful exercise because wave propagation outside the flow is well understood and modeled by the linear wave equation. The purpose of a Ffowcs Williams and Hawkings (FWH) surface is to provide a far field solution to the wave equation given accurate numerical calculations on a surface which bounds the source region. It is assumed that the calculations are accurate inside the source region and that they accurately capture the compressible flow fluctuations and the acoustic waves, so that the FWH surface may be arbitrarily located within the numerical domain. This is important because the numerical calculations at the edges of the computational domain may be adversely influenced by numerical boundary conditions, so the FWH surface is usually placed inside the numerical domain in a region where there is confidence in the calculations. However, it is preferable, but not absolutely necessary, to choose the surface so that the Lighthill stress tensor does not contribute to the far field in the region outside the FWH surface. Then Eq. (5.2.13) can be used without the quadrupole source terms to give the acoustic field as

$$\rho'(\mathbf{x},t)c_\infty^2 H_s = -\frac{\partial}{\partial x_i}\int_{S_o}\left[\frac{\left(\rho v_i\left(v_j - V_j\right) + p_{ij}\right)n_j}{4\pi r|1 - M_r|}\right]_{\tau = \tau^*}dS$$
$$+\frac{\partial}{\partial t}\int_{S_o}\left[\frac{\left(\rho v_j - \rho' V_j\right)n_j}{4\pi r|1 - M_r|}\right]_{\tau = \tau^*}dS \qquad (5.5.1)$$

This result implies that if we know the flow quantities on $S_o$ (which may be moving relative to the observer) then we can calculate the acoustic far field everywhere outside the computational domain. Obviously we need to check that the surface is in the right place and that the dropping of the quadrupole term is a good approximation.

As shown by Brentner and Farassat [2], this can be done by recalculating the results on a slightly different surface and checking that the far field is independent of surface location. Notice that in this formula we need to know the mass flux, the surface stress, and the momentum flux on the surface. These variables are available in an unsteady compressible flow calculation, but the incorrect result will be obtained from an incompressible flow calculation because the acoustic waves which couple with the far field will be missing unless the surface is very small compared to the acoustic wavelength. For a rotating source, such as a helicopter rotor, the steady flow in blade based coordinates can be used as the input to Eq. (5.5.1) to give a time varying far field. This can include detached shock waves that are local to the blade and inside the computational domain. When the blade encounters an unsteady flow event, such as a blade vortex interaction, then obviously the time varying surface quantities must be used.

To apply Eq. (5.5.1), it is necessary to calculate the integrand at the emission time for of each element of the FWH surface. This can require significantly more accurate time computations on the surface than would be required in the acoustic far field. It can therefore be advantageous to move the time derivatives in Eq. (5.5.1) inside the integrand. For the acoustic far field this can be done by using

$$\frac{\partial}{\partial x_i} \approx -\frac{x_i}{|\mathbf{x}|} \frac{1}{c_\infty} \frac{\partial}{\partial t}$$

and Eq. (5.2.16) to give the form used by Farassat [3]

$$\rho'(\mathbf{x}, t) c_\infty^2 H_s \approx \frac{1}{c_\infty} \int_{S_o} \left[ \frac{x_i}{4\pi |\mathbf{x}|^2 (1 - M_r)^2} \left\{ \frac{\partial L_i}{\partial \tau} + \frac{L_i}{(1 - M_r)} \frac{\partial M_r}{\partial \tau} \right\} \right]_{\tau = \tau^*} dS$$
$$+ \int_{S_o} \left[ \frac{1}{4\pi |\mathbf{x}| (1 - M_r)^2} \left\{ \frac{\partial q}{\partial \tau} + \frac{q}{(1 - M_r)} \frac{\partial M_r}{\partial \tau} \right\} \right]_{\tau = \tau^*} dS$$

(5.5.2)

with

$$L_i = \left( \rho v_i (v_j - V_j) + p_{ij} \right) n_j \qquad q = \left( \rho v_j - \rho' V_j \right) n_j$$

where terms of order $|\mathbf{x}|^{-2}$ have been dropped. The source terms on the surface can now be readily evaluated and differentiated in source time, and then propagated to the far field. There is still a requirement to match the observer time and the emission time, but this can be done with less restrictive requirements on numerical accuracy if the differentials are evaluated in source time.

In conclusion, the approach described above is very attractive because it overcomes one of the main limitations of Lighthill's theory. Specifically, it moves the application of Lighthill's equation to the region where it is unambiguously defined, and allows all the processes which are taking place to be determined by the numerical code inside the FWH surface. However, the accuracy of the result depends on the accuracy of the calculations on the surface. It is only the surface perturbations that couple to the acoustic field which are needed, and unfortunately these may be a small part of the net motion.

The signal to noise ratio of the integrands in Eq. (5.5.2) is therefore determined by the ratio of the propagating waves to the numerical noise, not the ratio of the absolute fluctuations to the numerical noise. In later chapters we will discuss how we can discriminate the propagating and nonpropagating parts of the surface source terms, and show that it is the disturbances with supersonic phase speeds on the FWH surface (relative to the observer) which are the most important.

## 5.6  Incompressible flow estimates of acoustic source terms

In the previous section we discussed the use of FWH surfaces to couple CFD calculations over a finite volume to the acoustic far field. This requires that the acoustic waves are correctly defined on the FWH surface and this will only be the case if they are computed using the compressible equations of motion. At low Mach numbers the CFD grid required for a compressible calculation stretches the limits of computational capabilities. In general, a FWH surface surrounding a computational domain of arbitrary size cannot be used with the output from an incompressible flow calculation. However, it has been shown by Wang et al. [4] that some valuable results can be obtained at low Mach numbers if the output of an incompressible flow calculation is used to specify the source terms in Lighthill's acoustic analogy, provided the volume integral is broken down into source regions that are very small compared to the acoustic wavelength.

With the flow computed incompressibly, the speed of sound used in the application of Lighthill's equation can be taken as constant so that acoustic density perturbations are directly related to the acoustic pressure fluctuations as $p - p_\infty = (\rho - \rho_\infty)c_\infty^2$. Lighthill's stress tensor can then be approximated by $\rho_0 w_i w_j$ where $w_i$ is the flow perturbation relative to a constant velocity $U_j^{(\infty)}$ at the inflow boundary of the computational domain (see Section 5.4). Acoustic radiation from low Mach number turbulent flows without surfaces present is normally of very low level, and so the surface source terms are usually of primary interest. In almost all cases the viscous terms on the surfaces are assumed to be negligible. The surface source term in Eq. (5.4.4) is then determined by the surface pressure. If the surface is acoustically compact and all parts of the turbulent flow are within a fraction of an acoustic wavelength from the surface, then the surface pressure can be approximated by the pressure fluctuations that are obtained from the incompressible CFD calculation. However, if the surface is large compared to the acoustic wavelength, then this approximation is invalid because the acoustic waves on the surface are not included. For example, the sound radiated by the trailing edge of an airfoil is often modeled by a semi-infinite flat plate that extends to the upstream limit of the computational domain. As we will show below, the far field sound from turbulence near the corner of a large flat plate will scale quite differently from the sound from an acoustically compact surface, and this limits the use of the surface hydrodynamic pressure for acoustic calculations.

An alternative approach, used by Wang et al. [4], is to combine the calculated Lighthill's stress tensor with a tailored Green's function, as in Eq. (4.5.2). This eliminates the need to evaluate the surface integral and the far field sound is calculated directly from the unsteady flow. Unfortunately, tailored Green's functions are only available for a limited set of idealized shapes. An alternative numerical approach, developed by Khalighi et al. [5], is to numerically calculate the surface pressure from the volume source terms using a boundary element method. This calculation is most easily done in the frequency domain and implies that the pressure term in Eq. (4.5.4) (or the Fourier transform of Eq. 5.4.4) is solved for the unknown surface pressure $\tilde{p}$ with the volume sources defined from an incompressible CFD calculation. The resulting integral equation is of course singular because the Green's function is singular when the source point and observer point merge, but the approach described in Section 4.6 can be used to eliminate the singularity. The size of the surface is in principle not limited, but boundary integral methods are only applicable to closed surfaces, and so some approximation must be used to close semi-infinite surfaces without introducing additional scattering. This is a three step approach: first an unsteady incompressible CFD code is used to calculate the turbulent flow, and then a compressible boundary element method is used to numerically calculate the surface pressure from the volume source terms in Lighthill's analogy. The final step is to calculate the acoustic far field from the surface pressure on the boundary and the volume sources. However, both these methods assume that Lighthill's analogy can be used to calculate the pressure perturbations within the flow and, as we will discuss below, this limits their application to low Mach number flows.

The physical element that is missing from these approaches is the back reaction of the acoustic waves on the unsteady flow near corners and edges. For example, acoustic feedback loops that occur in cavities and in laminar boundary layers will not be simulated because they are often caused by acoustic rather than hydrodynamic feedback. A classical example is a singing airfoil where the instabilities in a laminar boundary layer close to the leading edge of an airfoil grow with distance downstream and produce the regular shedding of eddies from the trailing edge. Potential flow and acoustic perturbations produced by the shedding are then felt upstream near the leading edge where they serve to originate and organize the boundary layer instability.

The process is controlled by the upstream feedback, which can be caused by either potential flow perturbations or acoustic waves that originate at the trailing edge. The amplitude of the potential flow perturbations scale with $1/r^2$, where $r$ is the distance from the source of the disturbance. In contrast acoustic perturbations, which are initially much weaker, scale with distance as $1/r$. At distances that are of the order of an acoustic wavelength, the acoustic disturbances can dominate because their decay with distance is much less than the decay of potential perturbations. If acoustic propagation is left out of the calculation, as it is in an incompressible flow code, then the feedback that disturbs the upstream laminar boundary layer will be much weaker and may not initiate a sufficient disturbance to trigger the feedback loop. This issue is not limited to low frequencies because the feedback loop is nonlinear, and so instabilities at very high frequencies can cause significant flow excursions at low frequencies that would not be modeled correctly by an incompressible flow calculation.

Finally, it must be pointed out that Lighthill's formulation was never intended for calculations within the flow, and this in itself can be a limitation when the mean flow Mach number is increased above about 0.3. To illustrate this point, consider Lighthill's source term when there is a significant distortion of the mean flow from its uniform inflow value. If we define the mean flow as $U_i = U_i^{(\infty)} + \Delta U_i$ where $U_i^{(\infty)}$ is constant then the velocity relative to the constant flow is $w_i = \Delta U_i + u_i$ and the source term in Lighthill's wave equation can be broken down into linear and nonlinear terms as follows:

$$\rho w_i w_j = \underbrace{\rho_\infty \Delta U_i \Delta U_j}_{\text{Steady flow}} + \underbrace{\rho_\infty \Delta U_i u_j + \rho_\infty u_i \Delta U_j + \rho' \Delta U_i \Delta U_j}_{\text{Linear terms}} + \underbrace{\rho_\infty u_i u_j + \rho' u_i u_j}_{\text{Nonlinear terms}}$$

$$(5.6.1)$$

The important point is that when there is significant distortion of the mean flow then terms that are linear in the perturbation velocity will dominate the source terms if $\Delta U \gg u$. For flows over obstructions that include stagnation points or for weakly turbulent inflows over streamlined shapes, the linear terms will dominate. However, if that is the case then we cannot ignore the term $\rho' \Delta U_i \Delta U_j$ which depends on the acoustic variable $\rho'$, and so is not a source term. It is also not included in incompressible flow calculations. This term represents the refraction of sound waves by the mean flow and should be part of the wave operator. This is one of the inherent limitations of Lighthill's analogy but will only have a small impact if $\rho' \Delta U_i \Delta U_j \ll \rho' c_\infty^2$, which suggests the limit $\Delta U / c_\infty < 0.3$. To address this limitation a theory is required that does not include the acoustic variable in the source term. In the next chapter we will address this issue by assuming that the turbulence is sufficiently weak that the linear terms in Eq. (5.6.1) dominate the production of sound.

# References

[1] J.E. Ffowcs Williams, D.L. Hawkings, Sound generation by turbulence and surfaces in arbitrary motion, Philos. Trans. R. Soc. Lond. A 264 (1969) 321–342.
[2] K. Brentner, F. Farassat, Modeling aerodynamically generated sound of helicopter rotors, Prog. Aeronaut. Sci. 39 (2003) 83–120.
[3] F. Farassat, Linear acoustic formulas for calculation of rotating blade noise, AIAA J. 19 (1981) 1122–1130.
[4] M. Wang, J.B. Freund, S.K. Lele, Computational prediction of flow-generated sound, Annu. Rev. Fluid Mech. 38 (2006) 483–512.
[5] Y. Khalighi, A. Mani, F. Ham, P. Moin, Prediction of sound generated by complex flows at low Mach numbers, AIAA J. 48 (2010) 306–316.

# The linearized Euler equations

## 6

In Chapters 4 and 5 we developed the basic equations for sound generation by fluid flow based on Lighthill's Acoustic Analogy. The solution to these equations gives the radiated sound in terms of sources in the flow and sources on surfaces that bound the flow. To obtain the surface source terms we need to solve a boundary value problem, but Lighthill's analogy is not ideally suited for this because it only describes acoustic waves propagating in a stationary medium. In practice, surfaces immersed in a turbulent flow will generate sound through their interaction with the turbulence. Predicting this type of sound generation may be thought of as involving two distinct physical mechanisms: the distortion of the turbulence by the flow around the body, and the generation of unsteady pressures on the surface as a response to that turbulence. To address these issues we will now discuss an alternate formulation based on the linearization of the Euler equations, which assume that the flow is isentropic and inviscid.

## 6.1 Goldstein's equation

Goldstein [1] derived a solution to the linearized Euler equations for small perturbations about a steady mean flow. The equations of motion for the mean flow, defined in terms of its density, pressure, and velocity $\rho_o$, $p_o$, and $\mathbf{U}$, were subtracted from the equations of motion for the unsteady flow and linearized, giving a set of linear equations in terms of the density, pressure, and velocity perturbations $\rho'$, $p'$, and $\mathbf{u}$. In this section we use a more general approach and take the time derivative of the equations of motion and identify the perturbation quantities as $\partial\rho/\partial t$, $\partial p/\partial t$, and $\partial\mathbf{v}/\partial t$. The theory assumes an inviscid flow with no heat conduction, which means that the unsteady flow is isentropic and so, from Eq. (2.6.7), $Ds/Dt = 0$. We also assume the mean flow to be homentropic. This restricts the approach to cases where boundary layer effects can be ignored and therefore excludes separated flows. It also assumes that the mean flow convects flow disturbances linearly, which implies that the turbulent eddies remain as coherent structures while they are convected from an upstream boundary (where they are defined) until they have completely passed the surfaces in the flow. This can represent a severe restriction in small-scale turbulent flows where the coherence length scale of the turbulence, in the convected frame, is small compared to the size of the body.

The starting point for the derivation of Goldstein's wave equation is to rewrite the continuity equation in terms of the density perturbation, or more generally, in terms of the rate of change of density $\partial\rho/\partial t$. This will enable us to separate out the fluctuating quantities from the mean quantities at a later stage. Taking the time derivative of the continuity Eq. (2.2.5) gives

$$\frac{\partial^2\rho}{\partial t^2} + \frac{\partial}{\partial t}\nabla\cdot(\rho\mathbf{v}) = 0$$

Aeroacoustics of Low Mach Number Flows. http://dx.doi.org/10.1016/B978-0-12-809651-2.00006-0

so

$$\frac{\partial^2 \rho}{\partial t^2} + \mathbf{v} \cdot \nabla \left( \frac{\partial \rho}{\partial t} \right) + \frac{\partial \rho}{\partial t} \nabla \cdot \mathbf{v} + \nabla \cdot \left( \rho \frac{\partial \mathbf{v}}{\partial t} \right) = 0 \tag{6.1.1}$$

The first two terms of the expansion combine to give the material derivative of $\partial \rho / \partial t$, and we can replace the divergence of the velocity in the third term by using the continuity equation in the form given by Eq. (2.2.12), $\rho D(1/\rho)/Dt = \nabla \cdot \mathbf{v}$, so, after multiplying and dividing by the density we have

$$\rho \left\{ \frac{1}{\rho} \frac{D}{Dt} \left( \frac{\partial \rho}{\partial t} \right) + \frac{\partial \rho}{\partial t} \frac{D}{Dt} \left( \frac{1}{\rho} \right) \right\} + \nabla \cdot \left( \rho \frac{\partial \mathbf{v}}{\partial t} \right) = 0 \tag{6.1.2}$$

Combining the terms in curly brackets then gives

$$\frac{D}{Dt} \left( \frac{1}{\rho} \frac{\partial \rho}{\partial t} \right) + \frac{1}{\rho} \nabla \cdot \left( \rho \frac{\partial \mathbf{v}}{\partial t} \right) = 0 \tag{6.1.3}$$

This is the continuity equation in terms of perturbation variables $\partial \rho / \partial t$ and $\partial \mathbf{v} / \partial t$. We will show later that this can be readily linearized about the mean time-invariant flow when the density perturbations are small.

Next we replace the density perturbation with the pressure perturbation by using the equation for entropy change in an ideal gas given by Eq. (2.4.9) in the form

$$\frac{1}{\rho} \frac{\partial \rho}{\partial t} = \frac{c_v}{p c_p} \frac{\partial p}{\partial t} - \frac{1}{c_p} \frac{\partial s}{\partial t} = \left( \frac{1}{\rho c^2} \frac{\partial p}{\partial t} \right) - \frac{1}{c_p} \frac{\partial s}{\partial t}$$

where $c^2 = \gamma p / \rho = p c_p / \rho c_v$ is the local *unsteady* speed of sound. The continuity equation is then

$$\frac{D}{Dt} \left( \frac{1}{\rho c^2} \frac{\partial p}{\partial t} \right) + \frac{1}{\rho} \nabla \cdot \left( \rho \frac{\partial \mathbf{v}}{\partial t} \right) = \frac{1}{c_p} \frac{D}{Dt} \left( \frac{\partial s}{\partial t} \right) \tag{6.1.4}$$

We linearize this equation about the mean flow by setting $\rho = \rho_o + \rho'$, $p = p_o + p'$ and assuming that terms of order $(\rho'/\rho_o)\partial p/\partial t$, $(p'/p_o)\partial p/\partial t$, and $\rho' \partial \mathbf{v}/\partial t$ are negligible compared to terms of order $\partial p/\partial t$ and $\rho_o \partial \mathbf{v}/\partial t$, which also impacts the definition of the speed of sound. (Note that in the present context the prime indicates fluctuation relative to the mean value, rather than disturbance relative to a static ambient as was the case for Lighthill's equation.) So, the left side of Eq. (6.1.4) becomes

$$\frac{D}{Dt} \left( \frac{1}{\rho_o c_o^2} \frac{\partial p}{\partial t} \right) + \frac{1}{\rho_o} \nabla \left( \rho_o \frac{\partial \mathbf{v}}{\partial t} \right)$$

For isentropic flow the right side becomes

$$\frac{D}{Dt}\left(\frac{\partial s}{\partial t}\right) = \frac{\partial}{\partial t}\left(\frac{Ds}{Dt}\right) - \left(\frac{\partial \mathbf{v}}{\partial t} \cdot \nabla\right)s$$

For isentropic flow $Ds/Dt = 0$, and since there is no mean entropy gradient, $((\partial \mathbf{v}/\partial t) \cdot \nabla)s$ is the product of two small quantities and so can also be ignored, and it follows that Eq. (6.1.4) may be reduced to

$$\frac{D}{Dt}\left(\frac{1}{\rho_o c_o^2}\frac{\partial p}{\partial t}\right) + \frac{1}{\rho_o}\nabla \cdot \left(\rho_o \frac{\partial \mathbf{v}}{\partial t}\right) = 0 \tag{6.1.5}$$

Additionally, we assume that the velocity perturbation is small compared to the mean velocity, so $|\mathbf{U}| \gg |\mathbf{u}|$. Thus for any fluctuating quantity $f$ we can approximate

$$\frac{Df}{Dt} = \frac{\partial f}{\partial t} + (\mathbf{U} + \mathbf{u}) \cdot \nabla f \approx \frac{\partial f}{\partial t} + \mathbf{U} \cdot \nabla f = \frac{D_o f}{Dt} \tag{6.1.6}$$

where $D_o/Dt$ is the substantial derivative relative to the mean flow. The assumption $|\mathbf{U}| \gg |\mathbf{u}|$ implies that the flow perturbation is only convected by the mean flow and does not convect itself. In turn this implies that the time scale of the turbulence in the moving frame of reference is much larger than the transit time of the turbulence through the region of interest. This is referred to as rapid distortion theory (RDT) and will be discussed in Section 6.3.

The linearized continuity equation and entropy equation then become

$$\frac{D_o}{Dt}\left(\frac{1}{\rho_o c_o^2}\frac{\partial p}{\partial t}\right) + \frac{1}{\rho_o}\nabla \cdot \left(\rho_o \frac{\partial \mathbf{v}}{\partial t}\right) = 0 \qquad \frac{D_o s}{Dt} = 0 \tag{6.1.7}$$

Following Goldstein [1] we decompose the velocity into three terms $\mathbf{v} = \mathbf{U} + \nabla\phi + \mathbf{u}^{(g)}$ where we define the velocity potential term $\phi$ to be directly related to the pressure fluctuations by

$$\rho_o \frac{D_o}{Dt}\left(\frac{\partial \phi}{\partial t}\right) = -\frac{\partial p}{\partial t} \tag{6.1.8}$$

Note that $\mathbf{u}^{(g)}$ can include both potential and rotational disturbances, leaving us free to choose $\phi$ as that part of the potential related directly to the pressure fluctuation. We can then obtain the wave equation in terms of this velocity potential by substituting for the velocity and pressure disturbances in Eq. (6.1.7) to give

$$\frac{D_o}{Dt}\left(\frac{1}{c_o^2}\frac{D_o \dot{\phi}}{Dt}\right) - \frac{1}{\rho_o}\nabla \cdot \left(\rho_o \nabla \dot{\phi}\right) = \frac{1}{\rho_o}\nabla \cdot \left(\rho_o \dot{\mathbf{u}}^{(g)}\right) \tag{6.1.9}$$

where the dot represents a derivative with respect to time. Since Eq. (6.1.9) is linear with respect to the potential and velocity disturbance terms, it applies to both the time derivatives and the fluctuating parts of the flow variables that were considered in the original derivation given by Goldstein [1]. In his formulation, therefore, the dots denoting time derivative do not appear.

The wave equation given by (6.1.9) includes a source term on the right side that still needs to be defined. To obtain this we will start with the momentum equation in its nonconservative form given by Eq. (2.3.10). To obtain this in terms of perturbation quantities, we take its time derivative and ignore viscous terms, giving

$$\frac{D}{Dt}\left(\frac{\partial \mathbf{v}}{\partial t}\right) + \left(\frac{\partial \mathbf{v}}{\partial t}\cdot\nabla\right)\mathbf{v} = -\frac{\partial}{\partial t}\left(\frac{1}{\rho}\nabla p\right)$$

Decomposing the velocity as above and expanding the right side of this equation we obtain

$$\frac{D}{Dt}\left(\dot{\mathbf{u}}^{(g)} + \nabla\dot{\phi}\right) + \left(\left(\dot{\mathbf{u}}^{(g)} + \nabla\dot{\phi}\right)\cdot\nabla\right)\mathbf{v} = -\frac{1}{\rho}\nabla\dot{p} + \frac{\dot{\rho}}{\rho^2}\nabla p$$

We then linearize this equation as we did continuity and use Eq. (6.1.8) to give

$$\frac{D_o}{Dt}\left(\dot{\mathbf{u}}^{(g)} + \nabla\dot{\phi}\right) + \left(\left(\dot{\mathbf{u}}^{(g)} + \nabla\dot{\phi}\right)\cdot\nabla\right)\mathbf{U} = \frac{1}{\rho_o}\nabla\left(\rho_o\frac{D_o\dot{\phi}}{Dt}\right) + \frac{\dot{\rho}}{\rho_o^2}\nabla p_o \qquad (6.1.10)$$

The first term on the right side of this equation expands as

$$\frac{1}{\rho_o}\nabla\left(\rho_o\frac{D_o\dot{\phi}}{Dt}\right) = \frac{\nabla\rho_o}{\rho_o}\left(\frac{D_o\dot{\phi}}{Dt}\right) + \frac{\partial(\nabla\dot{\phi})}{\partial t} + \nabla\left(\mathbf{U}\cdot\nabla\dot{\phi}\right)$$

Which, using the vector identity

$$\nabla(\mathbf{A}\cdot\mathbf{B}) = (\mathbf{A}\cdot\nabla)\mathbf{B} + (\mathbf{B}\cdot\nabla)\mathbf{A} + \mathbf{A}\times(\nabla\times\mathbf{B}) + \mathbf{B}\times(\nabla\times\mathbf{A})$$

can be rewritten as

$$\frac{1}{\rho_o}\nabla\left(\rho_o\frac{D_o\dot{\phi}}{Dt}\right) = \frac{\nabla\rho_o}{\rho_o}\left(\frac{D_o\dot{\phi}}{Dt}\right) + \frac{\partial(\nabla\dot{\phi})}{\partial t} + (\mathbf{U}\cdot\nabla)\nabla\dot{\phi} + (\nabla\dot{\phi}\cdot\nabla)\mathbf{U} + \nabla\dot{\phi}\times\boldsymbol{\omega}_o$$

where $\boldsymbol{\omega}_o = \nabla\times\mathbf{U}$ is the vorticity of the mean flow. The middle three terms in this expansion match the terms on the left-hand side of Eq. (6.1.10) that depend on the potential $\phi$, and so we obtain

$$\frac{D_o\dot{\mathbf{u}}^{(g)}}{Dt} + \left(\dot{\mathbf{u}}^{(g)}\cdot\nabla\right)\mathbf{U} = \nabla\dot{\phi}\times\boldsymbol{\omega}_o + \frac{\nabla\rho_o}{\rho_o}\left(\frac{D_o\dot{\phi}}{Dt}\right) + \frac{\dot{\rho}}{\rho_o^2}\nabla p_o \qquad (6.1.11)$$

then using Eq. (6.1.8) the last two terms on the right side of Eq. (6.1.11) become

$$\frac{\nabla \rho_o}{\rho_o}\left(\frac{D_o \dot{\phi}}{Dt}\right) + \frac{\dot{\rho}}{\rho_o^2}\nabla p_o = -\frac{\nabla \rho_o}{\rho_o^2}\dot{p} + \frac{\dot{\rho}}{\rho_o^2}\nabla p_o$$

Further we can relate these quantities to the entropy using Eq. (2.4.9), so

$$-\frac{\nabla \rho_o}{\rho_o^2}\dot{p} + \frac{\dot{\rho}}{\rho_o^2}\nabla p_o = -\frac{\nabla \rho_o}{\rho_o^2}\dot{p} + \frac{\nabla p_o}{\rho_o}\left(\frac{c_v \dot{p}}{c_p p_o} - \frac{\dot{s}}{c_p}\right)$$

Since the mean flow is assumed to be homentropic it follows that $(c_v/p_o)\nabla p_o = (c_p/\rho_o)\nabla \rho_o$, and so

$$-\frac{\nabla \rho_o}{\rho_o^2}\dot{p} + \frac{\dot{\rho}}{\rho_o^2}\nabla p_o = -\frac{\nabla p_o}{\rho_o}\left(\frac{\dot{s}}{c_p}\right)$$

making use of the momentum equation for the mean flow gives

$$-\frac{\nabla \rho_o}{\rho_o^2}\dot{p} + \frac{\dot{\rho}}{\rho_o^2}\nabla p_o = \left(\frac{\dot{s}}{c_p}\right)(U \cdot \nabla)U$$

Then, since $D_o s/Dt = 0$ we can write the right side of this equation as

$$\left(\frac{\dot{s}}{c_p}\right)(U \cdot \nabla)U = \frac{D_o}{Dt}\left(\frac{\dot{s}U}{2c_p}\right) + \left(\left(\frac{\dot{s}U}{2c_p}\right) \cdot \nabla\right)U$$

With these rather extensive manipulations we obtain Eq. (6.1.11) as

$$\frac{D_o \dot{u}^{(h)}}{Dt} + \left(\dot{u}^{(h)} \cdot \nabla\right)U = \nabla\dot{\phi} \times \omega_o \qquad \dot{u}^{(h)} = \dot{u}^{(g)} - \frac{U\dot{s}}{2c_p} \qquad \frac{D_o \dot{s}}{Dt} = 0 \qquad (6.1.12)$$

As in the wave Eq. (6.1.9), the dot represents a differentiation with respect to time and identifies fluctuating quantities. Eqs. (6.1.9), (6.1.12) are a set of linear partial differential equations for small perturbations in a homentropic mean flow. The perturbation can include an entropy gust defined by $U\dot{s}/2c_p$ which would be relevant to supersonic or heated flows. However, at low Mach number we are generally only interested in vortical disturbances, so $\dot{s} = 0$. We also note that the wave Eq. (6.1.9) has variable coefficients so that it allows for the diffraction of the sound by the mean flow and any variation in local sound speed. It applies both within the flow and in the acoustic far field, which makes it more valuable than Lighthill's equation in solving certain problems. However, it excludes the nonlinear turbulent terms that dominate in cases where there are no surfaces in the flow and so cannot be used when direct radiation from turbulence is important.

One of the most significant aspects of this derivation is that in a potential mean flow, $\omega_o$ is zero, and so Eqs. (6.1.9), (6.1.12) are uncoupled, and analytical solutions can be found for $\mathbf{u}^{(h)}$ and $\partial s/\partial t$. We will discuss these solutions in more detail in the next section and introduce the concept of drift coordinates that allow for the analytical solution to Eq. (6.1.12).

## 6.2 Drift coordinates

One of the most important applications of Goldstein's equation is to the sound radiation from unsteady flow over stationary rigid surfaces. The results from this application can also be used for moving surfaces by making use of the Ffowcs Williams and Hawkings surface as described in Chapter 5.

Consider the flow over a body, as shown in Fig. 6.1. The mean flow is defined as uniform at the inflow boundary, where a disturbance is defined and is convected or propagated downstream in accordance with Eqs. (6.1.9), (6.1.12).

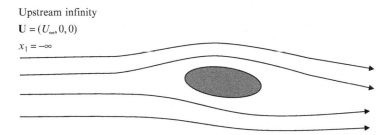

Upstream infinity
$\mathbf{U} = (U_\infty, 0, 0)$
$x_1 = -\infty$

**Fig. 6.1** The flow over a stationary object and the boundary conditions at upstream infinity.

The problem can be solved in terms of *drift coordinates* $X_1$, $X_2$, $X_3$ which are defined as the solutions to the first-order differential equations

$$\frac{D_o X_1}{Dt} = U_\infty \qquad \frac{D_o X_2}{Dt} = 0 \qquad \frac{D_o X_3}{Dt} = 0 \qquad (6.2.1)$$

with the upstream boundary conditions that $X_i = x_i$, and the flow speed is $U_\infty$ in the $x_1$ direction as $x_1$ tends to $-\infty$. The coordinate $X_1$ is described as the "drift" because surfaces of constant $X_1$ represent the locations of fluid particles after they have been convected by the mean flow for the same amount of time. For example, if a set of particles are on the surface $X_1 = 0$ at time $t = 0$ then the same particles will lie on the surface $X_1 = U_\infty \tau$ after they have been convected by the mean flow for a time $t = \tau$ (see Fig. 6.2). It follows then that

$$\frac{D_o X_1}{Dt} = \mathbf{U} \cdot \nabla X_1 = U_\infty \qquad \text{and} \qquad X_1 = U_\infty \int_{\text{streamline}} \frac{d\sigma}{U}$$

where $\sigma$ is the distance along a mean-flow streamline, and the integral is initiated at the upstream boundary where $X_1 = 0$.

The surfaces $X_2 = const$ and $X_3 = const$ represent stream surfaces, and the unit vectors normal to those surfaces are orthogonal to each other. The intersection of the two surfaces represents the streamlines of the mean flow (see Fig. 2.2).

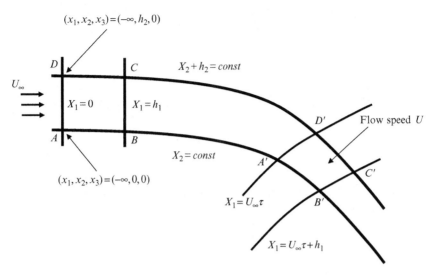

**Fig. 6.2** The evolution of drift coordinates. The surface $X_2 = const$ is a stream surface, and $X_1/U_\infty$ gives the time it takes for a particle to convect along a streamline from the point $X_1 = 0$ to the point at $X_1$.

The application of drift coordinates to the equations defined in Section 6.1 is readily seen because they give the solution to the differential equation

$$\frac{D_o f(x_i, t)}{Dt} = 0 \tag{6.2.2}$$

with the initial condition

$$[f]_{x_1 \to -\infty} = g(x_2, x_3, t - x_1/U_\infty)$$

as

$$f(x_i, t) = g(X_2, X_3, t - X_1/U_\infty)$$

It follows immediately that the evolution of an entropy perturbation $\dot{s}$ for which $D_o \dot{s}/Dt = 0$ is given by $\dot{s}(X_2, X_3, t - X_1/U_\infty)$, and any entropy perturbation defined at the upstream boundary will be convected downstream according to this behavior.

Finding a solution for $\partial \mathbf{u}^{(h)}/\partial t$ in Eq. (6.1.12) is not as straightforward. However, if the mean flow is irrotational then the equation for $\partial \mathbf{u}^{(h)}/\partial t$ is given by

$$\frac{D_o \dot{\mathbf{u}}^{(h)}}{Dt} + \left( \dot{\mathbf{u}}^{(h)} \cdot \nabla \right) \mathbf{U} = 0 \tag{6.2.3}$$

Since the flow is irrotational the vector expansion of $\nabla(\nabla X_i . \mathbf{U})$ reduces to

$$\nabla(\nabla X_i \cdot \mathbf{U}) = (\nabla X_i \cdot \nabla)\mathbf{U} + (\mathbf{U} \cdot \nabla)\nabla X_i$$

and so, using Eq. (6.2.1)

$$\frac{D_o(\nabla X_i)}{Dt} + (\nabla X_i \cdot \nabla)\mathbf{U} = \frac{\partial(\nabla X_i)}{\partial t} + \nabla(\mathbf{U} \cdot \nabla X_i) = \nabla(U_\infty) = 0 \qquad (6.2.4)$$

It follows that $\nabla X_i$ satisfies the same differential equation as $\partial \mathbf{u}^{(h)}/\partial t$ and that the solution to Eq. (6.2.3) for an irrotational mean flow is given by

$$\mathbf{u}^{(h)} = \nabla X_j u_j^{(\infty)}(X_2, X_3, t - X_1/U_\infty) \qquad (6.2.5)$$

This shows that the effect of drift is not to simply convect the velocity disturbance but to also modify its amplitude and direction. At the upstream boundary the surfaces defined by constant drift coordinates are aligned with the Cartesian coordinates $x_1$, $x_2$, $x_3$, and so $\partial X_j/\partial x_i = \delta_{ij}$ and

$$\left[ u_i^{(h)} \right]_{x_1 \to -\infty} = u_i^{(\infty)}(x_2, x_3, t - x_1/U_\infty) \qquad (6.2.6)$$

This represents the gust at the inflow boundary and Eq. (6.2.5) specifies how the gust propagates downstream.

It would appear that these solutions uniquely define the gust everywhere in the flow given the upstream boundary condition, but unfortunately, the solution is not valid for points that lie downstream of a stagnation point on a stationary body. Physically, a fluid particle on a stagnation streamline comes to rest at the stagnation point and can never be convected past that point. So, in principle, $X_1$ is not defined at any point on a surface downstream of a stagnation point. However, it should be remembered that this is a linear theory that assumes $|\mathbf{u}| \ll |\mathbf{U}|$, and this approximation is not valid when $|\mathbf{U}|$ tends to zero.

Solutions based on drift coordinates embody the stretching of the turbulent eddies by the mean flow. However, as the stretching takes place the gust is no longer divergence free. To demonstrate this we note that the divergence of Eq. (6.2.5) includes the terms $\left( \partial u_j^{(\infty)}/\partial X_i \right)(\nabla X_j \cdot \nabla X_i)$, and Fig. 6.2 shows that the surfaces of constant drift and the stream surfaces are not orthogonal after the flow has been stretched, so $\nabla X_i \cdot \nabla X_j \neq 0$ when $i \neq j$. The consequence of this is that the right side of the wave Eq. (6.1.9) is not zero, even if the gust is incompressible at the upstream boundary, and so there will be both volume source terms and surface source terms that result from the boundary conditions imposed on the solution to the wave Eq. (6.1.9).

For a two-dimensional potential mean flow, we can define unit vectors $\mathbf{s}$ and $\mathbf{n}$ that are in the direction of the flow and normal to the flow, respectively. In general, the drift coordinates $X_2$ and $X_3$ define the stream surfaces of the mean flow, and for a potential flow can be represented by stream functions with a suitable normalization. On this basis we can define the drift function gradients for two-dimensional flow as

$$\nabla X_1 = \frac{U_\infty}{U}\mathbf{s} + q\mathbf{n}, \quad \nabla X_2 = \frac{U}{U_\infty}\mathbf{n}$$

where $q$ is to be determined. An illustration of these functions is shown in Fig. 6.3.

The problem with drift coordinates is accurately calculating the distortion $q$ which requires an integration along adjacent streamlines to find the position of the drift surface $X_1 = const$. This is most readily achieved by considering the displacement vectors $\delta \mathbf{l}^{(i)}$ which are shown in Fig. 6.3. These join the corners of a material volume that is a rectangular box at the upstream boundary with sides of length $h_i$ (Fig. 6.2).

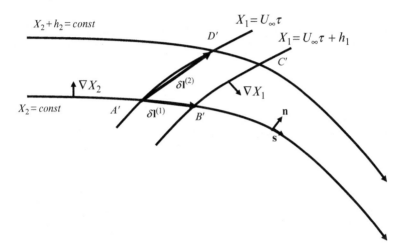

**Fig. 6.3** Displacement and drift coordinates.

The displacement coordinates $\delta \mathbf{l}^{(i)}$ and the drift coordinates are linked, and we can define one in terms of the other. First we note that the $\delta \mathbf{l}^{(1)}$ coordinate lies in the direction of the flow along the intersection of the stream surfaces $X_2 = const$ and $X_3 = const$. Similarly the $\delta \mathbf{l}^{(2)}$ coordinate lies along the intersection of the surfaces $X_1 = const$ and $X_3 = const$, while $\delta \mathbf{l}^{(3)}$ lies along the intersection of the surfaces $X_1 = const$ and $X_2 = const$. It follows that

$$\nabla X_i \cdot \delta \mathbf{l}^{(j)} = 0 \quad i \neq j$$

where we are using superscript notation to indicate that no summation is implied by the repeated index. We can also use a Taylor series expansion to show that, for differential small material volumes

$$X_i\left(\mathbf{x} + \delta \mathbf{l}^{(i)}\right) = X_i(\mathbf{x}) + \delta \mathbf{l}^{(i)} \cdot \nabla X_i$$

and since by definition $X_i(\mathbf{x} + \delta \mathbf{l}^{(i)}) - X_i(\mathbf{x}) = h_i$ we obtain the general result that

$$\delta \mathbf{l}^{(i)} \cdot \nabla X_j = h_i \delta_{ij} \tag{6.2.7}$$

where $\delta_{ij}$ is the Kroneker delta function. Furthermore, in a low Mach number flow where the mean density is constant, the material volume remains the same, so

$$\delta\mathbf{l}^{(1)} \cdot \left( \delta\mathbf{l}^{(2)} \times \delta\mathbf{l}^{(3)} \right) = \delta\mathbf{l}^{(2)} \cdot \left( \delta\mathbf{l}^{(3)} \times \delta\mathbf{l}^{(1)} \right) = \delta\mathbf{l}^{(3)} \cdot \left( \delta\mathbf{l}^{(1)} \times \delta\mathbf{l}^{(2)} \right) = h_1 h_2 h_3$$

which is written in tensor notation as

$$\delta\mathbf{l}^{(i)} \cdot \left( \delta\mathbf{l}^{(j)} \times \delta\mathbf{l}^{(k)} \right) = h_i h_j h_k \quad (i,j,k) = (1,2,3),(2,3,1), \text{ or } (3,1,2) \tag{6.2.8}$$

Since $\delta\mathbf{l}^{(j)}$ lies on the surfaces $X_i = const$ and $X_k = const$, and $\delta\mathbf{l}^{(k)}$ lies on the surfaces $X_i = const$ and $X_j = const$, it follows that $(\delta\mathbf{l}^{(j)} \times \delta\mathbf{l}^{(k)})$ lies normal to the plane $X_i = const$. Since these vectors are in the same direction, Eqs. (6.2.7), (6.2.8) show that

$$\nabla X_i = \frac{\delta\mathbf{l}^{(j)} \times \delta\mathbf{l}^{(k)}}{h_j h_k} \quad (i,j,k) = (1,2,3),(2,3,1), \text{ or } (3,1,2) \tag{6.2.9}$$

This gives a useful relationship between drift and displacement coordinates which we will find useful in the next section. In general $\delta\mathbf{l}^{(1)}$ lies in the direction of the mean flow and is equal to $\mathbf{U}$ times the time taken for the particle to move between two points defining the material volume in the direction of the flow. Since those two points were initially a distance $h_1$ apart in a region where the speed was $U_\infty$, continuity requires that $\delta\mathbf{l}^{(1)} = \mathbf{U}h_1/U_\infty$. For two-dimensional flows $\delta\mathbf{l}^{(3)}$ is constant, and so the only remaining unknown is $\delta\mathbf{l}^{(2)}$, which will depend on the particular flow, and can be calculated from the distortion of the material volume.

It is also noteworthy that this relationship is invertible using vector triple products. We find that

$$\nabla X_i \times \nabla X_s = -\nabla X_s \times \left( \frac{\delta\mathbf{l}^{(j)} \times \delta\mathbf{l}^{(k)}}{h_j h_k} \right) = -\left( \frac{\left(\nabla X_s \cdot \delta\mathbf{l}^{(k)}\right)\delta\mathbf{l}^{(j)} - \left(\nabla X_s \cdot \delta\mathbf{l}^{(j)}\right)\delta\mathbf{l}^{(k)}}{h_j h_k} \right)$$

$$= -\frac{\delta_{ks}\delta\mathbf{l}^{(j)}}{h_j} + \frac{\delta_{js}\delta\mathbf{l}^{(k)}}{h_k}$$

so

$$\nabla X_i \times \nabla X_j = \begin{cases} \dfrac{\delta\mathbf{l}^{(k)}}{h_k} & (i,j,k) = (1,2,3),(2,3,1), \text{ or } (3,1,2) \\ -\dfrac{\delta\mathbf{l}^{(k)}}{h_k} & (i,j,k) = (1,3,2),(2,1,3), \text{ or } (3,2,1) \end{cases} \tag{6.2.10}$$

## 6.3  Rapid distortion theory

In Section 6.1 we derived Goldstein's equation for perturbations to the flow over a body. The perturbations are defined on some upstream boundary where the flow is uniform and can be specified as either a perturbation of velocity or entropy. Entropy waves are only important in heated or high-speed flows, and so we will focus on velocity perturbations. Furthermore, we will consider low Mach number flows where the mean density and the speed of sound can be assumed to be constant. The unsteady velocity throughout the flow is then

$$\mathbf{v} = \mathbf{U} + \nabla\phi + \mathbf{u}^{(h)} \tag{6.3.1}$$

where the perturbation potential is the solution to Eq. (6.1.9), given in terms of the fluctuating parts, as

$$\frac{1}{c_\infty^2}\frac{D_o^2\phi}{Dt^2} - \nabla^2\phi = \nabla \cdot \mathbf{u}^{(h)} \tag{6.3.2}$$

To solve for the perturbation potential, we need to define the velocity term on the right side, which is the solution (6.1.12), and we need to specify the boundary conditions on all the surfaces bounding the flow.

The problem to which Eq. (6.3.2) is most readily applied is the sound radiation from a stationary rigid body in an irrotational mean flow. For this problem the boundary conditions are

$$[\phi]_{|\mathbf{x}|\to\infty} = 0, \quad \left[\mathbf{u}^{(h)}\right]_{x_1\to-\infty} = \mathbf{u}^{(\infty)}(x_2, x_3, t - x_1/U_\infty), \quad [\mathbf{u}\cdot\mathbf{n}]_{\text{surfaces}} = 0$$

The first step in finding a solution is to determine the evolution of the flow perturbation as a function of distance from the inflow boundary, and the second step is to solve the scattering problem for the potential $\phi$ that ensures the boundary conditions on the surface are met. In this section we focus on the evolution of the gust through the flow and show how particularly simple results can be obtained for high-frequency gusts. In the next section we will consider the scattering problem and show how acoustic waves are generated by the interaction of the gust with the surface.

In Section 6.2 we saw that if we define the velocity on the upstream boundary as $\mathbf{u}^{(\infty)}(x_2, x_3, t-x_1/U_\infty)$ then, in a potential mean flow, the gust velocity is given by

$$\mathbf{u}^{(h)} = \nabla X_j u_j^{(\infty)}(X_2, X_3, t - X_1/U_\infty) \tag{6.3.3}$$

These results assume that the flow perturbations are represented by the linearized equations of motion, which assume $|\mathbf{U}| \gg |\mathbf{u}|$. This implies that the evolution of the gust is determined by the mean flow and that nonlinear turbulent interactions are negligible. This will only be the case if the turbulent structure in the flow remains coherent during its passage through the domain being studied, which implies that the mean flow distortion is so rapid that it dominates the distortion of the turbulent eddies. This is RDT.

It has also been assumed that the gust velocity at the upstream boundary is incompressible and $\mathbf{u}^{(h)}$ is specified as a *vortical* gust. Due to the incompressibility condition, the divergence of $\mathbf{u}^{(h)}$ is zero at the upstream boundary, but its divergence does not remain zero when it is distorted by the mean flow, and so the right side of Eq. (6.3.2) is in general nonzero. The perturbation potential $\phi$ is related to the pressure perturbation by Eq. (6.1.8), is a solution to the acoustic wave Eq. (6.3.2), and is defined by Goldstein as the "acoustic" part of the gust, even if it does not propagate at the speed of sound as an acoustic wave.

Further insight is obtained by considering a harmonic wave at the inflow boundary defined such that

$$\mathbf{u}^{(\infty)}(x_2, x_3, t - x_1/U_\infty) = \mathrm{Re}(\hat{\mathbf{a}} \exp{(i\mathbf{k} \cdot \mathbf{x} - ik_1 U_\infty t)}) \tag{6.3.4}$$

If this gust is incompressible then the divergence of $\mathbf{u}^{(\infty)}$ is zero so that $\mathbf{k} \cdot \hat{\mathbf{a}} = 0$. By combining Eq. (6.3.4) with Eq. (6.3.3) for a potential mean flow we find

$$\hat{\mathbf{u}}^{(h)} = \nabla X_k \hat{a}_k e^{i\mathbf{k} \cdot \mathbf{X}} \tag{6.3.5}$$

where the hat represents the complex amplitude of a variable with implied harmonic time dependence, so

$$\mathbf{u}^{(h)} = \mathrm{Re}\left(\hat{\mathbf{u}}^{(h)} e^{-i\omega t}\right), \qquad \omega = k_1 U_\infty$$

The right side of Eq. (6.3.2) is then obtained as

$$\nabla \cdot \hat{\mathbf{u}}^{(h)} = \left(i\boldsymbol{\kappa} \cdot \nabla X_k + \nabla^2 X_k\right) \hat{a}_k e^{i\mathbf{k} \cdot \mathbf{X}}, \qquad \boldsymbol{\kappa} = \nabla X_j k_j \tag{6.3.6}$$

As was pointed out by Majumdar and Peake [2] this result is greatly simplified by considering a high-frequency gust where $|\mathbf{k}|$ is large. This implies that the lengthscale of the gust is much smaller than the scale of the mean flow distortion. In this case the first term on the right side of Eq. (6.3.6) is dominant, and this simplifies the solution to Eq. (6.3.2) considerably. However, it also implies that the scale of the gust is small, and this challenges the concept of a rapid distortion described earlier, so this approximation must be used selectively. Quantitatively, if the scale of the mean flow distortion is $L$, then the high-frequency gust approximation assumes that $k_1 L \gg 1$, or in terms of the gust frequency $\omega = k_1 U_\infty$, we require $\omega L/U_\infty \gg 1$. This condition must be used in conjunction with the assumption used above that $|\mathbf{u}| \ll U$ which is required by the constraints of RDT.

Next consider a solution to Eq. (6.3.2) that is of the form

$$\phi = \mathrm{Re}\left(\hat{Q} e^{i\mathbf{k} \cdot \mathbf{X} - i\omega t}\right)$$

so that $D_o \phi/Dt = 0$ is zero, and the wave Eq. (6.3.2) becomes

$$\left(|\boldsymbol{\kappa}|^2 - ik_j \nabla^2 X_j\right) \hat{Q} e^{i\mathbf{k} \cdot \mathbf{X}} = \left(i\boldsymbol{\kappa}.\nabla X_k + \nabla^2 X_k\right) \hat{a}_k e^{i\mathbf{k} \cdot \mathbf{X}}$$

In the high-frequency approximation the variations of $\boldsymbol{\kappa}$ and $X_i$ are assumed to be relatively slow compared to the term $\exp(i\mathbf{k} \cdot \mathbf{X})$. We can therefore find a local approximate solution to this equation that assumes that $\boldsymbol{\kappa}$ and $\nabla X_i$ are effectively constant over several wavelengths. The high-frequency approximation implies that $|\boldsymbol{\kappa}|^2 \gg k_j \nabla^2 X_j$ and $|\boldsymbol{\kappa} \cdot \nabla X_j| \gg \nabla^2 X_j$, so we have an approximate solution (6.3.2) given by

$$\hat{Q} \approx \frac{i\boldsymbol{\kappa} \cdot \nabla X_k \hat{a}_k}{|\boldsymbol{\kappa}|^2}$$

Then using Eq. (6.3.1) we find the unsteady velocity as

$$\hat{\mathbf{u}} = \nabla \hat{\phi} + \hat{\mathbf{u}}^{(h)} = \left( -\frac{\boldsymbol{\kappa}(\boldsymbol{\kappa} \cdot \nabla X_k)}{|\boldsymbol{\kappa}|^2} + \nabla X_k \right) \hat{a}_k e^{i\mathbf{k} \cdot \mathbf{X}} = \left( \hat{\mathbf{u}}^{(h)} - \frac{\boldsymbol{\kappa}\left(\boldsymbol{\kappa} \cdot \hat{\mathbf{u}}^{(h)}\right)}{|\boldsymbol{\kappa}|^2} \right) \quad (6.3.7)$$

which is the result given by Majumdar and Peake [2]. Using the vector triple product theorem this can be written as

$$\hat{\mathbf{u}} = -\frac{\boldsymbol{\kappa} \times \left( \boldsymbol{\kappa} \times \hat{\mathbf{u}}^{(h)} \right)}{|\boldsymbol{\kappa}|^2} \quad (6.3.8)$$

This shows that the gust is incompressible throughout the flow because $\boldsymbol{\kappa} \cdot \hat{\mathbf{u}} = 0$ and, at upstream infinity where $\boldsymbol{\kappa} \cdot \hat{\mathbf{u}}^{(h)}$ tends to $\mathbf{k} \cdot \hat{\mathbf{u}}^{(\infty)} = 0$, $\hat{\mathbf{u}}$ tends to $\hat{\mathbf{u}}^{(h)}$, and so the potential correction is zero on the upstream boundary. We also note that there is no pressure perturbation that is caused by this gust because $D_o \phi / Dt = 0$, so there are no propagating acoustic waves, in spite of the fact that the potential perturbation is sometimes referred to as the *acoustic* part.

Continuing with our analysis, the local vorticity fluctuation $\hat{\boldsymbol{\omega}} = \nabla \times \hat{\mathbf{u}}$ can be calculated from Eq. (6.3.5) as

$$\begin{aligned} \hat{\boldsymbol{\omega}} &= \nabla \times \left( \nabla X_k \hat{a}_k e^{i\mathbf{k} \cdot \mathbf{X}} \right) \\ &= \nabla \left( \hat{a}_k e^{i\mathbf{k} \cdot \mathbf{X}} \right) \times \nabla X_k = \left( i\boldsymbol{\kappa} \hat{a}_k e^{i\mathbf{k} \cdot \mathbf{X}} \right) \times \nabla X_k = i\boldsymbol{\kappa} \times \nabla X_k \hat{a}_k e^{i\mathbf{k} \cdot \mathbf{X}} \\ &= i\boldsymbol{\kappa} \times \hat{\mathbf{u}}^{(h)} \end{aligned} \quad (6.3.9)$$

It follows that the local gust velocity, the local vorticity, and the local gust wavenumber vector $\boldsymbol{\kappa}$ are orthogonal to each other at a point in the flow and that, from Eq. (6.3.8),

$$\hat{\mathbf{u}} = i \frac{\boldsymbol{\kappa} \times \hat{\boldsymbol{\omega}}}{|\boldsymbol{\kappa}|^2} \quad (6.3.10)$$

We denote the gust vorticity at the inflow boundary as $\boldsymbol{\omega}^{(\infty)} = \mathrm{Re}\left( \hat{\boldsymbol{\omega}}^{(\infty)} \exp\left( i\mathbf{k} \cdot \mathbf{x} - ikU_\infty t \right) \right)$, where $\hat{\boldsymbol{\omega}}^{(\infty)}$ is the complex amplitude of the inflow vorticity obtained by taking the curl of Eq. (6.3.4) to give $\hat{\boldsymbol{\omega}}^{(\infty)} = i\mathbf{k} \times \hat{\mathbf{a}}$.

We can use this amplitude to express the vorticity at any point in the flow by substituting $\hat{\mathbf{u}}^{(h)}$ given by Eq. (6.3.5) into Eq. (6.3.9) and using Eq. (6.2.10) to give,

$$\boldsymbol{\omega} = \mathrm{Re}\left( \hat{\omega}_i^{(\infty)} \frac{\delta l_i^{(i)}}{h_i} e^{i\mathbf{k}\cdot\mathbf{X} - ik_1 U_\infty t} \right) \tag{6.3.11}$$

where the sum over repeated indices is implied. This is a form Cauchy's theorem (see Batchelor [3]) in which the vorticity remains aligned with and proportional to the length of the fluid lines, represented here by the displacement coordinates.

To illustrate the modification of a gust in an accelerating flow we will consider the two-dimensional example shown in Fig. 6.4. This flow represents the contraction of a wind tunnel. The mean flow streamlines lie along radial lines that converge to an apparent sink shown on the right-hand side of the figure. The upstream boundary is on the surface $r=r_o$ in terms of the cylindrical polar coordinates $(r,\theta)$. The flow speed on the inflow boundary is $U_o$ and, at any point on a streamline, is given by $U=U_o(r_o/r)$. The unit vector $\mathbf{s}$ lies in the direction of the flow, and the unit vector $\mathbf{n}$ lies normal to the streamlines in the plane shown. In addition, there will be a unit vector $\mathbf{t}$ pointing out of the paper. We can calculate the drift from the time it takes for a particle to be convected from the inflow boundary to a point a distance $s=r_o-r$ downstream as

$$X_1 = U_o \int_0^s \frac{ds}{U} = \int_0^s \frac{rds}{r_o} = s - \frac{s^2}{2r_o} \tag{6.3.12}$$

The surfaces of constant $X_1$ are shown in Fig. 6.4 and are seen to move further apart as the flow accelerates along the contraction.

The first thing to notice from Fig. 6.4 is that the accelerating flow stretches the gust in the direction of the flow, and that there is a contraction of each volume element in the direction normal to the flow.

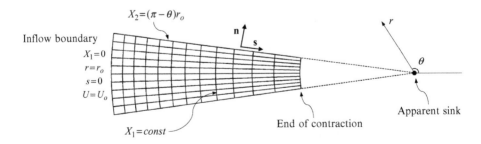

**Fig. 6.4** A wind tunnel contraction modeled by a converging two-dimensional flow.

We will define a vortical gust on the inflow boundary with components in the $\mathbf{s}, \mathbf{n}, \mathbf{t}$ directions given by $\hat{\mathbf{a}} \exp(ik_1 s + ik_2 \mu + ik_3 z - ik_1 U_o t)$, where $\mu = r_o(\pi - \theta)$ is the circumferential distance normal to the flow direction, and $\mathbf{k} \cdot \hat{\mathbf{a}} = 0$. We can readily calculate the drift function gradients and the displacement vectors as

$$\nabla X_1 = \frac{r}{r_o}\mathbf{s}, \quad \nabla X_2 = \frac{r_o}{r}\mathbf{n}, \quad \nabla X_3 = \mathbf{t}, \quad \frac{\delta\mathbf{l}^{(1)}}{h_1} = \frac{r_o}{r}\mathbf{s}, \quad \frac{\delta\mathbf{l}^{(2)}}{h_2} = \frac{r}{r_o}\mathbf{n}, \quad \frac{\delta\mathbf{l}^{(3)}}{h_3} = \mathbf{t}$$

so that the wavenumber vector and vortical gust are

$$\boldsymbol{\kappa} = \frac{k_1 r}{r_o}\mathbf{s} + \frac{k_2 r_o}{r}\mathbf{n} + k_3\mathbf{t}, \qquad \hat{\mathbf{u}}^{(h)} = \left(\frac{\hat{a}_1 r}{r_o}\mathbf{s} + \frac{\hat{a}_2 r_o}{r}\mathbf{n} + \hat{a}_3\mathbf{t}\right)e^{i\mathbf{k}\cdot\mathbf{X}}$$

and the local vorticity is given by

$$\hat{\boldsymbol{\omega}} = \left(\hat{\omega}_1^{(\infty)}\frac{r_o}{r}\mathbf{s} + \hat{\omega}_2^{(\infty)}\frac{r}{r_o}\mathbf{n} + \hat{\omega}_3^{(\infty)}\mathbf{t}\right)e^{i\mathbf{k}\cdot\mathbf{X}}$$

This shows that the effect of the accelerating flow is to reduce the streamwise component of the wavenumber vector and increase the streamwise vorticity. Similarly, the wavenumber vector normal to the flow is increased as the flow is stretched, and the vorticity component normal to the flow is reduced. The spanwise wavenumber and vorticity in the $\mathbf{t}$ direction is unaltered by the flow acceleration.

To complete the picture, we must also include the potential correction to the velocity to ensure that the gust remains incompressible. This is particularly simple to calculate for a two-dimensional gust for which $k_3 = 0$ and $a_3 = 0$. The only component of the local vorticity is then in the $\mathbf{t}$ direction and remains unaltered by the flow acceleration, giving from Eq. (6.3.10)

$$\hat{\mathbf{u}} = -\frac{\left(k_2\left(\frac{r_o}{r}\right)\mathbf{s} - k_1\left(\frac{r}{r_o}\right)\mathbf{n}\right)(k_1\hat{a}_2 - k_2\hat{a}_1)e^{i\mathbf{k}\cdot\mathbf{X}}}{(k_1 r/r_o)^2 + (k_2 r_o/r)^2} \tag{6.3.13}$$

Because the gust is incompressible we have $\hat{a}_2 = -k_1\hat{a}_1/k_2$, and so this reduces to

$$\frac{\hat{\mathbf{u}}}{|\hat{\mathbf{a}}|} = \frac{\left(k_2\frac{r_o}{r}\mathbf{s} - \frac{k_1 r}{r_o}\mathbf{n}\right)|\mathbf{k}|e^{i\mathbf{k}\cdot\mathbf{X}}}{(k_1 r/r_o)^2 + (k_2 r_o/r)^2}$$

The important conclusion that we can draw from this example is that as the flow is accelerated in the region where $r < r_o$ the streamwise gust becomes stronger than

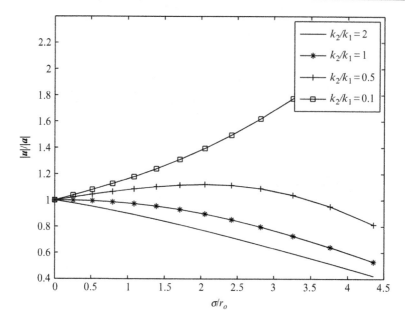

**Fig. 6.5** The magnitude of a two-dimensional gust in a contraction for $r_o = 7$.

the gust normal to the flow. When $k_2 < k_1$ the magnitude of the gust initially increases as it is stretched by the flow, as shown in Fig. 6.5, and reaches a maximum when the numerator of Eq. (6.3.13) reaches a maximum. However, the magnitude of the gust is always reduced by stretching when $k_2 > k_1$. It follows that the stretching tends to amplify the gusts that have a large scale ($k_2 < k_1$) in the cross stream direction.

## 6.4 Acoustically compact thin airfoils and the Kutta condition

In many applications in aero and hydroacoustics we are concerned with calculating the sound radiation from thin blades at relatively low frequencies, where the blade chord is much less than the acoustic wavelength. The case where the blade is moving at low Mach number through turbulence will be discussed in Chapter 14. Here we consider the case where the blade is fixed in a turbulent stream and the observer is outside the flow, as shown in Fig. 6.6.

For this case Curle's theorem applies, and we can use Eq. (4.4.7) to calculate the radiated sound, providing the flow Mach number is sufficiently small for the quadrupole sources in the flow to be negligible (as discussed in Section 4.4). Since the blade is assumed to be acoustically compact, differences in the propagation time from each point on the blade surface to the observer will be small enough for sound waves generated on different parts of the surface to arrive in phase at the observer. Eq. (4.4.7) then gives

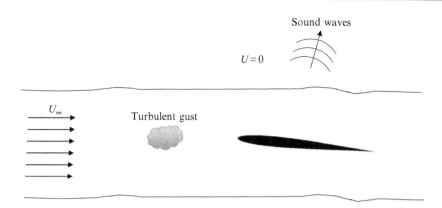

**Fig. 6.6** An airfoil in a uniform flow.

$$\left(\rho'(\mathbf{x},t)c_\infty^2\right)_{\text{dipole}} \approx \frac{x_i}{4\pi|\mathbf{x}|^2 c_\infty}\left[\frac{\partial F_i}{\partial \tau}\right]_{\tau=\tau^*} \quad F_i = \int_S p_{ij}n_j dS(\mathbf{y}) \tag{4.4.7}$$

where $F_i$ is the net unsteady loading on the blade from both lift and drag. To calculate the acoustic field, we need to determine the unsteady blade loading from the characteristics of the blade shape and the unsteady inflow.

To carry out this calculation the first step is to assume that at high Reynolds numbers over a streamlined blade the unsteady loads can be characterized using the inviscid flow approximation, so the linearized Euler equations are applicable. This excludes bluff bodies that have thick unsteady wakes and other cases where flow separation occurs. However, in well-designed systems, flow separation is avoided at all costs, and the blade drag is much smaller than the lift force. To meet this requirement blades are often very thin, with thickness to chord ratios that are $<10\%$, and the perturbation of the mean flow speed at most locations is small, so $|U-U_\infty| = \varepsilon U_\infty$, where $\varepsilon \ll 1$. In Section 6.1 the linearized Euler equations were introduced by assuming an unsteady turbulent inflow with velocity perturbation $|\mathbf{u}| \ll U$ so that terms of order $|\mathbf{u}|^2$ could be ignored compared to terms of order $|\mathbf{u}|U$. In unsteady thin airfoil theory, it is assumed that $|\mathbf{u}|$ is of order $\varepsilon U_\infty$, so the mean flow perturbation and the gust amplitude are of the same magnitude. The hierarchy of velocity terms for the thin airfoil problem is then illustrated by considering

$$(\mathbf{v} \cdot \nabla)\mathbf{v} = \left(\left(U_\infty \hat{\mathbf{i}} + \Delta\mathbf{U} + \mathbf{u}\right) \cdot \nabla\right)\left(U_\infty \hat{\mathbf{i}} + \Delta\mathbf{U} + \mathbf{u}\right)$$

$$= U_\infty \underbrace{\frac{\partial}{\partial y_1}(\Delta\mathbf{U} + \mathbf{u})}_{o(\varepsilon U_\infty^2)} + \underbrace{((\Delta\mathbf{U} + \mathbf{u}) \cdot \nabla)(\Delta\mathbf{U} + \mathbf{u})}_{o(\varepsilon^2 U_\infty^2)} \tag{6.4.1}$$

This shows that if we can ignore terms of order $\varepsilon^2$ compared to terms of order $\varepsilon$, then we can approximate $D_o/Dt$ by $D_\infty/Dt = \partial/\partial t + U_\infty \partial/\partial x_1$. The steady flow can be approximated by including a small correction $\Delta\mathbf{U}$ to the uniform flow, and the turbulence is modeled by a gust superimposed on the uniform flow $U_\infty$ in the $x_1$ direction. Given these approximations the linearized Euler equations (6.3.2) and (6.1.12), for the unsteady gust in a potential mean flow, reduce to

$$\frac{1}{c_\infty^2}\frac{D_\infty^2\phi}{Dt^2} - \nabla^2\phi = 0, \qquad \frac{D_\infty \mathbf{u}^{(h)}}{Dt} = 0, \qquad p' = -\rho_o\frac{D_\infty\phi}{Dt} \tag{6.4.2}$$

(Note that in this case the gust is undistorted and so remains divergence free, and the right side of the wave equation is zero). It follows that $\mathbf{u}^{(h)} = \mathbf{u}^{(\infty)}(x_2, x_3, t - x_1/U_\infty)$ and is defined by the unsteady vortical gust far upstream of the airfoil. The boundary condition on the surface requires that $\mathbf{u}\cdot\mathbf{n} = 0$, and the normal to the surface has components $n_1 = \sin\alpha$ and $n_2 = \cos\alpha$, where $\alpha$ is the angle (in radians) that the surface makes with the $x_1$ direction. In thin airfoil theory it is assumed that $\alpha \ll 1$, which is reasonable except near the blunt leading edge of a thick airfoil section. Given this assumption $n_1 \ll n_2$, the boundary condition on the surface reduces to $u_2 = 0$. Physically, this models the airfoil as an infinitely thin flat plate at zero angle of attack. This is a very useful approximation that is used extensively in aeroacoustics because it is correct to first order and facilitates many important solutions to challenging problems. Furthermore, the approximation is of the same order as thin airfoil theory for the mean lift and drag of the surface and so applies to lifting surfaces at small angles of attack. We will return to this issue in subsequent chapters, but first we will explore the implications of this first-order approximation.

The problem of an unsteady flow encountering flat plate airfoil of chord $c$, at zero angle of attack, was first investigated by von Kármán and Sears [4]. They considered the flow perturbation to be a two-dimensional harmonic upwash gust $\hat{a}_2 \exp(ik_1(x_1 - U_\infty t))$ convected by the mean flow toward the airfoil located at $-c/2 < x_1 < c/2$, $x_2 = 0$, as shown in Fig. 6.7. This gust causes a time varying angle of attack at the leading edge of the airfoil. It is assumed, however, that the flow continues to leave smoothly from the trailing edge. This assumption, based on empirical observations of real flows, is the Kutta condition discussed in Section 2.7. In the inviscid model it conveniently avoids the infinite velocities that would be produced if the flow were required to negotiate this sharp corner. To maintain the Kutta condition requires that the circulation around the airfoil fluctuates with time. Since, according to Kelvin's theorem, the overall circulation of the flow must be conserved, the change in circulation implies the shedding of vorticity into its wake, the strength of the vorticity matching, instant by instant, the change in circulation on the airfoil. This requirement was sufficient for von Kármán and Sears to calculate the unsteady lift for a harmonic gust in a two-dimensional ideal flow.

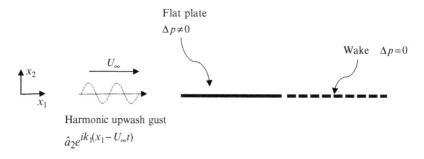

**Fig. 6.7** The Sears problem: a harmonic upwash gust incident on a flat plate.

The Kutta condition in unsteady flow has been the subject of many different modeling efforts, and in some cases it is assumed that the vorticity in the near wake, just downstream of the trailing edge, is convected at a speed that is less than the free stream velocity because of the reduced flow speed in the blade boundary layer and at the center of the near wake. von Kármán and Sears assumed that the vorticity was convected at the mean flow speed, and the correctness of this assumption has sometimes been challenged. The modeling issue was not resolved until Amiet [5] pointed out that, for the Sears problem, the flat plate and the wake are effectively replaced by a jump in velocity potential $\Delta\phi$, which is directly related to both the vortex sheet strength $\partial(\Delta\phi)/\partial x_1$ and the pressure jump $\Delta p = -\rho_o D_\infty(\Delta\phi)/Dt$ across the blade and wake, as shown in Fig. 6.7. Amiet argued that there could be no pressure jump across the wake because there is no structure in the fluid to support it. The correct boundary condition therefore is that $\Delta p = 0$ across the wake, and this will include all the effects of the Kutta condition, the blade boundary layer, and the formation of vorticity in the near wake. It follows that $D_\infty(\Delta\phi)/Dt = 0$ across the wake, and so the perturbations in the effective vortex sheet are convected downstream at the free stream velocity, which confirms the assumption used in the early work on unsteady blade loading.

The solution of Sears problem will be discussed in more detail in Chapters 7 and 14, and here we will give the result for future reference. The unsteady lift force on a compact airfoil located at $-c/2 < x_1 < c/2$, and with span $b$, was given by Sears as

$$
\begin{aligned}
-F_2(t) &= \mathrm{Re}\left(\pi\rho_o U_\infty \hat{a}_2 bc S(\sigma) e^{-ik_1 U_\infty t}\right) \\
S(\sigma) &= \frac{2}{\pi\sigma\left(H_o^{(1)}(\sigma) + iH_1^{(1)}(\sigma)\right)} \qquad \sigma = k_1 c/2
\end{aligned}
\tag{6.4.3}
$$

where $H_o^{(1)}(\sigma)$ and $H_1^{(1)}(\sigma)$ are Hankel functions of the first kind, and $\sigma$ is the nondimensional frequency. Note that the left-hand side of Eq. (6.4.3) is negative since $F_2$ denotes the force of the airfoil on the fluid.

The Sears function is plotted as a function of frequency in Fig. 6.8, and it is well approximated by

$$
|S(\sigma)| \approx \frac{1}{\sqrt{1 + 2\pi\sigma}}
\tag{6.4.4}
$$

These results are based on the assumption of an incompressible flow. This approximation was later shown by Amiet to be valid for frequencies where the acoustic wavelength is greater than four times the blade chord. Sears' result is therefore a low-frequency approximation, which is consistent with the compact chord assumption that is required to calculate the acoustic field using Eq. (4.4.7).

By combining Eq. (6.4.3) with Eq. (4.4.7) we obtain the acoustic far-field pressure in terms of the upwash gust amplitude as

$$
(p(\mathbf{x}, t))_{\mathrm{dipole}} \approx \mathrm{Re}\left(\frac{-i\omega x_2 e^{-i\omega(t-|\mathbf{x}|/c_\infty)}}{4\pi|\mathbf{x}|^2 c_\infty}\left(\pi\rho_o U_\infty \hat{a}_2 bc S(\sigma)\right)\right)
\tag{6.4.5}
$$

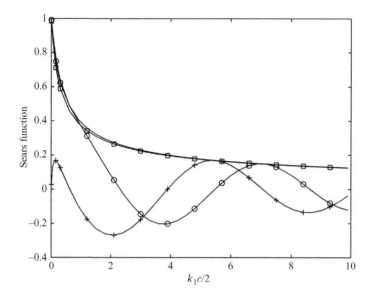

**Fig. 6.8** The Sears function plotted against nondimensional frequency $\sigma = k_1 c/2$. *Solid line,* $|S|$; o, Re($S$); +, Im($S$); *squares,* approx $|S|$.

where $\omega = k_1 U_\infty$ is the frequency. This represents an acoustic dipole that points in the direction of the unsteady lift. It has a cosine directionality with no sound in the direction of the flow. If the gust amplitude is taken to be proportional to $U_\infty$, then for fixed values of $k_1 c$ and $b/|\mathbf{x}|$, the acoustic pressure scales as $\rho_o U_\infty^3 / c_\infty$, as expected from the discussion in Chapter 3. The response function is reduced at higher frequencies as shown in Fig. 6.8, and the approximation of the Sears function implies that the acoustic pressure is proportional to $(\omega c/2U_\infty)^{1/2}$ at high frequencies. In practice, the amplitude of the gust also reduces with frequency, so the spectral level is reduced as the frequency is increased. We will discuss this characteristic in subsequent chapters when we introduce turbulence modeling.

## 6.5 The Prantl–Glauert transformation

In Section 6.4 we discussed the approximations of thin airfoil theory and showed that if the unsteady flow is linearized about the mean flow, and the mean flow perturbations are small, then the sound radiation from an airfoil in an unsteady flow can be reduced to the boundary value problem in which the perturbation potential satisfies the convected wave equation and the airfoil is modeled by a flat plate of zero thickness. The sound radiation from surfaces in a uniform flow was discussed in Section 5.4, and the acoustic field was given by Eq. (5.4.4) in terms of a Green's function that was the solution to the inhomogeneous wave equation

$$\frac{1}{c_\infty^2}\frac{D_\infty^2 G_e}{D\tau^2} - \nabla^2 G_e = -\delta(t-\tau)\delta(\mathbf{x}-\mathbf{y}) \tag{5.4.5}$$

To obtain a solution to this equation we can make use of the Prantl–Glauert transformation that conveniently transforms Eq. (5.4.5) into a regular wave equation for which we already have a solution. The transformation is defined as follows

$$G_e(\mathbf{x},t|\mathbf{y},\tau) = G_g\left(\boldsymbol{\xi},t_g|\boldsymbol{\zeta},\tau_g\right) \quad \boldsymbol{\xi} = (x_1,\beta x_2,\beta x_3) \quad t_g = t + Mx_1/\beta^2 c_\infty \\ \boldsymbol{\zeta} = (y_1,\beta y_2,\beta y_3) \quad \tau_g = \tau + My_1/\beta^2 c_\infty \tag{6.5.1}$$

where $\beta^2 = 1 - M^2$ and $M = U_\infty/c_\infty$. The inhomogeneous wave Eq. (5.4.5) is then reduced to

$$\frac{1}{\beta^4 c_\infty^2}\frac{\partial^2 G_g}{\partial t_g^2} - \frac{\partial^2 G_g}{\partial \xi_i^2} = -\delta(\boldsymbol{\xi}-\boldsymbol{\zeta})\delta(t_g - \tau_g) \tag{6.5.2}$$

The solution to this equation was discussed in Chapter 3, and from Eq. (3.9.17) we obtain for the free-field Green's function

$$G_g\left(\boldsymbol{\xi},t_g|\boldsymbol{\zeta},\tau_g\right) = \frac{\delta\left(t_g - |\boldsymbol{\xi}-\boldsymbol{\zeta}|/\beta^2 c_\infty - \tau_g\right)}{4\pi|\boldsymbol{\xi}-\boldsymbol{\zeta}|}$$

Converting this back to regular coordinates then gives

$$G_e(\mathbf{x},t|\mathbf{y},\tau) = \frac{\delta\left(t - r_g/\beta^2 c_\infty - \tau + M(x_1-y_1)/\beta^2 c_\infty\right)}{4\pi r_g} \\ r_g = \left((x_1-y_1)^2 + \beta^2(x_2-y_2)^2 + \beta^2(x_3-y_3)^2\right)^{1/2} \tag{6.5.3}$$

This provides the Green's function that is applicable to problems where the source and the observer are both in a uniform flow. Note how the across flow scale is reduced by the factor $\beta$ and the effect of mean flow convection on the wave propagation appears in the argument of the delta function. The relationship between the Green's function in the time domain and frequency domain is given by Eq. (3.10.7), and using inverse transforms we obtain the frequency domain Green's function

$$\widetilde{G}_e(\mathbf{x}|\mathbf{y}) = \frac{e^{i\omega r_g/\beta^2 c_\infty - i\omega M(x_1-y_1)/\beta^2 c_\infty}}{4\pi r_g} \tag{6.5.4}$$

For example, to calculate the sound radiation from a flat plate airfoil on the $y_1$ axis in a moving medium we could use the stationary medium result given by Eq. (4.5.4) and replace the Green's function using Eq. (6.5.4) as described in Section 5.4. Then using the far-field approximation, we obtain

$$\left(\widetilde{\rho}(\mathbf{x}, \omega)c_\infty^2\right)_{\text{dipole}} \approx$$

$$\frac{-i\omega x_2 e^{ik_o r_e}}{4\pi c_\infty r_e^2} \int\limits_{-c/2}^{c/2} \int\limits_{-b/2}^{b/2} \Delta\widetilde{p}(\mathbf{y}, \omega)e^{-ik_o x_1 y_1/r_e - ik_o x_3 y_3 \beta^2/r_e - ik_o M(x_1 - y_1)}dy_1 dy_3$$

$$k_o = \omega/\beta^2 c_\infty \qquad r_e = \sqrt{x_1^2 + \beta^2\left(x_2^2 + x_3^2\right)} \tag{6.5.5}$$

This result shows how we can calculate the far-field sound in the flow providing that we know the surface pressure distribution. It also shows how the wavenumber description of the surface pressure can be used to give the far field as in Eq. (4.7.11) with the wavenumbers

$$k_1^{(o)} = k_o\left(\frac{x_1}{r_e} - M\right), \qquad k_3^{(o)} = k_o\beta^2\left(\frac{x_3}{r_e}\right) \tag{6.5.6}$$

which allow for propagation enhanced by convection due to the steady flow, and the appropriate correction to the propagation distance.

# References

[1] M. Goldstein, Unsteady vortical and entropic distortions of potential flows round arbitrary obstacles, J. Fluid Mech. 89 (1978) 433–468.
[2] S.J. Majumdar, N. Peake, Noise generation by the interaction between ingested turbulence and a rotating fan, J. Fluid Mech. 359 (1998) 181–216.
[3] G.K. Batchelor, An Introduction to Fluid Dynamics, Cambridge University Press, Cambridge, 2000.
[4] T. von Kármán, W.R. Sears, Airfoil theory for non-uniform motion, J. Aeronaut. Sci. 5 (1938) 379–390.
[5] R.K. Amiet, Gust response for flat-plate airfoils and the Kutta condition, AIAA J. 28 (1990) 1718–1727.

# Vortex sound

# 7

One of the concerns with Lighthill's analogy is the physical interpretation of the source term $T_{ij}$ and relating it to easily recognized features of the flow. In many cases the flow may include relatively large coherent structures that are characterized by almost two-dimensional line vortices. Examples include coherent vortex shedding behind a cylinder or into a wake, and tip vortices interacting with helicopter rotor blades. These problems are often more readily understood if the sound is related to the vorticity in the flow rather than the Lighthill stress tensor. In this chapter we review the theory of vortex sound and give some examples of its application in low Mach number flows. In particular we will consider the sound radiation caused by the unsteady loading on rigid, acoustically compact surfaces in the presence of flows that can be modeled by coherent vortical structures.

## 7.1 Theory of vortex sound

The theory of vortex sound was first discussed by Powell [1] who manipulated the source term in Lighthill's wave equation to specifically include the vorticity. In this section we extend this concept and show how the linearized Euler equations developed in the previous chapter can be used to identify acoustic source terms that depend on vorticity [2]. In Section 6.1 we derived Goldstein's wave equation, which was given by Eq. (6.1.9) as

$$\frac{D_o}{Dt}\left(\frac{1}{c_o^2}\frac{D_o\dot{\phi}}{Dt}\right) - \frac{1}{\rho_o}\nabla\cdot\left(\rho_o\nabla\dot{\phi}\right) = \frac{1}{\rho_o}\nabla\cdot\left(\rho_o\dot{\mathbf{u}}^{(g)}\right) \tag{6.1.9}$$

where the dot represents a time derivative and the velocity and pressure perturbations were defined such that

$$\mathbf{v} = \mathbf{U} + \nabla\phi + \mathbf{u}^{(g)}, \qquad \rho_o\frac{D_o}{Dt}\left(\frac{\partial\phi}{\partial t}\right) = -\frac{\partial p}{\partial t}$$

The velocity perturbation can be specified using Crocco's equation (Eq. 2.5.4) for an inviscid flow

$$\frac{\partial\mathbf{v}}{\partial t} = -\nabla H + T_e\nabla s - \boldsymbol{\omega}\times\mathbf{v}$$

and so Eq. (6.1.9) can be written as

$$\frac{D_o}{Dt}\left(\frac{1}{c_o^2}\frac{D_o\dot{\phi}}{Dt}\right) + \frac{1}{\rho_o}\nabla\cdot(\rho_o\nabla H) = \frac{1}{\rho_o}\nabla\cdot(\rho_o T_e\nabla s - \rho_o\boldsymbol{\omega}\times\mathbf{v}) \tag{7.1.1}$$

Aeroacoustics of Low Mach Number Flows. http://dx.doi.org/10.1016/B978-0-12-809651-2.00007-2

We can then modify this equation to make $H$ the dependent variable of the wave operator on the left side. From the definition of the stagnation enthalpy $H$ we find that

$$\frac{\partial H}{\partial t} = T_e \frac{\partial s}{\partial t} + \frac{1}{\rho}\frac{\partial p}{\partial t} + \frac{1}{2}\frac{\partial v_i^2}{\partial t}$$

and taking the dot product of Crocco's equation with the velocity $\mathbf{v}$, and ignoring viscous terms, gives

$$\mathbf{v}\cdot\frac{\partial \mathbf{v}}{\partial t} = \frac{1}{2}\frac{\partial v_i^2}{\partial t} = -\mathbf{v}\cdot\nabla H + \mathbf{v}\cdot(T_e\nabla s)$$

Rearranging these equations gives

$$\frac{DH}{Dt} = \frac{1}{\rho}\frac{\partial p}{\partial t}$$

for an isentropic flow where $Ds/Dt=0$. If this result is linearized about the mean flow then $D_oH/Dt = -D_o\dot{\phi}/Dt$, which can be used in Eq. (7.1.1) to obtain Howe's wave equation [2]

$$\frac{D_o}{Dt}\left(\frac{1}{c_o^2}\frac{D_oH}{Dt}\right) - \frac{1}{\rho_o}\nabla\cdot(\rho_o\nabla H) = \frac{1}{\rho_o}\nabla\cdot(\rho_o\boldsymbol{\omega}\times\mathbf{v} - \rho_o T_e\nabla s) \tag{7.1.2}$$

The significant point here is that the acoustic variable is now defined by the stagnation enthalpy and the source terms are specified in terms of the Lamb vector $\boldsymbol{\omega}\times\mathbf{v}$, and the gradient of the entropy. The form of the equation is identical to Goldstein's equation but the source terms are specified in a more convenient form, if the vorticity is a well-defined quantity.

For the special case when the flow Mach number is very small, the mean density is constant, there are no entropy fluctuations, and the fluid is at rest outside a bounded region, we can simplify Howe's wave equation to

$$\frac{1}{c_\infty^2}\frac{\partial^2 H}{\partial t^2} - \frac{\partial^2 H}{\partial x_i^2} = \nabla\cdot(\boldsymbol{\omega}\times\mathbf{v}) \tag{7.1.3}$$

The solution to Eq. (7.1.3) can be obtained using the approaches described in Chapters 4 and 5. In the absence of scattering surfaces we can use the method of Green's functions to obtain

$$H(\mathbf{x},t) = -\int_{-T}^{T}\int_{V}(\boldsymbol{\omega}\times\mathbf{v})_i\frac{\partial G_o}{\partial y_i}dVd\tau \tag{7.1.4}$$

The source term on the right of Eq. (7.1.4) reveals the role of vorticity as a source of sound. It shows that there is no sound caused by vorticity that is aligned with the local flow. Furthermore, if the source term is linearized about an irrotational mean flow, we can use Eq. (6.3.11) to relate the local vorticity to the vorticity of a harmonic gust on some upstream inflow boundary, defined by $\boldsymbol{\omega} = \mathrm{Re}\left( \hat{\boldsymbol{\omega}}^{(\infty)} \exp\left(i\mathbf{k}\cdot\mathbf{x} - ikU_\infty t\right)\right)$, so that

$$\boldsymbol{\omega} \times \mathbf{U} = \mathrm{Re}\left( \hat{\omega}_i^{(\infty)} \left( \frac{\delta\mathbf{l}^{(i)}}{h_i} \times \mathbf{U} \right) e^{i\mathbf{k}\cdot\mathbf{X} - ik_1 U_\infty t} \right)$$

where $\delta\mathbf{l}^{(i)}$ are displacement coordinates shown in Fig. 6.3. Since $\delta\mathbf{l}^{(1)}/h_1 = \mathbf{U}/U_\infty$ we can use the relationship given by Eq. (6.2.9) to show that

$$\boldsymbol{\omega} \times \mathbf{U} = \mathrm{Re}\left( U_\infty \left( \hat{\omega}_3^{(\infty)} \nabla X_2 - \hat{\omega}_2^{(\infty)} \nabla X_3 \right) e^{i\mathbf{k}\cdot\mathbf{X} - ik_1 U_\infty t} \right)$$

where $X_2 = const$ and $X_3 = const$ are the stream surfaces of the mean flow. It follows that the source term in Eq. (7.1.4) is determined by the two components of the vorticity that are initially normal to the mean flow, and the streamwise component of the vorticity has no impact on the source term, even after distortion by the mean flow. The important consequence of this is that trailing tip vortices or other similar flows where the unsteady vorticity is initially aligned with the mean flow can only cause sound radiation because of nonlinear or self-induced unsteady motion.

## 7.2   Sound from two line vortices in free space

A simple canonical model of vortex sound [1] is given by two line vortices of equal strength $\Gamma$ that are separated by the distance $2d$ in an otherwise stationary fluid, as shown in Fig. 7.1. The vortex pair will spin about the point halfway between them because of the induced flow at each vortex. Because there is no background flow, the motion of each vortex is caused only by its convection by the other vortex in the pair, and so each moves with a speed of $U = \Gamma/4\pi d$.

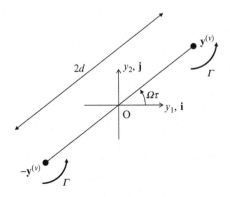

**Fig. 7.1** The spinning vortex pair.

The location of each vortex is given as $\mathbf{y} = \pm\mathbf{y}^{(v)}$ where $\mathbf{y}^{(v)}(\tau) = (d\cos(\Omega\tau),\ d\sin(\Omega\tau),\ 0)$, and the angular velocity of the system is $\Omega = \Gamma/4\pi d^2$ and $U = \Omega d$. The vorticity of each vortex is defined as

$$\boldsymbol{\omega} = \mathbf{k}\Gamma\delta\left(y_1 \pm y_1^{(v)}(\tau)\right)\delta\left(y_2 \pm y_2^{(v)}(\tau)\right) \tag{7.2.1}$$

where $\mathbf{k}$ is a unit vector pointing out of the page in Fig. 7.1. It follows that the Lamb vector in Eq. (7.1.4) is

$$\boldsymbol{\omega} \times \mathbf{v} = -\Omega\Gamma\mathbf{y}^{(v)}(\tau)\left\{\delta\left(y_1 - y_1^{(v)}(\tau)\right)\delta\left(y_2 - y_2^{(v)}(\tau)\right)\right.$$
$$\left. -\delta\left(y_1 + y_1^{(v)}(\tau)\right)\delta\left(y_2 + y_2^{(v)}(\tau)\right)\right\} \tag{7.2.2}$$

Note that, with the vorticity lumped into discrete vortices, the velocity field of the flow $\mathbf{v}$ can be replaced with the convection velocity of the vortices. Using Eq. (7.2.2) in Eq. (7.1.4) and integrating over $y_1$ and $y_2$ gives the sound field as

$$H(\mathbf{x}, t) = \Omega\Gamma\int_{-\infty}^{\infty}\int_{-T}^{T}\left[y_i^{(v)}(\tau)\frac{\partial G_o(\mathbf{x},t|\mathbf{y},\tau)}{\partial y_i}\right]_{\mathbf{y}=\mathbf{y}^{(v)}(\tau)+\mathbf{k}y_3}$$
$$-\left[y_i^{(v)}(\tau)\frac{\partial G_o(\mathbf{x},t|\mathbf{y},\tau)}{\partial y_i}\right]_{\mathbf{y}=-\mathbf{y}^{(v)}(\tau)+\mathbf{k}y_3}\ dy_3 d\tau$$

The integral is carried out over the infinite length of the two vortices because this is a two-dimensional problem. However, if we consider the sound radiation from a small linear segment of the vortices we can evaluate the far-field sound from that segment. This enables us to use the far-field approximation for the Green's function. The far-field sound from the segment of length $b$ centered on $y_3$ is then $\delta H$, where

$$\delta H(\mathbf{x}, t) = \Omega\Gamma b\int_{-T}^{T}\left[y_i^{(v)}(\tau)\frac{\partial G_o(\mathbf{x},t|\mathbf{y},\tau)}{\partial y_i}\right]_{\mathbf{y}=\mathbf{y}^{(v)}(\tau)+\mathbf{k}y_3}$$
$$-\left[y_i^{(v)}(\tau)\frac{\partial G_o(\mathbf{x},t|\mathbf{y},\tau)}{\partial y_i}\right]_{\mathbf{y}=-\mathbf{y}^{(v)}(\tau)+\mathbf{k}y_3}\ d\tau$$

We note from this result that if we were to ignore the effect of retarded time then the integrand would be zero, giving no far-field sound. However, if we expand the Green's function using a Taylor series expansion about the point $\mathbf{k}y_3$

$$\left[\frac{\partial G_o(\mathbf{x},t|\mathbf{y},\tau)}{\partial y_i}\right]_{\mathbf{y}=\pm \mathbf{y}^{(v)}(\tau)+\mathbf{k}y_3} = \left[\frac{\partial G_o(\mathbf{x},t|\mathbf{y},\tau)}{\partial y_i}\right]_{\mathbf{y}=\mathbf{k}y_3}$$

$$\pm y_j^{(v)}(\tau)\left[\frac{\partial^2 G_o(\mathbf{x},t|\mathbf{y},\tau)}{\partial y_i \partial y_j}\right]_{\mathbf{y}=\mathbf{k}y_3} + \cdots$$

and so the integral reduces to

$$\delta H(\mathbf{x},t) = 2\Omega\Gamma b \int_{-T}^{T} y_i^{(v)}(\tau)y_j^{(v)}(\tau)\frac{\partial^2 G_o(\mathbf{x},t|\mathbf{k}y_3,\tau)}{\partial y_i \partial y_j}d\tau \tag{7.2.3}$$

and we obtain an acoustic field that has the characteristics of a quadrupole. As we did in Section 4.4, in the acoustic far field we can shift the derivative of the free-space Green's function so that they are derivatives with respect to time, and then integrate over source time, so we obtain

$$\delta H(\mathbf{x},t) = \frac{2\Omega\Gamma b}{c_\infty^2}\frac{\partial^2}{\partial t^2}\left[\frac{\left(x_j y_j^{(v)}\right)\left(x_i y_i^{(v)}\right)}{4\pi|\mathbf{x}-\mathbf{k}y_3|^3}\right]_{\tau=\tau^*}$$

The analysis is simplified if we choose the observer location to be at $\mathbf{x}=(R_o,0,0)$ which gives $x_i y_i^{(v)}=R_o d\cos(\Omega\tau)$. We then obtain

$$\delta H(\mathbf{x},t) = \frac{\Omega\Gamma b d^2 R_o^2}{2\pi c_\infty^2}\frac{\partial^2}{\partial t^2}\left[\frac{\cos^2(\Omega\tau)}{\left(R_o^2+y_3^2\right)^{3/2}}\right]_{\tau=\tau^*} \tag{7.2.4}$$

Since $\cos^2(\Omega\tau)=(1+\cos(2\Omega\tau))/2$ we can simplify the integrand so that it only depends on $\cos(2\Omega\tau)$ because the constant term will not contribute to the acoustic field.

Then since the correct retarded time is

$$\tau^* = t - r/c_\infty, \qquad\qquad r = \left(R_o^2+y_3^2\right)^{1/2}$$

we obtain

$$\delta H(\mathbf{x},t) = -\frac{\Omega^3\Gamma b d^2 R_o^2}{\pi c_\infty^2 r^3}\cos\left(2\Omega(t-r/c_\infty)\right) \tag{7.2.5}$$

We can then evaluate the far-field pressure because $\partial H/\partial t$ is approximately equal to $(1/\rho_o)\partial p/\partial t$ in the acoustic far field. To obtain a scalable result we can define a Mach number of the vortex motion as $M=\Omega d/c_\infty$, and the speed of the vortex as $U=\Omega d=\Gamma/4\pi d$ so the far-field pressure signal is given by

$$\delta p(\mathbf{x},t) = -\frac{4\rho_o U^2 M^2 R_o^2 b\cos\left(2\Omega(t-r/c_\infty)\right)}{r^3} \tag{7.2.6}$$

This is a canonical example of two-dimensional flow noise and as such it is valuable to investigate the important features of the result. It shows that the far-field pressure from an elemental part of the vortex scales as $U^2 M^2$ which is the same as the result obtained in Chapter 4 for free turbulence. The acoustic intensity in the far field $I = |p|^2/(2\rho_o c_\infty)$ will scale as $M^4 U^4$ or the eighth power of the velocity at the vortex. Eq. (7.2.6) also shows that the acoustic radiation occurs at a frequency that is twice the angular speed of the vortex, which is to be expected since the flow field replicates itself after half a revolution of the vortex pair.

Some care needs to be taken when evaluating this result for a line vortex whose span exceeds the acoustic wavelength. The result given by Eq. (7.2.6) needs to be integrated along the length of the vortex and the motion is only specified by the model given above if the vortex is of infinite length. The infinite length vortex results in a two-dimensional problem and the far-field approximations used above will no longer be valid, and the result will scale differently, but the problem can be solved exactly [2]. However, if the vortices are of finite length, and their span is acoustically compact, but their motion is well approximated by the self-induced spinning motion used above, then the far-field sound is given by the integral of Eq. (7.2.6) over $y_3$.

## 7.3 Surface forces in incompressible flow

In Chapter 4 we showed that, for acoustically compact bodies in low Mach number flows, the flow noise is dominated by dipole sources whose strength is directly proportional to the unsteady loading on the body surface. In this case the acoustic wavelength is much larger than the size of the body, and so the unsteady loading will be dominated by the incompressible part of the flow perturbations. We can therefore separate the problem into two parts, the incompressible flow that causes the unsteady load and the sound radiation from the loading on the surfaces. When this approximation is made, the acoustic far field can be evaluated from Curle's equation (4.4.7) with the source term defined by the unsteady loading caused by incompressible flow perturbations.

In an incompressible flow the force applied to the fluid by a stationary surface can be calculated directly from the vorticity in the flow (see Ref. [2] for a more detailed derivation of the results that follow, in which the effect of viscosity is included). To obtain this relationship consider the problem illustrated in Fig. 7.2 where fluid is bounded by a surface $S_\infty$ at a large distance from the body and the mean flow tends to a constant on this surface. Since the fluid is assumed incompressible its density will be constant and the divergence of the velocity $\mathbf{v}$ will be zero. The surface of the body is assumed impenetrable and so $\mathbf{v} \cdot \mathbf{n}$ on $S$, and we will assume that $|\mathbf{v}|^2$ tends to $U_\infty^2$ on $S_\infty$ with a difference that tends to zero as $1/r^\alpha$, where $\alpha > 2$. This implies that there is no net flux of momentum across $S_\infty$ because in the limit that $r$ tends to infinity $(\rho \mathbf{v} \cdot \mathbf{n})$ $\mathbf{v} = \rho U_\infty^2 n_1$ which integrates to zero on $S_\infty$.

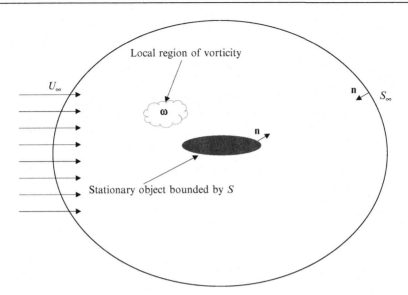

**Fig. 7.2** A stationary object in a steady uniform flow that includes a small region of vorticity and is bounded by a surface $S_\infty$.

The force applied to the fluid by the body $F_i$ can be equated to the fluid momentum and the pressure on $S_\infty$ using Eq. (2.3.3), with the assumption that there is no net flux of momentum across $S_\infty$ giving

$$F_i = \rho_o \frac{d}{dt} \int_V v_i dV - \int_{S_\infty} p n_i dS \qquad (7.3.1)$$

(where in this case the normal is chosen as $\mathbf{n} = -\mathbf{n}^{(o)}$ so it points into the enclosed volume). The volume integral can be changed to a surface integral by making use of the identity

$$v_i = \mathbf{v} \cdot \nabla(x_i) = \nabla \cdot (\mathbf{v} x_i) - (x_i \nabla \cdot \mathbf{v})$$

and noting that the last term is zero in an incompressible fluid. Substituting this into the volume integral in Eq. (7.3.1) and using the divergence theorem results in

$$\rho_o \frac{d}{dt} \int_V v_i dV = -\rho_o \frac{d}{dt} \int_{S_\infty} x_i \mathbf{v} \cdot \mathbf{n} dS - \rho_o \frac{d}{dt} \int_S x_i \mathbf{v} \cdot \mathbf{n} dS$$

The boundary condition on the surface of the body requires that $\mathbf{v} \cdot \mathbf{n} = 0$, so the surface integral over $S$ is zero and we can write Eq. (7.3.1) as

$$F_i = -\int_{S_\infty} \left( p n_i + x_i \rho_o \frac{\partial \mathbf{v}}{\partial t} \cdot \mathbf{n} \right) dS \tag{7.3.2}$$

(where $d/dt$ has been replaced by $\partial/\partial t$ inside the integral because the surface is stationary).

The right side of Eq. (7.3.2) can be related to the vorticity in the flow by making use of Kirchhoff coordinates and the form of the momentum equation given by Eq. (2.5.3). The Kirchhoff coordinates $Y_i$ are defined as equal to the potential of the flow around the body that results from an inflow on $S_\infty$ of unit amplitude in the $x_i$ direction. They have the properties that

$$\left. \begin{array}{c} Y_i = x_i \\ \mathbf{n} \cdot \nabla Y_i = n_i \end{array} \right\} \text{on } S_\infty, \qquad \nabla^2 Y_i = 0, \qquad \mathbf{n} \cdot \nabla Y_i = 0 \text{ on } S \tag{7.3.3}$$

If we take the dot product of $\nabla Y_i$ with the momentum equation expressed in the form given by Eq. (2.5.3) and integrate over the volume of the fluid then, ignoring viscous terms for a homentropic flow, we obtain

$$\int_V \nabla Y_i \cdot \left( \rho_o \frac{\partial \mathbf{v}}{\partial t} \right) dV + \int_V \nabla Y_i \cdot \nabla \left( p + \frac{1}{2} \rho_o v_i^2 \right) dV = -\rho_o \int_V \nabla Y_i \cdot (\boldsymbol{\omega} \times \mathbf{v}) dV$$

The first volume integral on the left of this equation can be turned into a surface integral over $S_\infty$ in exactly the same way as was done for Eq. (7.3.2). Similarly the integrand of the second volume integral can be turned into a divergence because $Y_i$ is a solution to Laplace's equation. Then making use of the boundary conditions on the surface defined in Eq. (7.3.3) we find that

$$\int_V \nabla Y_i \cdot \left( \rho_o \frac{\partial \mathbf{v}}{\partial t} \right) dV + \int_V \nabla Y_i \cdot \nabla \left( p + \frac{1}{2} \rho_o v_i^2 \right) dV$$

$$= \int_V \nabla \left( Y_i \rho_o \frac{\partial \mathbf{v}}{\partial t} \right) dV + \int_V \nabla \left( \nabla Y_i \left( p + \frac{1}{2} \rho_o v_i^2 \right) \right) dV$$

$$= -\int_{S_\infty + S} Y_i \rho_o \frac{\partial \mathbf{v}}{\partial t} \cdot \mathbf{n} dS - \int_{S_\infty + S} \mathbf{n} \cdot \nabla Y_i \left( p + \frac{1}{2} \rho_o v_i^2 \right) dS$$

$$= -\int_{S_\infty} Y_i \rho_o \frac{\partial \mathbf{v}}{\partial t} \cdot \mathbf{n} dS - \int_{S_\infty} \mathbf{n} \cdot \nabla Y \left( p + \frac{1}{2} \rho_o v_i^2 \right) dS$$

Since the surface lies at infinity we can use the limiting values given by Eq. (7.3.3) to obtain

$$-\int_{S_\infty} x_i \left( \rho_o \frac{\partial \mathbf{v}}{\partial t} \cdot \mathbf{n} \right) + n_i \left( p + \frac{1}{2} \rho_o v_i^2 \right) dS = -\rho_o \int_V \nabla Y_i \cdot (\boldsymbol{\omega} \times \mathbf{v}) dV$$

Then since the velocity tends to its value in the uniform flow with an error proportional to $1/r^\alpha$ where $\alpha > 2$ and $S_\infty$ increases as $r^2$ the integral of the squared velocity term becomes

$$\int_{S_\infty} n_i \left(\frac{1}{2}\rho_o v_i^2\right) dS \to \int_{S_\infty} n_i \left(\frac{1}{2}\rho_o U_\infty^2\right) dS = 0$$

By matching terms in Eq. (7.3.2) we obtain the force applied to the fluid by the body in terms of the volume distribution of the Lamb vector $\boldsymbol{\omega} \times \mathbf{v}$ as

$$F_i = -\rho_o \int_V \nabla Y_i \cdot (\boldsymbol{\omega} \times \mathbf{v}) dV \tag{7.3.4}$$

To evaluate this integral, we need to include the vorticity throughout the fluid, including any bound vorticity such as the vorticity in the surface boundary layer and the wake behind the body. However, since the mean vorticity and mean velocity in the boundary layer next to the surface of the body will be parallel to the surface, the Lamb vector near the surface will point in the direction of the surface normal which is orthogonal to $\nabla Y_i$. This greatly reduces the contribution from the blade boundary layer velocity fluctuations in this calculation. However, the vorticity in the wake behind the body, that is an extension of the surface boundary layers, cannot be ignored and may play a dominant role. The other important conclusion from this result is that stationary vorticity does not contribute to the unsteady loading, so we need to only consider vorticity that is convected when evaluating Eq. (7.3.4). Also we note that the streamwise force is given by the $F_1$ component $\nabla Y_1 \cdot (\boldsymbol{\omega} \times \mathbf{v})$. If the vorticity is convected by the potential mean flow around the body then $\mathbf{v} = U_\infty \nabla Y_1$, and there will be no streamwise force. The streamwise force is therefore determined by the deviation of the vortex path from the potential flow streamlines, which results from the influence of other regions of vorticity or the image vorticity inside the body that impacts the vortex trajectory in the flow.

When vorticity is close to a rigid surface it will be convected by both the local mean flow (which would still exist if the vorticity was zero) and a self-induced flow that is required to meet the non-penetration boundary condition on the surface. The self-induced flow is often characterized by equivalent image vorticity that is placed inside the surface but is a potential flow outside the surface. The image vorticity can have a big impact if the path of the vortex takes it close to the stagnation point, but little impact if the vortex is relatively weak and some distance from the surface. To assess the importance of this we will consider a line vortex near a surface and estimate the velocity induced by the image vortex to be $\Gamma/4\pi\delta$, where $\delta$ is the distance of the line vortex from the surface and $\Gamma$ is its circulation. Then if $\Gamma/4\pi U\delta \ll 1$ (where $U$ is the local mean flow speed) the motion induced by the image vortex will not be important. However, this criterion cannot be met close to the stagnation point at the front of the body where the mean flow speed tends to zero. In this case the motion of the vortex will be controlled by the induced flow and the amplitude of the unsteady loading will be of order $\rho_o\Gamma^2/4\pi\delta$. However this will only have a small impact on the unsteady loading pulse which will have a peak amplitude of order $\rho_o\Gamma U_\infty$. This leads to a criterion that requires $\Gamma/4\pi U_\infty\delta \ll 1$, which is less restrictive than the criterion based on the local flow speed. In real flows one might expect $\Gamma/2\pi \sim u_o L$, where $u_o$ is a typical gust velocity and $L$ is the lengthscale of the turbulence. Typically, $L \sim \delta$, so the importance of the self-induced motion of the vortex will depend on $u_o/4\pi U_\infty$ which is very small in most flows of interest for which $u_o \ll U_\infty$. Therefore, we will not consider the impact of the image vortex on the convection speed any further, and we will assume the vortex is simply convected by the mean flow.

## 7.4  Aeolian tones

A well-known phenomenon of sound caused by flow over solid bodies is the Aeolian tone. This describes the mechanism of sound generation from the wind blowing through boat rigging and over telephone wires, or any other cylindrical body. The mechanism responsible for Aeolian tones has been studied extensively and is a classic example of vortex sound. The flow is illustrated in Fig. 7.3, which shows a stationary cylinder in a uniform flow, which sheds well-defined vortical structures into its wake. The vortical structures form a von Kármán vortex street that consists of equally spaced vortices that are convected downstream at a constant speed. At Reynolds numbers $Re_d = U_\infty d/\nu$ (based on the flow speed $U_\infty$ and the cylinder diameter $d$) that lie in the range $40 < Re_d < 5 \times 10^4$ it is found that well-defined vortical structures are shed periodically, and the frequency of vortex shedding $f$ is given by a Strouhal number $St = fd/U_\infty$ of approximately 0.2. This is a weak function of Reynolds number and given by Goldstein [3] as

$$St = 0.198(1 - 19.7/Re_d)$$

Flow visualization of the wakes behind cylinders have shown that the formation of the wake vortices follows a consistent process. As shown in Fig. 7.3 the vortices are initiated just behind the cylinder on either the upper or lower side of the wake.

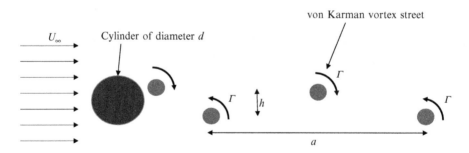

**Fig. 7.3** The vortex street that occurs behind a cylinder in a steady flow.

During the period of growth, the vortex is almost static as it spins up, extracting circulation from the boundary layer on the surface of the cylinder. Once it has reached a critical circulation, it detaches from its point of initiation and is swept downstream by the flow. The initial period of acceleration is quite fast, and once it reaches the wake region it is convected with constant speed $U_c$. The process is then repeated for a vortex in the lower part of the wake, and it grows and separates in exactly the same way as the vortex in the upper part of the wake. The process is then repeated resulting in a time varying, periodic flow.

The periodic vortex shedding causes an unsteady lift fluctuation on the cylinder that is readily understood by using the results given in Section 7.3. In terms of Eq. (7.3.4) the unsteady lift on the cylinder will be given by

$$F_2(t) = -\rho_o \int_V \nabla Y_2 \cdot (\boldsymbol{\omega} \times \mathbf{v}) dV \qquad (7.4.1)$$

where the function $Y_2$ is the velocity potential of a flow of unit free-stream velocity in the $i = 2$ direction around the cylinder and is given by Eq. (2.7.27) as

$$Y_2 = \mathrm{Re}\left(ze^{-i\pi/2} + \frac{d^2 e^{i\pi/2}}{4z}\right) = y_2\left(1 + \frac{d^2}{4\sqrt{y_1^2 + y_2^2}}\right) \quad \text{where} \quad z = y_1 + iy_2 \quad (7.4.2)$$

The first thing we note from Eq. (7.4.1) is that, since we are assuming that the vorticity is concentrated at a point in the flow, there will be no unsteady load while the vortex is stationary during its period of growth because $\mathbf{v} = 0$. However, once it reaches its critical strength and starts to move then it will contribute to the unsteady load. Vortex formation occurs at a point $l_o > d$ downstream from the center of the cylinder, and so when the vortex is in motion we can approximate $Y_2 \sim y_2$ and the trajectory is almost parallel to the $y_1$ axis, so we can write

$$\nabla Y_2 \cdot (\boldsymbol{\omega} \times \mathbf{v}) \approx \Gamma_n u(t - t_n)\delta(y_1 - l_o - s(t - t_n))\delta(y_2 \pm h/2)H(|y_3| - b/2)$$

where $u(t - t_n)$ is the speed of the vortex after it detaches at time $t = t_n$ and is zero for $t < t_n$. The location of the vortex in the downstream direction is given by $l_o + s(\tau)$, where $\tau$ is the time since the vortex was shed and $s(\tau)$ is the distance traveled. The vertical location of the vortex is given by $y_2 = \pm h/2$ depending on whether it is in the upper or lower row, and the spanwise extent of the vortex is given by $b$. Furthermore the vortices shed in the upper row will have the opposite direction of rotation to the vortices in the lower row, so $\Gamma_n = (-1)^n \Gamma$. Using this model in Eq. (7.4.1) for multiple vortices then gives

$$F_2(t) = -\rho_o \Gamma b \sum_{n=0}^{\infty} (-1)^n u(t - t_n) \qquad (7.4.3)$$

If the vortex convection speed increases linearly with time until it reaches a steady convection speed $U_c$ and the acceleration time is half a period, then we can model $u(\tau)$ such that

$$u(\tau) = \begin{cases} 0 & \tau < 0 \\ 2U_c\tau/T & 0 < \tau < T/2 \\ U_c & \tau > T/2 \end{cases}$$

The series in Eq. (7.4.3) then represents a triangular wave with amplitude $U_c$, and it can be expanded into a Fourier series so that

$$F_2(t) = -\rho_o \Gamma b U_c \left( \frac{1}{2} + \frac{4}{\pi^2} \cos(\Omega t) + \frac{4}{9\pi^2} \cos(3\Omega t) + \cdots \right) \tag{7.4.4}$$

where $\Omega = 2\pi/T$ is the fundamental frequency of the fluctuations. We can then define a lift coefficient for the first harmonic based on the frontal area of the cylinder as

$$C_L = \frac{|F_2|}{(1/2)\rho_o U_\infty^2 bd} = \frac{8}{\pi^2} \left( \frac{\Gamma}{U_\infty d} \right) \left( \frac{U_c}{U_\infty} \right) \tag{7.4.5}$$

Analysis of a von Kármán vortex street [4] shows that the convection speed of the vortices in the sheet relative to a fixed body in a uniform flow is given by

$$U_c = U_\infty - \frac{\Gamma}{2a} \tan h(\pi h/a) \tag{7.4.6}$$

where $a$ is the horizontal distance between vortices and is approximately $a = 4d$ for Reynolds numbers above 1000. The stability of the vortex street requires that $h/a = 0.281$. To obtain $\Gamma$ we note that the Strouhal number is directly related to the convection velocity, so $St = U_c d/U_\infty a$, and since $St = 0.2$ the convection velocity is approximately $U_c = 0.8 U_\infty$. However, there is also experimental evidence that the convection velocity is higher than this and has a value of $U_c = 0.9 U_\infty$ [5]. Solving Eq. (7.4.6), with $U_c = 0.9 U_\infty$, then gives

$$\frac{\Gamma}{U_\infty d} = \frac{2a}{d \tan h(\pi h/a)} \left( 1 - \frac{U_c}{U_\infty} \right) \approx 1.13$$

and a lift coefficient of approximately 0.81, which is in good agreement with the measured values given by Blake [5].

The sound radiation from the unsteady loading at the fundamental frequency is readily calculated using Eqs. (7.4.4), (4.4.7), and we obtain the far-field sound as

$$p \approx \left( \frac{\rho_o U_\infty^3 b C_L St}{4c_\infty} \right) \frac{x_2 \sin(\Omega(t - r/c_\infty))}{|\mathbf{x}|^2} \tag{7.4.7}$$

This shows the classic dipole scaling of the far-field sound in which the acoustic intensity $I = (p_{rms})^2/\rho_o c_\infty$ scales as $M^3 U_\infty{}^3$ and depends on the sixth power of the flow speed. Also note the directionality that has cosine dependence with a maximum in the direction of the unsteady lift.

## 7.5 Blade vortex interactions in incompressible flow

An important source of noise in aeroacoustics occurs when a blade moves past a line vortex. This is particularly relevant to helicopter noise when blade vortex interactions occur during maneuvers and cause a loud thumping sound. The source of the vortices in this case is the tips of the blades themselves. Using the formulas developed in Section 7.3 we can derive a simple incompressible flow model for the unsteady load caused by a blade vortex interaction from which we can calculate the acoustic field. More detailed analysis of this problem, which includes the effect of compressibility, will be given in Chapter 14, and here we will focus on the incompressible flow solution, using the approach given by Howe [2].

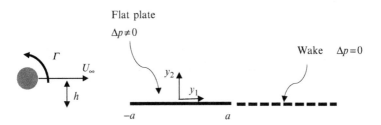

**Fig. 7.4** A blade vortex interaction in which the blade is modeled as a flat plate with zero thickness.

If we make the assumptions of thin airfoil theory then the blade can be modeled as a flat plate with zero thickness at zero angle of attack to the mean flow, as shown in Fig. 7.4. The vortex is convected past the plate with the mean flow $U_\infty$, and its path takes it a height $h$ above the plate. The coordinate system is chosen to be coincident with the center of the plate and the blade chord is $2a$. If we consider a line vortex with circulation $\Gamma$ then we can calculate the unsteady loading per unit span of the plate on the fluid using Eq. (7.3.4). Since the direction of the vorticity vector is out of the page, and the flow is in the $y_1$ direction the Lamb vector points in the $y_2$ direction and the unsteady load per unit span for a vortex located at $y_1 = U_\infty t$, $y_2 = h$ is

$$\frac{F_2}{b} = -\rho_o U_\infty \Gamma \left[\frac{\partial Y_2}{\partial y_2}\right]_{y_1=U_\infty t, y_2=h} \tag{7.5.1}$$

where $b$ is the span of the blade. The Kirchhoff coordinate $Y_2$ for a flat plate can be calculated from potential flow theory and is given by Eq. (2.7.29) with $\alpha = \pi/2$ as

$$Y_2 = \text{Re}\left(-i\sqrt{z^2 - a^2}\right), \qquad\qquad z = y_1 + iy_2 \tag{7.5.2}$$

The branch cut of the square root is chosen to lie between $z = \pm a$ so that there is no flow through the plate. Fig. 7.5 shows the streamlines of the flow defined by $\nabla Y_2$, and it is seen that the function contours around the plate, changing rapidly for locations near the leading and trailing edges. Using this in Eq. (7.5.1) then gives the unsteady loading as a function of time as

$$\frac{F_2}{b} = -\frac{1}{2}\rho_o U_\infty \Gamma \text{Re}\left\{ \sqrt{\frac{U_\infty t + ih - a}{U_\infty t + ih + a}} + \sqrt{\frac{U_\infty t + ih + a}{U_\infty t + ih - a}} \right\} \tag{7.5.3}$$

(note: this is the force applied to the fluid that tends to a constant when $t = \pm\infty$ because the vortex remains in the flow). The time history of the unsteady load therefore has two contributions. The first has a peak when $U_\infty t = -a$, which corresponds to the location of the vortex when it is closest to the leading edge of the blade. The second contribution occurs when $U_\infty t = a$, which corresponds to the location of the vortex when it is closest to the trailing edge. These pulses are shown in Fig. 7.6, and their peak levels are determined by the height of the vortex above the surface, so the results for different values of $h/a$ are shown. The closer the vortex is to the blade the sharper and larger is the unsteady loading pulse, and the leading and trailing edge pulses are of the same magnitude.

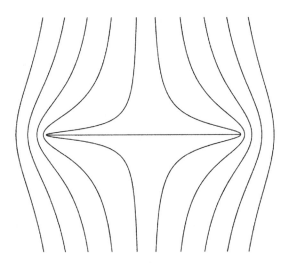

**Fig. 7.5** The streamlines of the flow defining the Kirchhoff coordinate $Y_2$ for a flat plate of zero thickness.

However, this model has not included the effect of vorticity in the blade wake, which is required to satisfy the Kutta condition. To fully calculate the impact of the wake vorticity we would need to solve the equations of motion to obtain the shed vorticity as a function of time. However, the impact of the wake vorticity is to eliminate the pressure jump at the trailing edge of the blade, and it was argued by Howe [2] that this reduces the trailing edge pulse. This leads to *Howe's approximation* which applies to the case when the vortex is close to the blade, so $h \ll a$. It approximates the net unsteady loading by the contribution from the leading edge pulse only, and sets the trailing edge pulse to zero, so the unsteady load is given by the first term in Eq. (7.5.3) in the limit that $U_\infty t = -a$, as

$$\frac{F_2}{b} = -\frac{1}{2}\rho_o U_\infty \Gamma \operatorname{Re}\left\{i\sqrt{\frac{2a}{U_\infty t + ih + a}}\right\} \qquad (7.5.4)$$

A comparison between Howe's approximation and the total pulse is shown in Fig. 7.7 for a vortex that passes above the blade at a standoff distance of $h/a = 0.05$.

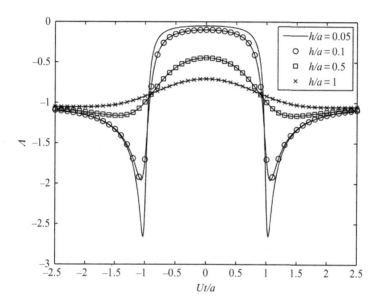

**Fig. 7.6** The leading and trailing edge loading pulses from a blade vortex interaction when there is no wake and the Kutta condition is not imposed, $\Lambda = F_2/\rho_o \Gamma U_\infty b$.

To justify this approximation, we need to compare the results using Howe's approximation to the solution that would have been obtained using an exact incompressible flow calculation that includes the Kutta condition. For a harmonic gust this is given by Sears function, which was discussed in Chapter 6. The comparison is most easily carried out by considering a gust that is caused by a harmonic vortex sheet at distance $h$ above the blade. This causes a velocity field given by

$$\hat{v}_1 e^{-i\omega t} = U_\infty - \hat{a}_1 \operatorname{sgn}(y_2 - h)e^{-k_1|y_2-h|+ik_1(y_1-U_\infty t)},$$
$$\hat{v}_2 e^{-i\omega t} = -i\hat{a}_1 e^{-k_1|y_2-h|+ik_1(y_1-U_\infty t)}$$

It is readily verifiable that this flow is incompressible, and the vorticity is only nonzero on the vortex sheet and has a single component in the $i=3$ direction given by $\hat{\omega}_3 e^{-i\omega t} = 2\hat{a}_1 \delta(y_2 - h)e^{ik_1(y_1-U_\infty t)}.$

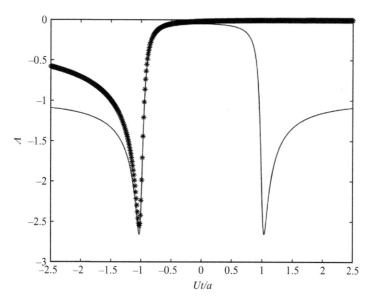

**Fig. 7.7** The comparison of Howe's approximation *(\*)* for the blade response to a passing vortex compared to the response without a wake *(line)*, $h/a = 0.05$.

Using this result in Eq. (7.3.4), we assume that the amplitude of the velocity fluctuations is very much less than the mean flow speed, giving the unsteady load as

$$\frac{\hat{F}_2}{b} e^{-i\omega t} = -2\rho_o U_\infty \hat{a}_1 \int_{-\infty}^{\infty} \left[ \frac{\partial Y_2}{\partial y_2} \right]_{y_2=h} e^{ik_1(y_1 - U_\infty t)} dy_1 \tag{7.5.5}$$

Using Howe's approximation this becomes

$$\frac{\hat{F}_2}{b} = -\rho_o U_\infty \hat{a}_1 \int_{-\infty}^{\infty} \text{Re} \left\{ i \sqrt{\frac{2a}{y_1 + ih + a}} \right\} e^{ik_1 y_1} dy_1$$

This integral can be evaluated using tables of Fourier transforms, and it is found that

$$\hat{F}_2 = -2\pi \rho_o U_\infty \hat{a}_1 ab \left( \sqrt{\frac{1}{2\pi k_1 a}} \right) e^{-ik_1 a - k_1 h - i\pi/4} \tag{7.5.6}$$

We can compare this with the high-frequency approximation to the unsteady loading calculated using Sears function using Eqs. (6.4.3), in the limit that $k_1 a = \sigma \gg 1$. In Eq. (6.4.3) the gust was defined relative to the upwash on the blade, and so the equivalence between the magnitude of the gust response given by Eqs. (6.4.3), (7.5.6) for

$\sigma \gg 1$ requires that the upwash gust amplitude is replaced by $\exp(-k_1 h)$ in Eq. (7.5.6). Comparing Eqs. (6.4.3), (7.5.6) shows that they give the same gust magnitude in the limit that $\sigma \gg 1$. Consequently Howe's approximation correctly accounts for the effect of vortex shedding in the wake and the Kutta condition and leads to the conclusion that the unsteady loading from a blade vortex interaction is dominated by the response of the blade as the vortex passes leading edge of the blade, hence the term *Leading Edge Noise* which we will discuss in more detail in subsequent chapters.

Another example that is of interest and also verifies Howe's approximation is the response of a flat plate to a step gust. In this case the incident disturbance is specified by the velocity and vorticity distribution where the gust reaches the leading edge at $t=0$. We assuming $w_o \ll U_\infty$, the unsteady loading per unit span is

$$v_1 = U_\infty, \quad v_2 = w_o H(U_\infty t - y_1 - a), \quad \omega_3 = -w_o \delta(y_1 + a - U_\infty t)$$

$$\frac{F_2}{b} = \rho_o U_\infty w_o \int_{-R}^{R} \left[ \frac{\partial Y_2}{\partial y_2} \right]_{y_1 = U_\infty t - a} dy_2$$

At first sight this integral would appear trivial and gives the load as $2\rho_o U_\infty w_o R$ which does not vary in time. However, there is an additional contribution when the step gust passes the leading edge of the blade because $Y_2$ has a discontinuity on $y_2 = 0$ between $y_1 = \pm a$. The contribution from the discontinuity is given by the jump in the value of $Y_2$ across the surface. Applying Howe's approximation to Eq. (7.5.2) gives the jump in $Y_2$ as

$$2\text{Re}\left( \sqrt{2a} \sqrt{y_1 + i\varepsilon + a} \right) H(y_1 + a)$$

where $\varepsilon$ tends to zero. The response to a step gust is then given by

$$\frac{F_2}{b} = -2\pi \rho_o U_\infty w_o a \left( \frac{\sqrt{2U_\infty t/a}}{\pi} \right) H(U_\infty t/a) \tag{7.5.7}$$

This can be compared to the approximate form of Kussner's function [6] for the response of a plate to a step gust, which is given by

$$\frac{F_2}{b} = -2\pi \rho_o U_\infty w_o a \left( \frac{(\tilde{\tau}^2 + \tilde{\tau})}{\tilde{\tau}^2 + 2.82\tilde{\tau} + 0.8} \right) H(U_\infty t/a), \quad \tilde{\tau} = \frac{U_\infty t}{a}$$

These results are compared in Fig. 7.8 and there is good agreement when the nondimensional time is less than one.

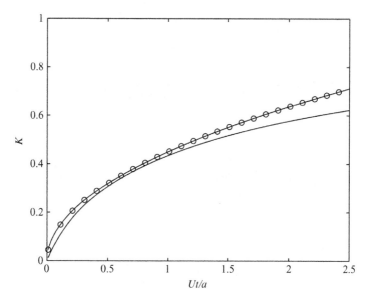

**Fig. 7.8** Howe's approximation *(circles)* compared to Kussner's function *(line)* $K = |F_2/2\pi\rho_o U_\infty w_o ab|$ for the response of a flat plate to a step gust in incompressible flow.

In conclusion the response of a blade to different types of gusts including a blade vortex interaction, a harmonic upwash gust and a step gust, is well approximated using Eq. (7.3.4) and Howe's approximation, which assumes the unsteady load is defined by the pulse that occurs when the disturbance is close to the leading edge of the blade.

## 7.6 The effect of angle of attack and blade thickness on unsteady loads

### 7.6.1 The effect of angle of attack

The calculation of the unsteady load produced by a two-dimensional body in an ideal potential flow can be carried out using Eq. (7.3.4) using a conformal mapping approach. This is particularly useful when considering the unsteady loading produced by airfoils at an angle of attack to the mean flow. This is not part of Sears theory since thin airfoil theory assumes that the effect of angle of attack is of second order on the unsteady loading and so is ignored. To evaluate its effect consider the unsteady loading caused by a line vortex with circulation $\Gamma$ that is convected by a two-dimensional steady flow with velocity $\mathbf{v} = (U, V, 0)$ past a flat plate airfoil as illustrated in Fig. 7.9. The vortex is defined as having its axis in the $y_3$ direction and so, by using Eq. (7.3.4), the unsteady loading per unit span is given by

$$\frac{F_i}{b} = -\rho_o \Gamma \left( U \frac{\partial Y_i}{\partial y_2} - V \frac{\partial Y_i}{\partial y_1} \right) \tag{7.6.1}$$

where $Y_i$ is the potential of a flow that has a speed of unit amplitude from the $i$ direction at infinity. Since the flow around the body is incompressible and irrotational it can be specified in terms of a complex potential $w(z) = \phi + i\psi$, where $z = y_1 + iy_2$, $\phi$ is the velocity potential, and $\psi$ is the stream function of the flow. The steady flow over the surface, and the flow defining $Y_i$ can then be specified as

$$U - iV = \frac{dw_o}{dz}, \qquad \frac{dY_i}{dy_1} - i \frac{dY_i}{dy_2} = \frac{dw_i}{dz} \tag{7.6.2}$$

and the unsteady loading can be written in terms of the complex potentials as

$$\frac{F_i}{b} = -\rho_o \Gamma \mathrm{Im} \left( \frac{dw_o}{dz} \left( \frac{dw_i}{dz} \right)^* \right) \tag{7.6.3}$$

where the $^*$ represents the complex conjugate.

To illustrate the application of this result we will consider a flat plate airfoil as

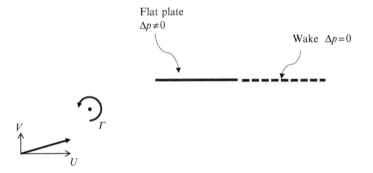

**Fig. 7.9** A vortex being convected past a flat plate with the flow at a finite angle of attack.

shown in Fig. 7.9, where the flow is incident at an angle of attack $\alpha$ to the chord line. The complex potential of the flow can be obtained using conformal mapping and is given by Eq. (2.7.30) as

$$w_o = U_\infty z \cos \alpha - i \left( \sqrt{z^2 - a^2} \right) \sin \alpha - \frac{i\Gamma_o}{2\pi} \ln \left( \frac{z + \sqrt{z^2 - a^2}}{2} \right) \tag{7.6.4}$$

The circulation $\Gamma_o$ defines the steady circulation around the airfoil, and to satisfy the Kutta condition at the blade trailing edge we require that $\Gamma_o = -2\pi a U_\infty \sin \alpha$ (Eq. 2.7.31). We can use this result along with the complex potentials that define the flow in the $y_1$ and $y_2$ directions, which are

$$w_1 = z \quad w_2 = -i\sqrt{z^2 - a^2}$$

to give the derivatives needed for Eq. (7.6.3) as,

$$\frac{dw_o}{dz} = U_\infty \left( \cos\alpha - i\sqrt{\frac{z-a}{z+a}}\sin\alpha \right) \quad \frac{dw_1}{dz} = 1 \quad \frac{dw_2}{dz} = \frac{-iz}{\sqrt{z^2 - a^2}} \quad (7.6.5)$$

However, the results are more easily interpreted if we use Howe's approximation and limit consideration to the leading edge pulse. This also corrects for the unsteady Kutta condition at low angles of attack and so is more likely to be accurate than a direct evaluation of Eq. (7.6.3). In this approximation the complex potential is evaluated in the limit that $z$ tends to $-a$ so that

$$\sqrt{z^2 - a^2} \approx i\sqrt{2a}\sqrt{z+a}$$

The two components of the unsteady load then separate out readily, and the evaluation of Eq. (7.6.3) gives

$$\frac{F_1}{b} = -\rho_o \Gamma U_\infty \mathrm{Im}\left( \sqrt{\frac{2a}{z+a}} \right) \sin\alpha, \quad \frac{F_2}{b} = \frac{1}{2}\rho_o \Gamma U_\infty \mathrm{Im}\left( \sqrt{\frac{2a}{z+a}} \right) \cos\alpha \quad (7.6.6)$$

The first important feature to note about this result is that the unsteady force normal to the blade $F_2$ is simply $\cos\alpha$ times the unsteady force given by Eq. (7.5.4) for a vortex located at $z = U_\infty t + ih$. In addition, there is a force $F_1$ in the direction of the blade chord, which is a suction force because it is negative. The dependence on vortex location as a function of $z$ is identical in each case, and the direction of the force vector is given by the vector components $(-2\sin\alpha, \cos\alpha, 0)$. For small angles of attack this corresponds to rotating the direction of the force anticlockwise through the angle $2\alpha$ as shown in Fig. 7.10.

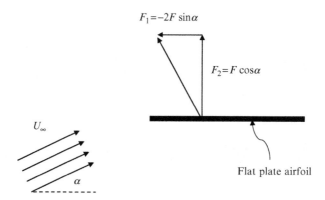

**Fig. 7.10** Flow incident at an angle of attack $\alpha$ on a flat plate airfoil and the rotation of the unsteady force in the anticlockwise direction.

The effect of the angle of attack on the amplitude of the pulse is relatively small and causes an increase in loading of $(1+3\sin^2\alpha)^{1/2}$ which is minimal for small $\alpha$. This result suggests that the effect of a small angle of attack on the unsteady loading is simply to rotate the direction of the load without affecting its amplitude.

However, the results given above are based on a fixed vortex displacement above the surface $h$, and we have not considered the effect of the rate of change of vortex position on the timescale of the pulse. If the vortex is convected by the mean flow along a streamline then its velocity is given by $(dw_o/dz)^*$ as defined in Eq. (7.6.5). In the vicinity of the leading edge where the pulse reaches its largest values we can approximate $z+a$ as equal to $h$, where $|h|$ is the closest distance of the vortex to the leading edge. The velocity of the vortex will depend on $(1+2a/h)^{1/2}$, and this will have only a minor impact on the vortex velocity if $(2a/|h|)^{1/2}\sin\alpha \ll 1$. For angles of attack of less than 6 degrees this implies that we require $(|h|/2a)^{1/2}>0.1$. It follows that if the vortex trajectory meets this criterion then the velocity of the vortex near the leading edge is not strongly affected by its proximity to the surface, and so the time scale of the pulse will remain the same as it was for zero angle of attack, and the leading order effect is simply the rotation of the direction of the force. However, in situations where the vortex passes closer to the leading edge than required by this limit then the vortex speed will be affected by the local flow speed, and some pulse distortion can be expected.

### 7.6.2    The effect of airfoil thickness

The same approach may be used to evaluate the effect of blade thickness on the unsteady loading. If we consider the Joukowski airfoil discussed in Section 2.7, then we can define the complex potential using Eqs. (2.7.32), (2.7.33). It will be assumed that the vortex is convected along a streamline, and so we need to take into account both the motion of the vortex and its location in drift coordinates.

To evaluate the streamlines of the flow and the drift coordinates around the airfoil we make use of the conformal mapping given in Table 2.1 so that

$$z=\zeta+a^2/4\zeta$$

and evaluate $\zeta$ from Eq. (2.7.32) for a blade of finite thickness and no circulation, so

$$(\zeta-\zeta_1)e^{-i\alpha}=\frac{W+\sqrt{W-2R}\sqrt{W+2R}}{2}, \qquad R=|\zeta_1-C|$$

and the branch cuts are chosen, so the real part of each square root is positive. The streamlines for the flow over an airfoil with $\zeta_1=-0.3C$ and $C=a/2$ are shown in Fig. 7.11A.

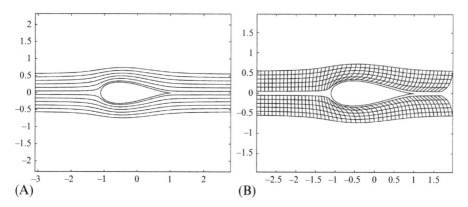

**Fig. 7.11** (A) The streamlines over a Joukowski airfoil with a thickness to chord ratio of 0.3. (B) The drift coordinates for the same airfoil.

The vortex will be convected by the mean flow, and if its circulation is sufficiently small that nonlinear convection effects can be ignored then the position of the vortex at any time will be determined by the drift coordinates of the mean flow. The drift coordinate $X_2 = \psi/U_\infty$ is simply the stream function of the mean flow normalized on the velocity, and the surfaces of constant drift along the streamline can be evaluated from the integral of the velocity along the streamline as

$$z = \int_{-t_o}^{X_1/U_\infty} \left(w'_o(z)\right)^* dt$$

where $w_o$ is given by Eq. (2.7.33) evaluated with $w_o = w$, $\alpha = 0$, and $\Gamma = 0$ for a blade at zero angle of attack.

The drift coordinates are shown in Fig. 7.11B and illustrate how the vortex that passes close to the airfoil will be retarded in comparison to the vortex that passes well above the airfoil. If the vortex is on the stagnation streamline then it comes to rest at the leading edge stagnation point and so is not convected over the surface unless it is ejected into the flow by a nonlinear interaction with the surface, caused by the image vortex required to match the non-penetration boundary condition. In a turbulent flow this ejection could cause the vortex to pass over either the upper or lower surface, so the net contribution from vorticity on the stagnation streamline is indeterminate.

Since the blade is symmetrical in this example, and the vortex is modeled as following a streamline, the unsteady loading caused by the passage of the vortex can be obtained from the complex velocity of the mean flow, as specified in Eq. (7.6.3). The terms required in this equation are obtained from Eq. (2.7.33) as

$$\frac{dw_o}{dz} = U_\infty \left\{ 1 - \frac{R^2}{(\zeta - \zeta_1)^2} \right\} \frac{1}{\left(1 - a^2/4\zeta^2\right)}$$

$$\frac{dw_2}{dz} = -i \left\{ 1 + \frac{R^2}{(\zeta - \zeta_1)^2} \right\} \frac{1}{\left(1 - a^2/4\zeta^2\right)}$$

Using these results in Eq. (7.6.3) then gives the unsteady loading for a vortex located at a point $\zeta$ on the streamline where

$$\zeta = \frac{z + \sqrt{z-a}\sqrt{z+a}}{2}$$

and as before the branch cuts are chosen so that square roots have a positive real part.

Fig. 7.12 shows the unsteady loading for a vortex passing at different distances from the blade shown in Fig. 7.11. The effect of the blade thickness on the leading edge pulse when compared to the results presented in Fig. 7.6 is significant. The sharpness of the leading edge pulse is smoothed, but the effect of thickness on the trailing edge pulse is relatively small. It is also noteworthy that when the vortex is on a streamline that is close to the airfoil the pulse is independent of the displacement of the vortex, indicating that for thick airfoils the proximity of the vortex to the stagnation streamline is unimportant.

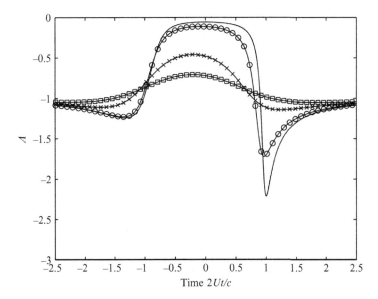

**Fig. 7.12** The nondimensional unsteady loading as a function of nondimensional time $2U_\infty t/c$ for a vortex passing a blade with thickness to chord ratio of 0.3. Displacement of the vortex from the stagnation streamline at upstream infinity: *(solid line)* $h/a = 0.05$, *(circles)* $h/a = 0.1$, *(crosses)* $h/a = 0.5$, and *(squares)* $h/a = 1$.

This example is for a very thick airfoil, and the effects of thickness for thin airfoils are better shown by considering a vortex at a fixed starting point being convected past an airfoil of different thicknesses, as shown in Fig. 7.13. The three cases shown are for an airfoil with thickness to chord ratios of 1.2%, 12%, and 30%, respectively. As before the effect of thickness is to reduce the leading edge pulse and has a much smaller effect on the trailing edge pulse. We also note that the effect of thickness is nonlinear and that for the thick blade (as in Fig. 7.12) the leading edge pulse is significantly smoothed.

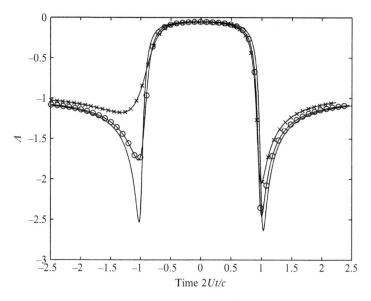

**Fig. 7.13** The nondimensional unsteady loading as a function of nondimensional time $2U_\infty t/c$ for a vortex passing a blade with thickness to chord ratios of 0.012 *(solid line)*, 0.12 *(circles)*, and 0.3 *(crosses)*. Displacement of the vortex from the stagnation streamline at upstream infinity is $2h/c = 0.05$.

Clearly the effect of thickness on the unsteady loading is to smooth out the leading edge pulse that occurs as the vortex interacts with the leading edge of the blade. It has less effect on the trailing edge pulse, but that is inconsequential because in these examples the Kutta condition has not been introduced. The smoothing of the leading edge pulse implies that the high-frequency content of the blade response function will be reduced by thickness effects. The effect of the blade thickness appears as if the vortex passes further from the blade than it would if the blade was a flat plate. This scales on the blade thickness to chord ratio, and Fig. 7.14 shows a comparison between the unsteady loading from airfoils of thicknesses of 1.2%, 6%, and 12% with a vortex initiated at the same point upstream and is compared to the unsteady loading calculated using a flat plate approximation with the vortex displacement increased by half the

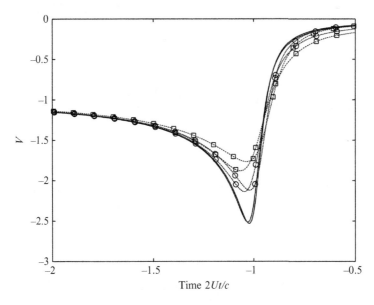

**Fig. 7.14** The nondimensional unsteady loading as a function of nondimensional time $2U_\infty t/c$ for a vortex passing a blade with thickness to chord ratios of 0.012 *(solid line)*, 0.06 *(circles)*, and 0.12 *(squares)*. The two curves show a calculation based on the blade of finite thickness and a flat plate with the vortex displaced by an additional half the blade thickness. Displacement of the vortex from the stagnation streamline at upstream infinity is $2h/c = 0.05$.

thickness of the blade. The curves are remarkably close, and this provides a first-order scaling on the effect of blade thickness on the unsteady loading.

We can also estimate the effect of blade thickness on the blade response function in the frequency domain by modifying Eq. (7.5.6) so that the displacement of the vortex $h$ is increased by half the thickness of the blade. This will reduce the high-frequency content of the blade response function by a factor of $\exp(-\omega t_{max}/2U_\infty)$, where $t_{max}$ is the thickness of the blade, and so the high-frequency content of the blade response function is reduced by an exponential factor that depends on the blade thickness. However, it should be noted that this analysis is for incompressible flow and assumes that the blade and wake respond instantaneously to an incoming disturbance. We will show in Chapter 14 that this approximation is only valid when the blade chord is less than one-quarter of the acoustic wavelength. The discussion outlined above shows that blade thickness is only important when $\omega t_{max}/U_\infty > 1$. The incompressibility criteria requires that $\omega < \pi c_o/2c$ and so for the effect of blade thickness to be modeled by an incompressible model we require $\pi t_{max}/2cM > 1$ or that the flow Mach number is less than 0.19 for a 12% thick blade. In most aeroacoustic applications the blade Mach number exceeds this criterion and so compressibility effects have to be considered.

# References

[1] A. Powell, Theory of vortex sound, J. Acoust. Soc. Am. 36 (1964) 177–195.

[2] M.S. Howe, Theory of Vortex Sound, Cambridge University Press, Cambridge, 2003.

[3] S. Goldstein (Ed.), Modern Developments in Fluid Dynamics, vol. 2, Dover, New York, 1965.

[4] G.K. Batchelor, An Introduction to Fluid Dynamics, Cambridge University Press, Cambridge, 1967.

[5] W.K. Blake, Mechanics of Flow Induced Sound and Vibration, Academic Press, Orlando, FL, 1986.

[6] J.G. Leishman, Principles of Helicopter Aerodynamics, Cambridge University Press, Cambridge, 2000.

# Turbulence and stochastic processes

# 8

This chapter discusses the stochastic nature of turbulence and the sound field produced by the interaction of a generic surface with a turbulent flow. The generation of turbulence at the largest scales, the energy cascade and dissipation are described, and the Reynolds averaged Navier Stokes (RANS) equations are introduced including the concept of a turbulence model. Computational methods based on these concepts are discussed. The chapter also introduces the necessary descriptors of turbulence for aeroacoustic analysis such as correlation functions integral scales frequency spectra, cross spectra cross correlations coherence phase wavenumber spectra in multiple dimensions in homogeneous and inhomogeneous flows.

## 8.1 The nature of turbulence

It is tempting to think of sound as an entirely deterministic phenomenon. The representation of sound fields as harmonic waves in space and time engenders an image of acoustic fields formed by the ordered propagation of entirely predictable periodic variations in density and pressure. There are many applications in aeroacoustics where this is the case, for example: the sound field generated by the thickness noise of a rotating propeller or the tones generated by a set of rotor blades as they cut through a nonuniform flow field. In many cases, however, sources of sound result from the unsteadiness of a turbulent flow and they, and the sound fields they produce, are stochastic.

Turbulence is chaotic, vortical motion found in the rotational regions of flows, such as boundary layers, wakes, and jets, where viscosity has influenced the motion. Turbulence in low Mach number flows is usually considered incompressible. Even though the mean motion may be compressible, the turbulent fluctuations in velocity are rarely more than 10–20% of the mean and thus do not have a significant Mach number.

Turbulence is initiated and maintained by viscous instability and characterized by eddying motions over a large and continuous range of scales. Sound produced by the action of turbulent flow therefore tends to be *broadband*, with energy over a continuous distribution of frequencies. On the largest scales the turbulent eddies, often also known as *coherent structures*, are formed by the instability and roll up of shear-layers associated with the large scale geometry of the flow, such as the von Kármán vortex street formed behind a circular cylinder (see Section 7.4). The scale of these largest structures $L$ is comparable to the overall dimension of the flow (e.g., the wake thickness), and the scale of their velocity fluctuations, $u$, varies directly with the overall velocity scale of the flow $U$.

*Aeroacoustics of Low Mach Number Flows.* http://dx.doi.org/10.1016/B978-0-12-809651-2.00008-4

The shearing motions associated with these largest structures are themselves unstable, however, and so they break down into smaller structures that are themselves subject to instability. This process continues until the structures formed are small enough to be directly slowed by the molecular viscosity of the fluid. Such a structure with a size $\eta$ and velocity $u_\eta$ experiences a viscous force that scales on $\mu u_\eta/\eta$ (from Newton's Law of Viscosity) multiplied by the eddy surface area that varies as $\eta^2$. It experiences a loss of momentum at a rate equal to its mass times the rate of change of its velocity, which will scale on $(\rho\eta^3)(u_\eta/\tau_\eta) = \rho u_\eta^2\eta^2$, where we have calculated the timescale for the deceleration $\tau_\eta$ as $\eta/u_\eta$. Given that the viscous force and rate of change of momentum must be in balance, we expect that

$$\rho u_\eta^2\eta^2 \cong (\mu u_\eta/\eta)(\eta^2)$$

and thus,

$$\eta u_\eta/\nu \cong 1$$

where $\nu$ the kinematic viscosity $\nu = \mu/\rho$. We see that the Reynolds number of the smallest scales in turbulence is of order 1. The scales $\eta, u_\eta,$ and $\tau_\eta$ are referred to as the Kolmogorov microscales after the great 20th century Russian mathematician Andrey Kolmogorov. These are most commonly expressed in terms of the rate of viscous dissipation of kinetic energy per unit mass $\varepsilon = u_\eta^2/\tau_\eta$. Since $\tau_\eta = \eta/u_\eta$ and $\eta = v/u_\eta$ we obtain

$$u_\eta = (\nu\varepsilon)^{1/4}, \ \eta = (\nu^3/\varepsilon)^{1/4}, \text{ and } \tau_\eta = (\nu/\varepsilon)^{1/2} \tag{8.1.1}$$

The flow of energy from large to small scales is referred to as the "energy cascade." Kinetic energy enters the cascade at the largest scales, at a rate that is determined by the largest motions based on the time scale $L/U$, i.e., at a rate, per unit mass, that scales with $U^2/(L/U) = U^3/L$. All this energy must be dissipated by viscosity at the smallest scales and so, counter-intuitively, this rate $\varepsilon \sim U^3/L$ is independent of viscosity. The ratio of the largest to smallest scales in the cascade is

$$L/\eta = L/(\nu^3/\varepsilon)^{1/4} \sim L\left/\left(\frac{\nu^3}{U^3/L}\right)^{1/4}\right. = \left(\frac{UL}{\nu}\right)^{3/4} = \text{Re}^{3/4} \tag{8.1.2}$$

where we have denoted the overall Reynolds number of the flow $UL/\nu$ as Re. We see, therefore, that the statement that turbulence is associated with a large range of scales is synonymous with the statement that turbulence is a high Reynolds number phenomenon. The ratio of largest to smallest velocity and time scales in the energy cascade is similarly shown to vary as $\text{Re}^{1/4}$ and $\text{Re}^{1/2}$ respectively.

Kolmogorov hypothesized that at sufficiently high Reynolds numbers, the smallest eddies in a turbulent flow depend only on viscosity and dissipation rate and thus are isotropic and universal between different flows. This is often referred to as the *dissipation range*. He also hypothesized universality in the mid-range scales, much larger than $\eta$ but much smaller than $L$, which should therefore be determined by the rate of energy flow through the cascade, which is equal to $\varepsilon$. This is referred to as the *inertial subrange*. The validity of Kolmogorov's hypotheses, and the extent to which turbulent flows exhibit a universal character, remain open questions in turbulence research.

Turbulence is a stochastic phenomenon. While it is completely described by the governing equations of fluid dynamics, the instantaneous details of flow are sensitive to its history and boundary conditions in such a complex and chaotic way that deterministic predictions are simply not possible. Fortunately, the instantaneous details are rarely important since, as engineers, we care about typical behavior. We wish to know the average behavior of the turbulence, and of the sound field with which it is associated.

## 8.2 Averaging and the expected value

In many aeroacoustic and fluid dynamic situations we find it helpful to take the mean value. For example consider an instantaneous acoustic or fluid dynamic variable, or a combination of variables, $a(\mathbf{y},t)$. We can, and often do, get the mean value of $a$ by averaging with respect to time

$$\bar{a}(\mathbf{y}) = \frac{1}{2T} \int_{-T}^{T} a(\mathbf{y}, t) dt$$

where we assume an averaging time $2T$ long enough compared to the flow processes that the result is independent of when the averaging period occurs or how long it is. Time averaging is intuitively simple but not appropriate when the typical behavior is a function of time itself, i.e., when the flow is not *time stationary*.

To understand this, consider the situation illustrated in Fig. 8.1 where an airfoil cuts through a turbulent wake. As this occurs, the airfoil encounters turbulence that produces an unsteady lift fluctuation and a burst of sound. The sound, the pressure fluctuations experienced on the airfoil and the eddy structures it encounters clearly will have typical characteristics, but these characteristics will depend on the position of the airfoil relative to the wake, which is itself a function of time.

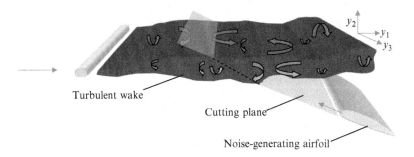

**Fig. 8.1** Airfoil cutting through a turbulent wake.

To handle this kind of situation we use the more general concept of the expected value. This is the mean of the value of a stochastic variable taken over many repeated realizations of the same flow. In each realization we imagine running the flow under conditions identical in all respects, except for the stochastic behavior. Thus we can imagine obtaining multiple independent samples of our flow quantity $a$ at the same defined position and time; e.g., for the same position of the airfoil relative to the wake. These independent samples can now be averaged to obtain a mean that remains dependent on time.

$$E[a(\mathbf{y},t)] = \lim_{N \to \infty} \frac{1}{N} \sum_{n=1}^{N} a^{(n)}(\mathbf{y},t)$$

where $a^{(n)}(\mathbf{y},t)$ is the $n$th sample of $a$ and $N$ is the total number of samples taken.

As already noted we apply averaging to many variables and combinations of variables. As we will see below, averaging of products of fluctuating velocity components at the same position and time, at different positions and times, of pressure fluctuations and of the Fourier transforms of these variables appear frequently in the analysis and results of aeroacoustic problems. When the terms "average" or "mean" are used, they will refer to the expected value, unless otherwise stated. In general we will denote averaged values using an overbar, e.g., $\overline{p'^2}$, unless a custom symbol has already been defined, such as in the case of mean velocity component $U_i$, or mean density $\rho_o$.

The particular types of statistical measures of turbulence that are most often required for aeroacoustic predictions become clear when the mechanisms of sound production are considered. Returning once more to the situation shown in Fig. 8.1, according to Curle's equation (4.3.9) for an impermeable stationary surface, the instantaneous acoustic pressure $p'(\mathbf{x}, t) = \rho'(\mathbf{x}, t) c_\infty^2$ depends on the time and area integral of the pressure over the airfoil surface, and a volume integral of quadrupole sources that is of second order. So, the first order approximation to the sound field is given by

$$p'(\mathbf{x},t) = \int_{-T}^{T} \int_{S} p'(\mathbf{y},\tau) \frac{\partial}{\partial y_i} (G(\mathbf{x},t|\mathbf{y},\tau)) n_i(\mathbf{y}) dS(\mathbf{y}) d\tau \qquad (8.2.1)$$

(note that the mean pressure produces no sound in this case, and viscous effects have been ignored so we can replace $p_{ij}$ with $p'$ in the integrand where $p' = p - p_\infty$). To characterize the typical character of the sound field, we typically measure in the mean square acoustic pressure $\overline{p'^2}(\mathbf{x}, t) = E\left[p'(\mathbf{x}, t)^2\right]$, for example, which is

$$\overline{p'^2}(\mathbf{x},t) = E\left[\left(\int_{-T}^{T} \int_{S} p'(\mathbf{y},\tau) \frac{\partial G(\mathbf{x},t|\mathbf{y},\tau)}{\partial n} dS(\mathbf{y}) d\tau\right)\left(\int_{-T}^{T} \int_{S} p'(\mathbf{y}',\tau') \frac{\partial G(\mathbf{x},t|\mathbf{y}',\tau')}{\partial n} dS(\mathbf{y}') d\tau'\right)\right]$$

$$= \int_{-T}^{T} \int_{S} \int_{-T}^{T} \int_{S} E[p'(\mathbf{y},\tau) p'(\mathbf{y}',\tau')] \frac{\partial G(\mathbf{x},t|\mathbf{y},\tau)}{\partial n} \frac{\partial G(\mathbf{x},t|\mathbf{y}',\tau')}{\partial n} dS(\mathbf{y}) dS(\mathbf{y}') d\tau d\tau'$$

where we have used $\partial/\partial n$ to represent the gradient in the direction normal to the surface. The expected value operator ends up containing only the multiplication of the pressures since pressure is the only stochastic variable on the right hand side. The sound is therefore a function of the correlation (i.e., the average of the product) of the pressure fluctuation at $\mathbf{y}$ and $\tau$, and that at $\mathbf{y}'$ and $\tau'$.

The important general point here is that average measures of the sound field will in general be functions of the two-point space-time correlations of the turbulent flow variables, and that simple averaging of a flow variable at a fixed position does not give all the information required to calculate the far field sound.

## 8.3 Averaging of the governing equations and computational approaches

Averaging plays a key role in how we approach the numerical solution to turbulent flow problems. At low Mach number a turbulent flow is, in principle, prescribed by the continuity and momentum equations

$$\frac{\partial \rho}{\partial t} + \frac{\partial(\rho v_i)}{\partial x_i} = 0 \tag{2.2.6}$$

$$\frac{\partial \rho v_i}{\partial t} + \frac{\partial(\rho v_i v_j + p_{ij})}{\partial x_j} = 0 \tag{2.3.9}$$

Numerical solution to these equations for turbulent flow is possible only in very simple situations and at Reynolds numbers that are low compared to most applications. These direct Navier Stokes (DNS) solutions are primarily useful for scientific research into the fundamental physics of a flow. Engineering solutions are not possible by this method because of the inherently large range of scales present in turbulent flows at practical conditions. At an overall Reynolds number Re of one million (which is typical as it applies to a blade of chord 20 cm in a flow of 75 m/s in air), the analysis of Section 8.1 tells us that we will need to resolve a range of length scales of about 30,000:1 ($\mathrm{Re}^{3/4}$) and a range of timescales of about 1000:1 ($\mathrm{Re}^{1/2}$). Numbers like these imply vast computational memory and speed requirements.

To compute turbulent flows at practical conditions and Reynolds numbers it is thus necessary to average the continuity and momentum equations. Splitting the dependent flow variables $\rho$, $v_i$, and $p_{ij}$ into their expected values and fluctuations about the mean, and taking the expected value of Eqs. (2.2.6), (2.3.9) we get, in turn

$$\frac{\partial \rho_o}{\partial t} + \frac{\partial\left(\rho_o U_i + \overline{\rho' u_i}\right)}{\partial x_i} = 0$$

$$\frac{\partial \rho_o U_i + \overline{\rho' u_i}}{\partial t} + \frac{\partial\left(\rho_o U_i U_j + \rho_o \overline{u_i u_j} + \overline{\rho'(U_i + u_i)(U_j + u_j)} + \overline{p_{ij}}\right)}{\partial x_j} = 0$$

Note that the $\rho'$ and $p'$ here denote fluctuations about the mean (expected) value, rather than changes from ambient conditions. At low Mach number, terms multiplied by $\rho'$ can be neglected since $\rho' \ll \rho_o$, and so we obtain

$$\frac{\partial \rho_o}{\partial t} + \frac{\partial (\rho_o U_i)}{\partial x_i} = 0 \tag{8.3.1}$$

and

$$\frac{\partial \rho_o U_i}{\partial t} + \frac{\partial \left( \rho_o U_i U_j + \rho_o \overline{u_i u_j} + \overline{p_{ij}} \right)}{\partial x_j} = 0 \tag{8.3.2}$$

We see that averaging of continuity equation merely returns the same expression in terms of the mean flow variables and, since the averaging process represents an expected value, the mean flow quantities can vary with time and so the time derivatives are not zero. Averaging of the momentum equation also returns a similar expression in terms of the mean variables, but with the additional terms $\rho_o \overline{u_i u_j}$ involving the velocity fluctuations. These terms (or, strictly speaking their negatives) are referred to as the *Reynolds* or *turbulent stresses*, since they appear in the equation in the same way as the mean compressive stress tensor $\overline{p_{ij}}$.

When attempting to solve Eqs. (8.3.1), (8.3.2) we can specify the mean compressive stress tensor $\overline{p_{ij}}$ from the sum of the mean pressure and the mean viscous stresses, which can be inferred from the mean velocity gradients and the viscosity. Also we can relate the pressure and density through the energy relations described in Chapter 2 (or for incompressible flows the density may be assumed constant). In the absence of the turbulent stresses there would then be four equations with four unknowns $p$, $U_1$, $U_2$, and $U_3$, and a solution is possible. However, when the turbulent stresses are included there is an additional set of six unknown terms,

$$\overline{u_1^2}, \overline{u_2^2}, \overline{u_3^2}, \overline{u_1 u_2}, \overline{u_1 u_3}, \overline{u_2 u_3}$$

and thus Eqs. (8.3.1), (8.3.2) do not form a closed set. Solving them requires that we introduce empirical expressions to relate the Reynolds stresses back to the mean flow variables. These relations are referred to as a "turbulence model." The first and most straightforward turbulence modeling concept is the Boussinesq eddy viscosity $\mu_t$, remarkably proposed in 1877 [1]. Boussinesq's hypothesis is that the turbulent stresses are related to the mean velocity gradients in almost the same way that the viscous stresses are related to the complete velocity gradients. That is, by near analogy with Eq. (2.3.11) (for incompressible flow), we write

$$-\rho_o \overline{u_i u_j} = \mu_t \left( \frac{\partial U_i}{\partial x_j} + \frac{\partial U_j}{\partial x_i} \right) - \frac{2}{3} \rho_o \kappa_e \delta_{ij} \tag{8.3.3}$$

where $\kappa_e \equiv \frac{1}{2} \overline{u_i^2} = \frac{1}{2} \left( \overline{u_1^2} + \overline{u_2^2} + \overline{u_3^2} \right)$ is the turbulence kinetic energy. The eddy viscosity is considered a property of the flow, rather than the fluid, and thus is a variable that must be modeled. One classic model is the "k-epsilon" model [2] for which $\mu_t$ is taken to be proportional to the mean scales $L$ and $U$ so $\mu_t \propto \rho_o \kappa_e L/U$ and since $\varepsilon$ scales as $U^3/L$ and $\kappa_e$ scales as $U^2$ it follows that $L/U$ scales as $\kappa_e/\varepsilon$ and $\mu_t$ scales as $\rho_o \kappa_e^2/\varepsilon$. The turbulence kinetic energy $\kappa_e$ and dissipation rate $\varepsilon$ are obtained from empirical differential equations designed to model the factors controlling the changes to a patch of turbulence as it is convected along by the mean flow. Much experimental, theoretical, and computational effort has gone into identifying and refining turbulence models, and there are many to choose from. All such models are, however, ultimately limited in application and accuracy since the models are not themselves solutions to the equations of motion.

In 1895 Osborne Reynolds [3] derived (in incompressible form) the time averaged version of Eqs. (8.3.1), (8.3.2),

$$\frac{\partial(\rho_o U_i)}{\partial x_i} = 0 \qquad \frac{\partial\left(\rho_o U_i U_j + \rho_o \overline{u_i u_j} + \overline{p_{ij}}\right)}{\partial x_j} = 0 \tag{8.3.4}$$

which are therefore referred to as the Reynolds averaged Navier Stokes (RANS) equations. Computational approaches that solve these equations for steady boundary conditions are referred to as RANS calculations. When the boundary conditions are unsteady (e.g., for the turbulent flow produced by an oscillating airfoil), Eq. (8.3.2) must be solved and the calculation is described as URANS. RANS and URANS calculations are feasible and regularly conducted for complete engineering configurations at full-scale conditions. Indeed, a number of commercial packages are available that perform such computations. The drawback of these methods is that a generic turbulence model is being asked to represent all the scales of the turbulence, including the largest scales that are expected to be characteristic of the specific flow conditions and geometry. The accuracy and reliability of such predictions can often be a concern. A further drawback from the aeroacoustic perspective is that the turbulence quantities computed as part of the solution are generally single-point statistics of velocity like the Reynolds stresses, $\kappa_e$, and $\varepsilon$, well short of the two-point quantities of velocity and pressure that are needed to completely define an acoustic source. Additional sweeping modeling assumptions are therefore normally needed to extrapolate RANS and URANS results to obtain the acoustic sources.

Time varying unsteady flows are more accurately computed using large eddy simulation (LES). Here the equations of motion, (2.2.6) and (2.3.9), are filtered to remove turbulence scales too small to be resolved by the computational grid or time step. Since the filtering is just another type of averaging, the result is equations that appear identical to Eqs. (8.3.1), (8.3.2) but where the averaged variables $\rho_o, U_i$ and $\overline{p_{ij}}$ now include the resolved turbulent fluctuations, and where the term $\rho_o \overline{u_i u_j}$ represents the unknown correlations between the small scale unresolved fluctuations. A turbulence model is required for these "subgrid" stresses. The LES approach was proposed by Smagorinsky [4] who also introduced the first subgrid turbulence model, which is an extension of Boussinesq's eddy viscosity hypothesis. More sophisticated modeling approaches have since been proposed including the "dynamic model" of Germano et al. [5] in which the parameters of the model are adjusted to match the local characteristics of the resolved turbulence. LES turbulence models are expected to be more generally applicable because the larger configuration-specific turbulence scales are computed directly. The model therefore only needs to account for the small scale turbulence that is hopefully more universal in character.

The grid resolution, and thus filter size, is an important factor in performing an LES calculation. Choosing a coarse filter and grid means a faster, smaller calculation but one that places greater reliance on the accuracy of the empirical turbulence model. Choosing a sufficiently fine filter can result in calculations that approach the fidelity of DNS. The need for both computational efficiency and accuracy has resulted in the development of hybrid RANS/URANS/LES methods where computational effort can be concentrated in those regions where it is most needed.

Better resolved LES calculations can be particularly useful for aeroacoustics because some fraction of the pressure and velocity correlations that form the acoustic source terms can be computed directly from the resolved scales. This fraction obviously increases as the resolution and expense of the calculation are increased. Integration of these sources over a surface bounding the flow and using a Ffowcs Williams Hawkings surface, can then be used to determine the sound heard by an observer in the far field, within the limitations discussed in Chapter 5.

Note that in principle, the origin and propagation of sound waves can be directly computed as part of any unsteady compressible turbulent flow simulation. However, this is rarely done for low Mach number flows because of the computational challenges posed by the large disparity in the scales and fluctuation levels of sound waves and of the turbulence producing them.

## 8.4  Descriptions of turbulence for aeroacoustic analysis

### 8.4.1  Time correlations and frequency spectra of a single variable

Consider the situation shown in Fig. 8.2 where noise is being radiated by the continuous passage of turbulence over the leading edge of an airfoil. Both, the time variations of the flow properties used to describe the turbulent source (whether they be velocity or pressure) and the sound waves that are heard by a far field observer are stochastic and time stationary. It is important that we have a quantitative measure of the typical frequency content of these signals because we need to assess the impact of the sound on a human listener, and because we are interested in separating out the contributions of different turbulence scales to the acoustic source. In Chapter 3 we introduced the Fourier transform as a way to extract the frequency components of a specific waveform. Applied to a generic flow or acoustic quantity $a(t)$ that varies with time and has zero mean, this definition is

$$\tilde{a}(\omega) = \frac{1}{2\pi} \int_{-T}^{T} a(t) e^{i\omega t} dt \qquad (8.4.1)$$

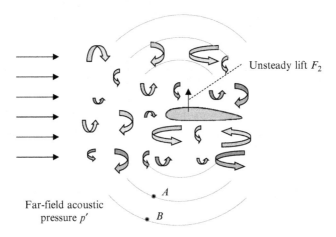

Far-field acoustic
pressure $p'$

Unsteady lift $F_2$

**Fig. 8.2** An airfoil in turbulence producing leading edge noise.

The Fourier transform $\tilde{a}\,(\omega)$ is not by itself a very useful measure since, just like $a(t)$, it will vary stochastically. The appropriate average measure of the frequency content is given by the autospectral density of $a$ (often referred to as the autospectrum, the power spectrum, or just the spectrum) defined as

$$S_{aa}(\omega) \equiv \frac{1}{2\pi}\int_{-T}^{T} R_{aa}(\tau)e^{i\omega\tau}d\tau \tag{8.4.2}$$

where $R_{aa}(\tau)$ is the time delay autocorrelation function, defined as

$$R_{aa}(\tau) = E[a(t)a(t+\tau)] \tag{8.4.3}$$

As can be seen from this expression $R_{aa}(\tau)$ is the average of the signal multiplied by itself at a later time. For a time stationary signal the expected value of $a(t)a(t+\tau)$ will not depend on $t$. The inverse Fourier transform relates the spectrum back to $R_{aa}(\tau)$.

$$R_{aa}(\tau) = \int_{-\infty}^{\infty} S_{aa}(\omega)e^{-i\omega\tau}d\omega \tag{8.4.4}$$

The definition (8.4.3) tells us that the autocorrelation function is even, i.e., it is symmetric about $\tau = 0$, because

$$E[a(t)a(t+\tau)] = E[a(t-\tau)a(t)] = E[a(t)a(t-\tau)]$$

This means that the spectrum is a real and even function of frequency because Eq. (8.4.2) can be expanded as

$$S_{aa}(\omega) = \frac{1}{2\pi}\int_{-T}^{T} R_{aa}(\tau)\cos(\omega\tau)d\tau + \frac{i}{2\pi}\int_{-T}^{T} R_{aa}(\tau)\sin(\omega\tau)d\tau$$

The imaginary term is zero because $R_{aa}(\tau)$ is even and $\sin(\omega\tau)$ is an odd function. It also follows that $S_{aa}(\omega)$ is an even function of frequency because $\cos(\omega\tau)$ is an even function of frequency.

To illustrate why the concepts of the spectrum and correlation are useful physically, consider the example shown in Fig. 8.3. Fig. 8.3A shows part of a velocity signal $u_1(t)$ measured in boundary layer turbulence near a wall where the mean flow velocity is close to 20 m/s. The signal appears quite random but, as we will see, does have a definite statistical character that reveals important information about the boundary layer turbulence. In Fig. 8.3B the time delay correlation of this signal is shown. The expected value required in Eq. (8.4.3) was obtained by calculating the product $u_1(t)u_1(t+\tau)$ for every time instant at which the signal was measured (about 400,000 times in this particular case) and then taking the mean value of those numbers.

The correlation function measures how similar the signal is to its time shifted copy. At zero time delay the signals are a perfect match and the correlation has its maximum value equal, by definition, to the velocity variance $\overline{u_1^2}$ (i.e., its mean square). For non-zero time delay, the correlation decays as the time shifted signal copy becomes less and less similar to the original, reaching 8% of $\overline{u_1^2}$ at about 0.02 s. The overall width of the correlation peak is a measure of the time scale of the largest turbulence and, indeed, the larger scales in the time signal of Fig. 8.3A do appear to be about 0.02 s. To precisely quantify this we introduce the concept of the integral timescale $\mathcal{T}$, defined as

$$\mathcal{T} \equiv \frac{1}{E[a^2(t)]} \int_0^\infty R_{aa}(\tau)d\tau = \int_0^\infty \rho_{aa}(\tau)d\tau \tag{8.4.5}$$

We also use this opportunity to introduce the correlation coefficient function $\rho_{aa}$, which is just $R_{aa}$ normalized on the variance, so that $\rho_{aa}(0) = 1$. In this example $\rho_{u_1u_1}(\tau)$ integrates to a time scale of 0.0064 s. The integral time scale is usually about one third of the overall half-width of the correlation peak. If we assume that all the turbulence is traveling with a mean speed of 20 m/s then this timescale can be used to estimate the streamwise lengthscale of the flow ($0.0064 \times 20 = 0.128$ m). In general, this type of time-to-space conversion is referred to as Taylor's *frozen flow* hypothesis. While it is commonly used, and probably reasonably accurate in this example, it is important to remember that it unrealistically treats the turbulence as though it didn't have velocities itself. It can therefore be misleading, particularly in flows where the turbulent velocity fluctuations are significant compared to the mean velocity.

It is clear that the correlation function also has information about smaller turbulence scales. For example, its initial rate of decay will reflect how much of the mean square velocity fluctuation is due to the smallest turbulence. Taking the Fourier transform to obtain the spectrum reveals this information in a more explicit way. Fig. 8.3C shows the spectrum of the velocity signal $S_{u_1u_1}(\omega)$ calculated according to Eq. (8.4.2). Both axes of the spectrum have been plotted on logarithmic scales to more clearly reveal the behavior in different frequency ranges. We see that the overall form of the spectrum is broken into three parts. At low frequencies, where it represents the

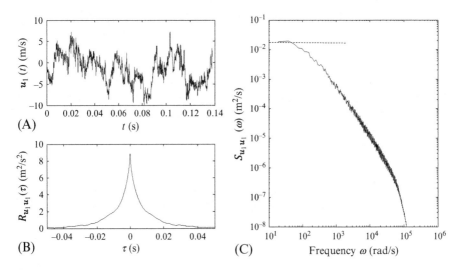

**Fig. 8.3** Velocity in a low-speed high Reynolds number turbulent boundary layer measured at a point that is 20% of the boundary layer thickness from the wall. (A) Time variation of the velocity fluctuation, (B) time delay autocorrelation function, and (C) autospectral density.

largest motions of the boundary layer, the spectrum curves downward from a plateau. The level of this plateau is characterized by the zero frequency value of the spectrum (indicated by the horizontal line in Fig. 8.3C) which, in this case gives $S_{u_1 u_1}(0) = 0.181 \, \text{m}^2/\text{s}^2/(\text{rad}/\text{s})$. Looking at Eq. (8.4.2) for zero frequency, we see that this value corresponds to the integral time scale, scaled as $T \overline{u_1^2}/\pi$ or, in general $S_{aa}(0) = T \overline{a^2}/\pi$. In the mid-frequency range we see that our example velocity spectrum becomes almost straight with a slope on the log-log scale of close to $-5/3$. As will be discussed in Chapter 9, this is indicative of an inertial subrange behavior in the mid-range scales of the boundary layer turbulence, as hypothesized by Kolmogorov. At the highest frequencies, which are assumed to be generated by the smallest eddies, the spectrum curves downward away from the $-5/3^{\text{rds}}$ slope as a result of the dissipation of these eddies by viscous action.

In general, the spectrum of a time history $a(t)$ can be physically interpreted as revealing the contributions to the mean square fluctuation $\overline{a^2}$ at each frequency. This can be demonstrated very simply using the inverse transform relationship, Eq. (8.4.4) which, for zero time delay $\tau$ becomes

$$R_{aa}(0) = \overline{a^2} = \int_{-\infty}^{\infty} S_{aa}(\omega) d\omega \qquad (8.4.6)$$

This is known as Parseval's theorem which states that the "power" in the time and frequency domains is the same, and that the spectrum divides up that power by frequency. (Note that the term "power" is loosely used in spectral analysis to refer to the mean square.) It is in this sense that $S_{aa}(\omega)$ is a spectral density function. We also see that the units of $S_{aa}(\omega)$ will be the units of $\overline{a^2}$ per radian-per-second.

One simple but critical detail here is that the mathematical definition of the spectral density includes both positive and negative frequencies. In the present context of one-dimensional spectra these mean the same thing. However, note from Eq. (8.4.6) that the energy is spread over both the positive and negative domains. We call the spectrum *double sided* and this is the norm for mathematical analysis. However, in many situations including presenting predictions or measurements of sound spectra, it is normal to consider them to exist only for positive frequencies and to double the spectral values. We then refer to the spectrum as *single sided* for which we introduce the symbol $G_{aa}(\omega)$.

The fact that $S_{aa}(\omega)$ (and by extension $G_{aa}(\omega)$) is a density function means that we expect it to integrate to a defined physical quantity. This limits the ways in which we may scale it and the frequency variable upon which it depends. For example, if we wanted to express our spectrum as a function of frequency $f$ in cycles per second or Hertz, then we would write $S_{aa}(f) = 2\pi S_{aa}(\omega)$ since we are expecting that

$$\overline{a^2} = \int_{-\infty}^{\infty} S_{aa}(f)df \tag{8.4.7}$$

and $df = d\omega/2\pi$. Another way of thinking about this is to envision $S_{aa}(f)$ as resulting in a Fourier transform defined in terms of frequency in Hertz as,

$$S_{aa}(f) = \int_{-T}^{T} R_{aa}(\tau)e^{2\pi i f \tau}d\tau$$

so comparing this with Eq. (8.4.2) we see that $S_{aa}(f) = 2\pi S_{aa}(\omega)$. A more complicated example is the scaled velocity spectrum, which we might reasonably want to normalize so that it integrates to $\overline{u_1^2}/U_\infty^2$ where $U_\infty$ is the free stream velocity. The integrand of Eq. (8.4.7) is then $S_{u_1 u_1}(\omega)/U_\infty^2$. However, at the same time it would be physically meaningful to nondimensionalize the frequency as $\sigma = \omega L/U_\infty$ where $L$ is a representative physical scale of the flow. To preserve the integral under the spectrum our spectral normalization must become $S_{u_1 u_1}(\sigma) = S_{u_1 u_1}(\omega)U_\infty/L$ so that

$$\frac{\overline{u_1^2}}{U_\infty^2} = \int_{-\infty}^{\infty} \frac{S_{u_1 u_1}(\sigma)}{U_\infty^2}d\sigma \qquad \sigma = \omega L/U_\infty$$

Some quite specific conventions exist for presenting broadband sound spectra. We present acoustic pressure spectra in decibels as using the "narrow band" sound pressure level (SPL), which for a spectrum is defined as

$$\text{SPL} = 10\log_{10}\left[G_{pp}(\sigma)/\left(p_{ref}^2/\Delta\sigma_{ref}\right)\right] \quad \text{dB re } p_{ref}/\Delta\sigma_{ref}^{1/2} \tag{8.4.8}$$

where $\Delta\sigma_{ref}$ refers to whatever frequency unit is being used for $\sigma$ (e.g., Hz, radians per second, or $U_\infty/L$) and $p_{ref}$ is the reference pressure, conventionally taken as 20 µPa for measurements in air. Note that the single sided spectrum $G_{pp}$ is used to express the

SPL. We refer to Eq. (8.4.8) as "narrow band" to distinguish it from one-third octave band SPL. As discussed in Chapter 1, the third octave spectrum divides the spectrum into frequency bands, with each band being $2^{1/3}$ times the size of the preceding band. In this case the SPL in the $n$th band is given by

$$\text{SPL}_n = 10\log_{10}\left(\frac{1}{p_{\text{ref}}^2}\int_{f_l^{(n)}}^{f_u^{(n)}} G_{pp}(f)df\right) \text{ dB re } p_{\text{ref}} \tag{8.4.9}$$

where the lower and upper frequency limits of each band are defined in terms of the mid-band frequency $f_n$ as

$$\begin{aligned}
f_l^{(n)} &= f_n/2^{1/6} \\
f_u^{(n)} &= f_n \times 2^{1/6}
\end{aligned} \tag{8.4.10}$$

The mid-band frequencies are calculated using Eq. (1.3.3). The third octave bands appear evenly spaced when plotted on a logarithmic scale, and the spectral levels represented are no longer densities but pure mean square contributions, the frequency dependence having been integrated out in Eq. (8.4.9).

To close this section, we return to the definition of the spectrum $S_{aa}(\omega)$ given in Eq. (8.4.2) and investigate the relationship between this average measure of the frequency content, and the Fourier transform of the instantaneous signal $\tilde{a}(\omega)$. Since we are restricting ourselves to a time stationary signal we can average the time delay correlation (Eq. 8.4.3) over time without changing its value, so

$$R_{aa}(\tau) = \frac{1}{2T}\int_{-T}^{T} E[a(t)a(t+\tau)]dt \tag{8.4.11}$$

Substituting this into Eq. (8.4.2) we obtain

$$\begin{aligned}
S_{aa}(\omega) &= \frac{1}{2\pi}\int_{-T}^{T}\frac{1}{2T}\int_{-T}^{T} E[a(t)a(t+\tau)]dt e^{i\omega\tau}d\tau \\
&= \frac{1}{4\pi T}E\left[\int_{-T}^{T}\int_{-T}^{T} a(t)a(t+\tau)e^{i\omega\tau}d\tau dt\right]
\end{aligned} \tag{8.4.12}$$

Now we make the substitution $t' = t+\tau$ and note that $dtd\tau = dtdt'$ to obtain

$$S_{aa}(\omega) = \frac{1}{4\pi T}E\left[\int_{-T}^{T}\int_{-T+t}^{T+t} a(t)a(t')e^{i\omega(t'-t)}dt'dt\right] \tag{8.4.13}$$

We next change integration limits from $(-T, -T+t)$ and $(T, T+t)$ to $(-T, -T)$ and $(T, T)$ as illustrated in Fig. 8.4. The expected value of the integrand $E[a(t)a(t')] = R_{aa}(t-t')$, which is only a function of $t'-t$, and is constant along diagonal lines such as the one shown. The two integration regions are shown by the solid and dashed boxes. We see that the change in limits is only valid if the expected value of the integrand decays to zero in a time $t'-t \ll T$ and that the time history itself is bounded by the limits $-T < t' < T$, as required by the definition of the Fourier transform. With this change, and rearranging Eq. (8.4.13), we obtain

$$
\begin{aligned}
S_{aa}(\omega) &= \frac{1}{4\pi T} E\left[\int_{-T}^{T} a(t)e^{-i\omega t}dt \int_{-T}^{T} a(t')e^{i\omega t'}dt'\right] \\
&= \frac{\pi}{T} E\left[\tilde{a}^*(\omega)\,\tilde{a}\,(\omega)\right]
\end{aligned}
\tag{8.4.14}
$$

where $\tilde{a}^*(\omega)$ denotes the conjugate of $\tilde{a}\,(\omega)$. An important observation here is that the definition of the Fourier transform that is used here does not give the spectrum for zero frequency and is limited to angular frequencies for which $\omega \gg \pi/T$.

Eq. (8.4.14) shows that the spectrum is also the expected value of the magnitude squared of the Fourier transform, supporting the physical interpretation given to it. The equivalence demonstrated in Eq. (8.4.14) is particularly useful when it comes to determining spectra from measured signals—as will be discussed further in Chapter 11. The fact that Eq. (8.4.14) does not apply at $\omega = 0$ means that spectra obtained using this method can only be used to estimate the integral scale by assuming that it is revealed by the asymptotic level of the spectrum as $\omega$ tends to zero.

The relationship in Eq. (8.4.14) can be cast in a somewhat different form that can be particularly useful in analysis. Specifically, consider the expected value of the right hand side but with two different frequency arguments, that is,

$$
E\left[\tilde{a}^*(\varpi)\,\tilde{a}\,(\omega)\right] = \frac{1}{(2\pi)^2}\int_{-T}^{T}\int_{-T}^{T} E[a(t)a(t')]e^{-i\varpi t}e^{i\omega t'}dtdt'
\tag{8.4.15}
$$

We make the substitution of $\tau$ for $t'$ with $\tau = t' - t$, and as before ignore the change in the limit by assuming large $T$ giving,

$$
\begin{aligned}
E\left[\tilde{a}^*(\varpi)\,\tilde{a}\,(\omega)\right] &= \frac{1}{(2\pi)^2}\int_{-T}^{T}\int_{-T}^{T} E[a(t)a(t+\tau)]e^{i\omega\tau}e^{i(\omega-\varpi)t}d\tau dt \\
&= \frac{1}{2\pi}\int_{-T}^{T} e^{i(\omega-\varpi)t}dt \frac{1}{2\pi}\int_{-T}^{T} R_{aa}(\tau)e^{i\omega\tau}d\tau
\end{aligned}
\tag{8.4.16}
$$

The first integral term on the right hand side is equal to $\delta(\varpi - \omega)$, whereas the second is simply the frequency spectrum. Our final result is therefore,

$$
E\left[\tilde{a}^*(\varpi)\,\tilde{a}\,(\omega)\right] = \delta(\varpi - \omega)S_{aa}(\omega)
\tag{8.4.17}
$$

This shows that for a stationary signal the fluctuations at each frequency are uncorrelated, which can be valuable in the analysis of some turbulent flow phenomena.

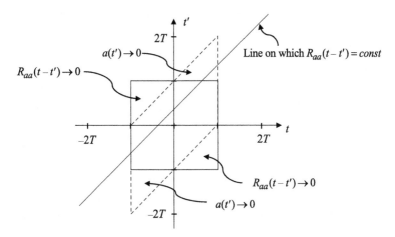

**Fig. 8.4** Integration regions in Eq. (8.4.13).

### 8.4.2 Time correlations and frequency spectra of two variables

Consider again the situation illustrated in Fig. 8.2. In addition to characterizing the typical behavior of the turbulent or acoustic variables, we are also interested in characterizing the typical relationships, for example between the sound pressure fluctuations at two points in acoustic the far field from the source. The way to do this is to define, by analogy with Eq. (8.4.3), a time delay cross correlation function

$$R_{ab}(\tau) \equiv E[a(t)b(t+\tau)] \qquad (8.4.18)$$

where $a$ and $b$ have zero mean. This measures the similarity of the stochastic variations in $a$ with those in $b$, as the time delay $\tau$ between $b$ and $a$ is varied. We are assuming that both variables are time stationary so that the cross correlation will only be a function of $\tau$. For zero time delay $R_{ab}(0)$ is equal to the covariance $\overline{ab}$. We define the cross correlation coefficient as

$$\rho_{ab}(\tau) \equiv \frac{R_{ab}(\tau)}{\sqrt{\overline{a^2}\,\overline{b^2}}} \qquad (8.4.19)$$

Defined this way $-1 \leq \rho_{ab}(\tau) \leq 1$ and the coefficient only reaches 1 or $-1$ if $a(t)$ and $b(t+\tau)$ are exact scaled copies of each other. Unlike the autocorrelation function, $R_{ab}(\tau)$ can be asymmetric and reach its maximum magnitude at a nonzero time delay. In the example of Fig. 8.2 the acoustic signal at $A$ would precede the acoustic signal at $B$ and thus $R_{AB}(\tau) = E[p_A(t)p_B(t+\tau)]$ would have a positive maximum at positive time delay $\tau$. Inverting the sign of the acoustic signal (say by shifting the point $A$ from below to above the airfoil if the sound is from a lift dipole only) would reverse the sign of $R_{AB}$ which would then have a negative peak.

Consistent with the autospectrum, we define the cross spectral density as the Fourier transform of the correlation function

$$S_{ab}(\omega) \equiv \frac{1}{2\pi} \int_{-T}^{T} R_{ab}(\tau) e^{i\omega\tau} d\tau \tag{8.4.20}$$

Expanding this we obtain

$$S_{ab}(\omega) = \frac{1}{2\pi} \int_{-T}^{T} R_{ab}(\tau) \cos(\omega\tau) d\tau + \frac{i}{2\pi} \int_{-T}^{T} R_{ab}(\tau) \sin(\omega\tau) d\tau$$

showing that $S_{ab}(\omega)$ will, in general, be a complex function because of the probable asymmetry of $R_{ab}(\tau)$. It is straightforward to verify that $S_{ab}$ will also be conjugate symmetric with frequency, so $S_{ab}(\omega) = S_{ab}^*(-\omega)$. The real and imaginary parts of $S_{ab}$ are referred to as the *cospectrum* $C_{ab}$ and the *quad-spectrum* $Q_{ab}$ respectively.

As one would expect, the inverse Fourier transform relates the cross spectrum back to the cross correlation

$$R_{ab}(\tau) = \int_{-\infty}^{\infty} S_{ab}(\omega) e^{-i\omega\tau} d\omega \tag{8.4.21}$$

For zero time delay we obtain an expression for the covariance in terms of the cross spectral density

$$\overline{ab} = R_{ab}(0) = \int_{-\infty}^{\infty} S_{ab}(\omega) d\omega = \int_{-\infty}^{\infty} C_{ab}(\omega) + i \cdot Q_{ab}(\omega) d\omega = \int_{-\infty}^{\infty} C_{ab}(\omega) d\omega \tag{8.4.22}$$

The conjugate symmetry of the cross spectrum ensures that quad-spectrum does not contribute to the covariance. The cospectrum may therefore be thought of as the covariance per unit frequency. Two more physically important expressions of the cross spectrum are the coherence and phase spectra, defined respectively as

$$\gamma_{ab}^2(\omega) = \frac{|S_{ab}(\omega)|^2}{S_{aa}(\omega) S_{bb}(\omega)} \tag{8.4.23}$$

and

$$\theta_{ab}(\omega) = \arctan\left(\frac{Q_{ab}(\omega)}{C_{ab}(\omega)}\right) \tag{8.4.24}$$

The coherence $\gamma_{ab}^2$ is a squared and normalized measure of the expected value of the product of the complex amplitude of the two quantities $a$ and $b$ at a given frequency. If, in multiple realizations of our airfoil example, the lift (assumed say to be a compact dipole source) and sound variations (which may include background noise) were randomly phased with respect to each other implying no linear connection, then the coherence between them would be zero. If the phasing were not random, but tended to prefer a particular value, then the coherence would be positive with a value that increases the more closely correlated the signals are at that frequency. For nonzero coherence, the phase spectrum $\theta_{ab}(\omega)$ gives the preferred value of the phase difference, and the phase is undefined for $\gamma_{ab}^2 = 0$.

The coherence is an even function of frequency and is normalized so that $0 \leq \gamma_{ab}^2 \leq 1$, a value of 1 being achieved when $b(t)$ is a constant multiple of $a(t)$. The phase is an odd function and has a straightforward relationship to the time delay between variables. In the case of the airfoil example we argued that the positive time delay between the sound fluctuations in the far field would lead to $R_{AB}(\tau)$ reaching its maximum magnitude at positive $\tau$. This situation corresponds to a positive gradient of the phase spectrum $\theta_{AB}(\omega)$ with frequency.

A review of the above material will make it clear that the auto correlation and auto spectrum functions are simply special cases of the cross correlation and cross spectrum functions with $b(t) = a(t)$. This is also true for the relationship between the cross spectral density and the stochastic Fourier transform. Using a derivation entirely analogous to that laid out in Eqs. (8.4.11)–(8.4.14), we can show that, for angular frequencies $\omega \gg \pi/T$,

$$S_{ab}(\omega) = \frac{\pi}{T} E\left[\tilde{a}^*(\omega)\, \tilde{b}(\omega)\right] \tag{8.4.25}$$

where $\tilde{a}(\omega)$ and $\tilde{b}(\omega)$ are the Fourier transforms of $a(t)$ and $b(t)$. Likewise we can show, analogously to the derivation of Eq. (8.4.17) that

$$E\left[\tilde{a}^*(\varpi)\, \tilde{b}(\omega)\right] = \delta(\varpi - \omega) S_{ab}(\omega) \tag{8.4.26}$$

Note that if one of the variables involved in the cross spectrum or correlation is a vector then these statistical measures are also vectors, e.g., the pressure velocity spectrum $S_{pu_i}$. If both are vectors then the resulting measures are tensors, e.g., the velocity correlation tensor $R_{u_i u_j}$, which is commonly abbreviated as $R_{ij}$.

### 8.4.3  Spatial correlation and the wavenumber spectrum

The correlation and spectral analysis techniques we have introduced to analyze the temporal behavior of a turbulent flow can equally well be applied to revealing its spatial structure. Consider turbulent velocity fluctuations $u_2$ as a function of position along a line $x_1$ through a turbulent flow (Fig. 8.5). The correlation between simultaneous fluctuations at two points on the line can in general be written as

$$R_{22}(x_1, x_1') = E\left[u_2(x_1) u_2(x_1')\right] \tag{8.4.27}$$

If the flow is assumed homogeneous along the line (i.e., its average properties are independent of $x_1$), then the correlation will only be a function of the separation between the two points $\Delta x_1 = x_1' - x_1$

$$R_{22}(\Delta x_1) = E\left[u_2(x_1)u_2(x_1')\right] \tag{8.4.28}$$

Homogeneity also allows us to define an *integral lengthscale* for the turbulence

$$L_{21} \equiv \frac{1}{\overline{u_2^2}} \int_0^\infty R_{22}(\Delta x_1)d\Delta x_1 \tag{8.4.29}$$

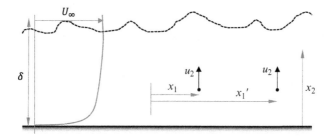

**Fig. 8.5** Velocity fluctuations in a two-dimensional turbulent boundary layer. Coordinate $x_3$ is out of the paper.

where $L_{ij}$ is defined as the integral lengthscale of velocity component $u_i$ in the direction $x_j$. We can additionally Fourier transform the correlation function in the homogeneous direction to give the wavenumber spectrum

$$\phi_{22}(k_1) \equiv \frac{1}{2\pi} \int_{-R_\infty}^{R_\infty} R_{22}(\Delta x_1)e^{-ik_1\Delta x_1}d\Delta x_1 \tag{8.4.30}$$

The inverse relationship giving the correlation function in terms of the wavenumber spectrum is

$$R_{22}(\Delta x_1) = \int_{-\infty}^{\infty} \phi_{22}(k_1)e^{ik_1\Delta x_1}dk_1$$

Note that, to be consistent with the Fourier transform convention introduced in Chapter 1, the signs of the exponents in the spatial Fourier transform and its inverse are reversed compared to those used with the time transform, Eqs. (8.4.1), (8.4.2). The wavenumber spectrum and spatial correlation function reveal the averaged structure of the turbulence in ways that are entirely analogous to the frequency spectrum and time correlation functions illustrated in Fig. 8.3.

The homogeneity of the turbulent flow is important in the same way that stationarity is important in time histories. Without it, the integral scale and spectrum, as well as properties derived from the spectrum, lose much of their physical meaning and mathematical utility. The assumption of homogeneity will be good if the average properties a flow vary on a scale much larger than the scale of its turbulence. For example, Fig. 8.5 depicts part of a two-dimensional turbulent boundary layer formed under a free stream in the $x_1$ direction. The largest turbulence scales will be of the order of the boundary layer thickness $\delta$, a distance much smaller than the scale on which the boundary layer is growing. Homogeneity in $x_1$ is therefore a good assumption in this case. Since the boundary layer is two dimensional it would be a good approximation in the spanwise $x_3$ direction, but not in direction $x_2$ perpendicular to the wall.

There is much scope in the correlation and spectrum descriptors that we have not yet acknowledged. It is clear from the boundary layer example that the spatial correlation and wavenumber spectrum will be functions of $x_2$, could also be taken in the $x_3$ direction, and could be taken of any velocity component combination. So, we would have $L_{in}(x_2), R_{ij}(x_2, \Delta x_n)$, and $\phi_{ij}(x_2, k_n)$, where $n = 1, 3$. Perhaps more interestingly we can expand correlations and spectra to more than one dimension. Most generally, we could consider the five-dimensional correlation between any two points in the boundary layer which has the form

$$R_{ij}\left(\Delta x_1, x_2, x_2', \Delta x_3, \tau\right) = E\left[u_i(x_1, x_2, x_3, t)u_j\left(x_1', x_2', x_3', t'\right)\right] \tag{8.4.31}$$

and we can envision taking the Fourier transform in $\Delta x_1$ and $\Delta x_3$ to yield the two-wavenumber spectrum

$$\phi_{ij}\left(x_2, x_2', k_1, k_3, \tau\right) = \frac{1}{(2\pi)^2} \int_{-R_\infty}^{R_\infty} \int_{-R_\infty}^{R_\infty} R_{ij}\left(\Delta x_1, x_2, x_2', \Delta x_3, \tau\right) e^{-ik_1\Delta x_1 - ik_3\Delta x_3} d\Delta x_1 d\Delta x_3 \tag{8.4.32}$$

and then the Fourier transform in time to obtain the wavenumber frequency spectrum

$$\Phi_{ij}\left(x_2, x_2', k_1, k_3, \omega\right) = \frac{1}{(2\pi)^3} \int_{-T}^{T} \int_{-R_\infty}^{R_\infty} \int_{-R_\infty}^{R_\infty} R_{ij}\left(\Delta x_1, x_2, x_2', \Delta x_3, \tau\right) e^{i\omega\tau - ik_1\Delta x_1 - ik_3\Delta x_3} d\Delta x_1 d\Delta x_3 d\tau \tag{8.4.33}$$

The spectral quantities can of course always be related back to correlations through inverse Fourier transforms. So, for example, the time cross spectrum of the velocities at two points in the flow, $(x_1, x_2, x_3)$ and $(x_1', x_2', x_3')$ can be obtained from the wavenumber frequency spectrum in Eq. (8.4.33) as

$$S_{ij}\left(\Delta x_1, x_2, x_2', \Delta x_3, \omega\right) = \int_{-\infty}^{-\infty} \int_{-\infty}^{\infty} \Phi_{ij}\left(x_2, x_2', k_1, k_3, \omega\right) e^{ik_1\Delta x_1 + ik_3\Delta x_3} dk_1 dk_3 \tag{8.4.34}$$

Descriptors like those of Eqs. (8.4.31)–(8.4.33) encompass all the possible second order statistics defined by a flow and are sufficiently rich to represent the complete acoustic source term in many linear aeroacoustic problems. Obviously, in any specific problem, the number of homogeneous directions in the flow will limit the dimensionality of the wavenumber spectrum that can be defined, just as time stationarity of the flow will determine if a physically useful frequency spectrum exists. Note that the two-point correlation, as written in the right hand side of Eq. (8.4.31), can always be defined.

Acoustic sources are often cast in terms of the fluctuating velocity field of the turbulence, but may also appear in terms of the fluctuating vorticity or the pressure on a surface, for example. In our boundary layer example, the fluctuating wall-pressure field will be characterized by the wavenumber frequency spectrum

$$\Phi_{pp}(k_1, k_3, \omega) = \frac{1}{(2\pi)^3} \int_{-T}^{T} \int_{-R_\infty}^{R_\infty} \int_{-R_\infty}^{R_\infty} R_{pp}(\Delta x_1, \Delta x_3, \tau) e^{i\omega\tau - ik_1\Delta x_1 - ik_3\Delta x_3} d\Delta x_1 d\Delta x_3 d\tau$$

(8.4.35)

where

$$R_{pp}(\Delta x_1, \Delta x_3, \tau) = E\left[p(x_1, x_3, t)p(x_1', x_3', t')\right]$$

(8.4.36)

Note that relationships exactly analogous to Eqs. (8.4.14), (8.4.17) exist for the wavenumber domain. Thus, for example, we can write for the wavenumber frequency spectrum of the wall-pressure,

$$\frac{\pi}{T}E\left[\tilde{p}^*(k_1, k_3, \omega)\, \tilde{p}\,(k_1', k_3', \omega)\right] = \Phi_{pp}(k_1, k_3, \omega)\delta(k_1 - k_1')\delta(k_3 - k_3')$$

(8.4.37)

where $\tilde{p}$ denotes the wavenumber frequency transform of $p'$. Also, we can relate the one-dimensional wavenumber spectrum to the one-dimensional wavenumber transform as,

$$\phi_{pp}(k_1) = \frac{1}{4\pi R_\infty}E\left[\int_{-R_\infty}^{R_\infty} p(\Delta x_1)e^{ik_1\Delta x_1} dt \int_{-R_\infty}^{R_\infty} p(\Delta x_1')e^{-ik_1\Delta x_1'} dt'\right] = \frac{\pi}{R_\infty}E\left[\tilde{p}^*(k_1)\,\tilde{p}\,(k_1)\right]$$

(8.4.38)

for example. In addition, by repeated application of Eq. (8.4.38) or (8.4.14) multi-dimensional relationships are obtained, such as,

$$\phi_{ij}(k_1 k_3) = \frac{\pi^2}{R_\infty^2}E\left[\tilde{u}_i^*(k_1, k_3)\tilde{u}_j(k_1, k_3)\right]$$

(8.4.39)

for the planar wavenumber spectrum of the velocity fluctuations.

The combination of space and time in the correlation function and in the wavenumber frequency spectrum in general captures information about the convection or phase velocity of fluid motions as well as their scale and intensity. We can always calculate a velocity component from $\Delta x_i/\tau$ or $\omega/k_i$. Whether such a component is physically meaningful depends on the form of the correlation or spectrum at that point. Where turbulence is carried by a mean flow, much of the energy of the flow will appear concentrated around a *convective ridge*.

Fig. 8.6 shows this type of feature in the pressure correlation coefficient function $R_{pp}(\Delta x_1, 0, \tau)/\overline{p^2}$ measured underneath a high Reynolds number turbulent boundary layer. The convective ridge defines the elongated form of the correlation function. The convection velocity $U_c$ is calculated as $\Delta x_1/\tau$ at the center of the ridge. For small separations $\Delta x_1$, the correlation levels on the ridge are high because the turbulence does not evolve much over short distances. The slope of the convective ridge and $U_c$ are relatively low here in the measurements (about $0.6\,U_\infty$) because the correlation includes a substantial contribution from small turbulence scales in the near-wall region where the flow speeds are slow. At larger separations the correlation values reduce as the turbulence has more distance to evolve. Smaller eddies evolve more quickly and so the correlation increasingly represents the larger eddies of the flow that are more likely to occupy faster flowing regions further from the wall. The slope of the convective ridge and $U_c$ therefore increase for larger separations. The convection velocity of wall-pressure fluctuations in a smooth wall boundary layer is usually observed to be between 60% and 80% of the free stream velocity. The boundary layer wall-pressure spectrum and correlation function will be discussed in more detail in the next chapter.

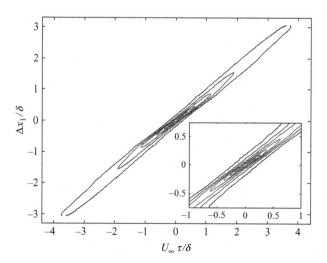

**Fig. 8.6** Streamwise space-time correlation coefficient function for the wall-pressure fluctuations $R_{pp}(\Delta x_1, 0, \tau)/\overline{p^2}$ under the boundary layer of Forest [6]. $U_\infty = 33.6$ m/s, boundary layer thickness $\delta = 231$ mm. Inset shows enlargement of region at small $\tau$ and $\Delta x_1$. Contours are in intervals of 0.05 starting at 0.05.

# References

[1] J. Boussinesq, Essai sur la theorie des eaux courantes, Memoires presentes par divers savants a l'Academie des Sciences, vol. XXIII, Impremerie Nationale, Paris, 1877, pp. 1–680.

[2] B.E. Launder, D.B. Spalding, The numerical computation of turbulent flows, Comput. Methods Appl. Mech. Eng. 3 (1974) 269–289.

[3] O. Reynolds, On the dynamical theory of incompressible viscous fluids and the determination of the criterion, Philos. Trans. R. Soc. Lond. A 186 (1895) 123–164.

[4] J. Smagorinsky, General circulation experiments with the primitive equations, Mon. Weather Rev. 91 (1963) 99–164.

[5] M. Germano, U. Piomelli, P. Moin, W.H. Cabot, A dynamic subgrid-scale eddy viscosity model, Phys. Fluids A 3 (1991) 1760–1765.

[6] J.B. Forest, The Wall Pressure Spectrum of High Reynolds Number Rough-Wall Turbulent Boundary Layers (M.S. thesis), Virginia Polytechnic Institute and State University, Blacksburg, VA, 2012.

# Turbulent flows

**9**

In Chapter 8 it was shown that predictions of sound radiated as a result of the unsteady motions of a turbulent flow require estimates of the two-point space-time correlations of those flows, or their spectral equivalents. In practice these estimates are usually obtained using models developed for homogeneous and isotropic turbulence, and by scaling or otherwise modifying those models to match the inhomogeneous and anisotropic reality of the application in question. The fundamental knowledge needed to do this, including the properties of homogeneous isotropic turbulence and of the most important turbulent shear flows for aeroacoustics at low Mach number, is the topic of this chapter.

## 9.1 Homogeneous isotropic turbulence

### 9.1.1 Mathematical description

Homogeneous isotropic turbulence is turbulence that has average properties that are both independent of position and direction. Either one considers there to be no mean flow, or the mean flow is uniform and the turbulence is viewed from a frame of reference moving with it. Homogeneous turbulence is an ideal that can be approximated by the uniform turbulence downstream of a grid, or by Direct Navier Stokes (DNS) simulations of turbulence in periodic boxes. It is the simplest kind of turbulence and therefore the most studied and understood.

Homogeneous turbulence is decaying turbulence. Its mean flow has no shear and therefore no mechanism for the generation of new eddies from instabilities in the mean flow. At its largest scales it is therefore missing the configuration-specific biases that are present in almost all other turbulent flows. However, it serves as a useful representation of the universal components of those flows. Given Kolmogorov's hypotheses we expect this representation to become increasingly accurate at smaller scales and higher Reynolds numbers.

The value of homogeneous turbulence to aeroacoustics is not just this universal character, but also the fact that it is simple enough for analytical models of the second order statistics of the turbulence to exist. Engineers often use these models to provide acoustic source terms for flows that can be approximated by homogeneous turbulence, that average in time to something that appears homogeneous, or that are clearly neither homogeneous nor isotropic but for which no better information exists.

In analyzing homogeneous turbulence, we will consider only spatial correlations and related quantities. This is because in aeroacoustic applications it is generally assumed that the decay rate of turbulence is negligible over the time taken for the turbulence to convect over noise-producing hardware and to generate sound. Thus Taylor's hypothesis can be used to infer time dependencies from the spatial description of the turbulence.

Aeroacoustics of Low Mach Number Flows. http://dx.doi.org/10.1016/B978-0-12-809651-2.00009-6

The forms of the spatial correlation function and wavenumber spectrum of the velocity fluctuations for homogeneous turbulence can be readily inferred from Eqs. (8.4.31), (8.4.32) to be

$$R_{ij}(\Delta x_1, \Delta x_2, \Delta x_3) = E\left[u_i(x_1, x_2, x_3)u_j\left(x_1', x_2', x_3'\right)\right]$$ (9.1.1)

and

$$\varphi_{ij}(k_1, k_2, k_3) = \frac{1}{(2\pi)^3} \int\limits_{-R_\infty}^{R_\infty} \int\limits_{-R_\infty}^{R_\infty} \int\limits_{-R_\infty}^{R_\infty} R_{ij}(\Delta x_1, \Delta x_2, \Delta x_3)$$
$$e^{-ik_1\Delta x_1 - ik_2\Delta x_2 - ik_3\Delta x_3} d\Delta x_1 d\Delta x_2 d\Delta x_3$$ (9.1.2)

The inverse relationship is simply given by the equivalent inverse Fourier transform

$$R_{ij}(\Delta x_1, \Delta x_2, \Delta x_3) = \int\limits_{-\infty}^{\infty} \int\limits_{-\infty}^{\infty} \int\limits_{-\infty}^{\infty} \varphi_{ij}(k_1, k_2, k_3) e^{ik_1\Delta x_1 + ik_2\Delta x_2 + ik_3\Delta x_3} dk_1 dk_2 dk_3$$

(9.1.3)

While both $R_{ij}$ and $\varphi_{ij}$ appear to be three-dimensional tensor functions they can, in fact, both be inferred from a single one-dimensional scalar function. This can be understood if we consider the velocity fluctuations at two points $A$ and $B$ in a homogeneous turbulent flow, as shown in Fig. 9.1. The points are separated by a distance $r$ and we define fluctuating velocity components $u_s$ and $u_n$ respectively parallel and perpendicular to the direction of separation. Isotropy means that the average flow properties must be invariant under rotation of the coordinate system used to describe them. The mean-square velocity in any direction must therefore be the same, and so $\overline{u_n^2} = u_s^2 \equiv \overline{u^2}$. The turbulent shear stress term $\overline{u_n u_s}$ must also be zero as any nonzero value would reverse the sign if we rotated our coordinates 180 degrees about the axis $AB$ in Fig. 9.1. Equivalently, we can argue that at any instant when $u_s$ is positive there

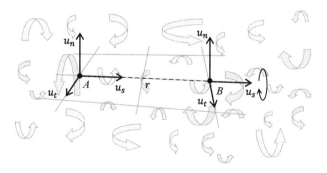

**Fig. 9.1** Velocity correlations in homogeneous turbulence.

is equal probability that $u_n$ is either positive or negative and so $E[u_s u_n] = 0$. Following the same arguments, the correlation between different components at different points $\overline{u_n(A)u_s(B)}$ must also be zero. With these restrictions the only distinct correlations we can make are $\overline{u_s(A)u_s(B)}$ and $\overline{u_n(A)u_n(B)}$, the value of the latter being independent of rotation about the axis AB and thus the same for the third velocity component $u_t$ directed out of the page.

Given that the velocity in any other direction can be composed from these components we conclude that no other independent correlations are possible. We therefore define the longitudinal and lateral correlation coefficient functions as, respectively

$$f(r) = \overline{u_s(A)u_s(B)}/\overline{u^2} \quad \text{and} \quad g(r) = \overline{u_n(A)u_n(B)}/\overline{u^2}$$

where $r = \sqrt{\Delta x_1^2 + \Delta x_2^2 + \Delta x_3^2}$, and the associated integral scales are

$$L_f = \int_0^\infty f(r)dr \quad \text{and} \quad L_g = \int_0^\infty g(r)dr \tag{9.1.4}$$

To obtain a general expression for the two-point correlation function $R_{ij}(\Delta x_1, \Delta x_2, \Delta x_3) = \overline{u_i(A)u_j(B)}$ we recognize that the Cartesian velocity components $u_i$ at $A$ and $B$ are simply a rotation of $u_s$, $u_n$, and $u_t$, as shown in Fig. 9.2, and thus

$$u_i = u_s(\mathbf{e}_s \cdot \mathbf{e}_i) + u_n(\mathbf{e}_n \cdot \mathbf{e}_i) + u_t(\mathbf{e}_t \cdot \mathbf{e}_i)$$

where $\mathbf{e}_i$ denotes the unit vector in the direction of $u_i$, and $\mathbf{e}_s$, $\mathbf{e}_n$, and $\mathbf{e}_t$ are the unit vectors defining the directions of $u_s$, $u_n$, and $u_t$. We will therefore have that

$$\overline{u_i(A)u_j(B)} = \overline{u_s(A)u_s(B)}(\mathbf{e}_s \cdot \mathbf{e}_i)(\mathbf{e}_s \cdot \mathbf{e}_j)$$
$$+\overline{u_n(A)u_n(B)}(\mathbf{e}_n \cdot \mathbf{e}_i)(\mathbf{e}_n \cdot \mathbf{e}_j) + \overline{u_t(A)u_t(B)}(\mathbf{e}_t \cdot \mathbf{e}_i)(\mathbf{e}_t \cdot \mathbf{e}_j)$$

Recognizing that $\overline{u_n(A)u_n(B)} = \overline{u_t(A)u_t(B)} = \overline{u^2}\, g(r)$ and $\overline{u_s(A)u_s(B)} = \overline{u^2} f(r)$, and that the direction cosine products can be written as

$$(\mathbf{e}_s \cdot \mathbf{e}_i)(\mathbf{e}_s \cdot \mathbf{e}_j) = \frac{\Delta x_i \Delta x_j}{r^2}$$

and

$$(\mathbf{e}_n \cdot \mathbf{e}_i)(\mathbf{e}_n \cdot \mathbf{e}_j) + (\mathbf{e}_t \cdot \mathbf{e}_i)(\mathbf{e}_t \cdot \mathbf{e}_j) = \delta_{ij} - \frac{\Delta x_i \Delta x_j}{r^2}$$

we obtain

$$R_{ij}(\Delta x_1, \Delta x_2, \Delta x_3) = \overline{u^2}\left[g(r)\delta_{ij} + \frac{f(r) - g(r)}{r^2}\Delta x_i \Delta x_j\right] \qquad (9.1.5)$$

This result is consistent with the requirement [1] that any second order isotropic tensor function must have the form $\alpha \Delta x_i \Delta x_j + \beta \delta_{ij}$ where $\alpha$ and $\beta$ are only functions of $r$. The additional requirement that $R_{ij}$ must be consistent with the incompressible continuity equation allows us to obtain a relationship between $f$ and $g$. Continuity requires that $\partial u_i / \partial x_i = 0$ and thus, from Eq. (9.1.1), $\partial R_{ij} / \partial \Delta x_i = \partial R_{1j} / \partial \Delta x_1 + \partial R_{2j} / \partial \Delta x_2 + \partial R_{3j} / \partial \Delta x_3 = 0$. Applying this to Eq. (9.1.5) (noting that $\partial \Delta x_i \Delta x_j / \partial \Delta x_i = 4\Delta x_j$) gives,

$$g = f + \frac{r}{2}\frac{\partial f}{\partial r} \qquad (9.1.6)$$

See also Batchelor [1].

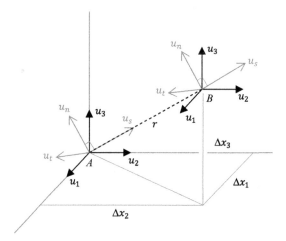

**Fig. 9.2** Coordinate systems for generalization of the two-point velocity correlation.

Substituting this expression into Eq. (9.1.4) for $L_g$ and integrating the second term by parts gives that $L_g = L_f / 2$. Also, from Eq. (9.1.5) it follows that the sum of the normal correlation function components

$$\frac{1}{2}R_{ii}(\mathbf{r}) = \frac{1}{2}[R_{11}(\mathbf{r}) + R_{22}(\mathbf{r}) + R_{33}(\mathbf{r})] = \overline{u^2}[g(r) + f(r)/2]$$

where $\mathbf{r} = (\Delta x_1, \Delta x_2, \Delta x_3)$ and $r = |\mathbf{r}|$. This function contains no reference to direction and, for $r = 0$, is equal to the turbulence kinetic energy $\kappa_e$.

To obtain a useful form of the wavenumber spectrum $\varphi_{ij}$ in terms of the one-dimensional description above, we first take the Fourier transform of $\frac{1}{2}R_{ii}(\mathbf{r})$ using Eq. (9.1.2).

$$\frac{1}{2}\varphi_{ii}(k_1,k_2,k_3) = \frac{1}{(2\pi)^3}\int\limits_{-R_\infty}^{R_\infty}\int\limits_{-R_\infty}^{R_\infty}\int\limits_{-R_\infty}^{R_\infty}\frac{1}{2}R_{ii}(\mathbf{r})e^{-i\mathbf{k}\cdot\mathbf{r}}d\Delta x_1 d\Delta x_2 d\Delta x_3$$

$$= \frac{\overline{u^2}}{(2\pi)^3}\int\limits_{-R_\infty}^{R_\infty}\int\limits_{-R_\infty}^{R_\infty}\int\limits_{-R_\infty}^{R_\infty}(g(r)+f(r)/2)e^{-ikr\cos\theta}d\Delta x_1 d\Delta x_2 d\Delta x_3$$

where $\mathbf{k}=(k_1,k_2,k_3)$ and $\theta$ is the angle between $\mathbf{r}$ and $\mathbf{k}$. The integral over volume $d\Delta x_1 d\Delta x_2 d\Delta x_3$ can instead be performed over the spherical surface defined by $r=const$ and then with respect to $r$ (Fig. 9.3), as

$$\frac{1}{2}\varphi_{ii}(k_1,k_2,k_3) = \frac{\overline{u^2}}{(2\pi)^3}\int\limits_0^{R_\infty}\int\limits_0^{2\pi}\int\limits_0^{\pi}(g(r)+f(r)/2)e^{-ikr\cos\theta}r^2\sin\theta d\theta d\varphi dr$$

$$= \frac{\overline{u^2}}{(2\pi)^2}\int\limits_0^{\infty}(g(r)+f(r)/2)\left(\int\limits_0^{\pi}\sin\theta e^{-ikr\cos\theta}d\theta\right)r^2 dr$$

$$= \frac{\overline{u^2}}{(2\pi)^2 k}\int\limits_0^{\infty}(2g(r)+f(r))\sin(kr)rdr$$

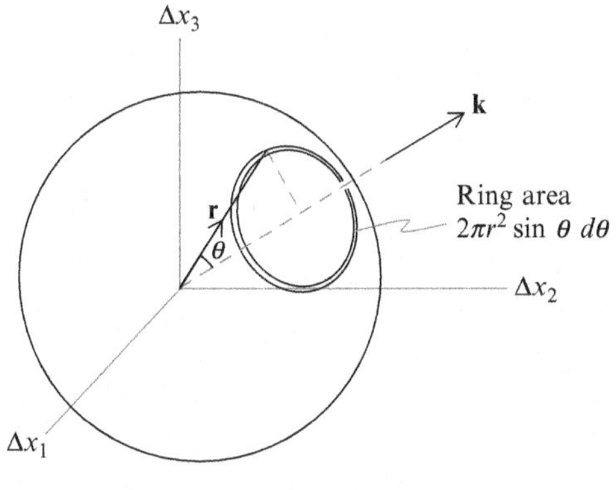

**Fig. 9.3** Integration over the spherical surface $r=const$.

We see that $\frac{1}{2}\varphi_{ii}(k_1,k_2,k_3)$ depends only on wavenumber magnitude $k$. The total energy at a given value of $k$ is therefore given by integrating $\frac{1}{2}\varphi_{ii}(k_1,k_2,k_3)$ over all possible wavenumber directions, which is simply equivalent to multiplying by the spherical surface area $4\pi k^2$. This result is called the *energy spectrum function* and is given the symbol $E(k)$. Thus,

$$E(k)\equiv 2\pi k^2\varphi_{ii}(k_1,k_2,k_3) = \frac{\overline{u^2}}{\pi}\int\limits_0^{\infty}[2g(r)+f(r)]kr\sin(kr)dr$$

The energy spectrum function represents the turbulence kinetic energy per wavenumber. It also can be used to express the full wavenumber spectrum $\varphi_{ij}$. Like $R_{ij}$ this is constrained to satisfy continuity and, as an isotropic tensor function, to have the form

$$\varphi_{ij}(k_1,k_2,k_3) = \alpha(k)k_ik_j + \beta(k)\delta_{ij} \tag{9.1.7}$$

where $\alpha(k)$ and $\beta(k)$ are functions to be determined. Continuity implies that $\partial R_{ij}/\partial \Delta x_i = 0$ and therefore, from Eq. (9.1.3), that $\varphi_{ij}k_i = 0$. With Eq. (9.1.7) for $\varphi_{ij}$ this gives

$$\beta(k) = -k^2\alpha(k)$$

Furthermore, Eq. (9.1.7) implies

$$\varphi_{ii}(k_1,k_2,k_3) = \frac{E(k)}{2\pi k^2} = k^2\alpha(k) + 3\beta(k) = 2\beta(k) = -2k^2\alpha(k)$$

Substituting $\alpha(k)$ and $\beta(k)$ back into Eq. (9.1.7) then gives the full wavenumber spectrum as

$$\varphi_{ij}(k_1,k_2,k_3) = \frac{E(k)}{4\pi k^2}\left(\delta_{ij} - \frac{k_ik_j}{k^2}\right) \tag{9.1.8}$$

In principle this result can also be obtained by taking the three-dimensional wavenumber transform of Eq. (9.1.5).

The energy spectrum function is conceptually useful because it is a one-dimensional function that can be used to divide the turbulence into its different scales and, as explained in Section 8.1, different physical mechanisms are expected to dominate in different scale ranges. It is therefore the natural function in which to express models of homogeneous turbulence. Below we detail and discuss two such models that are commonly used in aeroacoustics.

## 9.1.2   The von Kármán spectrum

In 1948, Theodore von Kármán [2] working at Caltech introduced a semiempirical model for the energy spectrum function of homogeneous turbulence based on two theoretical arguments. The first concerned Kolmogorov's hypothesis of a universal inertial subrange in turbulence where the statistical behavior of the flow is completely determined by the rate of energy flow through the cascade $\varepsilon$, as discussed in Section 8.1. This implies that $E(k) \sim k^{-5/3}$ in this range as a consequence of the fact that $E(k)k^{5/3}\varepsilon^{-2/3}$ is the only nondimensional group that can be formed from these variables. The second argument concerned the rate of decay of the turbulence at the largest scales and indicated that $E(k) \sim k^4$ at small wavenumbers. With these behaviors set he proposed the interpolation function

$$E(k) = const \frac{(k/k_e)^4}{\left[1+(k/k_e)^2\right]^{17/6}} \tag{9.1.9}$$

where $k_e$ is defined as the wavenumber scale of the largest eddies. The value of the constant can be fixed by requiring that $E(k)$ integrate to the turbulence kinetic energy, giving

$$E(k) = \frac{55}{9\sqrt{\pi}} \frac{\Gamma(5/6)}{\Gamma(1/3)} \frac{\overline{u^2}}{k_e} \frac{(k/k_e)^4}{\left[1+(k/k_e)^2\right]^{17/6}} \tag{9.1.10}$$

where $\Gamma()$ is the Gamma function, and $k_e$ can be related to the longitudinal integral scale as

$$k_e = \frac{\sqrt{\pi}}{L_f} \frac{\Gamma(5/6)}{\Gamma(1/3)} \tag{9.1.11}$$

Eq. (9.1.10) can be analytically integrated into many other spectral and correlation forms, both of fundamental interest and of use in aeroacoustics problems. First we consider the planar wavenumber spectrum, obtained by taking the inverse Fourier transform of $\varphi_{ij}(k_1,k_2,k_3)$ with respect to $k_2$ and evaluating it for $\Delta x_2 = 0$

$$\phi_{ij}(k_1, k_3) = \int_{-\infty}^{\infty} \varphi_{ij}(k_1,k_2,k_3)dk_2 \tag{9.1.12}$$

This, for example, is the wavenumber spectrum of relevance to the type of problem depicted in Fig. 8.2 where an aerodynamic surface is cutting a plane through a turbulent flow and we are interested in quantitatively evaluating the scales of the turbulence in that plane. Substituting Eqs. (9.1.10), (9.1.8) into Eq. (9.1.12), we obtain

$$\phi_{11}(k_1, k_3) = \frac{1}{18\pi} \frac{\overline{u^2}}{k_e^2} \frac{3+3(k_1/k_e)^2+11(k_3/k_e)^2}{\left[1+(k_1/k_e)^2+(k_3/k_e)^2\right]^{7/3}} \tag{9.1.13}$$

$$\phi_{22}(k_1, k_3) = \frac{4}{9\pi} \frac{\overline{u^2}}{k_e^2} \frac{(k_1/k_e)^2+(k_3/k_e)^2}{\left[1+(k_1/k_e)^2+(k_3/k_e)^2\right]^{7/3}} \tag{9.1.14}$$

$$\phi_{33}(k_1, k_3) = \frac{1}{18\pi} \frac{\overline{u^2}}{k_e^2} \frac{3 + 11(k_1/k_e)^2 + 3(k_3/k_e)^2}{\left[1 + (k_1/k_e)^2 + (k_3/k_e)^2\right]^{7/3}} \tag{9.1.15}$$

We can integrate one more time, this time with respect to $k_3$, to give the one-dimensional wavenumber spectra along a single line through the turbulence. For $\phi_{11}$ this gives

$$\phi_{11}(k_1) = \int_{-\infty}^{\infty} \phi_{11}(k_1, k_3)dk_3 = \frac{1}{\sqrt{\pi}} \frac{\Gamma(5/6)}{\Gamma(1/3)} \frac{\overline{u^2}}{k_e} \frac{1}{\left[1 + (k_1/k_e)^2\right]^{5/6}} \tag{9.1.16}$$

Applying the same integral to $\phi_{22}$ and $\phi_{33}$ we obtain

$$\phi_{22}(k_1) = \phi_{33}(k_1) = \frac{2}{27\sqrt{\pi}} \frac{\Gamma(5/6)}{\Gamma(7/3)} \frac{\overline{u^2}}{k_e} \frac{3 + 8(k_1/k_e)^2}{\left[1 + (k_1/k_e)^2\right]^{11/6}} \tag{9.1.17}$$

Note that these are double-sided spectra. The inverse Fourier transforms of $\phi_{11}(k_1)$ and $\phi_{22}(k_1)$ with respect to $k_1$ give the longitudinal and lateral correlation functions $f$ and $g$ multiplied by $\overline{u^2}$ as functions of $\Delta x_1$. Replacing $\Delta x_1$ with $r$ in those expressions gives their more general form:

$$f(r) = \frac{2^{2/3}}{\Gamma(1/3)} (k_e r)^{1/3} K_{1/3}(k_e r) \tag{9.1.18}$$

and

$$g(r) = \frac{2^{2/3}}{\Gamma(1/3)} (k_e r)^{1/3} \left( K_{1/3}(k_e r) - \frac{k_e r}{2} K_{-2/3}(k_e r) \right) \tag{9.1.19}$$

where $K$ is the modified Bessel function of the second kind.

One-dimensional wavenumber spectra are of particular interest because they are easily related to the frequency spectrum that would be measured if the turbulence was convecting in the $x_1$ direction past a fixed point. In Fig. 8.3C we plotted the turbulence frequency spectrum $S_{11}(\omega)$ produced by boundary layer turbulence moving at $U_1 = 20$m/s over a probe. Assuming Taylor's hypothesis we can estimate this curve using a homogeneous turbulence spectrum model as $S_{11}(\omega) = \phi_{11}(\omega/U_1)/U_1$. Note that the division by $U_1$ is necessary to ensure that $S_{11}(\omega)$ integrates to the mean square velocity fluctuation. Fig. 9.4A shows that this prediction, made using Eq. (9.1.16) and scaled with the actual mean-square velocity and the integral scale obtained from the measured correlation functions, is remarkably accurate. The von Kármán formula only departs from the measurement at the highest frequencies where dissipation begins to affect the spectral form. Figs. 9.4B and C show similar measurements and predictions for the normal to wall and spanwise velocity spectra $S_{22}(\omega)$ and $S_{33}(\omega)$ at the same position in the boundary layer. The scaling of the predictions is based on the mean-square velocity and integral scale associated with each component. We see that, except in the dissipation range, the von Kármán model is quite realistic.

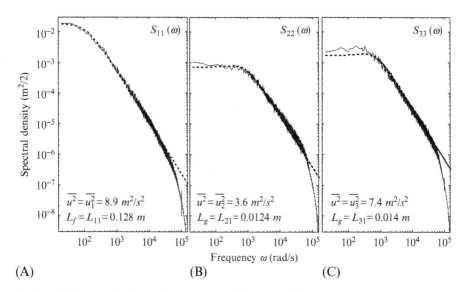

**Fig. 9.4** (A) Streamwise ($S_{11}$), (B) normal to wall ($S_{22}$), and (C) spanwise ($S_{33}$) velocity spectral density for the same turbulent boundary layer flow represented in Fig. 8.3 *(solid lines)* compared with curves estimated using the von Kármán formula with parameters adapted for each component *(dotted lines)*.

The agreement in Fig. 9.4 does not imply that the boundary layer turbulence is homogeneous and isotropic, merely that the von Kármán formula may realistically describe the form of the spectrum. Just how anisotropic the boundary layer actually is can be seen in the integral scale and velocity variance values listed in the figure. We see that $\overline{u_2^2}$ is only about 40% of $\overline{u_1^2}$ (instead of being equal), and that the lateral scale $L_g$ of the $u_2$ and $u_3$ components is only about 10% of longitudinal scale of the $u_1$ component $L_f$ (as opposed to half). However, the agreement seen in the spectra of Fig. 9.4 is still significant. It tells us that we can probably make a fair estimate of the acoustic source terms in inhomogeneous anisotropic turbulent flows by assuming a von Kármán spectrum scaled on the local integral scale and turbulence intensity of the velocity component of interest. Indeed, it is on this basis that many broadband noise predictions are made.

Pope [3] has extended the von Kármán interpolation formula to include a dissipation range model at high frequencies and to allow $E(k)$ to vary as $k^2$ at low frequencies in some situations. In its high Reynolds number form, Pope's high frequency dissipation range model involves multiplying the energy spectrum (Eq. 9.1.10) by the function

$$f_\eta(k\eta) = \exp\left(-\beta\left[(k\eta)^4 + c_\eta^4\right]^{\frac{1}{4}} - \beta c_\eta\right) \tag{9.1.20}$$

where $\eta$ is the Kolmogorov lengthscale, $\beta = 5.2$, and $c_\eta \cong 0.40$. Strictly speaking, since the effect of this multiplication is to lower spectral levels at the highest frequencies a slight rescaling of the numerical constant multiplying Eq. (9.1.10) then becomes necessary to ensure that the energy spectrum still integrates to the turbulence kinetic energy.

### 9.1.3 The Liepmann spectrum

Shortly after von Kármán introduced his interpolation function, his Caltech colleagues Hans Liepmann, John Laufer, and Kate Liepmann [4] proposed an alternative expression for the turbulence spectrum

$$E(k) = \frac{8}{\pi}\overline{u^2}L_f\frac{(kL_f)^4}{\left[1 + (kL_f)^2\right]^3} \tag{9.1.21}$$

At higher wavenumbers this model implies that $E(k)$ varies as $k^{-2}$ rather than the more fundamental $k^{-5/3}$. The difference can be small, however, and gives a function containing only integer powers of wavenumber that can be significantly easier to manipulate mathematically as part of aeroacoustic analyses. Integrating the Liepmann spectrum we obtain, for the planar wavenumber spectra,

$$\phi_{11}(k_1, k_3) = \frac{\overline{u^2}L_f^2}{4\pi}\frac{1 + (k_1L_f)^2 + 4(k_3L_f)^2}{\left[1 + (k_1L_f)^2 + (k_3L_f)^2\right]^{5/2}} \tag{9.1.22}$$

$$\phi_{22}(k_1, k_3) = \frac{3\overline{u^2}L_f^2}{4\pi}\frac{(k_1L_f)^2 + (k_3L_f)^2}{\left[1 + (k_1L_f)^2 + (k_3L_f)^2\right]^{5/2}} \tag{9.1.23}$$

$$\phi_{33}(k_1, k_3) = \frac{\overline{u^2}L_f^2}{4\pi}\frac{1 + 4(k_1L_f)^2 + (k_3L_f)^2}{\left[1 + (k_1L_f)^2 + (k_3L_f)^2\right]^{5/2}} \tag{9.1.24}$$

and for the one-dimensional spectra,

$$\phi_{11}(k_1) = \frac{\overline{u^2}L_f}{\pi}\frac{1}{1 + (k_1L_f)^2} \tag{9.1.25}$$

$$\phi_{22}(k_1) = \phi_{33}(k_1) = \frac{\overline{u^2}L_f}{2\pi} \frac{1 + 3\left(k_1 L_f\right)^2}{\left[1 + \left(k_1 L_f\right)^2\right]^2} \tag{9.1.26}$$

and, finally, for the longitudinal and lateral correlation functions

$$f(r) = e^{-r/L_f} \tag{9.1.27}$$

$$g(r) = e^{-r/L_f}\left(1 - \frac{r}{2L_f}\right) \tag{9.1.28}$$

Fig. 9.5 compares the Liepmann and von Kármán interpolation formulae. The one-dimensional wavenumber spectra (Fig. 9.5A) are almost identical up to a wavenumber $k/k_e$ of about 10. The longitudinal and lateral correlation functions (Fig. 9.5B and C) are very similar. Note that for both models $f(r)$ decays monotonically, but $g(r)$ has a shallow overshoot at larger $r$ so it dips below zero.

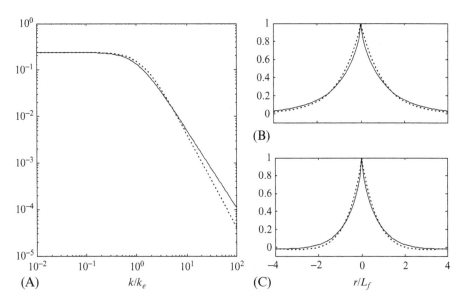

**Fig. 9.5** Comparison of Karman *(solid line)* and Liepmann *(dotted line)* models. (A) One-dimensional wavenumber spectrum $\phi_{11}k_e/\overline{u^2}$, (B) longitudinal correlation coefficient function $f(r)$. (C) Lateral correlation coefficient function $g(r)$.

## 9.2 Inhomogeneous turbulent flows

Homogeneous isotropic turbulence is rare in practical applications. Aeroacousticians are almost always dealing with turbulent flows with average properties that vary substantially across their width and that involve motions with a clear preference for direction. The specifics of these vary from configuration to configuration making a single prescription impossible. However, turbulence only forms in the rotational regions of a flow where viscous action has had an influence. Thus a substantial proportion of the turbulent flows of interest are wakes or boundary layers. In this section we describe these flows in their most idealized form; the fully developed plane wake and the zero-pressure gradient flat plate turbulent boundary layer. The goal here is to give the reader a qualitative and quantitative understanding of these canonical flows that they can then adapt and extend when faced with aeroacoustic sources generated by more configuration-specific turbulent shear flows.

It is important to state at the outset that no analytic interpolation formulae for the velocity correlations exist for these flows. Comprehensive numerical characterizations of the correlation functions (either measured or computed) exist in only a few cases and are too unwieldy to be used for routine aeroacoustic analysis. Instead, such analysis must usually rely on scaling of von Kármán or Liepmann spectra to the local or spatially averaged properties of these flows. Therefore it is these properties—the Reynolds stress fields, the velocity scales, and the lengthscales—that we will highlight in our discussion. Where visible, we also point out features of the large-scale turbulence structure that are unlikely to be modeled well with homogeneous turbulence spectra. Interpolation formulae do exist for the wavenumber-frequency spectrum of the wall-pressure fluctuations of the turbulent boundary layer. These models, which are central to the aeroacoustic analysis of noise from flow over surfaces, will be presented at the end of this section.

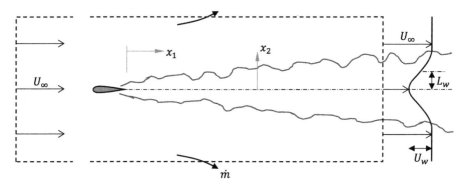

**Fig. 9.6** Control volume and nomenclature for a plane wake.

### 9.2.1 The fully developed plane wake

Consider a two-dimensional body, such as an airfoil or strut, placed in a free stream as shown in Fig. 9.6. The body disturbs the flow causing the formation of a wake that trails downstream. After an initial period of comparatively rapid evolution the wake settles down into a slender flow with almost parallel mean streamlines in which gradients in the flow direction $\partial/\partial x_1$ are much less than those across the wake $\partial/\partial x_2$. This fully developed *far wake* is the focus of our interest here. The mean-velocity profile of the far wake has a symmetric form characterized by $U_w$, the maximum velocity deficit reached on the wake centerline, and $L_w$, the distance measured from the wake centerplane to the point where the deficit is half $U_w$. It is usual to refer to $L_W$ as the *half wake width*.

At high Reynolds number, no scales beyond $U_w$ and $L_w$ are needed to characterize the mean flow or the important turbulence scales. Viscous scales are only important in controlling the dissipation of the energy of the smallest turbulence and have no direct influence on the overall flow structure. Thus the flow reaches a self-similar state in which the mean flow, turbulence stresses, spectra, and correlations can all be described by functions that are independent of streamwise position. For example, using the coordinate system of Fig. 9.6, we expect $(U_\infty - U_1)/U_w$, $\overline{u_i u_j}/U_w^2$, $L_{i1}/L_w$, and $L_{i3}/L_w$; all to be only functions of $\eta \equiv x_2/L_w$ (where $L_{ij}$ is defined as the integral lengthscale of velocity component $u_i$ in the direction $x_j$). The distances defining two point correlations (and wavenumbers) are also expected to scale on $L_w$. Frequencies seen at a fixed point will scale closely as $\omega L_w/U_\infty$ since fluctuations are produced by eddies being convected past the point almost at $U_\infty$. Note that $U_w/U_\infty \ll 1$ in the far wake.

The fact that the form of the wake becomes invariant with streamwise position does not imply that this flow is universal. Fully developed wakes generated by different bodies are qualitatively very similar, but the relative magnitudes of such things as the peak turbulence intensity $\sqrt{u_1^2}$ to the maximum deficit $U_w$ may differ by as much as 30% [5].

We can infer how the controlling scales $U_w$ and $L_w$ change as the wake grows. Consider first the control volume shown in Fig. 9.6. Assuming incompressible flow and requiring that the mass flow in and out of the volume be the same, we have

$$\dot{m} = 2\rho_o h U_\infty - \int_{-h}^{h} \rho_o U_1 dx_2 \tag{9.2.1}$$

where $-h$ and $h$ are the $x_2$ limits of the control volume and $\dot{m}$ is the mass flow per unit span out of the top and bottom faces of the volume. Equating the difference in the momentum flow into and out of the control volume in the $x_1$ direction to the drag force per unit span on the body as $D$, gives

$$2\rho_o U_\infty^2 h - \dot{m} U_\infty - \int_{-h}^{h} \rho_o U_1^2 dx_2 = D \tag{9.2.2}$$

where we are assuming that the mass flow $\dot{m}$ leaves the volume with a $U_1$ velocity component equal to the free-stream velocity—a good approximation if $h$ is large. Substituting for $\dot{m}$ and normalizing on $\rho_o U_\infty^2$ this expression simplifies to

$$\frac{D}{\rho_o U_\infty^2} = \int\limits_{-h}^{h} \frac{U_1}{U_\infty}\left(1 - \frac{U_1}{U_\infty}\right) dx_2 \tag{9.2.3}$$

The integral on the right-hand side has units of distance and is referred to as the *momentum thickness* $\theta$. Since it is equal to the normalized drag it must be invariant with streamwise position in the wake $x_1$. Note that the Eq. (9.2.3) can be written as $\theta/c = 2C_d$, where $C_d$ is the two-dimensional drag coefficient on the body and $c$ is the reference length used to normalize $C_d$, such as airfoil chord.

Writing the integrand of Eq. (9.2.3) as $\dfrac{U_1}{U_\infty}\left(\dfrac{U_\infty - U_1}{U_\infty}\right) dx_2$, we note that $(U_\infty - U_1)/U_\infty$ will scale with $U_w/U_\infty$, and $U_1/U_\infty \approx 1$ since $U_w/U_\infty \ll 1$. Similarly $x_2$ will scale with $L_w$. We must therefore have that $\theta \sim U_w L_w/U_\infty$, and thus the product of the velocity and length scales of the wake must be constant.

To determine how these parameters vary individually, we use the streamwise component of the Reynolds averaged Navier Stokes equations (8.3.4). For two-dimensional mean flow with $\partial/\partial x_3 = 0$ and ignoring viscous terms this is, for $i = 1$

$$U_1 \frac{\partial U_1}{\partial x_1} + U_2 \frac{\partial U_1}{\partial x_2} = -\frac{1}{\rho_o}\frac{\partial p_o}{\partial x_1} - \frac{\partial \overline{u_1^2}}{\partial x_1} - \frac{\partial \overline{u_1 u_2}}{\partial x_2} \tag{9.2.4}$$

A formal order of magnitude analysis [6] shows that only the first term on the left hand side and the last term on the right hand side are significant in the far wake, giving

$$U_\infty \frac{\partial U_1}{\partial x_1} = -\frac{\partial \overline{u_1 u_2}}{\partial x_2} \tag{9.2.5}$$

where the substitution of $U_\infty$ for $U_1$ on the left hand side follows from $U_w/U_\infty \ll 1$. Following Townsend [7], we can use the concept of self similarity to scale the flow in the wake. This requires the velocity profile to scale on the wake width $L_W$. Given that $\theta = U_W L_W/U_\infty$ is constant, this also requires scaling of the wake velocity deficit on $U_W$. Consequently, we can write the mean velocity and stress profiles as a functions of $\eta = x_2/L_W$ in the form

$$\frac{U_\infty - U_1}{U_w} = h_1(\eta) \quad \text{and} \quad \frac{\overline{u_1 u_2}}{U_w^2} = h_2(\eta) \tag{9.2.6}$$

It follows that

$$\frac{\partial U_1}{\partial x_1} = -\frac{\partial}{\partial x_1}(U_W h_1(\eta)) \quad \text{and} \quad \frac{\partial \overline{(u_1 u_2)}}{\partial x_2} = \frac{U_W^2}{L_W}\frac{\partial h_2}{\partial \eta}$$

Substituting these into Eq. (9.2.5), recognizing that $\dfrac{\partial \eta}{\partial x_1} = -\dfrac{\eta}{L_w}\dfrac{dL_w}{dx_1}$, and rearranging gives

$$\left(\frac{U_\infty}{U_w}\frac{dL_w}{dx_1}\right)\eta\frac{dh_1}{d\eta} - \left(\frac{U_\infty}{U_w^2}L_w\frac{dU_w}{dx_1}\right)h_1 = -\frac{dh_2}{d\eta}$$

Since the terms outside the parentheses are not functions of $x_1$, the terms inside the parentheses must be constants and this gives two simultaneous differential equations for $L_w$ and $U_w$. The solution to these is simply that $L_w \sim x_1^n$ and $U_w \sim x_1^{n-1}$ where $n$ is a constant. Since we already know that $\dfrac{U_w L_w}{U_\infty}$ is a constant then we must have that $L_w \sim x_1^{1/2}$ and $U_w \sim x_1^{-1/2}$.

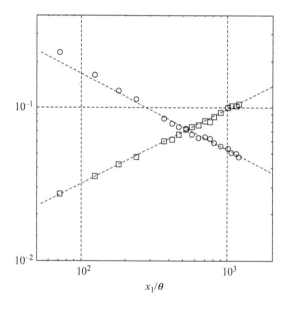

**Fig. 9.7** Variations of $U_w/U_\infty$ ($\circ$) and $L_w/\theta \div 100$ ($\square$) with streamwise distance for a plane wake. Data from Ref. [8].

This is realized in practice. For example, Fig. 9.7 shows data from the plane wake shed from a roughened airfoil in a uniform air flow [8], much as pictured in Fig. 9.6. The airfoil is 0.2 m in chord and the free-stream velocity $U_\infty$ is 20 m/s, for a Mach number of 0.059 and a Reynolds number $U_\infty c/\nu = 339{,}000$. The drag coefficient on the airfoil is 0.0187 implying $\theta/c = 0.093$. The scaling variables are shown as a function of $x_1/\theta$ where $x_1$ is measured downstream from the wake origin, very close to the airfoil trailing edge. After some slight initial adjustment, $U_w$ and $L_w$ closely follow square root variations predicted. The specific dependencies shown by the dotted lines in Fig. 9.12 are;

$$U_w/U_\infty = 1.66(x_1/\theta)^{-1/2} \tag{9.2.7}$$

$$L_w/\theta = 0.306(x_1/\theta)^{1/2}$$

The ratio between $\theta/L_w$ and $U_w/U_\infty$, of 1.97 is expected to be universally constant [5]. This wake develops a self-similar structure by about eight chord lengths ($860\theta$) downstream of the trailing edge. In Figs. 9.8–9.11 we show the self-similar form so as to illustrate the typical structure of a fully developed wake. Mean velocity and Reynolds-stress profiles are plotted in Fig. 9.8. The mean-velocity profile is an inverted bell shape that has an almost Gaussian form. The curve fit of Wygnanski et al. [5]

$$\frac{U - U_\infty}{U_w} = -\exp\left(-0.637\left(\frac{x_2}{L_w}\right)^2 - 0.056\left(\frac{x_2}{L_w}\right)^4\right) \tag{9.2.8}$$

provides a quite accurate model of the mean profile with the only effect of the $(x_2/L_w)^4$ term in the exponent being to slightly narrow the tails of the profile (Fig. 9.8A).

**Fig. 9.8** Profiles through the fully developed plane wake of Ref. [8]. (A) Mean velocity; $\circ$, $(U - U_\infty)/U_w$; —, Eq. (9.2.8); ---, Eq. (9.2.8) without the non-Gaussian term. (B) Turbulence normal stresses; $\square$, $\overline{u_1^2}/U_w^2$; $\triangle$, $\overline{u_2^2}/U_w^2$; $\diamond$, $\overline{u_3^2}/U_w^2$. (C) Turbulence shear stress; $\triangledown$, $\overline{u_1 u_2}/U_w^2$.

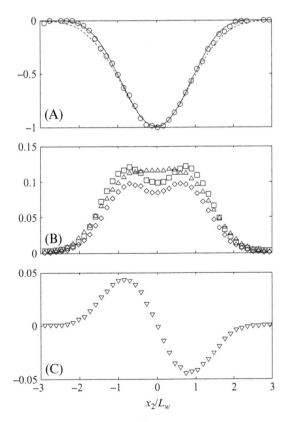

The Reynolds normal stresses (Fig. 9.8B) are almost constant and equal over the central portion of the wake $|x_2| < L_w$ where they reach values of about $0.1U_w^2$. They fall below 10% of this value just over two wake half-widths from the center. The Reynolds shear stress $\overline{u_1 u_2}/U_w^2$ forms an antisymmetric profile that roughly mirrors the negative of the gradient of the mean velocity. It reaches its peak magnitude of just over 0.04, near $|x_2| = 0.8L_w$. Note that two-dimensionality implies that the other Reynolds shear stresses $\overline{u_1 u_3}$ and $\overline{u_2 u_3}$ are zero.

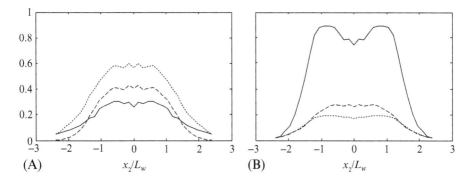

**Fig. 9.9** Variation of the integral lengthscales with distance $x_2$ across the wake of Ref. [8]. (A) Spanwise scales: —, $L_{13}/L_w$; - - -, $L_{23}/L_w$; – –, $L_{33}/L_w$. (B) Streamwise scales inferred from the integral timescale and Taylor's hypothesis: —, $L_{11}/L_w$; - - -, $L_{21}/L_w$; – –, $L_{31}/L_w$.

Fig. 9.9 shows the integral lengthscales in the flow. The flow is homogeneous in the spanwise $x_3$ direction and almost homogeneous in the streamwise $x_1$ direction, since it is slowly growing. We can therefore define meaningful integral scales of each of the three velocity components in $x_1$ and $x_3$, and these integral scales are functions of the position in the wake $x_2/L_w$. Recall that, in the convention established in Section 8.4, $L_{in}$ denotes the lengthscale of velocity component $u_i$ taken in direction $x_n$. The streamwise lengthscales in Fig. 9.9B were measured by determining the integral timescale from time-delay correlations, and then applying Taylor's hypothesis. We expect this to be a good assumption since $U_w/U_\infty \ll 1$ and so the mean velocity is always close to $U_\infty$ and the timescale on which the turbulence evolves $L_w/U_w$ is therefore much longer than the timescale for it to convect past a fixed point $L_w/U_\infty$.

In many ways the lengthscale profiles of Fig. 9.9 have similar shapes to those of the turbulence normal stresses, with roughly constant regions over the middle 50% or so of the wake flanked by regions of decay towards the edges. The lengthscales reveal strong anisotropy of the largest scale eddies in the wake. Near the wake center the longitudinal lengthscale in the streamwise direction $L_{11}$ (Fig. 9.9B) is about 80% of the half-wake width and about four times the two lateral scales $L_{21}$ and $L_{31}$ (recall that for homogeneous turbulence the longitudinal scale is twice the lateral). At the same position, the longitudinal scale in the spanwise direction $L_{33}$ (Fig. 9.9A) is about $0.4L_w$ and actually smaller (by about 30%) than the lateral scale $L_{23}$.

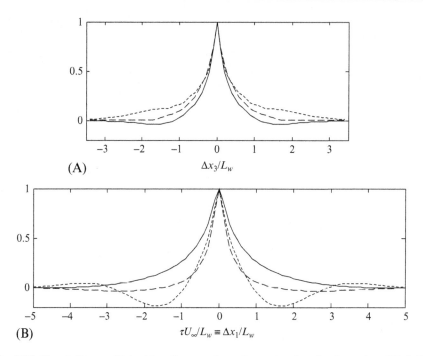

**Fig. 9.10** Correlation coefficient functions on the wake centerplane for the wake of Ref. [8]. (A) As a function of spanwise separation. (B) As a function of time delay and, through Taylor's hypothesis, streamwise separation. —, $\rho_{11}$; ---, $\rho_{22}$; − −, $\rho_{33}$.

The nature of the large-scale eddies producing this behavior is more apparent in the spanwise and streamwise correlation coefficient functions from which the length-scales are integrated. Fig. 9.10 shows these functions at the wake center, $x_2 = 0$. The streamwise correlation (Fig. 9.10B) of the vertical velocity component $\rho_{22}$ is seen to oscillate as it decays reaching a minimum at $|\Delta x_1| \approx 1.7L_w$ followed by a shallow maximum at $|\Delta x_1| \approx 3.5L_w$. The negative area produced by the overshoot results in the comparatively low integral lengthscale $L_{21}$ at the wake center. The oscillation indicates some regularity in the spacing of eddies in the streamwise direction, termed *quasi periodicity*. What we are seeing is organization associated with eddies generated directly from the roll up of the shear visible in the mean-velocity profile pictured in Fig. 9.8A, a feature that would not be represented well with a Liepmann or von Kármán model (Fig. 9.4C), even though such a model might still adequately describe the smaller scales. These large scale organized motions may also be responsible for the broad wings seen in the spanwise decay of $\rho_{22}$ visible in Fig. 9.10A.

The inhomogeneity of the wake in the $x_2$ direction means that the spatial correlation in this direction is a function of two positions $x_2$ and $x_2'$ and not just the distance between them. Fig. 9.11 shows this two-dimensional correlation function $R_{ij}(x_2, x_2')$, for zero $\Delta x_1$, $\Delta x_3$ and $\tau$, plotted in coefficient form by normalizing on the geometric average of the corresponding mean-square velocity fluctuations at $x_2$ and $x_2'$. In this

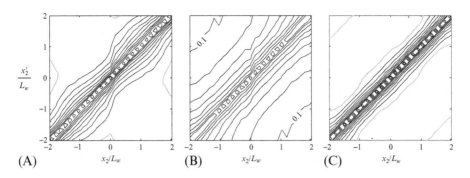

**Fig. 9.11** Zero-time delay correlation coefficient $R_{ij}\left(x_2,x_2'\right)/\left[\overline{u_i^2}\left(x_2\right)\overline{u_j^2}\left(x_2'\right)\right]^{1/2}$ for the wake. (A) Streamwise velocity $i=j=1$, (B) normal velocity $i=j=2$, (C) spanwise velocity $i=j=3$. Contours in steps of 0.1: —, Positive levels; ▭▭, zero level. Data from Ref. [8].

normalization the diagonals of the contour maps in Fig. 9.11 have a value of unity and the rate of decay to either side is an indication of the extent of the correlation perpendicular to the plane of the wake. For example, Fig. 9.11A shows that streamwise velocity fluctuations at the wake center correlate over about one half-wake width above and below the center. The vertical extent of the correlation in the region of highest mean-velocity gradient near $x_2/L_w = \pm 1$ is actually larger than at the centerline, giving the correlation map a waisted appearance. Otherwise the vertical extent of the correlations of all three velocity components is remarkably constant with the position in the wake, with the vertical velocity component $u_2$ correlating over the greatest distance, and the spanwise correlation $u_3$ the smallest. We can therefore define meaningful integral scales in the $x_2$ direction. Integrating the correlation coefficient function, as shown in Fig. 9.11 with respect to $x_2'$ for $x_2 = 0$ we obtain scales of $0.35L_w$, $0.82L_w$, and $0.17L_w$ for the $u_1$, $u_2$, and $u_3$ velocity components respectively.

## 9.2.2 The zero pressure gradient turbulent boundary layer

Any surface exposed to a high Reynolds number air stream will develop a boundary layer. The boundary layer is an expression of the no-slip condition imposed at the surface, which requires that the fluid layer immediately adjacent to a surface not move relative to it. At low Mach numbers the no-slip condition, and the boundary layer it produces, are the only sources of vorticity within a flow. Wakes, regions of separation and vortical flows, all originate from boundary layers.

Boundary layers are by definition thin. That is, thin relative to the streamwise distance over which the boundary layer grows significantly, and thin compared to the local radius of curvature of the surface. A boundary layer is therefore another example of a slender flow with almost parallel streamlines in which gradients of the average

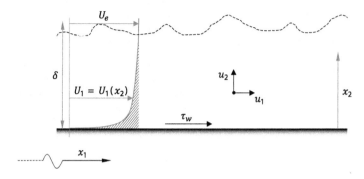

**Fig. 9.12** Turbulent boundary layer.

velocity properties in the flow direction are much less than those across the boundary layer. The almost parallel streamlines also ensure that the mean pressure is constant across the boundary layer, and that pressure variations in the streamwise direction are impressed by the overriding irrotational flow. We refer to the streamwise pressure gradient as *favorable* or *adverse* depending, respectively on whether it is tending to accelerate or decelerate the fluid.

Fig. 9.12 shows the coordinate set up for our boundary layer discussion, with $x_1$ measured streamwise from the origin of the boundary layer far upstream, $x_2$ perpendicular to the wall, and $x_3$ spanwise. The velocity of the irrotational flow just outside the boundary layer edge is $U_e$ and the time mean frictional stress exerted by the boundary layer on the wall is $\tau_w$. We define the local friction coefficient $C_f \equiv \tau_w / \frac{1}{2}\rho U_e^2$. We denote the boundary layer thickness as $\delta$ and this is defined as the distance from the wall to the point where the mean-velocity $U_1$ is 99% of $U_e$. We also define a boundary layer momentum thickness,

$$\theta = \int_0^\delta \frac{U_1}{U_e}\left(1 - \frac{U_1}{U_e}\right) dx_2 \tag{9.2.9}$$

Similar to the wake, the momentum thickness is related to viscous drag. Specifically, a control volume analysis similar to the analysis for the wake in Section 9.2.1 relates the normalized friction drag per unit length to twice the incremental increase in momentum thickness per unit length, $C_f = 2d\theta/dx_1$. A second integral measure of the boundary layer thickness is the displacement thickness $\delta^*$.

$$\delta^* = \int_0^\delta 1 - \frac{U_1}{U_e} dx_2 \tag{9.2.10}$$

This is a measure of the loss of volumetric flow rate in the boundary layer per unit span due to the frictional slowing of the flow. Graphically it is equivalent to the shaded area outside the mean-velocity profile shown in Fig. 9.12 divided by $U_e$. The slower flow rate in the boundary layer effectively pushes the overriding irrotational flow away from the wall by a distance equal to $\delta^*$.

A smooth surface exposed to a low-turbulence free stream will first develop a laminar boundary layer, but transition to a turbulent flow occurs relatively quickly at most Reynolds numbers of engineering relevance. A typical length Reynolds number $x_1 U_e/\nu$ for transition to turbulence with zero pressure gradient is about 600,000 implying, for example, transition within about 10 cm in an air flow of 100 m/s. Transition can occur much sooner in the presence of surface roughness, significant free-stream turbulence, or an adverse pressure gradient, or can be delayed substantially in a favorable pressure gradient or if the free stream is particularly clean.

At high Reynolds numbers a turbulent boundary layer contains a vast range of eddy sizes, from those that encompass the whole boundary layer to microscopic motions that dissipate energy directly to viscosity. Away from the wall the largest eddies are responsible for an intricately convoluted boundary layer edge, beneath which a full turbulent energy cascade is established. Close to the wall, however, the extreme velocity gradients associated with the no-slip condition produce instability and roll up of new turbulent eddies so small as to be comparable to that of the energy dissipating motions. The presence of the wall thus enforces some spatial sorting of the turbulence structure.

In a zero pressure gradient turbulent boundary layer two sets of scales are needed to describe the mean flow and average turbulence properties. Close to the wall the flow is determined by viscosity $\nu$ and the wall shear stress $\tau_w$ from which we can form the "inner" velocity and length scales $u_\tau \equiv \sqrt{\tau_w/\rho_o}$ and $\nu/u_\tau$. Note that $u_\tau$ is referred to as the *friction velocity*. The flow near the boundary layer edge is also determined by the friction (but with no direct dependence on viscosity) as well as the boundary layer thickness and so the "outer" scales here are $u_\tau$ and $\delta$. As we will see below, the edge velocity $U_e$ also plays some role throughout the boundary layer.

Accurate and fast computational methods for boundary layer calculation, with and without pressure gradient, are well established and readily available [9,10]. For zero pressure gradient turbulent boundary layers a number of empirical formulae exist as well. The skin-friction coefficient

$$C_f = 2\frac{u_\tau^2}{U_e^2} \tag{9.2.11}$$

can be estimated using the well-known Schulz-Grunow formula

$$C_f = 0.37(\log_{10}\mathrm{Re}_x)^{-2.584} \tag{9.2.12}$$

where $\mathrm{Re}_x \equiv U_e x_1/\nu$. Note that many other curve fits exist [11]. In Fig. 9.13 this expression has been used to plot the variation in $u_\tau/U_e$ with Reynolds number. The normalized friction velocity and the skin-friction coefficient gently decrease with an increase in Reynolds number, and $u_\tau/U_e$ is close to 4% for Reynolds numbers between about

1 and 4 million (this Reynolds number range encompasses most wind tunnel work and some full-scale applications, such as wind turbines). Simple calculations for the boundary layer thicknesses can be obtained by approximating the mean-velocity profile using a 1/7th power-law curve [12] to give:

$$\delta = 0.37 x_1 \left( \frac{U_e x_1}{\nu} \right)^{-1/5} \tag{9.2.13}$$

with $\theta/\delta = 7/72$ and $\delta^*/\delta = 1/8$. Eq. (9.2.13) implies that the boundary layer thickness grows almost linearly as $x_1^{4/5}$, and has a thickness of roughly equal to 2% of its running length. Favorable and adverse pressure gradients will result in thinner and substantially thicker boundary layer thicknesses, respectively.

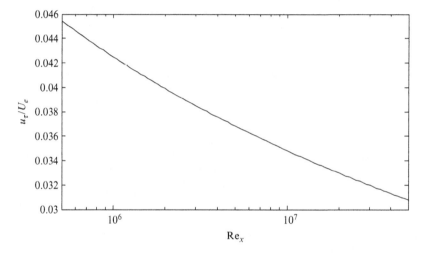

**Fig. 9.13** Schultz-Grunow formula for the skin-friction coefficient plotted as friction velocity.

As Eqs. (9.2.12), (9.2.13) imply, the ratios of the micro and macro turbulent boundary layer scales $u_\tau/U_e$ and $\nu/u_\tau\delta$ are not constant. This means that, unlike the wake, the turbulent boundary layer does not reach a self-similar state. Instead it continually sustains two distinct scaling regions. This is particularly apparent for the mean-velocity profile. In the outer region the slowing of the mean flow relative to the free-stream scales with the friction and distances scale on the boundary layer thickness, so

$$\frac{U_1 - U_e}{u_\tau} = f_o \left( \frac{x_2}{\delta} \right) \tag{9.2.14}$$

This is known as the "law of the wake." In the inner region, adjacent to the wall, only the friction and viscosity determine the flow:

$$u^+ = f_+ \left( x_2^+ \right) \tag{9.2.15}$$

where the inner variables are defined as $u^+ \equiv U_1/u_\tau$ and $x_2^+ \equiv x_2 u_\tau/\nu$. This is "the law of the wall." Obviously, these two descriptions of the profile must be consistent and presumably, must overlap. If so, the mean-velocity gradients implied by Eqs. (9.2.14), (9.2.15) must match in the overlap region and we have

$$\frac{\partial U_1}{\partial x_2} = \frac{u_\tau}{\delta} f_o' \left(\frac{x_2}{\delta}\right) = \frac{u_\tau^2}{\nu} f_+' \left(\frac{x_2 u_\tau}{\nu}\right) \tag{9.2.16}$$

and thus

$$\frac{x_2}{\delta} f_o' \left(\frac{x_2}{\delta}\right) = \frac{x_2 u_\tau}{\nu} f_+' \left(\frac{x_2 u_\tau}{\nu}\right) \tag{9.2.17}$$

Since they are functions of different nondimensional variables, the left and right hand sides of this equation can only be the same if they are both equal to a constant, defined as $1/\kappa$. Integrating $f_+'$ in Eq. (9.2.17) we a have

$$u^+ = \frac{1}{\kappa} \ln x_2^+ + C \tag{9.2.18}$$

where $\kappa$ is referred to as the von Kármán constant. We see that the overlap portion of the profile must have a semilogarithmic form and this is indeed observed in practice. There is no universal agreement as to the exact values of the constants $\kappa$ and $C$, though most authors choose values close to 0.40 and 4.9, respectively.

The structure of the entire mean-velocity profile is illustrated in Fig. 9.14 which includes experimental data from a flat plate boundary layer with a momentum-thickness Reynolds number $Re_\theta = \theta U_e/\nu$ of 15,500 [13]. Fig. 9.14A shows the profile plotted in terms of $x_2^+$ and on a semilogarithmic scale to more clearly show

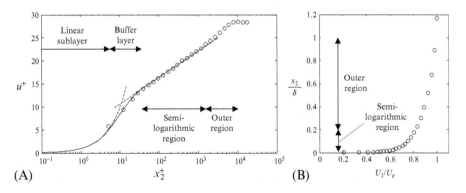

**Fig. 9.14** Mean-velocity profile in a turbulent boundary layer at $Re_\theta = 15,500$ plotted normalized on (A) inner variables, (B) outer variables. Data from Ref. [13].

the near-wall behavior. We see that the law of the wall consists of three regions. For $x_2^+$ less than about 5 the profile approaches the linear form $u^+ = x_2^+$ with a slope set by the velocity gradient at the wall. This is called the *linear sublayer*. In the *buffer layer*, between about $x_2^+ = 5$ and 30, the profile transitions to the semilogarithmic form of Eq. (9.2.18).

Both sublayer and buffer layer portions of the profile can be estimated by integrating the van Driest [14] formulation

$$\frac{du^+}{dx_2^+} = \frac{2}{1 + \sqrt{1 + 4\kappa^2 x_2^{+2}\left(1 - e^{-x_2^+/A^+}\right)}} \tag{9.2.19}$$

from the wall, with $A^+ \cong 26$. The semilogarithmic region, or *log layer*, is also part of the law of the wake and extends, approximately from $x_2^+ = 30$ to $x_2/\delta \cong 0.2$. In the outer region, beyond $x_2/\delta \cong 0.2$, the profile curves over to meet with the free stream. The form of the profile in the semilogarithmic and outer regions is often estimated by using Coles [15] extension of Eq. (9.2.18)

$$u^+ = \frac{1}{\kappa}\ln x_2^+ + C + 2\frac{\Pi}{\kappa}\sin^2\left(\frac{\pi}{2}\frac{x_2}{\delta}\right) \tag{9.2.20}$$

where $\Pi \cong 0.51$.

The profiles of the turbulence stresses also have two scaling regions, in the inner region varying as $x_2^+$ and as $x_2/\delta$ in the outer region. The stresses $\overline{u_1 u_2}$, $\overline{u_1^2}$, $\overline{u_2^2}$, and $\overline{u_3^2}$ are generally taken to scale with $u_\tau^2$ over the whole boundary layer, though this is a matter of some debate. In particular, it is fairly clear that the peak nondimensional streamwise turbulence normal stress $\overline{u_1^2}/u_\tau^2$ increases significantly with Reynolds number [16] and it has been proposed that $\overline{u_1^2}$ scales on $u_\tau U_e$ [17]. This is believed to result from "inactive" motion on the boundary layer—essentially velocity fluctuations in the $u_1$ component driven by large-scale irrotational motions in the outer part of the boundary layer [18].

Fig. 9.15 shows sample turbulence stress profiles for the same $Re_\theta = 15,500$ boundary layer represented in Fig. 9.14. The boundary layer thickness is based on a statistical average quantity and thus the instantaneous boundary layer edge often exceeds this height. As a result, the turbulence stresses (Fig. 9.15) are not completely zero outside the boundary layer edge. At $x_2 = \delta$ the normal stresses are about 0.004 $U_e^2$ (2% turbulence intensity). Turbulence levels intensify as the wall is approached, and become increasingly anisotropic, with the streamwise normal stress $\overline{u_1^2}$ becoming roughly twice the spanwise $\overline{u_3^2}$ and wall-normal $\overline{u_2^2}$ stresses. While not visible in the figure, $\overline{u_1^2}$ peaks very close to the wall towards the bottom of the buffer layer, whereas $-\overline{u_1 u_2}$ and $\overline{u_2^2}$ reach almost constant maximum values in the semilogarithmic region of the mean-velocity profile (Fig. 9.15). In particular, the shear-stress $-\overline{u_1 u_2}$ is constant here with a value equal to half the skin friction coefficient $C_f$.

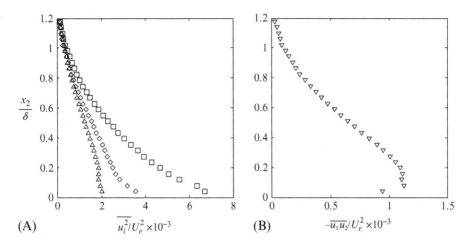

**Fig. 9.15** Turbulence stress profiles for a flat plate turbulent boundary layer at $Re_\theta = 15,500$. (A) Turbulence normal stresses: $\square$, $\overline{u_1^2}/U_e^2$; $\triangle$, $\overline{u_2^2}/U_e^2$; $\diamond$, $\overline{u_3^2}/U_e^2$. (B) Turbulence shear stress; $\triangledown$, $-\overline{u_1 u_2}/U_e^2$.
Data from Ref. [38].

This equivalence derives from the average streamwise momentum Eq. (8.3.4) which in the inner region can be shown [9] to reduce to the statement that the sum of the turbulent and viscous shear stresses in the inner region is constant and equal to the wall shear stress:

$$\tau_w = \mu \frac{\partial U_1}{\partial x_2} - \rho_o \overline{u_1 u_2} \tag{9.2.21}$$

Thus, $-\overline{u_1 u_2}$ is constant in the log layer, where viscous effects are insignificant, and decreases in the buffer layer to balance the increase in viscous shear.

The size and form of the largest eddies responsible for the velocity fluctuations are, to some extent, revealed in Figs. 9.16 and 9.17. Fig. 9.16A shows the integral time-scale associated with each of the three velocity components as a function of distance from the wall. Sample time delay correlation functions from which these scales were obtained are shown in Figs. 9.17A through C. Fig 9.16B shows integral lengthscales in the spanwise direction as a function of $x_2$ with example correlation functions appearing in Figs. 9.17D through F.

The integral timescales (Fig. 9.16A), are approximately constant or increase slowly as the boundary layer is traversed from top down. Close to the wall, in the region not shown in Fig. 9.16A, the timescales are expected to become proportional to distance from the wall [19] and thus reduce to zero. By far the largest timescale and therefore, through the qualitative application of Taylor's hypothesis, the largest streamwise

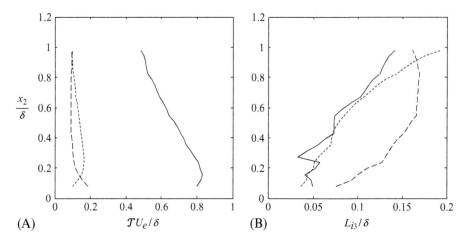

**Fig. 9.16** Integral scale profiles for a flat plate turbulent boundary layer at $Re_\theta = 15,500$. (A) Streamwise integral time scales $\mathcal{T}U_e/\delta$ for: —, $u_1$; ---, $u_2$; — —, $u_3$. (B) Spanwise integral lengthscales: —, $L_{13}/\delta$; ---, $L_{23}/\delta$; — —, $L_{33}/\delta$. Data from Ref. [38].

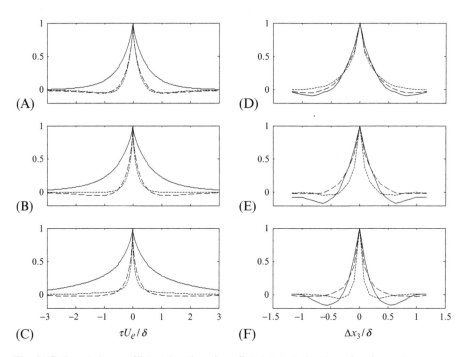

**Fig. 9.17** Correlation coefficient functions for a flat plate turbulent boundary layer at $Re_\theta = 15,500$. Time delay correlations at $x_2/\delta = $ (A) 0.98, (B) 0.51, and (C) 0.20. Spanwise correlations at $x_2/\delta = $ (D) 0.98, (E) 0.51, and (F) 0.20. Velocity components: —, $u_1$; ---, $u_2$; — —, $u_3$. Data from Ref. [38].

lengthscale is that associated with the streamwise velocity fluctuations $u_1$. This scale is about five times that associated with the spanwise and wall normal velocity fluctuations. The underlying time-delay correlation coefficient functions (Fig. 9.17A–C) show that this is because $\rho_{11}$ has very long positive tails, indicating significant correlation for time delays as large as $3\delta/U_e$ (the time taken for the flow at the boundary layer edge to move downstream by three boundary layer thicknesses). The $\rho_{22}$ and $\rho_{33}$ correlations have less pronounced tails and for $\tau > 0.5\delta/U_e$ those tails are negative and thus subtract from the corresponding integral scales.

The spanwise integral lengthscales (Fig. 9.16B) are quite small compared to the implied streamwise scales and are seen to increase with distance from the wall in the outer region. Over most of the boundary layer the lengthscale of the spanwise velocity fluctuations $L_{33}$ is largest and reaches a value of some 17% of $\delta$ over the top half of the boundary layer. However, the associated correlation coefficient functions (Fig. 9.17D–F) show that the streamwise velocity component $u_1$ actually correlates over larger spanwise distances than $u_3$. The streamwise velocity correlation function $\rho_{11}$ has negative lobes, however, for larger spanwise separations that reduce its integral scale. The negative lobes are more pronounced towards the bottom of the boundary layer.

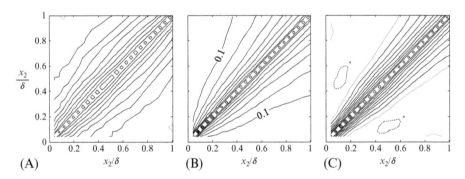

**Fig. 9.18** Zero-time delay correlation coefficient $R_{ij}(x_2, x_2') / \left[ \overline{u_i^2}(x_2) \, \overline{u_j^2}(x_2') \right]^{1/2}$ for a flat plate turbulent boundary layer at $\mathrm{Re}_\theta = 15{,}500$. (A) Streamwise velocity $i=j=1$, (B) Normal velocity $i=j=2$, (C) Spanwise velocity $i=j=3$. Contours in steps of 0.1; —, Positive levels; ▬, zero level; - - -, negative levels.
Data from Ref. [38].

As with the plane wake, the inhomogeneity of the boundary layer means that the spatial correlation in the vertical direction is a function of two positions $x_2$ and $x_2'$ and not just the distance between them. This correlation coefficient function $R_{ij}(x_2, x_2') / \left[ \overline{u_i^2}(x_2) \, \overline{u_j^2}(x_2') \right]^{1/2}$ is plotted in Fig. 9.18. This figure shows that streamwise velocity fluctuations in the log layer near the wall have about a 10%

correlation coefficient with those at the mid height of the boundary layer, and that streamwise velocity fluctuations at the mid height correlate measurably with those over almost the entire boundary layer thickness. The scale of the vertical correlation of $u_1$ appears almost independent of position in the boundary layer. This contrasts with the scale of the correlation of normal-to-wall fluctuations $u_2$ which grows approximately linear with distance from the wall (Fig. 9.18B). The spanwise velocity correlation also grows with distance from the wall (Fig. 9.18C), but this velocity component is the least well correlated in the vertical direction.

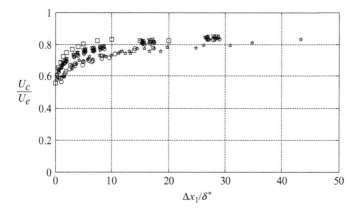

**Fig. 9.19** Smooth wall boundary layer convection velocities $U_c$ obtained from the center of the convective ridge in the wall-pressure space-time correlation, as a function of streamwise separation $\Delta x_1$. $\bigcirc$ , Blake [21], $Re_\theta = 8210$–13,200; $\star$, Bull [22], 10,500–35,200; $\square$, Willmarth and Wooldridge [23], 38,000; $\diamondsuit$ , $\triangleright$, $\triangleleft$, $\triangle$, $\triangledown$, Forest [24], 27,400, 37,700, 52,000, 64,300, 76,700, respectively.

Convection velocities of the velocity fluctuations of the turbulence in the boundary layer are a function of distance from the wall $x_2$. For most of the boundary layer, the convection velocity $U_c$ measured from the streamwise space-time correlation is close to the local mean-velocity [20] However, in the buffer layer and linear sublayer the convection velocity becomes almost constant at between 10 and 15 $u_\tau$ indicating, perhaps, that the streamwise velocity correlations in this region are predominantly generated by the overriding turbulent structures, rather than by eddies contained within these regions. A bulk convection velocity for the boundary layer turbulence, which is of particular relevance to aeroacoustic applications, can be defined using the streamwise space-time correlation wall-pressure fluctuations. As will be explained in the next section, the pressure fluctuations represent an integral over the boundary layer thickness. Fig. 9.19 shows convection velocities deduced from the wall-pressure correlations over streamwise separations $\Delta x_1$ for a range of momentum-thickness Reynolds numbers [20–23]. For the smallest streamwise separations, for which the correlation contains the greatest contribution from small scale eddies near the wall, $U_c$ is close to $0.6U_e$. At large separations, where the correlation is mostly determined by the largest eddies, $U_c$ is between about $0.8U_e$ and $0.85U_e$. Note that there is some scatter in the rate of increase with spacing between studies that may indicate a slight increasing trend with Reynolds number.

### 9.2.3  The turbulent boundary layer wall-pressure spectrum

Scattering of turbulent surface pressure fluctuations into sound is one of the dominant mechanisms behind some important aeroacoustic noise sources, such as trailing edge noise and rough-wall boundary layer noise. To relate the surface pressure fluctuations on a flat surface below a turbulent boundary layer to the velocity fluctuations in the boundary layer we can use Lighthill's analogy, provided that the flow is of sufficiently low Mach number that it may be regarded as locally incompressible. Since the distances between the local velocity fluctuations and a point on the surface are small compared to the acoustic wavelength, we need not be concerned with the effect of propagation time on the solution to Lighthill's wave equation given by Eq. (4.3.3). The unsteady surface pressure will be given by $p - p_\infty = \rho' c_\infty^2$ and, since the surface is rigid, the pressure gradient will be zero normal to the surface. We can then replace the Green's function in Eq. (4.3.3) with the tailored Green's function specified in Eq. (4.5.3) to eliminate the surface integral, and give the surface pressure on the surface as

$$p(\mathbf{x}, t) - p_\infty = \int_{-T}^{T} \int_V G_T(\mathbf{x},t|\mathbf{y},\tau) \frac{\partial^2 T_{ij}(\mathbf{y}, \tau)}{\partial y_i \partial y_j} dV(\mathbf{y}) d\tau \tag{9.2.22}$$

This simplifies considerably if we assume a completely incompressible flow because we can approximate $T_{ij}$ by $\rho_o v_i v_j$ (Section 4.2.1) and the tailored Green's function, for a point on the surface as $G_T = \delta(\tau - t)/2\pi|\mathbf{x} - \mathbf{y}|$, since the distance from the sources and their images to the surface is the same. These approximations are valid very close to the source where the fluctuations are dominated by the hydrodynamic part of the flow and the acoustic waves can be ignored because they are of a completely different scale. Using these approximations and carrying out the integral over time gives

$$p(\mathbf{x}, t) - p_\infty = \frac{\rho_o}{2\pi} \int_V \left[ \frac{\partial^2 \left( v_i v_j \right)}{\partial y_i \partial y_j} \right]_{(\mathbf{y}, t)} \frac{dV(\mathbf{y})}{|\mathbf{x} - \mathbf{y}|} \tag{9.2.23}$$

where the terms in square brackets are evaluated at $\mathbf{y}$ and $t$. The mean flow speed in the boundary layer varies with height and so we can split the velocity into its mean and unsteady parts as $v_i = U_i + u_i$ and so $v_i v_j = (U_i U_j + U_i u_j + u_i U_j + u_i u_j)$. It follows then that this formulation gives both a steady and unsteady surface pressure. The unsteady part of the surface pressure is obtained by subtracting the mean of the fluctuations and gives the nonlinear velocity term in Eq. (9.2.23) as

$$v_i v_j - \overline{v_i v_j} = U_i u_j + u_i U_j + u_i u_j - \overline{u_i u_j}$$

Then making use of the relationship for an incompressible flow that was used in Eq. (4.2.2), we find that, if the mean flow speed is parallel to the boundary in the $y_1$ direction, then

$$\frac{\partial^2}{\partial y_i \partial y_j}(U_i u_j + u_i U_j) = 2\frac{\partial U_1}{\partial y_2}\frac{\partial u_2}{\partial y_1}$$

Using these results in Eq. (9.2.23) then gives

$$p(\mathbf{x},t) - p_o = \frac{\rho_o}{2\pi}\int_V \left[2\frac{\partial U_1}{\partial y_2}\frac{\partial u_2}{\partial y_1} + \frac{\partial^2 \left(u_i u_j - \overline{u_i u_j}\right)}{\partial y_i \partial y_j}\right]_{(\mathbf{y},t)} \frac{dV(\mathbf{y})}{|\mathbf{x} - \mathbf{y}|} \tag{9.2.24}$$

The two terms of the integrand are referred to as the rapid and the slow term, respectively. The "rapid" term incorporates the mean-velocity gradient and this is thought of as responding immediately to changes in the mean flow, whereas the "slow" term responds only indirectly as a consequence of the influence of the mean flow on the turbulence structure. The rapid term is often assumed to dominate in boundary layers, and thus the usual focus of modeling, though DNS calculations show both contributions to be of similar magnitude [25]. In free turbulent flows, where mean-velocity gradients are small, the slow term dominates [26].

At a fundamental level, Eq. (9.2.24) shows that the pressure fluctuation at a point on the wall $\mathbf{x} = (x_1, 0, x_3)$ will be given by the integral of fluctuating velocities over the boundary layer, weighted by the inverse of the distance from that point. Thus the pressure fluctuation will tend to reflect contributions from small scale turbulent motions just above the wall as well as from larger scale motions from the outer part of the boundary layer that are coherent over a substantial volume of the flow. Scale decompositions of the pressure fluctuation into frequency or wavenumber-frequency spectra therefore reveal these different contributions.

Fig. 9.20 shows the wall-pressure frequency spectrum $G_{pp}(\omega)$ measured under a flat plate turbulent boundary layer flow as a function of momentum-thickness Reynolds number $Re_\theta$. At low frequencies pressure fluctuations are predominantly generated by the large structures in the outer part of the boundary layer with scales on the order of the boundary layer thickness $\delta$. The pressure fluctuations generated by these structures should scale on the velocity difference with the free stream that sustains them and thus, consistent with the law of the wake, should scale on $\rho_o u_\tau^2$. At the same time they are carried over the wall at a convection velocity that will likely be proportional to the edge velocity $U_e$ (Fig. 9.19) and so their passage frequency will scale on $U_e/\delta$. Thus we expect the spectrum in this region to have a fixed form when plotted as:

$$\frac{G_{pp}(\omega)}{\left(\rho_o u_\tau^2\right)^2}\frac{U_e}{\delta} = g_o\left(\frac{\omega\delta}{U_e}\right) \tag{9.2.25}$$

As demonstrated in Fig. 9.20A we see exactly this behavior, with pressure spectra measured over a 3:1 range of Reynolds numbers grouping into a narrow band that, in this case, extends up to $\omega\delta/U_e \approx 100$. Note that there are a number of (mostly minor) variations of the outer scaling that are commonly used, such as using $\delta^*$ as the distance scale, or assuming a convective velocity scaled on $u_\tau$.

The small near-wall eddies contributing to the pressure spectrum at high frequencies will have sizes determined by the same viscous scale $\nu/u_\tau$ that determines the mean-velocity profile at the bottom of the boundary layer. These structures will be moving at flow-speeds that vary with $u_\tau$ and produce turbulent velocity fluctuations that scale with $u_\tau$ and thus, presumably, produce pressure fluctuations that scale as $\rho_o u_\tau^2$. The high frequency part of the spectrum should therefore appear invariant when normalized as:

$$\frac{G_{pp}(\omega)\, u_\tau^2}{\left(\rho_o u_\tau^2\right)^2 \nu} = g_i\left(\frac{\omega\nu}{u_\tau^2}\right) \tag{9.2.26}$$

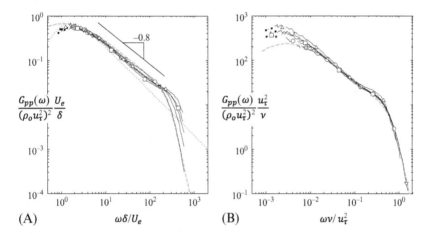

**Fig. 9.20** Flat plate turbulent boundary layer wall-pressure spectra scaled using (A) outer and (B) inner scales. Experimental data of Forest [24]: $\bigtriangledown$, $Re_\theta = 27{,}400$; $\circ$, 37,700; $\diamondsuit$, 52,000; $\square$, 64,300; $\triangle$, 76,700. Models scaled using boundary layer parameters for $Re_\theta = 27{,}400$; — — —, Goody [27]; · · · · ·, Howe [37].

Again this behavior is realized, as illustrated in Fig. 9.20B. It also appears from Fig. 9.20 that there is a mid-frequency range where both the above scalings exist simultaneously. In that case we must have that

$$\frac{\delta}{U_e} g_o\left(\frac{\omega\delta}{U_e}\right) = \frac{\nu}{u_\tau^2} g_i\left(\frac{\omega\nu}{u_\tau^2}\right) \tag{9.2.27}$$

This is only possible if

$$
g_o\left(\frac{\omega\delta}{U_e}\right) \propto \left(\frac{\omega\delta}{U_e}\right)^{-1} \quad \text{and} \quad g_i\left(\frac{\omega\nu}{u_\tau^2}\right) \propto \left(\frac{\omega\nu}{u_\tau^2}\right)^{-1} \tag{9.2.28}
$$

and thus we expect the pressure spectrum to have a $-1$ slope in the overlap region when plotted on a log-log scale. Mysteriously almost all turbulent boundary layer experiments, like that represented in Fig. 9.20 reveal a mid-frequency region with a slope of $-0.7$ to $-0.8$. A $-1$ region has been seen in atmospheric boundary layer measurements [29], a finding that may indicate that this behavior may be very slow to appear with increase in Reynolds number.

Analysis [30,31] of the fundamental constraints on the pressure spectrum and of the turbulent contributions to the integrand of Eq. (9.2.24), as well as measurements [32,33] have established other power-law regions in the wall-pressure time spectrum, including a $(\omega\delta/U_e)^2$ region at low frequencies, and a $\left(\omega\nu/u_\tau^2\right)^{-5}$ region at very high frequencies where the pressure fluctuations are generated by action in the viscous sublayer.

An empirical interpolation formula for the wall-pressure frequency spectrum, which takes advantage of these scaling regions, was developed by Goody [27] and has the form:

$$
\frac{G_{pp}(\omega)}{\left(\rho_o u_\tau^2\right)^2}\frac{U_e}{\delta} = \frac{C_2(\omega\delta/U_e)^2}{\left[(\omega\delta/U_e)^n + C_1\right]^{3.7} + \left[C_3 R_T^{-4/7}(\omega\delta/U_e)\right]^7} \tag{9.2.29}
$$

where $R_T$ is the ratio of the outer to inner layer timescales $(\delta/U_e)/(\nu/u_\tau^2)$. Goody recommends the constants:

$$
C_1 = 0.5, \ C_2 = 3.0, \ C_3 = 1.1 \ \text{and} \ n = 0.75 \tag{9.2.30}
$$

Goody's equation reduces at low frequencies $\omega\delta/U_e \ll 1$ to

$$
\frac{G_{pp}(\omega)}{\left(\rho_o u_\tau^2\right)^2}\frac{U_e}{\delta} = \frac{C_2}{C_2^{3.7}}\left(\frac{\omega\delta}{U_e}\right)^2 \tag{9.2.31}
$$

and at high frequencies $\omega\delta/U_e \gg \left(C_3 R_T^{4/7}\right)^{7/(7-3\cdot7n)}$ to

$$
\frac{G_{pp}(\omega)}{\left(\rho_o u_\tau^2\right)^2}\frac{u_\tau^2}{\nu} = \frac{C_2}{C_3^7}\left(\frac{\omega\nu}{u_\tau^2}\right)^{-5} \tag{9.2.32}
$$

If $R_T$ is sufficiently large then a mid-frequency "overlap" range will exist where

$$
\frac{G_{pp}(\omega)}{\left(\rho_o u_\tau^2\right)^2}\frac{U_e}{\delta} = C_2\left(\frac{\omega\delta}{U_e}\right)^{2-3\cdot7n} \tag{9.2.33}
$$

Goody chose $n = 0.75$ based on comparisons with lower-Reynolds number boundary layer measurements than those of Fig. 9.20 where the overlap region slope was observed to be about $-0.7$. Note that $R_T$ was not sufficient in those cases to realize the full implied slope in the overlap region of $-0.775$. Choosing $n = 0.79$ agrees very well with the higher Reynolds number, higher slope, and data of Fig. 9.20 and also comes close to satisfying the infinite Reynolds number limit that requires the overlap region slope of $-1$ predicted by dimensional analysis. The important take-away here is the need for adjustment of the parameter $n$ according to Reynolds number if accuracy in the mid-frequency range is important.

The full wavenumber-frequency spectrum of surface pressure fluctuations $\Phi_{pp}(k_1, k_3, \omega)$, as defined in Eq. (8.4.35), is of particular interest for aeroacoustic calculations. This spectrum captures the spatial scales on which the pressure fluctuations occur at each frequency and thus is the source term in surface and trailing edge noise applications. It also captures convection of turbulence over the wall at $U_c$ and thus includes a convective ridge that lies along $k_1 = \omega/U_c$. Because of its importance, considerable effort has gone into developing models for the wavenumber-frequency spectrum. We include here the details of a sophisticated model developed by Chase [28] and a more elemental model due to Corcos [34] with the intent that these span the range of need from accuracy to analytical simplicity. A number of other such models have been proposed which may be more easily applied in certain situations, many of which are reviewed by Graham [35], Liu and Dowling [36], and Howe [37]. The Chase model spectrum, in the incompressible limit, is given by the expression:

$$\Phi_{pp}(k_1, k_3, \omega) = \frac{\rho_o^2 u_\tau^3}{\left[\kappa_+^2 + (b\delta)^{-2}\right]^{5/2}} \left\{ C_T \kappa^2 \frac{\left[\kappa_+^2 + (b\delta)^{-2}\right]}{\left[\kappa^2 + (b\delta)^{-2}\right]} + C_M k_1^2 \right\} \tag{9.2.34}$$

where $\kappa$ is the magnitude of the in-plane wavenumber vector $\kappa^2 = k_1^2 + k_3^2$, and $\kappa_+$ is $\kappa$ modified according to the difference of $k_1$ and its value on the convective ridge, $\omega/U_c$, normalized on a multiple $h$ of the friction velocity, so

$$\kappa_+^2 = \kappa^2 + \left[\left(\frac{\omega}{U_c} - k_1\right)\frac{U_c}{h u_\tau}\right]^2 \tag{9.2.35}$$

Chase recommended that the empirical constants take the values

$$h = 3, \ C_T h = 0.014, \ C_M h = 0.466, \ b = 0.75 \tag{9.2.36}$$

and the range of validity is $U_e/c_o \ll 1$, $\kappa \gg |\omega|/c_o$, and $\omega\delta/U_e > 1$.

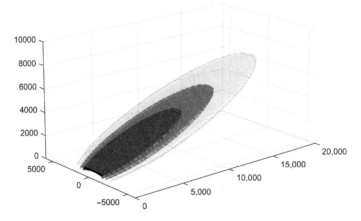

**Fig. 9.21** Wavenumber-frequency spectrum from Eq. (9.2.34) at conditions corresponding to the model Chase frequency spectrum in Fig. 9.20A. Contour spacing is 4 dB.

Fig. 9.21 shows a perspective view of the Chase spectrum in wavenumber frequency space which forms a thin tongue aligned with the convective ridge. The spectrum extends further in spanwise wavenumber ($k_3$) than streamwise ($k_1$) because the boundary layer turbulence correlates over significantly greater distances in the streamwise than spanwise direction. The exact frequency spectrum corresponding to Eq. (9.2.34) can be derived [35], but a simpler, approximate form due to Chase [31] and given by Howe [37] is,

$$\frac{G_{pp}(\omega)}{(\rho_o u_\tau^2)^2} \frac{U_e}{\delta^*} = \frac{2(\omega\delta^*/U_e)^2}{\left[(\omega\delta^*/U_e)^2 + 0.0144\right]^{3/2}} \tag{9.2.37}$$

This spectral form is compared with data and Goody's model in Fig. 9.20A. We see that the Chase model includes no viscous range at high frequencies, and at mid-frequencies takes on the $-1$ slope expected at very high Reynolds number.

The Corcos spectrum provides an algebraically simpler wavenumber-frequency spectrum model with the form

$$\Phi_{pp}(k_1, k_3, \omega) = \frac{S_{pp}(\omega)U_c^2}{\pi^2\omega^2} \frac{\alpha_1\alpha_3}{\left[\alpha_1^2 + (U_c k_1/\omega - 1)^2\right]\left[\alpha_3^2 + U_c^2 k_3^2/\omega^2\right]} \tag{9.2.38}$$

with empirical constants $\alpha_1 = 0.1$ and $\alpha_3 = 0.77$. Note that Corcos' formulation leaves the frequency spectrum $S_{pp}(\omega)$ unspecified.

# References

[1] G.K. Batchelor, The Theory of Homogeneous Turbulence, Cambridge University Press, Cambridge, 1953.

[2] T. von Kármán, Progress in the statistical theory of turbulence, Proc. Natl. Acad. Sci. U. S. A. 34 (1948) 530–539.

[3] S.B. Pope, Turbulent Flows, Cambridge University Press, Cambridge, 2000.

[4] H.W. Liepmann, J. Laufer, K. Liepmann, On the Spectrum of Isotropic Turbulence, NACA TN 2473, NACA, Washington, DC, 1951.

[5] I.J. Wygnanski, F. Champagne, B. Marasli, On the large-scale structures in two-dimensional, small-deficit, turbulent wakes, J. Fluid Mech. 168 (1986) 31–71.

[6] H. Tennekes, J.L. Lumley, A First Course in Turbulence, The MIT Press, Cambridge, MA, 1972.

[7] A.A. Townsend, The Structure of Turbulent Shear Flow, Cambridge University Press, Cambridge, 1956.

[8] W.J. Devenport, C. Muthanna, R.L. Ma, S. Glegg, Two-point descriptions of wake turbulence with application to noise prediction, AIAA J. 39 (2001) 2302–2307.

[9] J. Schetz, R.D.W. Bowersox, Boundary Layer Analysis, second ed., AIAA, Reston, VA, 2011.

[10] M. Drela, Xfoil, 2001–2013. From http://web.mit.edu/drela/Public/web/xfoil/.

[11] H.M. Nagib, K.A. Chauhan, P.A. Monkewitz, Approach to an asymptotic state for zero pressure gradient turbulent boundary layers, Philos. Trans. A Math. Phys. Eng. Sci. 365 (2007) 755–770.

[12] H. Schlichting, Boundary-Layer Theory, seventh ed., McGraw-Hill, New York, 1979.

[13] M. Awasthi, High Reynolds Number Turbulent Boundary Layer Flow Over Small Forward Facing Steps (M.S. thesis), Virginia Polytechnic Institute and State University, Blacksburg, VA, 2012.

[14] E.R. van Driest, On turbulent flow near a wall, J. Aeronaut. Sci. 23 (1956) 1007–1011.

[15] D. Coles, The law of the wake in the turbulent boundary layer, J. Fluid Mech. 1 (1956) 191–226.

[16] A.J. Smits, B.J. McKeon, I. Marusic, High-Reynolds number wall turbulence, Annu. Rev. Fluid Mech. 43 (2011) 353–375.

[17] D.B. DeGraaff, J. Eaton, Reynolds-number scaling of the flat-plate turbulent boundary layer, J. Fluid Mech. 422 (2000) 319–346.

[18] P. Bradshaw, 'Inactive' motion and pressure fluctuations in turbulent boundary layers, J. Fluid Mech. 30 (1967) 241–258.

[19] P.-A. Krogstad, R.A. Antonia, Structure of turbulent boundary layers on smooth and rough walls, J. Fluid Mech. 277 (1994) 1–21.

[20] J.M. Österlund, B. Lindgren, A.V. Johansson, Flow structures in zero pressure-gradient turbulent boundary layers at high Reynolds numbers, Eur. J. Mech. B Fluids 22 (2003) 379–390.

[21] W. Blake, Turbulent boundary-layer wall-pressure fluctuations on smooth and rough walls, J. Fluid Mech. 44 (1970) 637–660.

[22] M.K. Bull, Wall-pressure fluctuations associated with subsonic turbulent boundary layer flow, J. Fluid Mech. 28 (1967) 719–754.

[23] W.W. Willmarth, C.E. Woolridge, Measurements of the fluctuating pressure at the wall beneath a thick turbulent boundary layer, J. Fluid Mech. 14 (1962) 187–210.

[24] J.B. Forest, The Wall Pressure Spectrum of High Reynolds Number Rough-Wall Turbulent Boundary Layers (M.S. thesis), Virginia Polytechnic Institute and State University, Blacksburg, VA, 2012.

[25] J. Kim, On the structure of pressure fluctuations in simulated turbulent channel flow, J. Fluid Mech. 205 (1989) 421–451.

[26] W.K. George, P.D. Beuther, R.E.A. Arndt, Pressure spectra in turbulent free shear flows, J. Fluid Mech. 148 (1984) 155–191.

[27] M. Goody, Empirical spectral model of surface pressure fluctuations, AIAA J. 42 (2004) 1788–1794.

[28] D.M. Chase, The character of the turbulent wall pressure spectrum at subconvective wavenumbers and a suggested comprehensive model, J. Sound Vib. 112 (1987) 125–147.

[29] J.C. Klewicki, P.J.A. Priyadarshana, M.M. Metzger, Statistical structure of the fluctuating wall pressure and its in-plane gradients at high Reynolds number, J. Fluid Mech. 609 (2008) 195–220.

[30] W. Blake, Mechanics of Flow Induced Sound and Vibration, Academic Press, Orlando, FL, 1986.

[31] D.M. Chase, Modeling the wavevector-frequency spectrum of turbulent boundary layer wall pressure, J. Sound Vib. 70 (1980) 29–67.

[32] T.M. Farabee, M.J. Casarella, Spectral features of wall pressure fluctuations beneath turbulent boundary layers, Phys. Fluids A 3 (1991) 2410–2420.

[33] S.P. Gravante, A.M. Naguib, C.E. Wark, H.M. Nagib, Characterization of the pressure fluctuations under a fully developed turbulent boundary layer, AIAA J. 36 (1998) 1808–1816.

[34] G.M. Corcos, The structure of the turbulent pressure field in boundary-layer flows, J. Fluid Mech. 18 (1964) 353–378.

[35] W.R. Graham, A comparison of models for the wavenumber-frequency spectrum of turbulent boundary layer pressures, J. Sound Vib. 206 (1997) 541–565.

[36] Y. Liu, A.P. Dowling, Assessment of the contribution of surface roughness to airframe noise, AIAA J. 45 (2007) 855–869.

[37] M.S. Howe, Acoustics of Fluid-structure Interactions, Cambridge University Press, Cambridge, 1998.

[38] M.A. Morton, W.J. Devenport, S. Glegg, Rotor inflow noise caused by a boundary layer: inflow measurements and noise predictions, in: 18th AIAA/CEAS Aeroacoustics Conference, Colorado Springs, Colorado, 2012.

# Part 2

# Experimental approaches

# Aeroacoustic testing and instrumentation

**10**

Aeroacoustic predictions must depend to some degree on assumptions or simplifications concerning the physics of sound sources and details of their modeling. Experimental testing provides the most direct way to observe sources, infer their physics, and provide quantitative results against which prediction methods can be validated.

Most experimental testing in aeroacoustics is carried out using wind tunnels. Unless information is required at full scale about a large vehicle or application, running a wind tunnel is usually far less expensive than conducting a field test. It also provides a more controlled environment that can be particularly useful when insight into the fundamental physics is desired. Wind tunnel testing is the focus of this chapter. Some of the different wind tunnel configurations used for low Mach number experimental work are described, as are the acoustic corrections that must be applied to most wind tunnel measurements. This chapter concludes with an overview of some pressure and velocity measurement techniques often used in aeroacoustic testing.

## 10.1 Aeroacoustic wind tunnels

The ideal aeroacoustic wind tunnel is one that can accurately reproduce the aerodynamics of the device or flow configuration of interest while providing capability for the measurement of the far-field sound that it generates with the minimum of background noise. Accuracy in aerodynamics at low Mach number implies a low-turbulence and closely uniform free stream, model Reynolds numbers that realistically represent the application of interest (which are usually high), and small predictable interference corrections. Aerodynamic interference refers to the differences between the desired flow and the modeled flow which result from the finite extent of the wind tunnel stream. Small corrections allow larger models to be used and thus lead to higher achievable Reynolds numbers.

The standard wind tunnel configuration for aerodynamic studies is the closed test section. Parallel, or nearly parallel, rigid test section walls guide the flow over the model. This type of test section has been in use since at least the time of the Wright brothers and is extremely well understood. Interference corrections are well known and comparatively small and allow for the use of quite large models compared to the test section size. Such wind tunnels can be used for acoustic measurements. Microphones may be placed in the flow using aerodynamic mounts and nose cones or mounted on, or recessed within, the wind tunnel walls. The QinetiQ 5 m wind tunnel at Farnborough in the United Kingdom is a classic example of one such facility. The 5 m × 4.2 m test section, with a model under test, is shown in Fig. 10.1. Acoustic measurements in this facility have included the use of arrays of microphones embedded in the hard surface of the test section ceiling. This facility, built in the 1970s, is used for

Aeroacoustics of Low Mach Number Flows. http://dx.doi.org/10.1016/B978-0-12-809651-2.00010-2

testing up to a Mach number of 0.33 and can be pressurized up to 3 atm. This allows for independent control of test Mach number and Reynolds number.

**Fig. 10.1** Hard-wall test section of the QinetiQ 5 m wind tunnel at Farnborough, United Kingdom, with a model of an A300 aircraft under test.
Image provided by Ian Smith, QinetiQ.

To reduce acoustic reflections and reverberation, the walls of a closed test section wind tunnel may be treated, such as has been done in the NASA Glenn 9- by 15-Foot Low-Speed Wind Tunnel [1]. Fig. 10.2 shows the test section of this tunnel which is used extensively for testing the aeroacoustics of aircraft propulsion systems, including model turbofans and propellers. The photograph shows a number of in-flow microphones surrounding a model engine as well as, to the left of the picture, a traversable microphone rake. Treatment applied to the test section walls to reduce acoustic reflections, covered by metal perforate panels, is also visible.

The performance of microphones or a microphone array in a closed test section facility can be substantially improved by shielding it from the flow to minimize pressure fluctuations produced by turbulence which can overwhelm the acoustic pressure signal. A common arrangement is to recess the microphone array into a shallow cavity in one of the test section walls, covered with an acoustically transparent membrane. This approach was pioneered by Jaeger and coworkers at NASA Ames [2] who found that very light, plain weave Kevlar fabric provided a suitable covering, being both acoustically transparent and strong enough to stand up to extended exposure to the flow. Fig. 10.3 shows an example of such a recessed array in the closed test section of the Virginia Tech Stability Wind Tunnel.

Useful acoustic measurements in closed test section wind tunnels are easier in large facilities, where the acoustic far field can be reached within the test section, and in the

**Fig. 10.2** Test section of the NASA Glenn 9- by 15-Foot Low-Speed Wind Tunnel. Image provided by David Stephens, NASA Glenn Research Center.

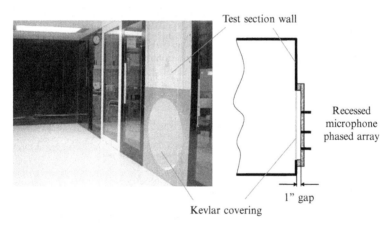

**Fig. 10.3** Thirty-two microphone phased array mounted in the closed test section of the Virginia Tech Stability Wind Tunnel.

investigation of louder sources that are not overwhelmed by reverberation or background facility noise. Dominant sources of background noise are usually the wind tunnel fan, turning vanes, and the roughness noise from walls both up- and downstream of the test section. Microphones placed in the flow, or on the test section walls, may be directly exposed to such parasitic sources.

A much more optimal arrangement from an acoustic perspective is the open-jet test section. Here the test section walls are partially or completely eliminated, and the free stream is projected as a jet across an anechoic chamber. The model is then placed in the jet so that the sound it produces radiates through the jet shear layer into the chamber where microphone instrumentation is placed. Free-jet tunnels come in a variety of arrangements. One example is the Quiet Flow Facility (QFF) at NASA Langley

Research Center [3], shown in Figs. 10.4 and 10.5. Air is supplied from a low-pressure fan through a diffuser, including a set of acoustic splitters to suppress noise from the fan, and into a plenum. Air from the plenum is accelerated to test speed through a contraction and exhausted into a large anechoic chamber through a vertical $2 \times 3$-foot nozzle (pictured in Fig. 10.4). For the experiment shown in Fig. 10.4 [4], an airfoil

**Fig. 10.4** The Quiet Flow Facility at NASA LaRC. Test-section area showing an airfoil model under test and microphone phased array in the foreground.
Image provided by Florence Hutcheson and Michael Doty, NASA Langley Research Center.

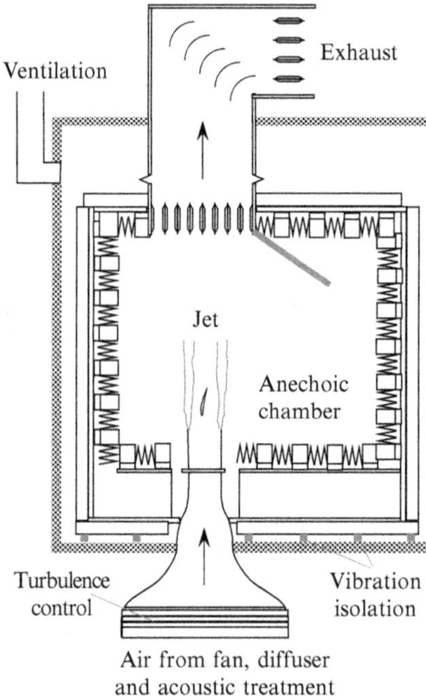

Ventilation

Exhaust

Jet

Anechoic
chamber

Turbulence
control

Vibration
isolation

Air from fan, diffuser
and acoustic treatment

**Fig. 10.5** Schematic of the Quiet Flow Facility at NASA LaRC.

model is mounted in the jet between two large end plates attached to the nozzle exit. The jet is allowed to diffuse over the height of the anechoic chamber where it is exhausted to atmosphere through a large duct (Fig. 10.5).

This type of open-circuit arrangement can produce extremely quiet flow conditions and is especially suited to smaller facilities. Flow losses associated with the acoustic treatment used to remove the fan noise can be minimized by arranging for low flow speeds in the bulk of the air delivery system. The exhaust system can be placed sufficiently far from the nozzle to keep speeds low here also and thus minimize parasitic noise generation. The NASA facility also has vibration isolation and ventilation of the anechoic chamber to minimize structure-borne noise and to prevent instability as the flow exhausts from the chamber.

**Fig. 10.6** Test section of the DNW-NWB low-speed wind tunnel showing 3.25 m × 2.8 m nozzle to the left and the collector to right.
Image provided by Andreas Bergmann, DNW.

For larger facilities it becomes more important to recover the kinetic energy of the jet. Such facilities thus usually have a closed circuit arrangement and a *collector* designed to receive the jet back into the tunnel duct with minimum loss. Figs. 10.6 and 10.7 show the test section and layout of the DNW-NWB low-speed wind tunnel with its open-jet configuration [5]. The 3.25 m × 2.80 m nozzle that launches the jet is visible to the left of this photograph. Six meters downstream the somewhat larger collector is visible to the right of the picture. This is designed to operate well even with lifting models in the jet [6]. The entire arrangement is enclosed within a large anechoic chamber.

In general, the design of this type of facility requires some care. Impingement of the jet shear layer on the collector surfaces may produce significant noise and, through resonance with the rest of the tunnel circuit, can generate large-scale flow pulsations known as *pumping*. Test models that substantially deflect the jet can exacerbate these problems. Pumping has been eliminated at the DNW-NWB [7] by introducing a sudden change in the cross-sectional area of the flow path just ahead of the drive fan (see Fig. 10.7). This produces a discontinuity in the acoustic impedance which acts to suppress resonances.

While the open-jet test section arrangement can provide a near optimal solution to the acoustic goals of a wind tunnel test, this configuration can suffer from large

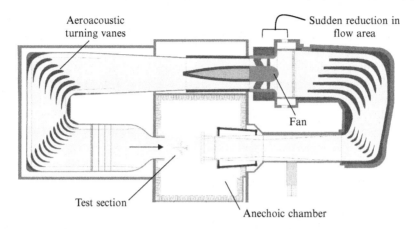

**Fig. 10.7** Plan view schematic of the DNW-NWB low-speed wind tunnel. Image provided by Andreas Bergmann, DNW.

aerodynamic interference effects depending on the size of the model compared to the test section. These may limit the accuracy of the aerodynamic data obtained so that separate test runs in different wind tunnels or test sections may be needed to adequately characterize both aerodynamic and acoustic performance. In airfoil testing the aerodynamic interference is primarily produced by the deflection of the jet. To a first level of approximation the effect of this is to reduce the angle of attack experienced by the airfoil from the geometric value implied by the airfoil chord line and the test section axis to an angle that accounts for the distortion of the wind tunnel flow. In closed wall tunnels, the dominant correction is referred to as the *blockage*. This is characterized as a change in the effective free stream velocity $U_\infty$ compared to the actual speed of the on-coming flow. Blockage also produces a smaller angle of attack error.

**Fig. 10.8** Schematics and nomenclature for interference calculations for (A) a closed test section, (B) an open-jet test section, and (C) a hybrid test section.

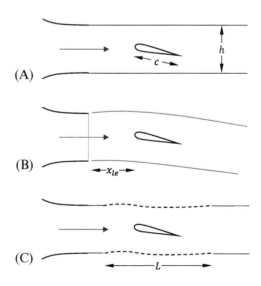

To predict corrections for an open-jet test section usually requires a numerical calculation of some type. Figs. 10.8 through 10.10 show example calculations. Fig. 10.8B is a schematic of a NACA 0012 airfoil in an open jet wind tunnel test section. The airfoil is placed at a geometric angle of attack of 5 degrees in the center of a free jet with its leading edge a distance $x_{le}$ downstream of the nozzle exit of height $h$. A two-dimensional panel method [8] is used to calculate the flow around the airfoil and the associated deflection of the jet boundaries, these being modeled as surfaces of constant pressure. The best match between the computed airfoil pressure distribution and a parallel free flight calculation is then used to determine the corrected angle of attack and free stream velocity.

Fig. 10.9 shows the corrections to angle of attack, $(\alpha' - \alpha)/\alpha'$, and free stream velocity $(U'_\infty - U_\infty)/U'_\infty$ computed as functions of chord length to test section height ratio $c/h$ for a free jet with $x_{le}/h = 1$. Also shown are these corrections computed using a similar method for a closed test section (Fig. 10.8A). In these ratios $\alpha$ and $U_\infty$ denote the effective angle of attack and free stream velocity, whereas $\alpha'$ and $U'_\infty$ denote the geometric angle of attack and nominal facility free stream. Note that closed test section corrections for airfoils can be accurately estimated analytically using well-established methods [9].

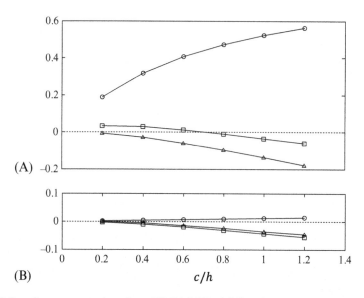

**Fig. 10.9** Interference corrections for a NACA 0012 airfoil at 5 degrees geometric angle of attack. Corrections for (A) angle of attack $(\alpha' - \alpha)/\alpha'$, and (B) free stream velocity $(U'_\infty - U_\infty)/U'_\infty$. Symbols: $\triangle$, closed test section; $\circ$, free-jet test section with $x_{le}/h = 1$; and $\square$, hybrid test section with $L/h = 2$.

The airfoil in the open-jet test section experiences almost no blockage effect (Fig. 10.9B), compared to a correction of a few percent in the closed test section. The interference effect on angle of attack is, however, several times that of the closed test section and, in an absolute sense, very large. For example, an airfoil placed in a free jet requires a correction to angle of attack of over 30% if the chord length is half of the test section width, or larger. Unfortunately, this error is also a function of the positioning of the model relative to the nozzle exit, as shown in Fig. 10.10 where it has been plotted as a function of both $x_{le}$ and $c/h$. Changing the airfoil profile significantly from a NACA 0012 airfoil section also noticeably affects the correction.

**Fig. 10.10** Corrections for angle of attack $(\alpha' - \alpha)/\alpha'$ for a NACA 0012 placed in a free jet at $\alpha' = 5$ degrees as a function of streamwise position $x_{le}/h$ and chord length $c/h$.

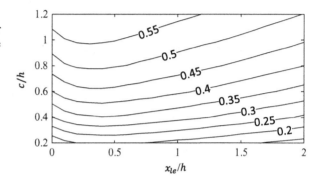

A significant problem with the size of the angle of attack effect is that it can change the character of the flow and thus render the fundamental assumptions of the aerodynamic correction invalid, namely, that the effects of the test flow boundaries are inviscid and in the aerodynamic far field. If this happens, then no effective angle of attack can be found where the pressure distribution resembles that produced in free flight [10]. In such a circumstance there is no other option than to reduce the size of the model relative to the jet or abandon the quantitative relationship of the test to conditions independent of the wind tunnel configuration. Such a test can still be of value, of course, if the goal of the experiment is fundamental scientific insight into phenomena still present in the wind tunnel flow or the validation of a computational method that is used to simulate both the free jet and the model mounted within it.

A third type of configuration used in aeroacoustic testing is the hybrid anechoic tunnel (Fig. 10.11). Here the hard walls of the closed test section are replaced with acoustically transparent walls (termed *acoustic windows*), generally made from Kevlar fabric. Acoustic instrumentation is placed in an anechoic chamber or chambers external to the acoustic windows. The acoustic windows contain the flow and thus limit the aerodynamic interference, at the same time as enabling sound measurements in the acoustic far field in much the same way as for a free jet.

Fig. 10.11 shows the layout of the hybrid acoustic test section of the Virginia Tech Stability Wind Tunnel [11]. This facility has a square test section 1.83 m on edge. The side walls of the test section consist of 4.21-m long 1.83-m high acoustic windows. Sound generated in the flow passes through these windows into two large anechoic chambers that sit to either side of the test section. The floor and ceiling of the test section include treated flow surfaces designed to minimize acoustic reflections. Fig. 10.12 shows an airfoil model mounted vertically in this test section, as well

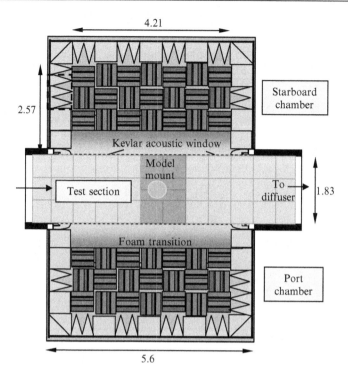

**Fig. 10.11** Plan view schematic of the anechoic test section system of the Virginia Tech Stability Wind Tunnel.
From W.J. Devenport, R.A. Burdisso, A. Borgoltz, P.A. Ravetta, M.F. Barone, K.A. Brown, M.A. Morton, The Kevlar-walled anechoic wind tunnel, J. Sound Vib. 332 (2013) 3971–3991.

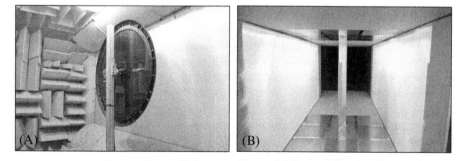

**Fig. 10.12** The anechoic test section system of the Virginia Tech Stability Wind Tunnel configured for an airfoil trailing edge noise measurement. (A) Phased array system in the port-side anechoic chamber and (B) airfoil in the test section.

as a phased microphone array system placed in one of the anechoic chambers to measure the airfoil trailing edge noise. Figs. 10.13 and 10.14 show other implementations of this concept at the 2 m × 2 m wind tunnel at the Japan Aerospace Agency (JAXA) in Tokyo, and the Anechoic Flow Facility at the Naval Surface Warfare Center, in Carderock, Maryland. The JAXA facility is configured similarly to the Virginia Tech Stability Tunnel, with the exception that only the starboard-side Kevlar acoustic window is backed by a full anechoic chamber. The comparatively shallow space-saving chamber on the port side serves, primarily, as an acoustic absorber.

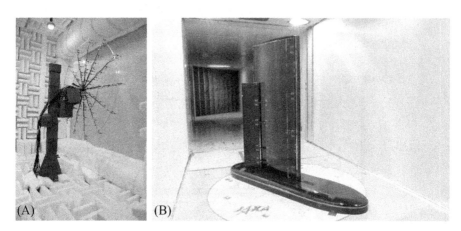

**Fig. 10.13** The hybrid anechoic test section system of the JAXA 2 m × 2 m Wind Tunnel. (A) Phased array system in the starboard-side anechoic chamber and (B) high-lift OTOMO model installed in the test section.
Images provided by Hiroki Ura, JAXA.

The Carderock facility represents a somewhat different arrangement used to contain the 8-foot diameter jet of this otherwise conventional open-jet wind tunnel. The setup shown is for a surface flow noise test on a large model that divides the test section into left and right halves. The facility has an octagonal nozzle necessitating segmented acoustic windows.

The basic idea behind the hybrid configuration is to combine the better features of closed and free-jet test-section configurations for aeroacoustic testing. From the aerodynamic perspective, enclosing the jet in Kevlar walls reduces interference corrections to closed-test section levels. That is not to say that closed-test section methods can be used without modification to estimate corrections, as the porosity and flexibility of the acoustic windows are both significant factors in determining these. Corrections require a test section flow simulation that incorporates these effects. Fig. 10.9 shows example calculations of interference effects for a NACA 0012 airfoil (Fig. 10.8C) computed using an inviscid panel method coupled with a membrane solver and porosity model [11]. Even for large $c/h$ the angle of attack corrections remain small. Free-stream velocity corrections are the same size as those for a closed

Kevlar
acoustic
windows

Foil used for surface flow noise study

**Fig. 10.14** Photographs of the Anechoic Flow Facility at NSWC Carderock Division with acoustic windows installed.
Image provided by Jason Anderson, NSWCCD.

test section. As evidenced in Fig. 10.15, these corrections are almost independent of the length of the acoustic windows relative to the test section height $L/h$, for $L/h$ greater than about 2. Note that the angle of attack correction actually passes through zero for $c/h \cong 0.5$. This is because there are two contrary influences on this parameter—the porosity of the acoustic windows (which tends to lower angle of attack) and their flexibility (which tends to increase blockage and thus angle of attack). The much lower interference effects compared to a free jet permit testing of substantially larger models, and thus a higher range of Reynolds numbers.

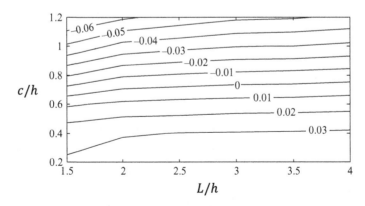

**Fig. 10.15** Corrections for angle of attack $(\alpha' - \alpha)/\alpha'$ for a NACA 0012 placed at $\alpha' = 5$ degrees in a hybrid acoustic test section with Kevlar 120 acoustic windows tensioned at 1500 N/m as a function of acoustic window length $L/h$ and chord length $c/h$.

The acoustic windows of a hybrid test section eliminate the need for a jet collector (and eliminate any noise it might generate) and permit a much longer test section compared to its width. Acoustic instrumentation can also be placed much closer to the flow, and thus the sources of interest, than it can be for a jet (Fig. 10.12A). The net effect of these two attributes is that a phased microphone array placed close to the outside of an acoustic window receives sound from a comparatively long stretch of the flow making it easier to distinguish the sounds produced from a model placed at the center of the test section from the parasitic noise sources of the wind tunnel, which appear at the ends of the test section. These acoustic benefits are balanced against the necessity of accounting for acoustic losses for sound transmission through the windows (see Section 10.2.4) and, at frequencies greater than about 15 kHz, the presence of some parasitic noise generated by the flow over the acoustic window material.

## 10.2 Wind tunnel acoustic corrections

When sound measurements are made in the anechoic chamber surrounding the test section of an open-jet or hybrid anechoic wind tunnel, the sound produced in the flow must cross a shear-layer as it propagates from model to microphone. The shear layer reflects some of the sound and refracts the remainder, bending the otherwise direct path of propagation to the microphone. This changes the amplitude of the sound and its apparent directivity. Some of the amplitude change results from the increased distance traveled by the refracted waves and from the distortion of the wave field produced by the refraction. In a hybrid anechoic tunnel additional amplitude attenuation is produced by the insertion loss of the acoustic window.

We begin by ignoring the shear layer thickness and a possible acoustic window, deferring discussion of these effects to later in this section. Without these complications this situation is amenable to mathematical analysis which provides formulae that can be used to correct sound measurements back to what would have been measured in the absence of the shear layer.

### 10.2.1 Shear layer refraction

In preparation for addressing this problem, consider a plane sound wave propagating through still air. The wave has a complex pressure amplitude $\hat{p}$, a frequency $\omega$ and thus a wavenumber, in the direction of propagation, of $k = \omega/c_o$. The wave will produce a small in-phase velocity fluctuation in the direction of propagation with an amplitude $\hat{u}$, referred to as the particle velocity. In terms of the acoustic momentum equation, Eq. (3.3.5), we have

$$i\omega\rho_o\hat{u} = \frac{\partial\hat{p}}{\partial s} = ik\hat{p} \quad \text{where} \quad \hat{p} = Ae^{iks} \tag{10.2.1}$$

and $s$ is distance along the path of propagation. Thus the amplitude of the particle velocity fluctuation is

$$\hat{u} = \hat{p}/\rho_o c_o \tag{10.2.2}$$

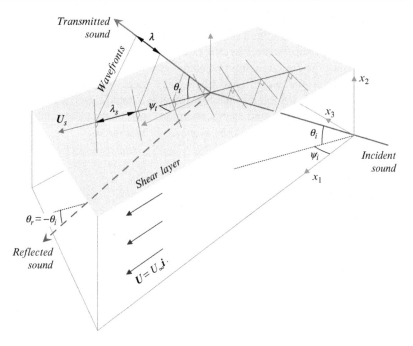

**Fig. 10.16** Interaction of a plane wave with a thin shear layer.

Now consider plane waves encountering an infinitely thin shear layer, as shown schematically in Fig. 10.16. The shear layer is perpendicular to the $x_2$ direction and separates a uniform flow of velocity $\mathbf{U}=U_\infty\mathbf{i}$ in the $x_1$ direction from stationary air. The sound waves incident on the shear layer originate from within the flow and have wavefronts defined by the spherical polar angles $\theta_i$ and $\psi_i$. The sound propagates perpendicular to these fronts at speed $c_o$ at the same time as being swept downstream by the flow at $U_\infty\mathbf{i}$. A portion of the sound is reflected from the shear layer back into the flow. The reflected sound field has the same form as the incident field except that its component of propagation perpendicular to the shear layer is reversed and thus $\theta_r=-\theta_i$. The remaining sound is transmitted through the shear layer and appears on the other side as plane wavefronts propagating along the direction defined by the angles $\theta_t$ and $\psi_t$. There is no convection on this side of the shear layer.

The relationship between the transmitted, reflected, and incident sound is determined by ensuring that these sound fields are consistent where they meet at the shear layer. The sound waves impose pressure variations on the shear layer and associated out-of-plane motion due to the vertical component of the acoustic particle velocity. These variations take the form of a surface wave that moves across the shear layer tracking the acoustic wavefronts that produce it. The reflection mechanism requires that the incident and reflected waves be in phase at the shear layer. The transmitted and incident/reflected waves must also be in phase here, since otherwise there would be no way for the pressure and surface motions to be consistent across the shear layer.

First we will match the velocity of propagation of the surface wave with the transmitted sound wave. The transmitted sound wave propagates at the sound speed $c_o$, and the distance between successive wavefronts in the direction of propagation is the wavelength $\lambda$. Since the direction of propagation makes an angle $\theta_t$ to the shear layer, the distance between the wavefronts where they meet with the shear layer is $\lambda_s = \lambda/\cos\theta_t$. The surface wave must therefore travel at a speed $U_s = c_o/\cos\theta_t$ to match the propagation of the transmitted sound, and thus the velocity of the surface wave expressed in terms of components in the $x_1$ and $x_3$ directions is

$$\mathbf{U}_s = \frac{c_o}{\cos\theta_t}\cos\psi_t \mathbf{i} + \frac{c_o}{\cos\theta_t}\sin\psi_t \mathbf{k} \tag{10.2.3}$$

For the incident wave we must also add the convection by the mean flow and so, in terms of the incident wave angles,

$$\mathbf{U}_s = \left(\frac{c_o}{\cos\theta_i} + U_\infty\cos\psi_i\right)\cos\psi_i \mathbf{i} + \left(\frac{c_o}{\cos\theta_i} + U_\infty\cos\psi_i\right)\sin\psi_i \mathbf{k} \tag{10.2.4}$$

We see that the surface wave speed can also be expressed as $U_s = c_o/\cos\theta_t + U_\infty\cos\psi_i$. Matching components in Eqs. (10.2.3), (10.2.4), we have

$$\frac{\cos\psi_t}{\cos\theta_t} = \frac{\cos\psi_i}{\cos\theta_i} + M\cos^2\psi_i \tag{10.2.5}$$

and

$$\frac{\sin\psi_t}{\cos\theta_t} = \frac{\sin\psi_i}{\cos\theta_i} + M\cos\psi_i\sin\psi_i \tag{10.2.6}$$

where $M = U_\infty/c_o$. Solving these equations gives the angles of the transmitted sound wave in terms of those of the incident sound wave

$$\cos\theta_t = \frac{\cos\theta_i}{1 + M\cos\psi_i\cos\theta_i} \qquad \psi_t = \psi_i \tag{10.2.7}$$

and, conversely, for the angles of the incident sound wave

$$\cos\theta_i = \frac{\cos\theta_t}{1 - M\cos\psi_t\cos\theta_t} \qquad \psi_i = \psi_t \tag{10.2.8}$$

These equations constitute Snell's law for shear layer refraction. To determine the amplitude of the transmitted sound compared to the incident we use the fact that the pressure fluctuations and deflections of the shear layer implied by the incident and reflected wave must be consistent with those implied by the transmitted wave.

Using subscripts "$i$," "$r$," and "$t$" to denote incident, reflected, and transmitted we have that, for the pressure,

$$\hat{p}_i = Ae^{ik\cos\theta_t(x_1\cos\psi_t + x_3\sin\psi_t) + ik(x_2-h)\sin\theta_i}$$
$$\hat{p}_r = Be^{ik\cos\theta_t(x_1\cos\psi_t + x_3\sin\psi_t) - ik(x_2-h)\sin\theta_i}$$
$$\hat{p}_t = Ce^{ik\cos\theta_t(x_1\cos\psi_t + x_3\sin\psi_t) + ik(x_2-h)\sin\theta_t}$$

where $k = \omega/c_\infty$, $x_2 = h$ identifies the shear layer, and we have matched the wavenumbers at the shear-layer interface. At the interface the pressure must match so

$$[\hat{p}_t]_{\text{shear layer}} = [\hat{p}_i]_{\text{shear layer}} + [\hat{p}_r]_{\text{shear layer}} \tag{10.2.9}$$

and thus $C = A + B$. Defining the reflection and transmission coefficients as $R = B/A$ and $T = C/A$ Eq. (10.2.9) becomes

$$T = 1 + R \tag{10.2.10}$$

We must also consider the displacement of the shear layer $\xi$ normal to the flow associated with the surface wave. For a harmonic sound wave, the displacement of the shear layer will be sinusoidal and have the form

$$\xi = \text{Re}\left\{De^{-i\omega t + ik\cos\theta_t(x_1\cos\psi_t + x_3\sin\psi_t)}\right\} \tag{10.2.11}$$

This must match the acoustic particle velocity on either side of the layer in the $x_2$ direction. In the stationary air outside the flow we have, from Eq. (10.2.2),

$$\frac{\partial\xi}{\partial t} = \text{Re}\left(\frac{[\hat{p}_t]_{\text{shear layer}}e^{-i\omega t}\sin\theta_t}{\rho_o c_o}\right) \quad \text{so} \quad -i\omega D = \frac{C\sin\theta_t}{\rho_o c_o}$$

Inside the flow, the sound wave is being convected at speed $U_\infty$, and so we must use the convective derivative to get the displacement velocity. We also have to account for the fact that the acoustic particle velocity of the reflected wave is reversed compared to that of the incident wave. Thus the displacement velocity is,

$$\frac{D_\infty\xi}{Dt} = \text{Re}\left(\frac{[\hat{p}_i - \hat{p}_r]_{\text{shear layer}}e^{-i\omega t}\sin\theta_i}{\rho_o c_o}\right)$$

so

$$-i\omega D(1 - M\cos\theta_t\cos\psi_t) = \frac{(A-B)\sin\theta_i}{\rho_o c_o}$$

Using Snell's law (Eq. 10.2.8) and combining the above expressions to eliminate $D$ gives

$$(A-B)\sin\theta_i\cos\theta_i = C\sin\theta_t\cos\theta_t \tag{10.2.12}$$

Dividing through by $A$ to evaluate $R$ and $T$, we obtain,

$$T = (1 - R)\sin 2\theta_i / \sin 2\theta_t \qquad\qquad (10.2.13)$$

which in combination with Eq. (10.2.10) implies

$$T = \frac{2}{1 + \sin 2\theta_t / \sin 2\theta_i} \qquad\qquad (10.2.14)$$

and

$$R = \frac{1 - \sin 2\theta_t / \sin 2\theta_i}{1 + \sin 2\theta_t / \sin 2\theta_i} \qquad\qquad (10.2.15)$$

Eqs. (10.2.14), (10.2.15), along with Eqs. (10.2.7), (10.2.8), first derived in two-dimensions by Ribner [12], completely define refraction effects for the case of a thin shear layer. These expressions have been derived for plane waves and a flat shear layer. However, they can equally well be taken to represent arbitrarily small portions of a nonplanar sound field or curved shear layer, and thus the results can also be applied locally to much more general situations. Exactly how the above relations are employed to correct measured noise data will depend on the specifics of an experimental setup.

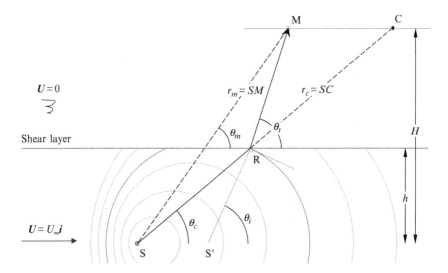

**Fig. 10.17** Sound measurement in a wind tunnel.

## 10.2.2   Corrections for a two-dimensional planar jet

Consider the arrangement shown in Fig. 10.17. The figure shows an acoustic source at location S in a wind tunnel flow of velocity $U_\infty$ located a distance $h$ inside the shear layer. The sound is being measured at location M, outside the shear layer at a distance and angle $r_m$ and $\theta_m$ from the source. Because of refraction, the measured sound arrives indirectly at M via the point R. Note that the angles $\theta_i$ and $\theta_t$ in Fig. 10.17 correspond to those in Fig. 10.16. The plane of the shear layer is perpendicular to the plane MRS. We wish to correct the measurement for the effects of the shear layer. In other words, we wish to know the true directivity angle $\theta_c$ (which also gives the corrected observer distance $r_c = H/\sin\theta_c$), the amplitude of the acoustic pressure $\hat{p}_c$ that would have been measured at point C if the flow had been unbounded, and the difference in the corresponding arrival time of the sound waves $\tau_m - \tau_c$. We need to know these in terms of $\theta_m$ and the measured pressure amplitude $\hat{p}_m$ and the geometrical parameters $h$ and $H$.

To obtain the true directivity angle $\theta_c$ we note first that

$$H \cot\theta_m = h\cot\theta_c + (H-h)\cot\theta_t \tag{10.2.16}$$

To eliminate $\theta_t$ we first use Snell's law (Eq. 10.2.7) for $\psi_i = 0$ to give

$$\frac{1}{\cos\theta_t} = \frac{1}{\cos\theta_i} + M \tag{10.2.17}$$

Next, we relate $\theta_i$ and $\theta_c$ using the kinematics of the sound propagation in the flow. Consider the point S′ (Fig. 10.17) located at the apparent center of the spherical wave arriving at point R on the shear layer. The time taken for this wave to propagate to R from the source is $h/(c_o \sin\theta_i)$, and in this time its center has convected downstream a distance $hM/\sin\theta_i$ between S and S′. We therefore have

$$h\cot\theta_c - h\cot\theta_i = hM/\sin\theta_i \tag{10.2.18}$$

which is solved using tangent half-angle identities to give,

$$\tan\theta_i/2 = \frac{1}{1-M}\left(\sqrt{\csc^2\theta_c - M^2} - \cot\theta_c\right) \tag{10.2.19}$$

Eqs. (10.2.19), (10.2.17), (10.2.16) are simple to evaluate analytically in reverse, i.e., to determine $\theta_m$ from $\theta_c$ given $M$ and $h/H$. An iterative method is needed to determine $\theta_c$ from $\theta_m$. For these calculations it is useful to rearrange Eq. (10.2.17) as

$$\cot\theta_t = 1/\sqrt{(\sec\theta_i + M)^2 - 1}$$

where the square root takes the sign of $\sec\theta_i$. Simpler relationships exist for limiting cases. In the limit of $H/h \to \infty$, $\theta_m = \theta_t$ and the relationship becomes

$$\tan\theta_c = \frac{\sqrt{(1-M\cos\theta_m)^2 - \cos^2\theta_m}}{(1-M^2)\cos\theta_m + M} \tag{10.2.20}$$

and, as $H/h \to 1$ we simply recover that $\theta_c = \theta_m$.

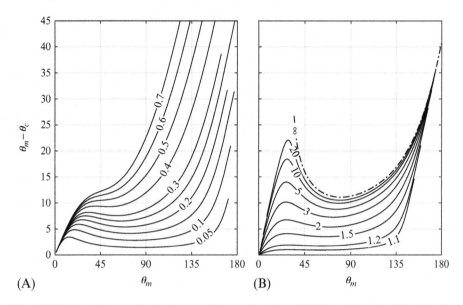

**Fig. 10.18** Corrections to $\theta_m$ in degrees; (A) as a function of Mach number for $H/h = 2$ and (B) as a function of $H/h$ for $M = 0.2$.

The correction to the receiving angle is plotted as a function of Mach number in Fig. 10.18A for $H/h = 2$. Corrections become particularly large in the forward arc and as the Mach number is increased. However, for $M \leq 0.2$ they are less than 10 degrees and roughly constant with $\theta_m$ over the central portion of the arc between $\theta_m$ of 45 and 135 degrees where most measurements are made. Fig. 10.18B shows the same correction plotted as a function of $H/h$ for $M = 0.2$. The correction increases with increasing distance of the observer from the flow. Eq. (10.2.20) representing the limit $H/h \to \infty$ only serves as an accurate approximation for $H/h$ greater than about 5.

With $\theta_c$ determined, the difference in the time of arrival of the measured signal at M and its hypothetical arrival across the flow at C is given by differencing the propagation times from point R

$$\tau_m - \tau_c = \frac{H - h}{c_o} \left( \frac{1}{\sin \theta_t} - \frac{1}{\sin \theta_i} \right) \tag{10.2.21}$$

where $\theta_t$ and $\theta_i$ can be determined from Eqs. (10.2.16), (10.2.17).

To determine $|\hat{p}_c/\hat{p}_m|$ we must account for both the loss of pressure amplitude in transmission through the shear layer (Eq. 10.2.14) and the difference in the spreading of the sound field between R and M as compared to R and C. In the absence of the shear layer the acoustic wave-fronts remain spherical, and so the square of the pressure

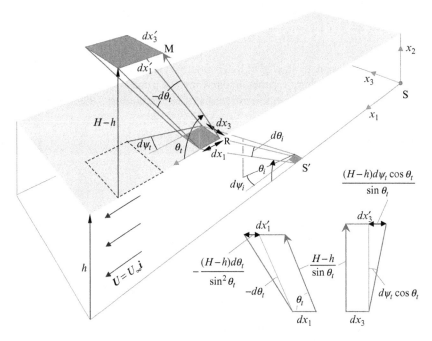

**Fig. 10.19** Ray tube connecting R and M. Insets (bottom right) show views of the ray tube from perpendicular to the $dx_1dx'_1$ and $dx_3dx'_3$ planes.

amplitude is inversely proportional to the square of the propagation distance from the source, i.e.,

$$\left|\frac{\hat{p}_c}{\hat{p}_i}\right|^2 = \left(\frac{h \sin \theta_i}{H \sin \theta_i}\right)^2 = \frac{h^2}{H^2} \qquad (10.2.22)$$

Where, consistent with the Snell's law analysis above, we are using $\hat{p}_i$ and $\hat{p}_t$ to denote the amplitude of the incident and transmitted sound pressure for the shear layer at R, respectively. Determining $|\hat{p}_m/\hat{p}_t|$ is more involved because we must account for the change in the spreading that occurs with refraction. Fig. 10.19 illustrates the method. We consider the sound passing through an elemental area $dx_1dx_3$ at R expanding out the area $dx'_1dx'_3$ at M. From Fig. 10.19,

$$dx'_1 dx'_3 = \left(dx_1 - \frac{H-h}{\sin^2 \theta_t} d\theta_t\right)\left(dx_3 + \frac{H-h}{\sin \theta_t} \cos \theta_t d\psi_t\right) \qquad (10.2.23)$$

Note that $d\theta_t$ as drawn in Fig. 10.19 is negative, resulting in the minus sign that appears inside the first bracket. Since the square of the sound pressure varies inversely with the cross-sectional area, and since the elemental areas at R and M make equal angles to the ray RM, then

$$\left|\frac{\hat{p}_t}{\hat{p}_m}\right|^2 = \frac{dx_1' dx_3'}{dx_1 dx_3} = \left(1 - \frac{H-h}{\sin^2\theta_t}\frac{\partial\theta_t}{\partial x_1}\right)\left(1 + \frac{H-h}{\sin\theta_t}\cos\theta_t\frac{\partial\psi_t}{\partial x_3}\right) \tag{10.2.24}$$

where the derivatives are evaluated at R. To determine the derivatives of $\theta_t$ and $\psi_t$, we first consider position where a general sound ray (i.e., one not confined to the $x_1 x_2$ plane) passes through the shear layer in terms of $\theta_i$ and $\psi_i$, Fig. 10.20, which is

$$x_1 = h\cot\theta_i\cos\psi_i + \frac{hM}{\sin\theta_i}$$
$$x_3 = h\cot\theta_i\sin\psi_i \tag{10.2.25}$$

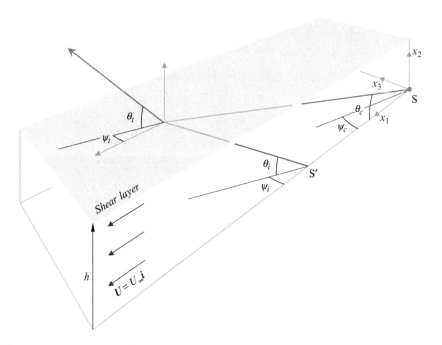

**Fig. 10.20** Path of an out-of-plane sound ray.

Differentiating these expressions with respect to $\theta_i$ and $\psi_i$, differentiating Snell's law for these angles (Eq. 10.2.8) with respect to $\theta_t$ and $\psi_t$, using the chain rule, and evaluating for $\psi_t = 0$ give

$$\frac{\partial x_1}{\partial\theta_t} = -\frac{h\sin\theta_t}{\zeta^3} \tag{10.2.26}$$

$$\frac{\partial x_3}{\partial\psi_t} = \frac{h\cos\theta_t}{\zeta} \tag{10.2.27}$$

where $\zeta^2 \equiv (1 - M \cos \theta_t)^2 - \cos^2 \theta_t$. Since the same process yields $\partial x_3 / \partial \theta_t = \partial x_1 / \partial \psi_t = 0$, the derivatives of $\theta_t$ and $\psi_t$ needed for Eq. (10.2.24) are simply the reciprocals of Eqs. (10.2.26), (10.2.27), and thus

$$\left| \frac{\hat{p}_t}{\hat{p}_m} \right|^2 = \left( 1 + \frac{H-h}{h} \frac{\zeta^3}{\sin^3 \theta_t} \right) \left( 1 + \frac{H-h}{h} \frac{\zeta}{\sin \theta_t} \right) \tag{10.2.28}$$

Finally, from Eqs. (10.2.28), (10.2.22), (10.2.14) we obtain

$$\begin{aligned} \left| \frac{\hat{p}_c}{\hat{p}_m} \right|^2 &= \left| \frac{\hat{p}_t}{\hat{p}_m} \right|^2 \left| \frac{\hat{p}_i}{\hat{p}_t} \right|^2 \left| \frac{\hat{p}_c}{\hat{p}_i} \right|^2 = \frac{1}{T^2} \left| \frac{\hat{p}_t}{\hat{p}_m} \right|^2 \left| \frac{\hat{p}_c}{\hat{p}_i} \right|^2 \\ &= \frac{1}{4} \left( 1 + \frac{H-h}{h} \frac{\zeta^3}{\sin^3 \theta_t} \right) \left( 1 + \frac{H-h}{h} \frac{\zeta}{\sin \theta_t} \right) \frac{h^2}{H^2} (1 + \sin 2\theta_t / \sin 2\theta_i)^2 \end{aligned} \tag{10.2.29}$$

which can be rewritten as

$$\left| \frac{\hat{p}_c}{\hat{p}_m} \right|^2 = \frac{1}{4\zeta^2} \frac{h^2}{H^2} \left( 1 + \frac{H-h}{h} \frac{\zeta^3}{\sin^3 \theta_t} \right) \left( 1 + \frac{H-h}{h} \frac{\zeta}{\sin \theta_t} \right) \left( \zeta + \sin \theta_t (1 - M \cos \theta_t)^2 \right)^2 \tag{10.2.30}$$

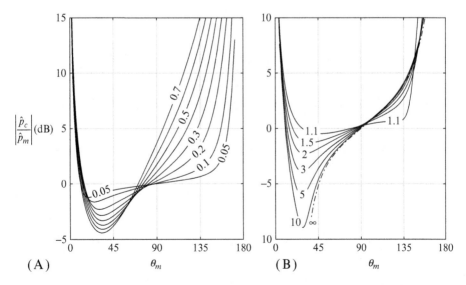

**Fig. 10.21** Corrections the acoustic pressure fluctuation amplitude (A) as a function of Mach number for $H/h = 2$ and (B) as a function of $H/h$ for $M = 0.2$.

This amplitude correction is plotted in Fig. 10.21A as a function of the angle of the sound measurement $\theta_m$ and Mach number for $H/h = 2$. Over much of the rearward arc $|\hat{p}_c/\hat{p}_m|$ is $<1$ (i.e., negative in terms of dB), primarily because the shear layer transmission coefficient $T$ is greater than 1 here. Over the forward arc, however, the spreading effect dominates making the corrected pressure amplitude significantly greater than that measured, particularly as the Mach number is increased above 0.2. Increasing the distance of the microphone from the shear layer, i.e., increasing $H/h$, tends to magnify the corrections as shown for $M = 0.2$ in Fig. 10.21B. At Mach numbers of 0.3 and less, microphone measurements made close to $\theta_m = 90$ degrees require no significant amplitude correction. In interpreting Figs. 10.18 and 10.21, it is important to note that space limitations and the extent of the flow often constrain acoustic measurements to directivity angles between $\theta_m$ of 45 and 135 degrees. Note that Eq. (10.2.30) and the directivity corrections introduced starting at Eq. (10.2.16) were first derived by Amiet [13,14].

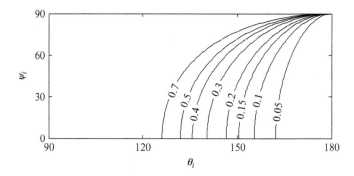

**Fig. 10.22** Boundary between refraction (to the left) and total internal reflection (to the right) as a function of Mach number.

The results of the above analysis allow for two important physical observations concerning the refraction phenomenon that we have not yet commented on. First, for incident wave angles $\theta_i$ sufficiently close to 180 degrees (i.e., directed toward the oncoming flow), the sound can suffer total internal reflection and no transmitted wave is produced. In this situation the magnitude of the right hand side of Eq. (10.2.7) is $>1$ so that there is no solution for $\theta_t$. Indeed, we can infer from this expression that for transmission to occur we must have

$$\frac{-1}{1 + M \cos \psi_i} \leq \cos \theta_i \leq \frac{1}{1 - M \cos \psi_i} \tag{10.2.31}$$

Fig. 10.22 shows this bound plotted as a function of Mach number, total internal reflection occurring for angles to the right of the lines drawn. The effect clearly grows with Mach number, but even for a Mach number of 0.15 it is not possible to receive sound outside the flow propagating into the wind, closer than 30 degrees from the oncoming flow direction. This may be a more significant concern for hybrid anechoic tunnels since it is more feasible to place instrumentation close to the shear layer in this configuration.

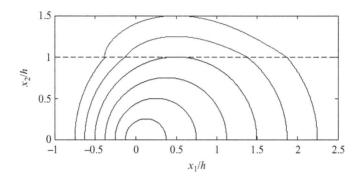

**Fig. 10.23** Wavefronts propagating from a point source at the origin through a shear layer for a flow of $M = 0.5$.

A second observation becomes apparent when Eq. (10.2.17) in combination with the trigonometry apparent in Fig. 10.17 is used to plot the wavefronts of sound propagating through a shear layer from a point source, as has been done in Fig. 10.23. Here the sound source is located at the origin. The refracted wavefronts are clearly noncircular (or nonspherical in the three-dimensional view), and their asymmetry becomes more pronounced as they propagate further from the shear layer. This means that there is no virtual image location for the sound. Thus if, as discussed in Chapter 12, we were to use a microphone phased array to construct an image of the source from the diffracted sound, then that image will be blurred unless we account for the wavefront distortion using, for example, the propagation time correction of Eq. (10.2.21).

### 10.2.3 Effects of shear layer thickness and curvature

The effects of shear layer thickness are discussed by Amiet [13,14], through comparison of measured corrections with the predictions made by the above formulae. He argues that shear-layer thickness effects are generally most significant for emission angles closest to those producing total internal reflection and thus are not often important in measurements. Additionally, Amiet [14] argues that Eq. (10.2.30) will still apply in the limit of a thick shear layer, with the exception that there will be no loss of acoustic energy in the shear layer since no reflected wave is produced. This has the effect of multiplying the measured pressure amplitude by the factor $(1 - R^2)^{-1/2}$, which, using Eqs. (10.2.15), (10.2.17), can be rewritten as

$$\left(1 - R^2\right)^{-1/2} = \frac{2(1 - M\cos\theta_t)\sqrt{\zeta\sin\theta_t}}{\zeta + \sin\theta_t(1 - M\cos\theta_t)^2} \tag{10.2.32}$$

Amiet [14] also gives results that incorporate the effects of shear layer curvature, as in the round shear layer formed around a circular free jet.

## 10.2.4  Considerations for hybrid anechoic tunnels

In hybrid anechoic wind tunnels the Kevlar acoustic windows used to contain the flow produce an attenuation of the sound additional to that associated with refraction. Otherwise, the effects of refraction are expected to be identical to those derived in Sections 10.2.1 and 10.2.2. Indeed, the presence of the Kevlar results in a much thinner and flatter shear layer than will be realized in most open-jet tunnel situations, and so some of the assumptions of the analysis are more accurate. Snell's law (Eqs. 10.2.7, 10.2.8) and the geometric relations derived in Eqs. (10.2.16)–(10.2.28) all apply. The transmission and reflection result (Eqs. 10.2.14, 10.2.15) and the pressure amplitude result that depends on it (Eq. 10.2.30), however, need to be modified to account for the insertion loss associated with the Kevlar barrier. Note that losses in transmission through an acoustic window do not usually impact the signal-to-noise ratio of a sound measurement, since both signal and the parasitic noise of the facility must pass through the Kevlar and are thus attenuated by the same amount.

The attenuation produced by the Kevlar depends not only on its physical characteristics (porosity and flexibility) but also on the speed of the flow that it is exposed to. This is because the flow influences the characteristics of the motion through the pores and the movement of the fabric that couples the sound waves on the incident and transmitted sides. Fortunately, such losses can be quite easily measured using, for example, the setup shown in Fig. 10.24.

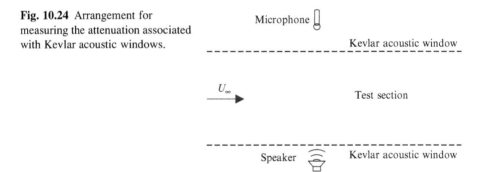

**Fig. 10.24** Arrangement for measuring the attenuation associated with Kevlar acoustic windows.

A speaker and a microphone are positioned facing each other across the empty test section. Both are mounted in anechoic chamber areas outside the test section, and thus the sound from the speaker must traverse both Kevlar windows and the flow in the test section to be recorded by the microphone. The sound attenuation between the speaker and microphone, as compared to that which would occur in a free-field environment, reveals the insertion loss associated with traversing two acoustic windows subject to flow. The loss associated with a single window, that is needed to adjust far-field noise measurements of a source in the test section, is simply half of this value when expressed in decibels. Measuring the attenuation associated with transmission through both windows and shear layers avoids the problems associated with placing speakers or microphones in the flow.

Most hybrid wind tunnels to date have used Kevlar 120 scrim as the acoustic window material. This fabric weighs 58 g/m$^2$ and has a plain weave with 34 threads per inch in both directions made from Kevlar 49 fiber. Insertion losses associated with one batch of this material [11], measured using the method depicted in Fig. 10.24, and then curve fitted are shown in Fig. 10.25. The curve fits are given by the expressions:

$$\Delta_{\text{Kevlar}}(\text{dB}) = 0.0059 \left(\frac{f}{1000\,\text{Hz}}\right)^2 + 0.0145 \left(\frac{f}{1000\,\text{Hz}}\right) \tag{10.2.33}$$

representing the transmission loss through the Kevlar with no flow, and

$$\Delta_{\text{Flow}}(\text{dB}) = \left[1 - e^{-1.057(f/1000\,\text{Hz})}\right] (5.4316M + 88.95M^2) \tag{10.2.34}$$

for the additional flow-related losses. As can be seen in Fig. 10.25, the total correction needed for the presence of the Kevlar ($\Delta_{\text{Kevlar}} + \Delta_{\text{Flow}}$) increases with both frequency and flow speed. While Fig. 10.25 and Eqs. (10.2.33), (10.2.34) are perhaps useful in general for providing rough estimates of Kevlar losses to be used in experiment or facility design, a measured calibration specific to the acoustic windows used in a particular test or facility is important for measurement accuracy. Different batches of the same material made to identical specifications can differ sufficiently (e.g., in porosity) to have a significant impact upon loss characteristics.

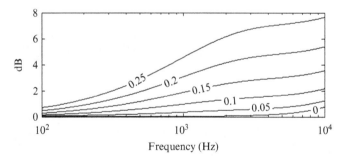

**Fig. 10.25** Losses ($\Delta_{\text{Kevlar}} + \Delta_{\text{Flow}}$) for a Kevlar 120 acoustic window as a function of frequency for different flow Mach numbers.

## 10.3 Sound measurement

Sound measurements at conditions likely to be generated by low Mach number flows are commonly made using condenser microphones—microphones that use the change in capacitance between a membrane and a backing plate. The membrane vibrates in response to the sound, and the capacitance changes as the distance between the membrane and the backing plate fluctuates as a function of time. Conventional condenser microphones require a power source to maintain a charge across the membrane and the backing plate, so a change in capacitance appears as a change in voltage. In electret microphones either the diaphragm or the fixed plate of the capacitor is made from a ferro-electric material that carries a permanent electric charge (referred to as *prepolarized*). This eliminates the need for a voltage source for the microphone, though power is usually still required because an integrated preamplifier is commonly part of these devices.

Instrumentation microphones designed for scientific work are often the most expensive but can provide well-defined, accurately documented, and stable characteristics. Examples are the 1/2-in. diameter B&K 4190 illustrated in Fig. 10.26A and the 1/4-in. G.R.A.S. 40-PH-S5-1 shown in Fig. 10.26B. Microphones of this type may be provided with documentation of their frequency response when exposed to plane sound waves parallel to the diaphragm, as well as deviations from that response for off-axis sound (e.g., Fig. 10.27). Within its operating range (6.3 Hz–20 kHz for the B&K 4190) this type of microphone will generally have an almost constant amplitude response to sources ahead of the diaphragm so that in this range only a single value of the microphone sensitivity is needed to make a quantitative sound measurement. Precise measurement of this sensitivity (to account for environmental conditions at the time of a measurement) can be made using a pistonphone. This is a handheld device produces pressure fluctuations of known amplitude by using the mechanical motion of a vibrating piston at a fixed frequency to simulate a sound wave by compressing a fixed volume of air to which the microphone diaphragm is exposed.

**Fig. 10.26** Examples of microphones for sound measurement. All are shown on the same scale, which may be inferred from the 1/2-in. diameter microphone in part (A) B&K model 4190, (B) G.R.A.S. 40-PH-S5-1, (C) B&K model 4138, (D) Sennheiser KE 4-211-2, and (E) Panasonic WM-64PNT.

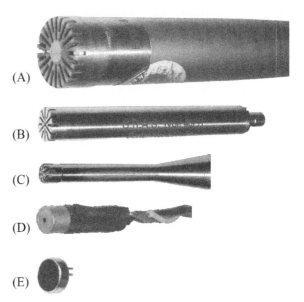

Instrumentation microphones come in a range of diameters. Large diameter micro-phones (e.g., 1/2 or 1 in.) are more sensitive to acoustic pressure fluctuations because those fluctuations are integrated over a larger diaphragm area. Such microphones are useful for measuring sources that are particularly quiet at laboratory scale, such as roughness noise, where good electrical signal-to-noise ratio is needed. At the same time, the larger diaphragm has greater inertia limiting the frequency response. While suffering from lower sensitivity, smaller diameter microphones (e.g., 1/4 and 1/8 in.) may have a greater range, both in terms of the intensity of the sound and the frequencies they can measure. For example, the B&K model 4138 1/8 in. microphone pictured in Fig. 10.26C can measure sounds up to 168 dB and 140 kHz.

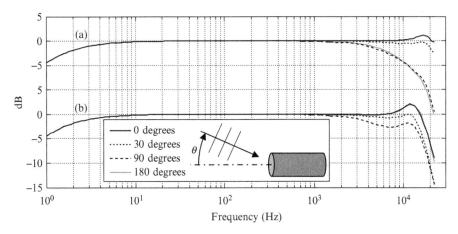

**Fig. 10.27** Amplitude response as a function of frequency and direction for a B&K 4190 1/2-in. microphone with (a) standard protection grid and (b) B&K model UA 0386 nose cone (curves for 0 and 180 degrees coincident).
Data provided by Brüel and Kjaer.

The size of the diaphragm is also important compared to the wavelength of the sound. When the wavelength becomes comparable to the diameter then the scattering of the sound field around the end of the microphone and the spatial distribution of the diaphragm sensitivity become important in determining the response. Fig. 10.27 shows, for example, the deviations in the response of the B&K 4190 with the direction of the incident sound. This microphone has a 1/2-in. diaphragm, and we see that the directionality of the sound becomes a substantial factor at around 6 kHz, where the quarter wavelength is approximately equal to the diaphragm diameter. At 20 kHz, the microphone is 10 dB less sensitive to sound directed at the side of the microphone, or from behind, than it is to sound originating from in front. Instrumentation micro-phones are usually designated as free-field microphones (like the B&K model 4190), optimized as far as possible to measure sound incident on the microphone from any direction, or pressure-field microphones designed to measure sound (or fluid dynamic pressure fluctuations) at a wall. Pressure-field microphones, such as the B&K model

4138 in Fig. 10.26C, usually include a vent designed to equalize the mean pressure on the two sides of the microphone for situations where the face of the microphone is exposed to a pressure significantly different from ambient.

At the lower end of the cost spectrum are devices designed for mass market applications such as lavalier microphones, cell-phones, hearing aids, and units designed for the recording and performance industry. Many of these devices are electret microphones, and a subset can be used for sound measurement in low Mach number applications as long as the experimentalist is willing to take the time to select and calibrate these sensors to the precision needed for scientific work. It is common practice to perform frequency response calibrations of such microphones using an instrumentation microphone as a reference, by subjecting both microphones to a broadband sound field generated by a loud speaker. Such a calibration needs to be done in an environment with a well-defined acoustic character, such as a sealed cavity, a pipe, or an anechoic chamber. Consistent placement and configuration of the test and reference microphones is usually critical to ensure that they are exposed to exactly the same sound field. Examples of such microphones include the Sennheiser KE 4-211-2 shown in Fig. 10.26D and the Panasonic WM-64PNT of Fig. 10.26E. Stability, low noise, and adequate amplitude and frequency range are important considerations in selecting a low-cost microphone for an aeroacoustic test. For situations where multiple microphones are to be used as part of a system (such as a phased microphone array) measuring and matching the phase calibrations of the microphones are often crucial.

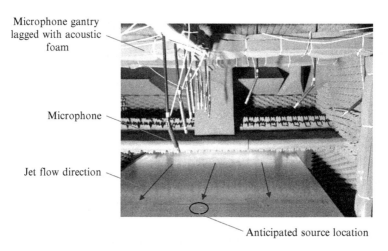

**Fig. 10.28** Microphone mounting in an anechoic wall jet wind tunnel.

The placement and mounting of microphones outside of the test flow involves a number of considerations. It is usually desirable for the microphone to be as close as possible to the source to maximize signal-to-noise ratio. At the same time, keeping the microphone in the acoustic far field (at least one wavelength from the source) is often desirable to simplify the interpretation of the measurement. Also, placing the microphone too close to the free jet or acoustic window can result in contamination by near-field pressure fluctuations associated with the turbulent shear layer. It is usual to orient the microphone to point as directly as possible at the source. This aligns the wavefronts parallel to the diaphragm and makes use of the most favorable microphone response. To exploit the best characteristics of the anechoic chamber, microphones are usually placed at least a quarter wavelength away from the walls, defined by the wedge tips. To avoid unpredictable scattering effects microphones should either use slender mounts and be held from behind or be mounted on a solid surface designed to cleanly reflect the incoming acoustic waves (in which case the measured sound pressure amplitude is doubled because of the addition of the incident and reflected waves). Fig. 10.28 shows single microphones mounted with these considerations in mind in an open-jet wind tunnel. Note that acoustic foam is wrapped around the support beams of the microphone gantry in order to reduce acoustic reflections. Two different mounting strategies for a microphone array are shown in Figs. 10.29 and 10.30. In Fig. 10.29 the microphones are mounted flush in the face of a circular carbon fiber disk designed to reflect the sound, and in Fig. 10.30 the microphones are supported from behind using an open lattice designed to transmit the sound at wavelengths of interest.

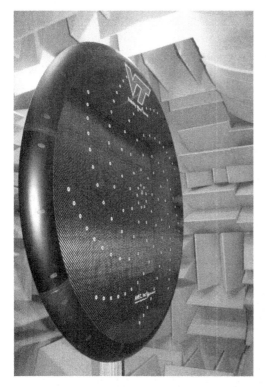

**Fig. 10.29** Microphones mounted in a carbon fiber disk to form a 117-microphone phased array.

Supporting lattice
structure

Microphone
diaphragm

**Fig. 10.30** Microphones mounted from behind using an open lattice in the anechoic chamber of a hybrid anechoic wind tunnel.

It is often of interest to obtain sound measurements at angles that require placing microphones inside the flow. A good example here is rotor testing, where sound radiated on or near the rotor axis is of particular scientific interest. When a microphone is placed directly in the flow its signal will be contaminated with turbulent and sound pressure fluctuations resulting from the interaction of the microphone and its support with the flow. To minimize the contamination an aerodynamic fore-body is used with the microphone, and a streamlined fairing is placed over the supporting strut. The typical microphone fairing shown in Fig. 10.31 is a 1/2-in. diameter B&K model UA 0386 nose cone. This consists of a bullet-shaped housing with a circumferential opening covered by a porous screen. When mounted to the front of a microphone the space interior to the screen forms a cylindrical cavity with one face formed by the microphone diaphragm. Regardless of the orientation of the acoustic source of interest, the microphone is mounted with the nose-cone facing directly into the flow. In this position its streamlined shape minimizes the generation of turbulence, and the screen keeps any flow generated pressure fluctuations separated from the diaphragm. Furthermore, the screen tends to average out turbulent pressure fluctuations that are incoherent around its circumference. The principal drawback of this arrangement is that the nose cone changes the response of the microphone to sound at wavelengths of comparable size. Fig. 10.27 shows the effect of this nose cone on the free-field response of the 1/2-in. B&K 4190 microphone. For sound waves directed roughly at the face of the microphone $\theta = 0$ degrees and 30 degrees the nose cone considerably attenuates the intensity of the sound measured at frequencies above about 15 kHz. The forebody actually amplifies sound coming from the side and behind in this frequency range. Obviously, these effects need to be corrected if sound at these frequencies is to be measured accurately. Note that other longer forebody designs can further reduce flow noise contamination [15].

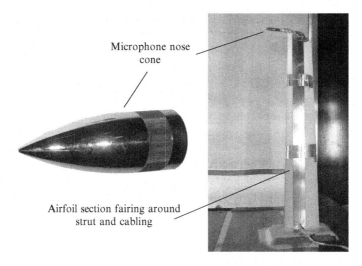

**Fig. 10.31** Microphone with nose cone mounted in the flow using a faired strut.

The design of the strut supporting the microphone in the flow is equally critical. Even as a streamlined airfoil, the strut will be an acoustic source that competes with the sound produced by the model under test. Making the strut as quiet as possible, maximizing the distance of the microphone from the strut (by using a long sting support) can be simple and effective measures. (Note that Fig. 10.31 shows a microphone being used to measure relatively intense sound from a rotor system, and thus a short microphone sting was adequate in this case.) Ideally, the airfoil section chosen for the strut needs to be quiet. In particular, it must not generate vortex shedding tones while being thick enough to provide rigid structural support for the microphone. The McMasters-Henderson airfoil [16], a symmetric 28% thick section illustrated in Fig. 10.32, works well in this role [15].

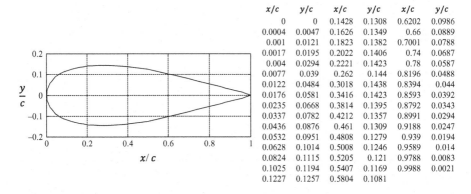

| $x/c$ | $y/c$ | $x/c$ | $y/c$ | $x/c$ | $y/c$ |
|---|---|---|---|---|---|
| 0 | 0 | 0.1428 | 0.1308 | 0.6202 | 0.0986 |
| 0.0004 | 0.0047 | 0.1626 | 0.1349 | 0.66 | 0.0889 |
| 0.001 | 0.0121 | 0.1823 | 0.1382 | 0.7001 | 0.0788 |
| 0.0017 | 0.0195 | 0.2022 | 0.1406 | 0.74 | 0.0687 |
| 0.004 | 0.0294 | 0.2221 | 0.1423 | 0.78 | 0.0587 |
| 0.0077 | 0.039 | 0.262 | 0.144 | 0.8196 | 0.0488 |
| 0.0122 | 0.0484 | 0.3018 | 0.1438 | 0.8394 | 0.044 |
| 0.0176 | 0.0581 | 0.3416 | 0.1423 | 0.8593 | 0.0392 |
| 0.0235 | 0.0668 | 0.3814 | 0.1395 | 0.8792 | 0.0343 |
| 0.0337 | 0.0782 | 0.4212 | 0.1357 | 0.8991 | 0.0294 |
| 0.0436 | 0.0876 | 0.461 | 0.1309 | 0.9188 | 0.0247 |
| 0.0532 | 0.0951 | 0.4808 | 0.1279 | 0.939 | 0.0194 |
| 0.0628 | 0.1014 | 0.5008 | 0.1246 | 0.9589 | 0.014 |
| 0.0824 | 0.1115 | 0.5205 | 0.121 | 0.9788 | 0.0083 |
| 0.1025 | 0.1194 | 0.5407 | 0.1169 | 0.9988 | 0.0021 |
| 0.1227 | 0.1257 | 0.5804 | 0.1081 | | |

**Fig. 10.32** The McMasters Henderson airfoil section [16], with coordinates.

## 10.4   The measurement of turbulent pressure fluctuations

Microphones respond to pressure fluctuations regardless of whether those fluctuations arise from sound waves or flow features. The same basic devices are thus used for measuring the fluid dynamic pressure fluctuations at a surface immersed in a turbulent flow. Such pressure fluctuations are often of interest since they represent an acoustic source (through Curle's equation) or because they help reveal the physical characteristics of a turbulent flow that is otherwise responsible for the aerodynamic noise produced.

The requirements of fluid dynamic pressure fluctuation measurements are quite different from those of sound measurements, and thus the details of the devices used, and how they are used, can be significantly different. The considerations that lead to these requirements concern the scale, intensity, and the location of the pressure fluctuations.

At low Mach number the scale of the turbulent eddies $L$ in a flow is much smaller than the wavelength of the sound they might produce $\lambda$, the ratio being roughly the Mach number. (This is because we expect the passing frequency of the eddies $U/L$ to be the same as the frequency of the sound $c_o/\lambda$.) Suppose, for example, we are concerned with characterizing the pressure fluctuations underneath a turbulent boundary layer on the surface of an airfoil. Such pressure fluctuations define the basic source terms for trailing edge noise and roughness noise to be discussed in Chapter 15. As discussed in Section 9.2.3 and shown in Fig. 9.20B the highest frequency pressure fluctuations in the boundary layer occur at a normalized angular frequency of about $\omega \nu / u_\tau^2 = 1$. The friction velocity $u_\tau$ is expected to be about 5% of the free stream $U_\infty$, so for an airfoil traveling through air at 50 m/s, $\omega \cong 4.3 \times 10^5$ or 68 kHz assuming sea-level conditions. If we take the convection velocity of the smallest eddies to be $0.6U_\infty$ (Fig. 9.19), then this implies that these eddies are a little under half a millimeter in size. If we are interested in completely resolving the pointwise pressure, then we must be able to resolve pressure scales at least this small. This is a major challenge. Studies using different size transducers [17] have shown that the measurement of the pressure spectrum becomes independent of transducer size when the effective transducer diameter, normalized on the inner boundary layer variables as $d^+ = d u_\tau / \nu$, is less than about 18. For the airfoil example, this implies a diameter $d$, of only 0.1 mm. In most cases it is not possible to meet this requirement—sufficiently small transducers simply do not exist. However, making the effective transducer size small is still important in minimizing the underresolution if we are interested in characterizing the pressure at a point. In doing so it usually makes sense to match the dynamic response of the microphone and its spatial response. If we need to use a transducer to measure the pressure fluctuations on our airfoil which has a dynamic response that dies off above 17 kHz, say, then there is no real purpose for its effective diameter to be less than about 0.4 mm.

At low Mach numbers, the intensity of pressure fluctuations in a turbulent flow is usually many orders of magnitude larger than the far-field sound the flow produces. Transducer dynamic range is therefore a consideration. In an equilibrium turbulent boundary layer the root-mean-square pressure fluctuation at the wall is about 1% of the free stream dynamic pressure $\frac{1}{2}\rho_\infty U_\infty^2$ and approximately three times the wall shear stress $\tau_w$, corresponding to 20 Pa or 120 dB (relative to 20 μPa) in the above airfoil example. If it were produced by sound waves, a level of 120 dB would be close to the threshold of pain and thus is already near the upper limit of many commercially available microphones. Furthermore, simple perturbations to the boundary layer, such as produced by a step in the surface, can increase the RMS pressure fluctuation by more than an order of magnitude, meaning that a transducer with a 140 dB range would be inadequate in this case. One benefit of the high range of aerodynamic pressure fluctuations is that it does permit the use of solid-state transducers that are normally too insensitive for sound measurement. Such transducers are usually piezo-resistive (electrical resistance proportional to pressure) or piezo electric (charge proportional to pressure).

The final factor in the selection of pressure transducers for flow measurement is the space available. In some applications this may be the dominant constraint. Trailing edge noise is again a convenient example. The radiated noise is nominally dependent on the surface pressure fluctuation difference between the two sides of the airfoil at the trailing edge itself. There is therefore a desire to measure as close as possible to this point so that the measured characteristics are accurately representative of the source term. Consider the commonly used NACA 0012 airfoil, for example. At a point 5% of the chord length $c$ upstream of the trailing edge (a reasonable place for a trailing edge pressure measurement), this airfoil has a thickness equivalent to 1.4% $c$. For a half-meter chord length model this implies 7 mm thickness to accommodate one or two surface pressure sensors (if the difference is being measured) and their wiring.

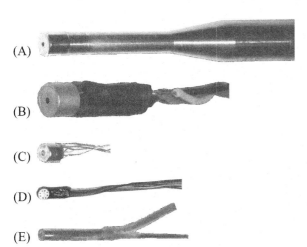

(A)

(B)

(C)

(D)

(E)

**Fig. 10.33** Examples of microphones for pointwise measurement of pressure in a flow. All are shown on the same scale, which may be inferred from the 1/8-in. diameter microphone in part (A) B&K model 4138, (B) Sennheiser KE 4-211-2, (C) Knowles FG-23742-C05, (D) Kulite LQ-062-5D, and (E) Endevco 8514-10.

Fig. 10.33 shows some examples of microphones suitable for turbulent pressure fluctuation measurement at low Mach number. We use these examples to illustrate the typical trade-offs made in selecting suitable transducers, rather than recommending any specific device. The two electret microphones, the Knowles FG-23742 and the Sennheiser KE-4-211-2, are supplied with fairly small openings (0.75 and 1 mm diameter, respectively) and thus have an inherent spatial resolution that may be sufficient for some applications. These microphones are quite sensitive (6.7 and 10 mV/Pa, respectively) implying measurements relatively free of electrical noise, have a useful frequency response (to 10 and 20 kHz), and are compact enough for installation within a confined space (particularly the Knowles for which the casing is only 2.5 mm in diameter and depth). On the other side, these devices have limited dynamic ranges of about 115 dB for the Knowles and 140 dB (with some distortion) for the Sennheiser. Much higher ranges can be achieved using the solid-state Kulite LQ-062-5D and the Endevco 8514-10 both of which have pressure fluctuation ranges that exceed 170 dB, have frequency response characteristics from DC to over 100 kHz, and come in relatively compact packages with effective sensor diameters of approximately 0.7 and 1.7 mm, respectively. The caveat here is that the large range implies a low sensitivity, of about 4 and 3 $\mu$V/Pa.

A different option is to use a small instrumentation microphone, such as the 1/8-in. B&K 4138 pictured in Fig. 10.33A. This microphone combines a range (168 dB) and sensitivity (1 mV/Pa) that are well suited to most low Mach number flows. If device size is not a limitation, then the dominant shortcoming of this transducer is the large diameter of its diaphragm (3 mm). A simple and common way to improve the spatial resolution of microphones of this type is to place a pinhole over the diaphragm (compare Figs. 10.33A and 10.26C). However, it is important to recognize and account for the fact that this substantially impacts the frequency response.

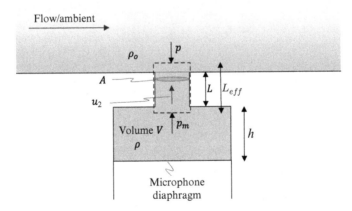

**Fig. 10.34** Nomenclature for analysis of the Helmholtz resonator formed by mounting a microphone behind a pinhole in a surface. The dashed line shows the control volume analyzed.

The cavity and pinhole aperture above the microphone together form what is known as a *Helmholtz resonator* as shown in Fig. 10.34. As long as the dimensions of the cavity and pinhole are small compared to the acoustic wavelength, then the pressure sensed in the cavity by the microphone $p_m$ can be determined from the pressure experienced at the top of the pinhole $p$ using a one-dimensional momentum balance. The rate of change of momentum of the fluid oscillating through the pinhole is generated by the pressure difference between the cavity and the ambient $(p_m - p)$ acting over the pinhole area $A$. This is opposed by the frictional resistance to flow through the pinhole which we will assume is proportional to the flow velocity with a damping coefficient $RA$, where $R$ is the termed the acoustic resistance. Thus,

$$M\frac{\partial u_2}{\partial t} = (p_m - p)A - u_2 RA \tag{10.4.1}$$

Here $M$ is the mass of fluid in motion and $u_2$ is its velocity. Assuming harmonic fluctuations, i.e., $u_2$, $p$, and $p_m$ are given by their amplitudes $\hat{u}_2$, $\hat{p}$ and $\hat{p}_m$ multiplied by $\exp(-i\omega t)$, this becomes

$$-i\omega M\hat{u}_2 = (\hat{p}_m - \hat{p})A - RA\hat{u}_2 \tag{10.4.2}$$

The movement of the flow out of the pinhole produces a rate of change of the density of the air in the cavity, given by the associated mass flow rate and the cavity volume

$$\frac{\partial \rho}{\partial t} = -\frac{\rho_o u_2 A}{V} = \frac{1}{c_o^2}\frac{\partial p_m}{\partial t} \tag{10.4.3}$$

where this has been equated to the pressure in the cavity $p_m$ assuming isentropic compression. In terms of amplitudes this relationship can be rewritten as

$$\hat{u}_2 = \frac{i\omega V}{\rho_o c_o^2 A}\hat{p}_m \tag{10.4.4}$$

Substituting into Eq. (10.4.2) and rearranging we obtain

$$\frac{\hat{p}_m}{\hat{p}} = \frac{\rho_o c_o^2 A^2 / V}{-\omega^2 M - i\omega RA + \rho_o c_o^2 A^2 / V} \tag{10.4.5}$$

We see that the pinhole cavity behaves as a simple second-order system with a stiffness given by $\rho_o c_o A^2 / V$. The mass of the fluid in motion $M$ can be computed from the product of the air density, the pinhole area, and an effective pinhole depth $L_{eff}$. This is generally different from the physical depth $L$ to account for the additional fluid just above and below the pinhole aperture that is involved in the motion.

The appropriate values for $L_{eff}$ and $R$ are dependent upon the specifics of the situation. With no mean motion in the ambient and a circular pinhole that is shallow compared to its diameter $d$, potential flow modeling [18] indicates that $L_{eff} = \pi d / 4$. When the pinhole depth is significant this estimate can be modified to

$$L_{eff} = L + \frac{\pi}{4}d \tag{10.4.6}$$

Different formulas for the acoustic resistance exist, a common one being [19]

$$R = \rho_o \sqrt{8\nu\omega}(1 + L/d)$$                                          (10.4.7)

where $\nu$ is the kinematic viscosity of the air. Behaving as a second-order system the response of the cavity $\hat{p}_m/\hat{p}$ will have a natural frequency

$$\omega_n = \sqrt{\frac{\rho_o c_o^2 A^2}{MV}} = \sqrt{\frac{c_o^2 A}{L_{eff} V}} = \frac{c_o d}{D\sqrt{hL_{eff}}}$$     (10.4.8)

where $D$ denotes the microphone diaphragm diameter and $h$ the distance between the diaphragm and the bottom of the pinhole. Obviously maximizing the frequency response means minimizing the depth of the diaphragm below the surface, given a fixed microphone and pinhole diameter. The amplitude response at the natural frequency is

$$\left|\frac{\hat{p}_m}{\hat{p}}\right| = \frac{\dfrac{\rho_o c_o^2 A^2}{V}}{\omega_n RA} = \frac{L_{eff}\sqrt{\omega_n}}{\sqrt{8\nu}(1 + L/d)}$$     (10.4.9)

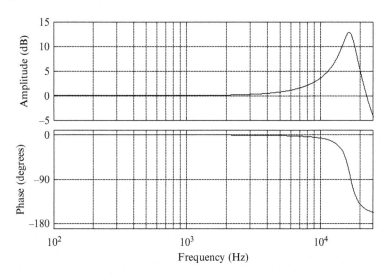

**Fig. 10.35** Measured response of a B&K 4138 microphone with 0.5-mm pinhole cap.

using the resistance formula of Eq. (10.4.7). The measured response function of the B&K 4138 microphone with half-millimeter diameter pinhole pictured in Fig. 10.33A as shown in Fig. 10.35. Using known geometry ($D = 3.2$ mm, $d = 0.5$ mm) and environmental conditions ($c_o = 340$ m/s, $\nu = 1.6 \times 10^{-5}$ m²/s) and taking $h$ and $L$ to be 0.5

and 0.2 mm, respectively, Eqs. (10.4.8), (10.4.9) give a natural frequency of 16.6 kHz and an amplitude response at this frequency of 21 dB. While the predicted natural frequency appears accurate, the amplitude is not, reflecting the fact that the resistance formula of Eq. (10.4.7) is quite uncertain. In particular, when using a microphone with pinhole to measure the fluctuating pressure under a flow, one would expect shear flow over the top of the pinhole to have a substantial effect on the acoustic resistance. Past measurements have indicated both an increase [19] and a decrease [20] in effective resistance in the presence of a grazing flow, in different circumstances.

Overall, the important observation here is that enhancing the spatial resolution of a microphone using a pinhole has a substantial effect on its dynamic response that is likely to impact the frequency range of interest in most low Mach number flows. Given the magnitude of the effects, and the uncertainties in their theoretical estimation, dynamic calibration of the microphone with pinhole is essential. This is often done by comparing the output with that of an unmodified instrumentation microphone when both are exposed to a broadband acoustic source.

**Fig. 10.36** Array of 24, 1/2-in. instrumentation microphones used by Awasthi [21] to measure the unsteady force on a step face. Note that microphone caps have been removed so that the diaphragms can be mounted flush with the flow surface.

Resolving the pointwise pressure is not always the goal of a surface pressure measurement in a flow. Sometimes we are interested in deliberate spatial averaging of the surface pressure fluctuations, typically if we are interested in revealing the portion of the pressure field that is directly related to the near-field of an acoustic response. For example, if we are considering the response of an airfoil in turbulence and the leading edge noise it produces (e.g., Fig. 8.2), then at low Mach number the source term is characterized by the zero spanwise wavenumber component of the surface pressure fluctuation (i.e., its spanwise average) produced on the airfoil leading edge by the incident turbulence. Likewise, at low Mach number the sound produced by flow over a step is determined by the unsteady force on the step faces and thus the spatially averaged pressures there. Fig. 10.36 shows a linear array of 24 half-inch microphones developed precisely for this application. The microphones are used with their diaphragms exposed and mounted flush with the surface, so as not to disturb the flow and to take advantage of the averaging of the pressure fluctuations over each diaphragm as well as the array as a whole.

The spatial sensitivity distribution of an instrumentation microphone diaphragm was measured by Brüel and Rasmussen [22] and can be predicted by the function [23,24]

$$S(r/r_{max}) = \frac{J_o(\alpha) - J_o(\alpha r/r_{max})}{J_o(\alpha) - 1} \qquad (10.4.10)$$

where $\alpha = 2.675$, $J_o$ denotes the Bessel function of the first kind of zero order, $r$ is the distance from the center of the microphone, and $r_{max}$ is the diaphragm radius. This function, plotted in Fig. 10.37, shows how the microphone is less sensitive at its edges and thus has somewhat narrower spatial resolution than implied by its physical diameter.

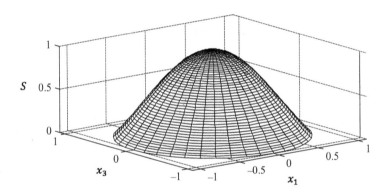

**Fig. 10.37** Sensitivity as a function of position within a microphone diaphragm.

The challenge with this type of measurement system is in establishing the extent to which the array produces an acceptable approximation to the ideal intended result (usually defined by a parallel aeroacoustic analysis). For the array of Fig. 10.36 the ideal result was the uniform averaging of the pressure over an infinitely extending spanwise strip. One way this can be done when the turbulence is generated by a boundary-layer like flow is to exploit the models for the wave-number frequency spectrum of the wall pressure $\Phi_{pp}(k_1,k_3,\omega)$ introduced at the end of Chapter 9. Specifically, given the sensitivity distribution of each microphone, any array result can be written as a wavenumber filter function $F(k_1,k_3)$. In our example, this is obtained by Fourier transforming the sum of the 24 spanwise distributed microphone response functions representing the spatial sensitivity map of the array. The model spectrum (scaled according to boundary layer parameters appropriate to the experiment) is then integrated, weighted by this filter, to estimate the frequency spectrum of the array output, i.e.,

$$G_{pp}(\omega) = \int\limits_{-\infty}^{\infty} \int\limits_{-\infty}^{\infty} \Phi_{pp}(k_1, k_3, \omega) F(k_1, k_3) dk_1 dk_3 \tag{10.4.11}$$

This can then be compared to the same integration carried out using a filter function $F_I(k_1, k_3)$ representing the ideal intended measurement $G_{pp}^I(\omega)$. By comparing $G_{pp}(\omega)$ and $G_{pp}^I(\omega)$ the frequency range within which the measurement is likely to be accurate can be inferred.

## 10.5  Velocity measurement

Measuring the velocity field of a flow is often a central component of aeroacoustics experiments. First, velocity measurements are needed to quantitatively reveal the form of the flow in which the noise is being produced. An example here would be measuring mean flow field around an airfoil that is a source of trailing edge noise. Such measurements would reveal the thickness of the boundary layer and the velocity distribution within it—important contextual and scaling information for the noise produced. Second, velocity measurements are often made to directly quantify the acoustic source, such as measuring the upwash fluctuations produced by turbulence approaching a noise-producing leading edge. In this section we focus our discussion on instrumentation suitable for this second purpose and the associated issues. For a more general review of instrumentation for aerodynamic velocity and turbulence measurements the reader is referred to a dedicated text such as Tropea et al. [25].

Hot-wire anemometry serves as a basic tool in many aeroacoustics experiments. Hot wires provide continuous measurement of the fluctuating velocity at a fixed point as in a flow as turbulence is convected past, in much the same way that a noise generating device experiences the turbulence. The fluctuating signals generated by a hot-wire probe can be readily analyzed to determine velocity spectra and, if pairs of probes are used, space time correlations. These are capabilities that match well with the scientific challenge of characterizing sound generated as a result of interaction with the turbulence. Hot-wire anemometry is long established technology that is well understood. One of its drawbacks is that it can only be used accurately in regions where the turbulent fluctuations are significantly smaller than the mean velocity and thus is restricted to flows outside regions of separated flow. A second limitation is that it is intrusive, requiring introduction of a solid probe to the flow. In most unseparated flows, it is possible to arrange the mounting of probes so that the introduction of the hot wire does not significantly disturb the flow at the measurement point. However, it is almost inevitable that any probe arrangement will generate significant sound or otherwise disturb the aeroacoustic interaction under study. Hot-wire measurements therefore cannot usually be made simultaneously with acoustic measurements, and duplicate runs of an experiment are usually required to collect both types of information.

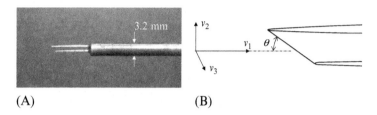

(A)                                          (B)

**Fig. 10.38** A hot-wire probe with a single-slanted sensor (A) photo and (B) schematic.

Hot wires measure velocity by sensing the rate at which a fine wire is cooled by the flow. Wires are typically 2.5 or 5 μm in diameter (at the limit of visibility for most people) and 1–2 mm in length. Tungsten is a common wire material. Fig. 10.38 shows a simple single hot-wire sensor supported on two needle-like prongs. The hot-wire sensor is operated using an electrical feedback circuit incorporating the sensor as one arm of a Wheatstone bridge. The circuit provides current to electrically heat the sensor, so its cooling rate can be measured. The absolute temperature to which the wire is heated as a multiple of the ambient temperature is referred to as the *overheat ratio*. The feedback loop is designed to maintain sensor conditions in the presence of the fluctuating cooling rate imposed by the turbulent velocity fluctuations to which the sensor is exposed. The most widely used scheme is referred to as *constant temperature anemometry*, or CTA. Here the voltage across the sensor is continually adjusted to maintain the sensor at the same temperature and thus the same resistance. The voltage required to do this varies with the velocity to which the sensor is exposed. Since the sensor remains at a fixed temperature the only inertia in the system that limits the response of the probe is electrical and thus it is quite possible to achieve the frequency response (usually in the 10s of kHz) necessary to characterize the turbulence. Overheat ratios of between 1.6 and 1.8 are commonly used in this scheme.

In CTA the relationship between the velocity experienced by the sensor $v_{eff}$ and voltage output by the bridge $E$ is known as King's law:

$$E^2 = A + B v_{eff}^n \tag{10.5.1}$$

This equation comes from the relationship between the nondimensional heat transfer rate from a circular cylinder in a flow and the Reynolds number of the flow. The exponent $n$ is generally taken as 0.45, whereas the constants $A$ and $B$ are determined by calibration. Calibration is performed by placing the probe in a known flow (such as a wind tunnel free stream) and measuring the voltage output of the anemometer over a sequence of several different flow speeds (usually about 10) covering the range of speeds that the sensor is expected to be exposed to during the measurement.

Hot-wire calibrations are notoriously sensitive, particularly to temperature of the flow but also to humidity and to any contaminants in the air that tend to coat the sensor and change its thermal or electrical properties. It is thus not uncommon to need to recalibrate for every hour or two of operation. In facilities where the temperature is not controlled to within a fraction of a degree, methods to correct for temperature drift [26,27] are necessary to maintain calibration and measurement accuracy.

As a first approximation the effective cooling velocity experienced by the sensor can be taken as the component of the velocity perpendicular to the sensor. Consider, for example, the single sensor arrangement shown in Fig. 10.38. We choose our coordinate system so that the sensor is in the $v_1 - v_2$ plane and so that the mean flow is predominantly in the $v_1$ direction. With the sensor sitting at an angle $\theta$ to this direction, the effective velocity will be:

$$v_{eff}^2 = (v_1 \sin \theta + v_2 \cos \theta)^2 + v_3^2 \tag{10.5.2}$$

Note that this relationship guarantees that $v_{eff}$ can only ever be determined as a positive quantity, reflecting the fact that flow can only ever increase the heat transfer from the wire, regardless of its direction. This rectification effect is what prevents hot wires being used accurately in highly turbulent flows, where there is no dominant mean flow direction and direction of the flow may reverse on the sensor. For the same reason, it usually makes little sense to pick probe arrangements that place a sensor anywhere near tangent to the flow direction. The angle $\theta$ in Fig. 10.38 would thus not usually be chosen to be less than about 45 degrees. To make Eq. (10.5.2) more useable, we linearize it by first breaking the velocity components into their mean and fluctuating components:

$$
\begin{aligned}
U_{eff} + u_{eff} &= \sqrt{[(U_1 + u_1)\sin\theta + (U_2 + u_2)\cos\theta]^2 + (U_3 + u_3)^2} \\
&= U_1 \sin\theta \sqrt{\left[1 + \frac{u_1}{U_1} + \left(\frac{U_2}{U_1} + \frac{u_2}{U_1}\right)\cot\theta\right]^2 + \left(\frac{U_3}{U_1} + \frac{u_3}{U_1}\right)^2 \csc^2\theta} \\
&= U_1 \sin\theta \sqrt{1 + 2\frac{u_1}{U_1} + 2\left(\frac{U_2}{U_1} + \frac{u_2}{U_1}\right)\cot\theta + O(2)} \\
&\approx (U_1 + u_1)\sin\theta + (U_2 + u_2)\cos\theta \tag{10.5.3}
\end{aligned}
$$

where we have neglected all terms involving the square or product of the ratio of a mean or fluctuating component to $U_1$ in order to expand the square root. So, in the approximation of Eq. (10.5.2) we see that a hot-wire sensor placed perpendicular to a mean flow ($\theta = 90$ degrees) senses the fluctuating component of velocity in the mean flow direction. A sensor placed at an angle to the mean flow can be thought of as sensitive to a linear combination of the mean and fluctuating velocity components that lie in the plane formed by the sensor and mean flow direction. This second observation allows us to design a hot-wire probe with multiple sensors placed at different angles to the flow so as to measure more than one velocity component. Figs. 10.39 and 10.40 show examples of such probes in the form of an X-wire probe, used for two-component velocity measurement, and a quad-wire probe used for three-component measurements.

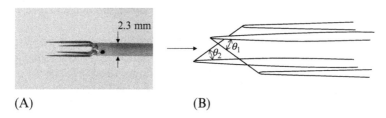

**Fig. 10.39** An X-array hot-wire probe (A) photo and (B) schematic.

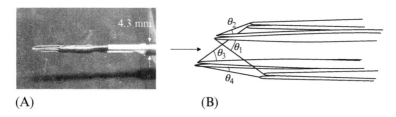

**Fig. 10.40** A quad hot-wire probe (A) photo and (B) schematic.

Attention to detail is necessary to successfully make hot-wire measurements, particularly with multiple sensor probes. First, it is usually not safe to rely on Eq. (10.5.3) and the geometric angles of the sensors to characterize the angle response of a probe. Cooling effects at the ends of the wires, curvature in the wires, interference between adjacent sensors and prongs, and the assumptions made in deriving Eq. (10.5.3) make it wise to calibrate probes for flow angle. This would involve, for example, pitching the X-wire probe of Fig. 10.39 over an arc of angles in a known flow and recording the effective velocities measured at each angle. For the three-component probe of Fig. 10.40 a complete cone of angle measurements is required. Modeling of the measured angle response can then be accomplished using methods of different fidelity depending on the application and the accuracy required. At the simplest level Eq. (10.5.3) may still be used, but with the sensor angles chosen so as to best fit the calibration data. More sophisticated methods include, in the case of the X-wire, using the calibration data to establish a look up table for the velocity and angle of the flow in terms of the voltages of the two sensors [28]. In the case of the quad-wire, methods include using Eq. (10.5.3) with each sensor to obtain rough estimates of the velocity components that are then used to address a look-up table of the errors in those estimates established using the calibration data [29]. Note that such schemes can be questionable if the sensors of a probe are exposed to different flows. In particular, hot-wire probes become unusable when placed in flow gradients that result in significant changes in the flow properties along or between sensors (such as at the bottom of a boundary layer). Equivalently, probes cease to be accurate at frequencies where the typical scale of the turbulence is comparable to the sensor dimensions or spacing. For these reasons probes tend to be made as small as possible (e.g., about 1 mm$^3$ in the case of the quad-wire probe).

As noted at the beginning of this section, hot wires are well suited to the measurement of single- and two-point velocity spectra that are of particular interest to aeroacoustics. It is therefore particularly important that the dynamic response of the hot-wire anemometers used is adequate to resolve the highest frequency of fluctuations of interest. Optimizing the dynamic response of a constant temperature anemometer is, in principle, a simple matter of making sure that the capacitance and inductance of the hot-wire sensor and its cabling are balanced in the Wheatstone bridge used to operate it. Commercial anemometer bridges generally permit the user to adjust this balance. Most bridges also incorporate the ability to impose a square wave voltage signal across the bridge so as to simulate the effect of impulsive changes in velocity at the hot-wire sensor. Alternatively, the anemometer response can be excited using a pulsed laser directed at the hot-wire sensor. Visual inspection of the impulse response can be enough to get an idea of the dynamic response limit within a few kHz. However, recording and Fourier transforming of the response to quantitatively document the phase and amplitude response of the sensor as a function of frequency is generally preferable since then the resulting dynamic calibration can be accounted for in the processing of velocity measurements. Quantitative documentation of the response is particularly important in multisensor probes since phase matching or correction of the different sensors is necessary if the velocity components are to be correctly inferred from the effective velocities at high frequencies.

Optical flow diagnostics also play a major part in aeroacoustic testing. At low Mach number, the dominant technique is particle image velocimetry (PIV). PIV can provide instantaneous cross sections through a flow field to directly reveal the turbulent eddies and their interactions with flow hardware. PIV requires that the flow be seeded with particles small enough to accurately follow the flow. Seeding materials include DEHS (di-ethyl-hexyl-sebacate), olive oil, dioctal phthalate, and poly-latex spheres that are dispersed into particles typically 1 μm in diameter and injected into the flow far enough upstream of the test region for them to be evenly distributed throughout the flow regions of interest. Seeding can be a problem in many aeroacoustic wind tunnels where the seed material may be absorbed into the pores of wedges and other acoustic treatment, compromising their performance. This problem often requires that optical measurements be made in a different facility than that in which the acoustics is being measured, sometimes complicating the comparison of acoustics and aerodynamics. This has been less of a problem in hybrid acoustic wind tunnels where the Kevlar acoustic windows tend to reduce the amount of seed material reaching the most critical acoustic treatment.

PIV involves directing a sheet of laser light through a flow to illuminate the motion of seed particles in the region of interest. A camera or cameras are then used to image the motion of the seed particles and thus infer a cross-sectional view of the flow field. A single camera can be used to infer two velocity components. Adding a second camera allows the out of plane component to be measured as particles pass through the thickness of the light sheet. The general approach is to use a double pulsed laser to produce the light sheet, and synchronized cameras to record images of the particles separated by a short time interval. Velocities are then determined by cross correlating small portions of the pairs of images, termed the interrogation domain, to infer the movement of the local distribution particles. This distance normalized on the time interval between the images gives the velocity vector of the flow at that location. One of the great attributes of this method is it provides direct measurements of the coherent structures present in the turbulence and, through differentiation, of the vorticity field with which they are associated.

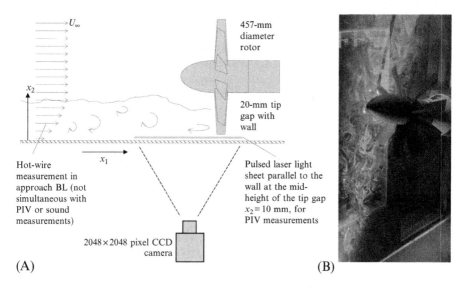

**Fig. 10.41** The experimental setup of Alexander et al. [30] and Murray [31]. (A) Side-view schematic showing the positioning of the PIV light sheet and (B) photograph showing the rotor and light sheet from inside the test section. Experiment included far-field sound measurements.

Depending on the arrangement and conditions, measurement accuracy of about 1% or better [25] can be obtained. As an example of a system that achieves this in a low Mach number environment, consider the PIV setup show in Fig. 10.41. Here a light is a sheet being used to interrogate the flow structure between a rotor immersed in a boundary layer and the adjacent wall on which the boundary layer is growing. The light sheet, generated by an Nd:YAG laser, is oriented parallel to the wall. This allows the camera to be placed behind the (transparent) wall with its image plane parallel to the light sheet. This is the optimum orientation since it allows for the best focusing across the entire field of view. Where other camera orientations are necessary (such as in using a second camera to get the third velocity component) it is necessary for the camera lenses to be tilted with respect to their CCD sensors so that the images can be properly focused. In this experiment, the flow speed was close to 20 m/s, and the time delay between laser pulse pairs was set to 50 μs, implying a particle displacement of about 1 mm. A camera with a resolution of $2048 \times 2048$ pixels was used to image a square portion of the illuminated cross section 28 cm on edge. This implies a particle displacement of about 8 pixels. Images were analyzed using a $32 \times 32$-pixel interrogation area. That is, data from the two images in each $32 \times 32$-pixel subdomain of the image plane were cross correlated in order to identify the spatial correlation peak and thus the local movement of the flow between the images. The fact that one can achieve 1% accuracy with this setup implies that the particle movement can be determined to an accuracy of better than 10% of the pixel spacing. This is done by curve fitting the data defining the correlation peak, usually to a Gaussian surface. Fig. 10.42B shows instantaneous velocity and vorticity fields measured in this experiment. These were used to identify and quantify the eddies interacting with the blade tips and thus characterize a major component of the rotor noise source.

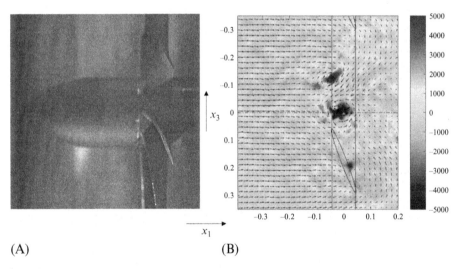

(A)                                                            (B)

**Fig. 10.42** Raw PIV image (A) and an instantaneous velocity vectors and vorticity contours (B) from the experiment of Murray [31]. Distances are normalized on the rotor radius and the origin of $x_1$ is at the rotor mid chord plane; vorticity levels are in radians per second.

Once set up, a PIV system can be very effective in collecting large quantities of flow measurements in a comparatively small amount of wind tunnel time. Postprocessing images to extract those measurements can be quite time-consuming, however, particularly if the images contain imperfections (as is common). Imperfections can include inhomogeneities in seeding or illumination, or nonuniform or varying background. An example of the latter can be seen in a sample image from the rotor experiment (Fig. 10.42A), where the rotor itself can clearly be seen in the background and the fine haze distributed across the picture are the particle images. Since the rotor is moving, separating correlations associated with the displacement of its image, as opposed to those of the seed particles, is necessary to extract useful measurements. The effects of a static nonuniform background can be mitigated, for example, by subtracting the background image, obtained by subtracting the minimum intensity recorded at each pixel over the course of the experiment. For the rotor experiment, it was necessary to establish a background as a function of rotor position to eliminate its effect.

Most off-the-shelf PIV systems provide relatively low image pair acquisition rates, typically of 15–30 Hz and thus do not yield information about the time variation of the flow that would be useful for most aeroacoustic applications. Time-resolved PIV is possible, however, and systems capable of several thousand frames per second have been applied successfully to aeroacoustic problems, particularly in jet flows where optical access is unrestricted, see, for example, Refs. [32,33].

# References

[1] D.B. Stephens, The acoustic environment of the NASA Glenn 9- by 15-Foot Low-Speed Wind Tunnel, in: 21st AIAA/CEAS Aeroacoustics Conference, Dallas, TX, 2015. AIAA paper 2015-2684.

[2] S.M. Jaeger, W.C. Horne, C.S. Allen, Effect of surface treatment on array microphone self-noise, in: 6th AIAA/CEAS Aeroacoustics Conference, Lahaina, HI, 2000. AIAA paper 2000-1937.

[3] H.H. Hubbard, J.C. Manning, Aeroacoustic research facilities at NASA Langley Research Center, NASA TM 84585, 1983.

[4] F.V. Hutcheson, T.F. Brooks, W.M.J. Humphreys, Noise radiation from a continuous mold-line link flap configuration, in: 14th AIAA/CEAS Aeroacoustics Conference, Vancouver, BC, 2008.

[5] A. Bergmann, The aeroacoustic wind tunnel DNW-NWB, in: 18th AIAA/CEAS Aeroacoustics Conference (33rd AIAA Aeroacoustics Conference), Colorado Springs, CO, 2012.

[6] V.S. Ciobaca, S. Melber-Wikending, G. Wichmann, A. Bergmann, A. Kupper, Experimental and numerical studies of the low speed wind tunnel DNW-NWB's open test section towards an aeroacoustic facility, in: 18th AIAA/CEAS Aeroacoustics Conference (33rd AIAA Aeroacoustics Conference), Colorado Springs, CO, 2012.

[7] M. Pott-Pollenske, W. von Heesen, A. Bergmann, Acoustical preexamination work and characterization of the low noise wind tunnel DNW-NWB, in: 18th AIAA/CEAS Aeroacoustics Conference (33rd AIAA Aeroacoustics Conference), Colorado Springs, CO, 2012.

[8] J. Katz, A. Plotkin, Low-Speed Aerodynamics, second ed., Cambridge University Press, Cambridge, 2001.

[9] H.J. Allen, W.G. Vincente, Wall interference in a two-dimensional-flow wind tunnel, with consideration of the effect of compressibility, NACA report 782, 1947.

[10] S. Moreau, M. Henner, G. Iaccarino, M. Wang, M. Roger, Analysis of flow conditions in free jet experiments for studying airfoil self-noise, AIAA J. 41 (2003) 1895–1905.

[11] W.J. Devenport, R.A. Burdisso, A. Borgoltz, P.A. Ravetta, M.F. Barone, K.A. Brown, M. A. Morton, The Kevlar-walled anechoic wind tunnel, J. Sound Vib. 332 (2013) 3971–3991.

[12] H.S. Ribner, Reflection, transmission, and amplification of sound by a moving medium, J. Acoust. Soc. Am. 39 (1957) 435–441.

[13] R.K. Amiet, Correction of open jet wind tunnel measurements for shear layer refraction, in: AIAA 2nd Aeroacoustics Conference, Hampton VA, 1975. AIAA Paper 75-532.

[14] R.K. Amiet, Refraction of sound by a shear layer, J. Sound Vib. 58 (1978) 467–482.

[15] T. Mueller, Aeroacoustic Measurements, Springer, New York, NY, 2002.

[16] J.H. McMasters, R.H. Nordvik, M.L. Henderson, J.H. Sandvig, Two airfoil sections designed for low Reynolds number, Tech. Soar. VI (1981) 2–24.

[17] S.P. Gravante, A.M. Naguib, C.E. Wark, H.M. Nagib, Characterization of the pressure fluctuations under a fully developed turbulent boundary layer, AIAA J. 36 (1998) 1808–1816.

[18] P.M. Morse, K.U. Ingard, Theoretical Acoustics, Princeton University Press, Princeton, NJ, 1987.

[19] N.S. Dickey, A. Selamet, M.S. Ciray, An experimental study of the impedance of perforated plates with grazing flow, J. Acoust. Soc. Am. 110 (2001) 2360.

[20] T. Meyers, J.B. Forest, W.J. Devenport, The wall-pressure spectrum of high-Reynolds-number turbulent boundary-layer flows over rough surfaces, J. Fluid Mech. 768 (2015) 261–293.

[21] M. Awasthi, Sound Radiated From Turbulent Flow Over Two and Three-Dimensional Surface Discontinuities, (Ph.D. dissertation), Department of Aerospace and Ocean Engineering, Virginia Tech, Blacksburg, VA, 2015.

[22] P.V. Bruel, G. Rasmussen, Free field response of condenser microphones, Bruel Kjaer Tech. Rev. 2 (1959) 1–19.

[23] W. Blake, D.M. Chase, Wavenumber-frequency spectra of turbulent-boundary-layer pressure fluctuations measured by microphone arrays, report 1769, Bolt, Beranek and Newman Inc., Cambridge, MA, 1969.

[24] T.M. Farabee, F.E.J. Geib, Measurements of boundary layer pressure fluctuations at low wavenumbers on smooth and rough walls, in: T.M. Farabee, W.L. Keith, R.M. Lueptow (Eds.), NCA—Flow Noise Modeling, Measurement, and Control, vol. 11, ASME, New York, 1991, pp. 55–68.

[25] C. Tropea, A. Yarin, J. Foss, Handbook of Experimental Fluid Mechanics, Springer Verlag, Berlin, 2007.

[26] P.W. Bearman, Corrections for the effect of ambient temperature drift on hot-wire measurements in incompressible flow, DISA Inf. 11 (1971) 25–30.

[27] M. Hultmark, A.J. Smits, Temperature corrections for constant temperature and constant current hot-wire anemometers, Meas. Sci. Technol. 21 (2010) 105404.

[28] A. Abdel-Rahman, G.J. Hitchman, P.R. Slawson, A.B. Strong, An X-array hot-wire technique for heated turbulent flows of low velocity, J. Phys. E Sci. Instrum. 22 (1989) 638–644.

[29] K.S. Wittmer, W.J. Devenport, J.S. Zsoldos, A four-sensor hot-wire probe system for three-component velocity measurements, Exp. Fluids 24 (1998) 416–423.

[30] W.N. Alexander, W.J. Devenport, D. Wisda, M.A. Morton, S.A.L. Glegg, Sound radiated from a rotor and its relation to rotating frame measurements of ingested turbulence, in: 20th AIAA/CEAS Aeroacoustics Conference, Atlanta, GA, 2014.

[31] H.D. Murray, Turbulence and Sound Generated by a Rotor Operating Near a Wall, (M.S. thesis), Department of Aerospace and Ocean Engineering, Virginia Tech, Blacksburg, VA, 2016.

[32] V. Lorenzoni, M. Tuinstra, F. Scarano, On the use of time-resolved particle image velocimetry for the investigation of rod–airfoil aeroacoustics, J. Sound Vib. 331 (2012) 5012–5027.

[33] M.P. Wernet, Temporally resolved PIV for space–time correlations in both cold and hot jet flows, Meas. Sci. Technol. 18 (2007) 1387–1403.

# Measurement, signal processing, and uncertainty

## 11

In this chapter we describe how measured data can be used to estimate the types of statistical and spectral quantities of importance to most aeroacoustic problems and the errors and limitations of that process. The chapter includes discussions of the inherent limitations in measurements and the concept of measurement uncertainty. The methods used to estimate uncertainties in raw measurements, results derived from those measurements, and statistical quantities are also explained. A large part of this chapter is dedicated to explaining how spectra and correlation functions are estimated. The numerical Fourier transform is introduced, and the distinctions with the continuous Fourier transform explained. Fundamental issues that limit the accuracy of spectral estimates, specifically aliasing, broadening, windowing, and the convergence of spectral averages, are discussed in detail.

Fig. 10.41 depicts an example of an aeroacoustics experiment [1,2]. A thrusting rotor is partially immersed in a turbulent boundary layer. As it ingests the layer turbulence, it generates noise that is measured in the far field using a set of microphones. Since the experiment is directed at examining how this sound is produced, the turbulent fluctuations in the boundary layer are also being recorded using hot-wire probes and particle image velocimetry (PIV). The signals produced by the microphones and hot-wire sensors are digitally sampled, as are the outputs of the reference sensors monitoring the free-stream velocity and the properties of the air. The data will need to be analyzed to estimate the quantities of interest, from simple averages of the flow conditions to space-time correlations of the hot-wire and from PIV-measured velocity fluctuations to cross spectra of the far field sound at the different microphone positions.

This task is more complex than it sounds. To successfully and accurately document the experiment and characterize the aeroacoustics it is necessary to consider many factors not immediately apparent in the mathematical definitions of the quantities we are measuring. In essence we must consider in detail how the reality of the experiment and the experimental data differ from the ideal of the mathematical descriptors we adopt for analysis. To be successful requires that we understand the errors that these differences produce and that we design our experiment to minimize their impact.

## 11.1 Limitations of measured data

The very act of measurement places fundamental limitations on the information we can extract. Whatever recording system we use will have a limited range and resolution, and therefore a limited dynamic range. In a digital measurement system, the

*Aeroacoustics of Low Mach Number Flows.* http://dx.doi.org/10.1016/B978-0-12-809651-2.00011-4

dynamic range is stated in bits. For example, a 12-bit system is one that can sense $2^{12}$ (4096) distinct signal levels, implying a dynamic range of $20\log_{10} 2^{12} = 72$ dB. Common also are 16- and 24-bit systems which imply dynamic ranges of 96 and 144 dB, respectively. Obviously the bit-resolution places a fundamental limit on the accuracy of a measurement, and the relative accuracy is optimized if a signal is amplified or attenuated so that its fluctuations span, as nearly as possible, the full range of a measurement system.

A further limitation is that the signal will have to be sampled at a finite rate $f_s = 1/\Delta t$, where $\Delta t$ is the time between successive samples and is referred to as the sampling period. To define a sinusoidal waveform requires at least two samples per period (e.g., one at each successive peak and trough). Thus the highest frequency that can be inferred from a sampled signal is limited to half the sampling rate. This is referred to as the *Nyquist frequency* or *Nyquist criterion*, after the Swedish-American electronics engineer Harry Nyquist [3]. Finally, we can only measure the signal for a finite period of time $T_o$. As such the measurement cannot be used to infer the presence of any frequency with a period greater than the sampling time, at least not without making assumptions about the behavior of the signal before we started or after we finished measuring. Thus the lowest frequency we can unambiguously infer from our measurement is $2\pi/T_o$ rad/s, regardless of whether lower frequencies are present.

It is particularly important to be aware of the consequences of not meeting the Nyquist criterion if one is interested in extracting spectral or time information about a signal. Consider, for example, the broadband turbulent velocity signal shown in Fig. 11.1A measured in a boundary layer much like that in our rotor example. We know from our analysis of the spectrum of this same signal in Chapter 8 (see Fig. 8.3C) that it contains little, if any, power at frequencies above 20 kHz. Sampling at 50 kHz, as shown by the black trace in Fig. 11.1A, therefore meets the Nyquist criterion, and the sampled signal contains a complete representation of the highest frequency fluctuations. Sampling at the much lower frequency of 2 kHz, shown with the gray trace, not only misrepresents the content of the signal at frequencies greater than the Nyquist but also corrupts the information we can obtain at lower frequencies.

Exactly how this occurs becomes clearer if we consider a simple sine wave. The 1900 Hz wave of Fig. 11.1B is well defined when sampled at 50 kHz. Sampling at 2 kHz, however, produces data points that combine to form a spurious and much lower frequency waveform of 100 Hz even though each sample still accurately represents the level of the signal at the instant it was taken. This misidentification of frequency content, called *aliasing*, is a serious problem. With only the 2 kHz sampled data we will have no way of knowing whether the 100 Hz signal was actually present or not. Furthermore, if we are measuring a broadband signal, like the turbulent velocity in Fig. 11.1A, the contribution at 100 Hz that comes from aliasing may mask any actual content we have at that frequency.

If all we are interested in is simple statistical information (such as the mean-square) which does not involve the signal sequence, then aliasing is of no concern. All that

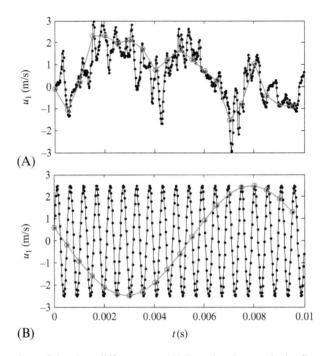

**Fig. 11.1** Sampling of signals at different rates (A) Boundary layer velocity fluctuations (taken from the measurement of Fig. 8.3) and (B) sinusoidal fluctuations.

matters is that we take sufficient independent samples to accurately estimate the expected value. We can see this in Fig. 11.1B where the amplitude, and therefore the mean square, of the two sampled signals is the same. If we are interested in frequency content or time-delay information, then avoiding aliasing is critical. Even if one can estimate the maximum frequency of a physical quantity to be measured, the aliasing of high-frequency interference unintentionally present in the sensor signal may doom a measurement or produce misleading results. For this reason, most digital measurement systems include analog low-pass filters that are applied to a signal before it is measured to ensure that the Nyquist criterion is satisfied.

In the above discussion we have made reference to time signals and frequency analysis, but the same limitations apply when the variation is in a signal that exists in space and we want to know its wavenumber content. For example, Fig. 11.2 shows an instantaneous velocity vector field measured in the bottom of the boundary layer parallel to the wall under the tips of the rotor blade in Fig. 10.41. Consider the fluctuating velocities seen along $x_3$ at $x_1 = 0$ as shown in Fig. 11.2B. These are measured every 2.23 mm over a total length of 240 mm. The Nyquist criterion therefore restricts us to inferring information at wavenumbers less than $2\pi/0.00223 = 2818$ rad/m, and smaller spatial scales will cause aliasing. At the same time we can only infer information at wavenumbers down to $2\pi/0.24 = 26$ rad/m.

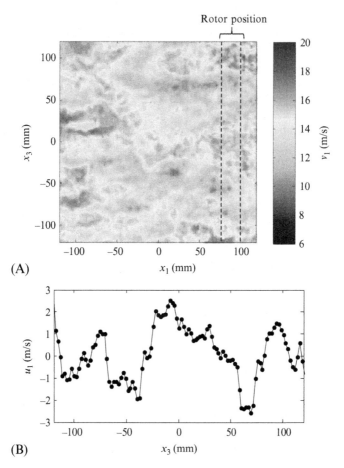

**Fig. 11.2** Measurements using the PIV in Fig. 10.41 for $U_\infty = 20$ m/s and with the rotor spinning at 1500 rpm. (A) Instantaneous velocities $v_1$ in a plane parallel to the wall at $x_2 = 10$ mm. (B) Instantaneous velocity variation $u_1$ extracted along the line $x_1 = 0$.

## 11.2 Uncertainty

All experimental measurements are subject to some degree of error. The errors may be *random*, in that they contribute to the scatter of a measurement about the actual value, or *bias* where the average of a sequence of measurements of the same quantity does not converge to its actual value. In order for an experiment to be useful we need to have a feel for the typical size of the errors in its results. *Uncertainty analysis* is the term given to the methods used to estimate errors and then track their propagation through the measurement and data analysis process to a final derived result. We denote the error in terms of an *uncertainty interval* or *band*. The interval is chosen so that there is a specified probability of the true value lying within it. The specified probability is generally taken to be 95%, unless other considerations (such as safety) dictate a more stringent standard.

An uncertainty interval is commonly represented using the $\pm$ sign as in "the Mach number was measured to be $0.06 \pm 0.001$" meaning that there is a 95% chance that the true Mach number was between 0.059 and 0.061. Symbolically, uncertainties are indicated using the $\delta[]$ notation and may be stated in absolute (e.g., $\delta[M] = 0.001$) or relative terms (e.g., $\delta[M]/M = 0.017$, or 1.7%). Note that we use square brackets here to distinguish the notation used for uncertainty from that used for the Dirac delta function, $\delta()$.

In general, any experimental result $R_o$ will be a function of some number of raw measurements $m_n$, i.e., measurements not derived from any other. Each measurement will be subject to an unknown error $\varepsilon_n$. If we assume the errors are small, then the error in the result $\varepsilon_R$ will be given by

$$\varepsilon_R \cong \sum_{n=1} \varepsilon_n \frac{\partial R_o}{\partial m_n} \tag{11.2.1}$$

where we are ignoring second- and higher-order terms in the Taylor expansion of $R_o$. In uncertainty analysis we are interested in estimating the typical error, which can be obtained by taking the mean square of Eq. (11.2.1) to give

$$\overline{\varepsilon_R^2} \cong \sum_{n=1} \overline{\varepsilon_n^2} \left(\frac{\partial R_o}{\partial m_n}\right)^2 \tag{11.2.2}$$

where we have assumed that the errors in different measurements are uncorrelated. If the errors are all normally distributed then the uncertainty at 95% probability is proportional to the root mean square error, where the constant of proportionality of 1.96 is usually rounded to 2. So, we are 95% certain that the actual value of $R_o$ lies between $-\delta[R_o]$ and $\delta[R_o]$ of the measured value, where

$$\delta[R_o] \cong \sqrt{\sum_{n=1} \left[\delta[m_n] \frac{\partial R_o}{\partial m_n}\right]^2} \tag{11.2.3}$$

Eq. (11.2.3) is a broadly useful tool for tracking the propagation of uncertainties through the analysis process. However, it is important to keep in mind that it rests on assumptions that may not be valid in some situations. Indeed, we will meet such a situation in the next section.

As an example of the application of Eq. (11.2.3), consider estimating an uncertainty interval for the measurement of free stream velocity in the rotor experiment of Fig. 10.41. The raw measurements contributing to this result are the dynamic pressure of the free stream $p_{dyn} = \frac{1}{2}\rho_o U_\infty^2$ measured using a Pitot static tube and its absolute temperature and pressure, $T_e$ and $p_o$, used to determine the density ($\rho_o = p_o/RT_e$). We have that

$$U_\infty = \sqrt{\frac{2RT_e p_{dyn}}{p_o}} \tag{11.2.4}$$

and thus, since the errors in these measurements will be independent,

$$\delta[U_\infty] = \sqrt{\left[\delta[p_{dyn}]\left(\frac{1}{2}\frac{U_\infty}{p_{dyn}}\right)\right]^2 + \left[\delta[T_e]\left(\frac{1}{2}\frac{U_\infty}{T_e}\right)\right]^2 + \left[\delta[p_o]\left(-\frac{1}{2}\frac{U_\infty}{p_o}\right)\right]^2} \tag{11.2.5}$$

In this particular experiment typical values were $U_\infty = 20$ m/s, $T_e = 290$ K, $p_o = 945$ mBar, and $p_{dyn} = 227$ Pa. The raw measurement uncertainties $\delta[p_{dyn}]$, $\delta[T_e]$, and $\delta[p_o]$ are obtained from the specifications and/or calibrations of the transducers used, the acquisition hardware, and the judgment of the experimentalist. Reasonable values in this case might be 2.5 Pa, 0.2 K, and 0.5 mBar, respectively, giving,

$$\delta[U_\infty] = \sqrt{0.0121 + 0.00005 + 0.00003} = 0.11 \, \text{m/s} \tag{11.2.6}$$

where the terms under the square root are written in the same order as in Eq. (11.2.5). We see that the uncertainty is dominated by the dynamic pressure measurement, and thus, if we wanted to improve our accuracy then we would invest time or money here. Note that we did not consider uncertainty in the gas constant $R$. This would only be necessary if we were expecting to achieve a relative uncertainty comparable to that with which $R$ is known.

Several general observations can be made here. First, it is sometimes easier and more reliable to calculate the derivatives needed for the uncertainty calculation numerically. In particular, with measurements processed using a computer program this can be done by perturbing each of the raw measurements input to the program, in turn, by its uncertainty. The change in the result produced by each trial will approximate $\delta[m_n]\partial R_o/\partial m_n$. Second, informed guesswork by the experimentalist is often necessary to estimate the uncertainty in a raw measurement (and, if necessary, the correlation between raw measurements). This is fine. The experimentalist still has a better idea of this error than anyone else and, if the measurement is worth making, the uncertainty will be a small proportion and so possible inaccuracy within the uncertainty estimate itself is not the main concern. Finally, the root mean square addition of uncertainties ensures that we will have a tendency to underestimate the uncertainty in a result, since any neglected contribution would only have increased its value. As such, it is the experimentalist's responsibility to substitute a credible value for the uncertainty if they feel that a computed estimate is unreasonably small (though any such estimate should of course be recorded).

## 11.3 Averaging and convergence

Averaging is a common feature of aeroacoustic measurement since many of the properties in which we are interested are defined as expected values. This means we need to know how many averages are necessary to determine the desired result to within a certain accuracy, i.e., the uncertainty band. Consider, as an example, the uncertainty of a mean value that is being measured from a series of $N$ samples of a quantity $a$:

$$\langle \bar{a} \rangle = \frac{1}{N} \sum_{n=1}^{N} a_n \tag{11.3.1}$$

The notation $\langle \bar{a} \rangle$ is used to recognize that, being determined from a limited number of samples, this is only an estimate of the true mean value $\bar{a}$. Since Eq. (11.3.1) describes the functional relationship between the mean estimator and the samples from which it is calculated, it appears quite possible to determine the uncertainty in $\langle \bar{a} \rangle$ from the range of fluctuations in $a_n$ using the error propagation equation (11.2.3). However, for almost all statistical quantities the nonlinear and/or correlation terms ignored in this equation are significant, and the analysis becomes unwieldy if these are included. We therefore use a different approach, directly deriving an equation for the variance in the error. For example, for the mean value we determine the variance of its estimated value as

$$\overline{\varepsilon^2(\langle \bar{a} \rangle)} = E\left[(\langle \bar{a} \rangle - \bar{a})^2\right] = E\left[\left(\left\{\frac{1}{N}\sum_{n=1}^{N} a_n\right\} - \bar{a}\right)^2\right] = E\left[\left(\frac{1}{N}\sum_{n=1}^{N}(a_n - \bar{a})\right)^2\right]$$
$$= E\left[\left(\frac{1}{N}\sum_{n=1}^{N} a'_n\right)^2\right] \tag{11.3.2}$$

where the prime denotes the fluctuating part. Replacing the square with a double sum we can write this as

$$\overline{\varepsilon^2(\langle \bar{a} \rangle)} = E\left[\frac{1}{N^2}\sum_{n=1}^{N}\sum_{p=1}^{N} a'_n a'_p\right] = \frac{1}{N^2}\sum_{n=1}^{N}\sum_{p=1}^{N} E\left[a'_n a'_p\right] \tag{11.3.3}$$

At this point we can make one of two choices. The first is to assume that the samples are all independent of each other, such as would be the case if they were collected by occasionally checking the value of some reference quantity, in which case $E\left[a'_n a'_p\right] = \overline{a'^2}\delta_{np}$ (where $\delta_{np}$ is the Kronecker delta) and Eq. (11.3.3) becomes

$$\overline{\varepsilon^2(\langle \bar{a} \rangle)} = \frac{\overline{a'^2}}{N} \tag{11.3.4}$$

Assuming the error in the mean is normally distributed, the uncertainty interval at 95% odds is determined as twice its standard deviation.

$$\delta[\langle \bar{a} \rangle] = \frac{2\sqrt{\overline{a'^2}}}{\sqrt{N}} \tag{11.3.5}$$

So, uncertainty decreases as the inverse square root of the number of independent samples used in forming the average.

The second option in evaluating Eq. (11.3.3) is to assume that the samples are taken in rapid succession at equal intervals $\Delta t$ as they would be if they were part of a time-resolved measurement. In this case $E\left[a'_n a'_p\right] = \overline{a'^2} \rho_{aa}([n-p]\Delta t)$, where $\rho_{aa}$ is the correlation coefficient function introduced in Eq. (8.4.5), and so we have

$$\overline{\varepsilon^2(\langle \bar{a} \rangle)} = E\left[\frac{1}{N^2}\sum_{n=1}^{N}\sum_{p=1}^{N}a'_n a'_p\right] = \frac{1}{N^2}\sum_{n=1}^{N}\sum_{p=1}^{N}\overline{a'^2}\rho_{aa}([n-p]\Delta t) \tag{11.3.6}$$

Introducing the index $q = n - p$ we can reorganize the summations as

$$\overline{\varepsilon^2(\langle \bar{a} \rangle)} = \frac{\overline{a'^2}}{N^2\Delta t}\sum_{q=1-N}^{N-1}(N - |q|)\rho_{aa}(q\Delta t)\Delta t \tag{11.3.7}$$

Now, from Eq. (8.4.5), $\displaystyle\sum_{q=1-N}^{N-1}\rho(q\Delta t)\Delta t$ is equal to twice the integral timescale $\mathcal{T}$, assuming $\Delta t \ll \mathcal{T}$ and the total sampling time $T_o = N\Delta t \gg \mathcal{T}$. Applying these assumptions to Eq. (11.3.7) gives,

$$\sum_{q=1-N}^{N-1}(N - |q|)\rho_{aa}(q\Delta t)\Delta t \cong \sum_{q=1-N}^{N-1}N\rho_{aa}(q\Delta t)\Delta t = 2N\mathcal{T} \tag{11.3.8}$$

and so we have

$$\overline{\varepsilon^2(\langle \bar{a} \rangle)} \cong \frac{\overline{a'^2}}{T_o}2\mathcal{T} \tag{11.3.9}$$

and an uncertainty interval of

$$\delta[\langle \bar{a} \rangle] = \frac{2\sqrt{\overline{a'^2}}}{\sqrt{T_o/2\mathcal{T}}} \tag{11.3.10}$$

In this case the uncertainty is independent of the number of samples we take if we are measuring for a fixed time. This can be visualized in terms of the rapidly sampled signal (black symbols) of Fig. 11.1A. Averaging the 100 samples of this signal between $t = 0.004$ and $0.006$ gives a poor estimate of $\overline{u_1}$. However, since the data taken already fully define the signal, we cannot improve this estimate by sampling faster to take more data over this period. Comparing Eqs. (11.3.10), (11.3.5) shows that for time-resolved data the effective number of independent samples is given by the total sampling time divided by twice the integral time scale. A safe rule in general is to take the effective number of independent samples as the minimum of $N$ and $T_o/2\mathcal{T}$.

In a similar fashion we can derive uncertainty relations for other statistics including the mean-square and the cross correlation [4]. These are summarized in Table 11.1. Note that these formulae assume independent samples, and that the cross correlation and coefficient expressions can equally well be applied to space or time-delay correlation function estimates of single variables or pairs of variables simply by appropriately assigning $a$ and $b$. Technically, the effective number of independent samples for the second-order statistics depends on the integral scale of the square or product of correlation functions. As a practical matter, however, this distinction is often ignored, and the minimum of $N$ and $T_o/2\mathcal{T}$ is still used.

**Table 11.1 Averaging uncertainties**

| Quantity | Estimator | Averaging uncertainty |
|---|---|---|
| Mean | $\langle \bar{a} \rangle = \dfrac{1}{N}\sum_{n=1}^{N} a_n$ | $\delta[\langle \bar{a} \rangle] = \dfrac{2\sqrt{\overline{a'^2}}}{\sqrt{N}}$ |
| Mean square fluctuation | $\left\langle \overline{a'^2} \right\rangle = \dfrac{1}{N}\sum_{n=1}^{N} a'^2_n$ | $\delta\left[\left\langle \overline{a'^2} \right\rangle\right] = \dfrac{2\sqrt{2}\,\overline{a'^2}}{\sqrt{N}}$ |
| RMS | $\sqrt{\left\langle \overline{a'^2} \right\rangle} = \sqrt{\dfrac{1}{N}\sum_{n=1}^{N} a'^2_n}$ | $\delta\left[\sqrt{\left\langle \overline{a'^2} \right\rangle}\right] = \dfrac{\sqrt{2}\sqrt{\overline{a'^2}}}{\sqrt{N}}$ |
| Cross correlation | $\left\langle \overline{a'b'} \right\rangle = \dfrac{1}{N}\sum_{n=1}^{N} a'_n b'_n$ | $\delta[\langle \overline{a'b'} \rangle] = \dfrac{2\sqrt{\overline{a'^2}}\sqrt{\overline{b'^2}}}{\sqrt{N}}\sqrt{\left(1+\rho_{ab}^2\right)}$ |
| Cross correlation coefficient | $\langle \rho_{ab} \rangle = \dfrac{\langle \overline{a'b'} \rangle}{\sqrt{\left\langle \overline{a'^2} \right\rangle \left\langle \overline{b^2} \right\rangle}}$ | $\delta[\langle \rho_{ab} \rangle] = \dfrac{2}{\sqrt{N}}\left(1-\rho_{ab}^2\right)$ |

## 11.4 Numerically estimating Fourier transforms

In Chapter 8 we introduced the autospectrum and the cross spectrum as important measures of turbulent flows and the acoustic fields that they generate. The measurement of these functions, and the quantities that can be inferred from them, is therefore often the central focus of aeroacoustics experiments. In this and the following sections we develop the tools needed to best estimate these functions and related quantities using sampled data. As a first step we introduce in this section the discrete Fourier transform (DFT) and its inverse discrete Fourier transform (IDFT), as approximations to the continuous Fourier transform we use in analysis. The DFT is defined as

$$A_m \equiv \sum_{n=1}^{N} a_n e^{-\frac{2\pi i (n-1)(m-1)}{N}} \equiv \mathrm{DFT}(a_n, m) \tag{11.4.1}$$

for $m = 1$ to $N$. The IDFT is defined as

$$a_n \equiv \frac{1}{N} \sum_{m=1}^{N} A_m e^{\frac{2\pi i (n-1)(m-1)}{N}} \equiv \mathrm{IDFT}(A_m, n) \tag{11.4.2}$$

for $n = 1$ to $N$. Note that these specific definitions are the same as those used by the Matlab programming environment[1] and other popular software tools, and that the IDFT is the inverse of the DFT so that

$$a_n = \mathrm{IDFT}(\mathrm{DFT}(a_n, m), n)$$

Eqs. (11.4.1), (11.4.2) are rarely computed explicitly as written since this is an expensive calculation requiring $O(N^2)$ operations. Instead the *Fast Fourier Transform* algorithm [5] is used which takes advantage of efficiencies which become possible when $N$ is a composite number, and particularly when it is a power of 2. This reduces the computational effort to $O(N \log N)$—a huge saving when large data sets are involved.

Consider now the definition of the continuous Fourier transform in time, Eq. (8.4.1), applied to a signal that lasts a finite time of $T_o$ seconds.

$$\tilde{a}^{(s)}(\omega) = \frac{1}{2\pi} \int_{0}^{T_o} a(t) e^{i\omega t} dt \tag{11.4.3}$$

We use the symbol $\tilde{a}^{(s)}(\omega)$ since this is slightly different than Eq. (8.4.1) where the time range of the integral extends from $-T$ to $T$. Suppose that the signal is formed from $N$ samples of the quantity $a$ taken at the beginning of a series of regular time intervals

---

[1] The DFT and IDFT are implemented by the Matlab functions fft() and ifft() respectively.

$\Delta t$, i.e., $a_n = a([n-1]\Delta t)$ for $n = 1$ to $N$ so that $T_o = N\Delta t$. To apply the definition to this signal we replace the integral by a summation

$$\langle \tilde{a}(\omega) \rangle = \frac{1}{2\pi} \sum_{n=1}^{N} a_n e^{i\omega(n-1)\Delta t} \Delta t \tag{11.4.4}$$

where $\langle \tilde{a}(\omega) \rangle$ indicates that we are using this discretization of Eq. (11.4.3) to provide an estimate of the continuous Fourier transform $\tilde{a}(\omega)$ as defined in Eq. (8.4.1), the function we are really interested in. We will address exactly how $\langle \tilde{a}(\omega) \rangle$ is different than $\tilde{a}(\omega)$ in Section 11.5. To apply Eq. (11.4.4) numerically we also need to discretize the result of the Fourier transform as $\langle \tilde{a}_m \rangle = \langle \tilde{a}((m-1)\Delta\omega) \rangle$ for $m = 1$ to $N$. Note that frequency resolution $\Delta\omega$ is also the smallest nonzero frequency at which we are going to get a result, so we choose this to be equal to the lowest frequency that can be unambiguously determined from the sampled time signal, $2\pi/T_o$. Thus,

$$\langle \tilde{a}_m \rangle = \frac{1}{2\pi} \sum_{n=1}^{N} a_n e^{i(m-1)\Delta\omega(n-1)\Delta t} \Delta t \tag{11.4.5}$$

Now $\Delta\omega\Delta t = 2\pi\Delta t/T_o = 2\pi/N$, and so,

$$\langle \tilde{a}_m \rangle = \frac{1}{2\pi} \sum_{n=1}^{N} a_n e^{\frac{2\pi i(n-1)(m-1)}{N}} \Delta t \tag{11.4.6}$$

Comparing this with Eq. (11.4.1), we see that the numerical estimate of the Fourier transform can be calculated as

$$\langle \tilde{a}_m \rangle = \frac{\Delta t}{2\pi} DFT^*(a_n, m) \tag{11.4.7}$$

where the asterisk denotes the complex conjugate of the DFT. If we restrict ourselves to times $t$ from 0 to $T_o$ then the inverse of the continuous Fourier transform of Eq. (11.4.3) is given by,

$$a(t) = \int_{-\infty}^{\infty} \tilde{a}^{(s)}(\omega) e^{-i\omega t} d\omega \tag{11.4.8}$$

and the equivalent inverse numerical transform to Eq. (11.4.7) is

$$a_n = \frac{2\pi}{N\Delta t} \sum_{m=1}^{N} \langle \tilde{a}_m \rangle e^{-\frac{2\pi i(m-1)(n-1)}{N}} = \frac{2\pi}{\Delta t} IDFT^*\left(\langle \tilde{a}_m \rangle^*, n\right) \tag{11.4.9}$$

where $\Delta t = 2\pi/N\Delta\omega$.

Consider the example shown in Fig. 11.3 of a time domain signal, represented by samples $a_n$, and the values of the transform $\langle \tilde{a}_m \rangle$ it implies. The signal is a simple square pulse defined by 32 samples. The sampling period $\Delta t$ is 1 s so that the frequency spacing in the Fourier domain is $2\pi/32$ rad/s. Both $a_n$ and $\langle \tilde{a}_m \rangle$ are plotted against their indices.

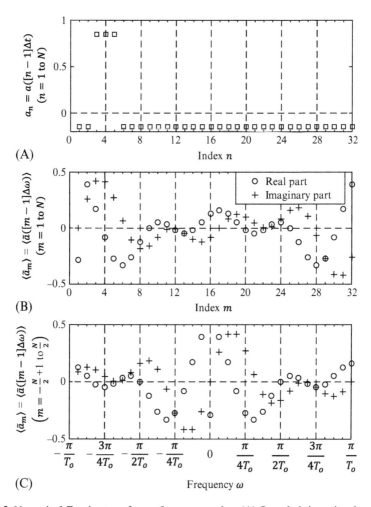

**Fig. 11.3** Numerical Fourier transform of a square pulse. (A) Sampled time signal, (B) transform vs index, and (C) transform vs frequency with conjugate reflection shifted to negative frequencies.

Before discussing its individual elements, it is important to note that transform defined by Eq. (11.4.6) is actually periodic. Since adding or subtracting any integer multiple of $2\pi$ to the exponent does not change its value, then $\langle \tilde{a}_m \rangle = \langle \tilde{a}_{m+N} \rangle = \langle \tilde{a}_{m-N} \rangle = \langle \tilde{a}_{m+2N} \rangle$, and so on. The frequency values for which we had derived this expression, and that are plotted in Fig. 11.3B, thus represent a single period of an infinitely repeating function. Unexpectedly, the same is also true for the inverse transform of Eq. (11.4.9) which tells us that $a_n = a_{n+pN}$, where $p$ is any integer. In other words, we have inadvertently assumed in our formulation of the numerical Fourier transform that the samples we measure form precisely one period of a periodic time variation. Since this is not usually the case, it can be a source of significant bias error, termed *broadening*. As will be discussed later, the source of this assumption was the discretization of the transform into frequency intervals.

Consider now the elements of $\langle \tilde{a}_m \rangle$ for $m = 1$ to $N$ shown in Fig. 11.3B. The first element $\langle \tilde{a}_1 \rangle$ is simply $T_o/2\pi$ times the mean value of the sampled signal since, from Eq. (11.4.6),

$$\langle \tilde{a}_1 \rangle = \frac{1}{2\pi} \sum_{n=1}^{N} a_n \Delta t \tag{11.4.10}$$

This element is therefore always real. Another element with no imaginary part is that corresponding to the Nyquist frequency ($\pi/\Delta t$ in terms of angular frequency) which occurs at index $m = N/2 + 1$ (equal to 17 in the example). Here Eq. (11.4.6) reduces to

$$\langle \tilde{a}_{N/2+1} \rangle = \frac{1}{2\pi} \sum_{n=1}^{N} a_n e^{\pi i (n-1)} \Delta t = \frac{1}{2\pi} \sum_{n=1}^{N} a_n (-1)^{n-1} \Delta t \tag{11.4.11}$$

Since this is always real we can conclude that the numerical Fourier transform extracts no useful phase information at the Nyquist frequency.

Fig. 11.3B shows that the remainder of the transform has a conjugate symmetry with $\langle \tilde{a}_2 \rangle = \langle \tilde{a}_N \rangle^*, \langle \tilde{a}_3 \rangle = \langle \tilde{a}_{N-1} \rangle^*, \dots \langle \tilde{a}_m \rangle = \langle \tilde{a}_{N-m+2} \rangle^*$. This is a consequence of the fact that the signal $a_n$ is real since, from Eq. (11.4.6),

$$\langle \tilde{a}_{N-m+2} \rangle = \frac{1}{2\pi} \sum_{n=1}^{N} a_n e^{\frac{2\pi i (n-1)(N-m+2-1)}{N}} \Delta t = \frac{1}{2\pi} \sum_{n=1}^{N} a_n e^{\frac{2\pi i (n-1)(-m+1)}{N}} \Delta t$$

$$= \frac{1}{2\pi} \sum_{n=1}^{N} a_n e^{-\frac{2\pi i (n-1)(m-1)}{N}} \Delta t \tag{11.4.12}$$

which is the conjugate of $\langle \tilde{a}_m \rangle$ for real $a_n$. The axis of symmetry axis occurs when $m = N - m + 2$, corresponding to $m = N/2 + 1$, the Nyquist frequency. Since the numerical transform is periodic we can, and usually do, ascribe the reflected spectral values from $m = N/2 + 2$ to $N$ to negative frequencies corresponding to $m = -N/2 + 2$ to 0, as illustrated in Fig. 11.4C. In this way the numerical transform is put in a form that is explicitly seen to match the conjugate symmetry of the continuous Fourier transform about zero frequency.

The above discussion demonstrates that the numerical Fourier transform contains no new information at frequencies greater than the Nyquist, as we would expect given our discussion in Section 11.1. Furthermore we see that $\langle \tilde{a}_m \rangle$ is defined by $N$ unique values (2 real numbers and $N/2 - 1$ complex numbers requiring 2 values each) matching the information content of the original sampled signal $a_n$. Table 11.2 summarizes the structure of the Fourier transform vector $\langle \tilde{a}_m \rangle$. Note that any vector that does not match this structure will imply that the time signal samples $a_n$ are complex numbers.

**Table 11.2 Structure of the numerical Fourier transform $\langle \tilde{a}_m \rangle$**

| Index | Description |
|-------|-------------|
| $m = 1$ | Mean value (real) |
| $2 \leq m \leq \dfrac{N}{2}$ | Numerical Fourier transform values for frequencies $\omega = (m-1)\dfrac{2\pi}{T_o}$ |
| $\dfrac{N}{2} + 1$ | Nyquist frequency value (real) |
| $\dfrac{N}{2} + 2 \leq m \leq N$ | Duplicate of values for $2 \leq m \leq \dfrac{N}{2}$ conjugated and in reverse order so that $\langle \tilde{a}_m \rangle = \langle \tilde{a}_{N-m+2} \rangle^*$ |

# 11.5 Measurement as seen from the frequency domain

A complete understanding of the measurement process and its impact on Fourier transform estimates requires that we model that process mathematically. To do this we introduce the *convolution theorem*. This says that multiplying together two functions in the time domain has the same effect as taking the convolution of (*convolving*) their Fourier transforms. Conversely, multiplying two transforms is equivalent to convolving their time functions and dividing by $2\pi$. The convolution of two functions $a(t)$ and $b(t)$ denoted as $a(t)*b(t)$ is defined as

$$c(t) = a(t)*b(t) \equiv \int_{-\infty}^{\infty} a(\tau)b(t-\tau)d\tau \tag{11.5.1}$$

To prove the convolution theorem, we take the Fourier transform of Eq. (11.5.1) over infinite limits.

$$\tilde{c}(\omega) = \frac{1}{2\pi} \int_{-\infty}^{\infty} \int_{-\infty}^{\infty} a(\tau)b(t-\tau)d\tau e^{i\omega t} dt \tag{11.5.2}$$

Reversing the order of integration gives

$$\tilde{c}(\omega) = \frac{1}{2\pi} \int_{-\infty}^{\infty} a(\tau) \int_{-\infty}^{\infty} b(t-\tau)e^{i\omega t} dt d\tau \tag{11.5.3}$$

With the substitution $t' = t - \tau$ we obtain

$$\tilde{c}(\omega) = \frac{1}{2\pi} \int_{-\infty}^{\infty} a(\tau) \int_{-\infty}^{\infty} b(t')e^{i\omega(t'+\tau)} dt' d\tau$$

$$= \frac{1}{2\pi} \int_{-\infty}^{\infty} a(\tau)e^{i\omega\tau} d\tau \int_{-\infty}^{\infty} b(t')e^{i\omega t'} dt' = 2\pi \tilde{a}(\omega) \tilde{b}(\omega) \tag{11.5.4}$$

So, using the operators $[\mathcal{F}]\{\}$ and $[\text{Fscr}]^{-1}\{\}$ to denote the Fourier transform and its inverse we have that

$$\tilde{a}(\omega) \tilde{b}(\omega) = \frac{1}{2\pi}[\text{Fscr}]\left\{ \int_{-\infty}^{\infty} a(\tau)b(t-\tau)d\tau \right\} \tag{11.5.5}$$

A nearly identical proof shows that

$$a(t)b(t) = [\text{Fscr}]^{-1}\left\{ \int_{-\infty}^{\infty} \tilde{a}(\varpi) \tilde{b}(\omega-\varpi)d\varpi \right\} \tag{11.5.6}$$

Note that it is simple to show that convolution is commutative, i.e., $a(t)*b(t) = b(t)*a(t)$ and $\tilde{a}(\omega)*\tilde{b}(\omega) = \tilde{b}(\omega)*\tilde{a}(\omega)$. To understand what the convolution operation does, suppose that $b$ is a delta function occurring at a time $t_1$, i.e., $b(t) = \delta(t-t_1)$. Convolving $a$ with $b$ simply shifts $a$ by the time $t_1$ since

$$a(t)*b(t) = \int_{-\infty}^{\infty} a(\tau)\delta(t-t_1-\tau)d\tau = a(t-t_1) \tag{11.5.7}$$

By extension, convolving $a$ with a sum of two delta functions at $t_1$ and $t_2$ produces a function that consists of two copies of $a$ shifted by $t_1$ and $t_2$ and added together, i.e., $a(t - t_1) + a(t - t_2)$. In general, we can think of the convolution as smoothing the signal $a(t)$ using the kernel $b(t)$, or vice versa. Exactly analogous examples and interpretation apply to convolution in the frequency domain.

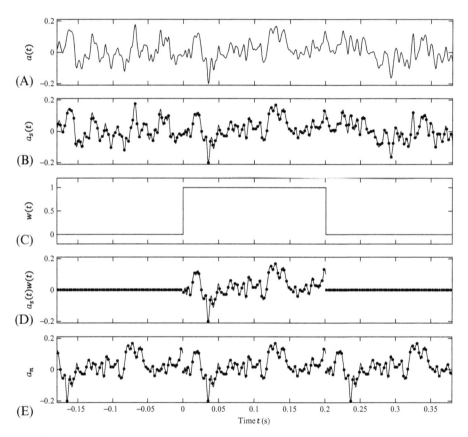

**Fig. 11.4** The measurement process as seen in the time domain. (A) Variation of physical quantity $a(t)$, (B) sampled signal, (C) window function, (D) sampled and windowed data, and (E) periodic sampled data implied by frequency sampling.

Now consider the measurement process illustrated in Figs. 11.4 and 11.5. We begin with the continuous time variation of a physical quantity $a(t)$ that effectively extends forever, Fig. 11.4A. This signal also has a continuous Fourier transform that in principal extends over an infinite frequency domain, Fig. 11.5A. To measure the signal, we sample it at regular intervals $\Delta t$, Fig. 11.4B. This can be modeled as multiplication of the signal by a train of delta functions, one delta function when each sample is taken:

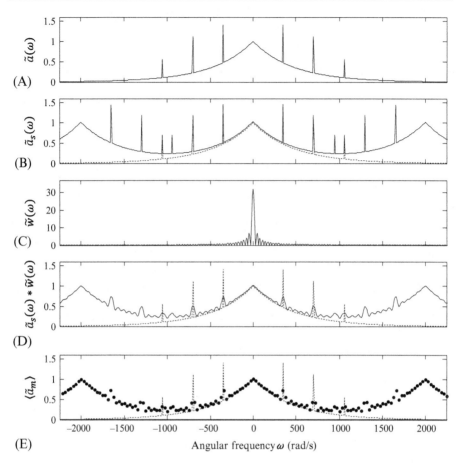

**Fig. 11.5** The measurement process as seen in the frequency domain (in terms of Fourier magnitudes). (A) Continuous Fourier transform of physical quantity $a(t)$ (also shown as the dashed line in parts B, D, and E), (B) effect of time sampling, (C) transform of the window function, (D) effect of time sampling and windowing, and (E) frequency sampling of final result.

$$f(t) = \sum_{n=-\infty}^{\infty} \delta(t - n\Delta t)\Delta t \tag{11.5.8}$$

This has a Fourier transform that is also a train of delta functions since,

$$\tilde{\tilde{f}}(\omega) = \frac{1}{2\pi} \sum_{n=-\infty}^{\infty} \int_{-T}^{T} \delta(t - n\Delta t)\Delta t\, e^{i\omega t} dt = \frac{\Delta t}{2\pi} \sum_{n=-\infty}^{\infty} e^{i\omega n\Delta t}$$

$$= \sum_{n=-\infty}^{\infty} \delta(\omega - n\Omega_o) \tag{11.5.9}$$

where $\Omega_o = 2\pi/\Delta t$. In the example of Figs. 11.4 and 11.5 the sampling period $\Delta t = 3.14$ ms, and so $\Omega_o = 2000$ rad/s. Thus the effect of sampling is to sum together a series of copies of the Fourier transform $\tilde{a}(\omega)$ each shifted by a different integer multiple of $\Omega_o$, Fig. 11.6B. The Fourier transform is now periodic with the form

$$\tilde{a}_s(\omega) = \int_{-\infty}^{\infty} \tilde{a}(\varpi) \sum_{n=-\infty}^{\infty} \delta(\varpi - \omega + n\Omega_o) d\varpi = \sum_{n=-\infty}^{\infty} \tilde{a}(\omega - n\Omega_o) \qquad (11.5.10)$$

This copying is the origin of aliasing. As shown in Fig. 11.5B, for example, $\tilde{a}(\omega)$ overlaps with the first copy to its right $\tilde{a}(\omega - \Omega_o)$, and the resulting sum produces an aliased curve that is different from both. The overlapping is centered on the Nyquist frequency $\pi/\Delta t$ at 1000 rad/s, and the tonal spike at $\omega = 1055$ in the original spectrum is aliased down to $\omega = 945$. We also see that although aliasing has its greatest effect around the Nyquist frequency, it can influence the entire shape of the Fourier transform. Obviously this overlapping and the associated error do not occur if the original signal has no content above the Nyquist frequency.

The fact that we can only measure the signal for a finite time is equivalent to multiplying it by a function that is 1 while we are measuring and 0 when we are not. This rectangular *window* function is shown in Fig. 11.4C, and its effect on the time domain data is illustrated in Fig. 11.4D. The window function can be written as

$$w(t) = H(t)H(T_o - t) \qquad (11.5.11)$$

where $H$ is the Heaviside step function and $T_o$ is the window length. This has the Fourier transform

$$\tilde{w}(\omega) = \frac{1}{2\pi} \int_0^{T_o} e^{i\omega t} dt = \frac{e^{i\omega T_o} - 1}{2\pi i \omega} = \frac{T_o}{2\pi} \frac{\sin(\omega T_o/2)}{\omega T_o/2} e^{i\omega T_o/2} \qquad (11.5.12)$$

The $\dfrac{\sin x}{x}$, or sinc, function that forms the magnitude of $\tilde{w}(\omega)$ is dominated by a central peak surrounded by decaying sidelobes with intervening nulls at frequencies of $2\pi n/T_o$, where $|n| \geq 1$, as illustrated in Fig. 11.5C. So, multiplying the time series by the window function convolves its Fourier transform with this sinc function. The resultant smoothing by $\tilde{w}(\omega)$, seen in Fig. 11.5D, substantially diminishes the sharp peaks in the transform associated with the tonal components of the signal and also generates sidelobes around those peaks mirroring those present in the window function. This convolution with the window function is the source of broadening.

The final measurement step, implicit in the numerical Fourier transform is the sampling of the Fourier transform in the frequency domain, Fig. 11.5E. The implied multiplication by a delta function train in frequency is equivalent to convolution in the

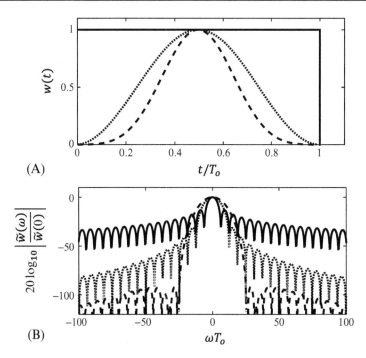

**Fig. 11.6** Window functions plotted in (A) the time domain and (B) frequency domain. Solid line, rectangular; dotted, Hanning; dashed, Blackman Harris.

time domain by a delta function train with a spacing equal to $T_o$, thereby replicating our sampled time signal and making it periodic with a period equal to the window length, Fig. 11.4E.

Unfortunately, we cannot completely mitigate the broadening error since it is a fundamental result of sampling the signal for a finite period of time. The worst effects can be reduced, however, if we window our sampled signal (with its mean value subtracted out) with a function that varies more smoothly to zero at the ends of the window than the default rectangular function of Fig. 11.4C. Numerous options are available, two of which are plotted in Fig. 11.6. These are the Hanning and Blackman Harris window functions which are all given, respectively, by the expressions:

$$w(t) = 0.5 - 0.5 \cos\left(\frac{2\pi t}{T_o}\right)$$

$$w(t) = 0.35875 - 0.48829 \cos\left(\frac{2\pi t}{T_o}\right)$$

$$+ 0.14128 \cos\left(\frac{4\pi t}{T_o}\right) - 0.01168 \cos\left(\frac{6\pi t}{T_o}\right)$$

(11.5.13)

The Fourier transforms of these functions reveal sidelobes that are greatly reduced compared to the rectangular window, though at the expense of increasing the width of the central lobe, Fig. 11.6B. The Hanning window is a satisfactory default choice for most applications. The application of a window reduces the average amplitude of the signal since $w(t) \leq 1$. This must be corrected by dividing by the RMS of the window function to ensure that the Fourier transform has the correct level, i.e., by dividing by $\sqrt{\overline{w^2}}$ where

$$\overline{w^2} = \frac{1}{T_o} \int_0^{T_o} w(t)^2 dt$$

For the Hanning window $\overline{w^2} = 3/8$.

## 11.6   Calculating time spectra and correlations

### 11.6.1   Calculating spectra

The efficiency of the fast Fourier transform means that it is far quicker to compute estimates of power and cross spectra directly from the Fourier transform of time series than it is to apply the definitions in terms of the auto and cross correlation functions Eqs. (8.4.2, 8.4.20 respectively). Specifically, for the power spectral density we make use of Eq. (8.4.14)

$$S_{aa}(\omega) = \frac{\pi}{T} E\left[\widetilde{a}^*(\omega)\, \widetilde{a}\,(\omega)\right] \tag{8.4.13}$$

To estimate the spectrum using this relation we first note that:

- The Fourier transform in Eq. (8.4.2) is taken from $-T$ to $T$, whereas the numerical transform is taken over a record lasting $T_o$ seconds (Eq. 11.4.3). Therefore, we set $T = T_o/2 = N\Delta t/2$.
- To estimate the Fourier transform $\widetilde{a}\,(\omega)$, the numerical transform is used, and thus the spectrum is estimated as

$$\left\langle S_{aa}^{(m)} \right\rangle = \langle S_{aa}([m-1]\Delta\omega)\rangle = \frac{2\pi}{T_0} E\left[\langle \widetilde{a}_m \rangle^* \langle \widetilde{a}_m \rangle\right] \tag{11.6.1}$$

- The numerical Fourier transform is calculated using Eq. (11.4.7) after first windowing the mean-subtracted measured data to avoid broadening. Thus for each measured record $a_n$ we will calculate

$$\langle \tilde{a}_m \rangle = \frac{\Delta t}{2\pi\sqrt{\overline{w^2}}} \text{DFT}^*(a_n w_n, m) \tag{11.6.2}$$

where $a_n w_n = a([n-1]\Delta t)w([n-1]\Delta t)$, and no summation is implied.

- The expected value operator is estimated by averaging the magnitude squared of the Fourier transform calculated from multiple time records.

The estimated spectrum is therefore given by

$$\langle S_{aa}^{(m)} \rangle = \langle S_{aa}([m-1]\Delta\omega) \rangle$$
$$= \frac{\Delta t}{2\pi N N_{rec}\overline{w^2}} \sum_{p=1}^{N_{rec}} \text{DFT}^*\left(a_n^{(p)} w_n, m\right) \text{DFT}\left(a_n^{(p)} w_n, m\right) \tag{11.6.3}$$

where $a_n^{(p)}$ denotes the $p$th record of $N$ samples $a_n$. Identical considerations lead to the cross-spectral density being estimated as:

$$\langle S_{ab}^{(m)} \rangle = \langle S_{ab}([m-1]\Delta\omega) \rangle$$
$$= \frac{\Delta t}{2\pi N N_{rec}\overline{w^2}} \sum_{p=1}^{N_{rec}} \text{DFT}^*\left(b_n^{(p)} w_n, m\right) \text{DFT}\left(a_n^{(p)} w_n, m\right) \tag{11.6.4}$$

Doubling these expressions gives the single-sided spectral estimates $\langle G_{aa}^{(m)} \rangle$ and $\langle G_{ab}^{(m)} \rangle$, of course. As noted in Section 8.4.1, Eq. (8.4.13) does not apply at zero frequency. Zero frequency spectral estimates (revealing, e.g., integral scales) can be obtained by assuming that they are given by the asymptotic level of the spectrum as $\omega$ tends to zero, e.g., by assuming $\langle G_{aa}^{(1)} \rangle = \langle G_{aa}^{(2)} \rangle$ or that $\langle G_{ab}^{(1)} \rangle$ is given by the real part of $\langle G_{ab}^{(2)} \rangle$.

In many cases signals are measured as single sequences of samples of length $N_{tot}$ rather than as multiple records. In this case we must decide how to break up these sequences into records. This decision is controlled by two competing factors. First, we want to choose the length of each record $T_o$ (and thus the number of samples in it $N$) so that the lowest frequency in the spectrum and its frequency resolution, $2\pi/T_o$, will be sufficiently small to capture and distinguish all the phenomena in which we are interested. Second, we wish to maximize the number of records $N_{rec}$ over which the averaging is performed to minimize the averaging uncertainty. At first sight it appears that these choices are constrained by the total number of samples, i.e., $N_{rec} = N_{tot}/N$. However, when a tapered window function is used, each numerical Fourier transform is only weakly dependent on the samples near the beginning and end of each record, and thus records are overlapped to make the best use of the data. The overlap ratio $\Lambda$, defined as the number of samples common to two adjacent records divided by $N$, increases the number of records that are averaged to

$$N_{rec} = \text{int} \left( \frac{N_{tot}/N - 1}{1 - \Lambda} \right) + 1 \tag{11.6.5}$$

where "int" indicates that only the integer part of the division result is retained. It is common to use an overlap ratio of 50%, at least with a Hanning window.

Once a spectrum has been computed we can give up some resolution to obtain a smoother result. This process, called *frequency averaging*, is done by defining a new frequency spacing that is an integer multiple of that for which the spectrum was calculated. Spectral density values for the intervals with this new spacing are then obtained by averaging the values computed for the corresponding original intervals. For example, a spectral estimate $\langle G_{aa} \rangle$ calculated with frequency spacing $\Delta \omega$ can be converted into a smoother estimate $\langle G_{aa1} \rangle$ with a frequency spacing $\Delta \omega_1 = B \Delta \omega$, where $B$ is a positive integer, as

$$\langle G_{aa1}([m_1 - 1]\Delta \omega_1) \rangle = \frac{1}{B} \left[ \frac{1}{2} \left\langle G_{aa}^{\left(M - \frac{B}{2}\right)} \right\rangle + \frac{1}{2} \left\langle G_{aa}^{\left(M + \frac{B}{2}\right)} \right\rangle + \sum_{m=M-\frac{B}{2}+1}^{M+\frac{B}{2}-1} \left\langle G_{aa}^{(m)} \right\rangle \right] \tag{11.6.6}$$

for even $B$, where $M = B(m_1 - 1) + 1$, and

$$\langle G_{aa1}([m_1 - 1]\Delta \omega_1) \rangle = \frac{1}{B} \sum_{m=M-\frac{B-1}{2}}^{M+\frac{B-1}{2}} \left\langle G_{aa}^{(m)} \right\rangle \tag{11.6.7}$$

for odd $B$. In effect these operations replicate what we would have obtained by reducing our original record length by the factor $B$ and increasing the number of averages $N_{rec}$ by the same factor. So, in choosing how to break up a single sequence of samples in order to compute the spectrum, it is better to err on the side of choosing fewer, longer records since once the spectrum has been computed we can trade resolution for a less uncertain spectrum using frequency averaging, though not vice versa.

## 11.6.2  Uncertainty estimates

Uncertainties in spectral estimates associated with statistical convergence can be determined by extending the methods introduced in Section 11.3 [6]. Estimates at the 95% confidence level, based on the effective number of independent spectral averages $N_{rec}$ taken, are listed in Table 11.3.

**Table 11.3** **Averaging uncertainties in spectral quantities**

| Quantity | Estimator | Averaging uncertainty |
|---|---|---|
| Power spectral density | $\langle G_{aa} \rangle$ | $\delta[\langle G_{aa} \rangle] = \dfrac{2G_{aa}}{\sqrt{N_{rec}}}$ |
| Cross-spectral density magnitude | $|\langle G_{ab} \rangle|$ | $\delta[|\langle G_{ab} \rangle|] = \dfrac{2\sqrt{G_{aa}G_{bb}}}{\sqrt{N_{rec}}}$ |
| Co-spectrum | $\langle C_{ab} \rangle = \mathrm{real}(\langle G_{ab} \rangle)$ | $\delta[\langle C_{ab} \rangle] = \dfrac{\sqrt{2}\sqrt{G_{aa}G_{bb} + C_{ab}^2 - Q_{ab}^2}}{\sqrt{N_{rec}}}$ |
| Quad-spectrum | $\langle Q_{ab} \rangle = \mathrm{imag}(\langle G_{ab} \rangle)$ | $\delta[\langle Q_{ab} \rangle] = \dfrac{\sqrt{2}\sqrt{G_{aa}G_{bb} - C_{ab}^2 + Q_{ab}^2}}{\sqrt{N_{rec}}}$ |
| Coherence | $\langle \gamma_{ab}^2 \rangle = \dfrac{|\langle G_{ab} \rangle|^2}{\langle G_{aa} \rangle \langle G_{bb} \rangle}$ | $\delta[\langle \gamma_{ab}^2 \rangle] = \dfrac{2\sqrt{2}\gamma_{ab}}{\sqrt{N_{rec}}}(1 - \gamma_{ab}^2)$ |
| Phase | $\langle \theta_{ab} \rangle = \arctan \dfrac{\langle Q_{ab} \rangle}{\langle C_{ab} \rangle}$ | $\delta[\langle \theta_{ab} \rangle] = \dfrac{\sqrt{2}\sqrt{1 - \gamma_{ab}^2}}{\gamma_{ab}\sqrt{N_{rec}}}$ (radians) |

## 11.6.3 Phase spectra

Estimates of the cross and auto-spectral density can be used to determine coherence and phase spectra. Particular care needs to be taken in interpreting phase. Superficially, a positive phase $\langle \theta_{ab}(\omega) \rangle$ can be thought of as implying a time delay, on average, between the sinusoidal components of $a$ and $b$ at frequency $\omega$ (with positive $\langle \theta_{ab} \rangle$ implying that $b$ is lagging). However, a phase shift between two sine waves can only be uniquely determined over a range of one wavelength ($2\pi$ radians) with the result that the relationship between the phase and the time delay has the form:

$$\langle \theta_{ab} \rangle = [(\tau\omega + \pi) \bmod 2\pi] - \pi \tag{11.6.8}$$

where "mod" refers to the modulo operation so that the item in square brackets is equal to the remainder of $(\tau\omega + \pi)$ after division by $2\pi$, and we have chosen this function so that $-\pi \leq \langle \theta_{ab} \rangle < \pi$. This causes a jump in phase whenever it is equal to $\pm\pi$ and is called *phase wrapping*. To illustrate this, consider a phase spectrum of the form shown in Fig. 11.7A. This shows the phase for a cross spectrum between two signals with a fixed time delay between them, so $a(t) = Cb(t + \tau)$. This results in the linear increase in phase with frequency and consequent discontinuities where $\theta_{ab}$ jumps from positive to negative value across the branch cut at $\pm\pi$, shown in Fig. 11.7A.

It is possible to unwrap the phase and determine the absolute time delay associated with it if the phase variation is assumed to be continuous with frequency, if that variation is sufficiently resolved in $\langle \theta_{ab} \rangle$, and if the phase is defined continuously down to

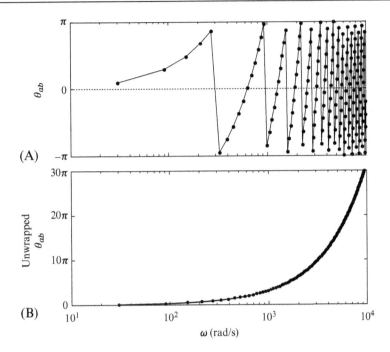

**Fig. 11.7** Phase spectrum implied by a 10 ms delay between signals $a(t)$ and $b(t)$. (A) Raw spectrum. (B) Unwrapped spectrum.

a frequency where the time delay is less than half the period. In that case we will be able to unambiguously identify and undo the $2\pi$ jumps in phase between adjacent frequencies associated with the phase ambiguity. In Fig. 11.7B we show the data of Fig. 11.7A unwrapped in this way so that the uninterrupted linear increase in phase with frequency is visible.

A second concern is that the estimated phase spectrum will adopt a value even when the coherence is insignificant and no meaningful phase exists. Such values should, of course, be ignored. The uncertainty relation in Table 11.3 can be used to identify when there is significant coherence. Specifically, if we consider the borderline case as that when the coherence is equal to its uncertainty, we obtain

$$N_{rec} \geq 8 \frac{\left(1 - \gamma_{ab}^2\right)^2}{\gamma_{ab}^2} \tag{11.6.9}$$

for the coherence to be significant. This relationship is plotted in Fig. 11.8. We see, for example, that 25 averages are sufficient to identify significant coherence values down to about $\gamma_{ab}^2 = 0.2$, and 100 and 1000 averages to identify coherence values as small as 0.07 and 0.008, respectively.

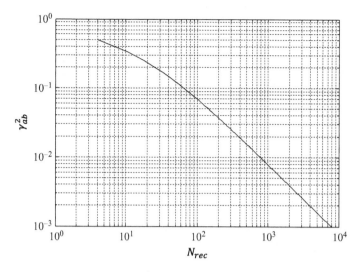

**Fig. 11.8** Minimum significant phase $\gamma_{ab}^2$ as a function of number of independent spectral averages $N_{rec}$.

## 11.6.4 Correlation functions

The fast Fourier transform also provides a computationally efficient route for estimating the auto and cross correlation functions. In principle, once we have $\langle S_{aa} \rangle$ and $\langle S_{ab} \rangle$ obtaining the correlations should just be a matter of applying an inverse Fourier transform consistent with Eqs. (8.4.4), (8.4.21). In practice, this assumes that the record length used in calculating $\langle S_{aa} \rangle$ and $\langle S_{ab} \rangle$ is much larger than any correlation time in the signals. Since this is often not the case some special care is required.

Consider first, for example, the calculation of the auto spectrum using Eq. (11.6.1). Substituting Eq. (11.4.6) for $\langle \tilde{a}_m \rangle$ (i.e., not using a tapered window function) we obtain

$$\left\langle S_{aa}^{(m)} \right\rangle = \frac{\Delta t}{2\pi N} \sum_{n=1}^{N} \sum_{n'=1}^{N} E[a_n a_{n'}] e^{2\pi i (m-1)(n-n')/N} \tag{11.6.10}$$

Now, since $a_n = a([n-1]\Delta t)$ then $E[a_n a_{n'}] = R_{aa}([n-n']\Delta t)$. Introducing the index $q = n - n'$ Eq. (11.6.10) can be rewritten as

$$\left\langle S_{aa}^{(m)} \right\rangle = \frac{\Delta t}{2\pi N} \sum_{n=1}^{N} \sum_{n'=1}^{N} R_{aa}((n-n')\Delta t) e^{2\pi i (m-1)(n-n')/N}$$

$$= \frac{\Delta t}{2\pi N} \sum_{n=1}^{N} \sum_{q=n-N}^{n-1} R_{aa}(q\Delta t) e^{2\pi i (m-1)q/N}$$

The summation over $n$ then has $N - |q|$ identical terms for $1 - N < q < N - 1$, giving

$$\left\langle S_{aa}^{(m)} \right\rangle = \frac{\Delta t}{2\pi N} \sum_{q=1-N}^{N-1} (N - |q|) R_{aa}(q\Delta t) e^{2\pi i (m-1)q/N} \tag{11.6.11}$$

To estimate the correlation function, we now apply the numerical inverse Fourier transform (Eq. 11.4.9) to $\left\langle S_{aa}^{(m)} \right\rangle$, giving

$$\left\langle R_{aa}^{(p)} \right\rangle = \frac{2\pi}{N\Delta t} \frac{\Delta t}{2\pi N} \sum_{m=1}^{N} \sum_{q=1-N}^{N-1} (N - |q|) R_{aa}(q\Delta t) e^{2\pi i (m-1)q/N} e^{-2\pi i (m-1)(p-1)/N}$$

$$= \frac{1}{N^2} \sum_{q=1-N}^{N-1} (N - |q|) R_{aa}(q\Delta t) \sum_{m=1}^{N} e^{2\pi i (m-1)(q-p+1)/N}$$

$$\tag{11.6.12}$$

for $p$ varying from 1 to $N$. Now, the inner summation is zero unless $q - p + 1$ is an integer multiple of $N$ in which case it sums to $N$. Since $q$ is limited to the range $|q| \le N - 1$, we find for $p > 0$ that $q - p + 1 = -N$ for each $q$ less than zero, and $q - p + 1 = 0$ for each $q$ greater than or equal to zero enabling us to also eliminate the summation over $q$ to give

$$\left\langle R_{aa}^{(p)} \right\rangle = \frac{N - (p-1)}{N} R_{aa}([p-1]\Delta t) + \frac{p-1}{N} R_{aa}([p-1-N]\Delta t) \tag{11.6.13}$$

So we see that the correlation estimate is actually formed from the scaled combination of two correlation values. Thus estimates $\left\langle R_{aa}^{(p)} \right\rangle$ for short positive time delays (from zero to $N\Delta t/2$) are contaminated with those for large negative time delays (from $-N\Delta t$ to $-N\Delta t/2$).

If the time period over which the correlation is nonzero is insignificant compared to $N\Delta t/2$ then Eq. (11.6.13) reduces to approximately $\left\langle R_{aa}^{(p)} \right\rangle = R_{aa}([p-1]\Delta t)$. If not, we can obtain an uncontaminated correlation estimate of the second half of our original series is set to zero. In other words, we *zero pad* each sample record by appending it with an equal number of trailing zeros. This ensures that $R_{aa}(n\Delta t) = 0$ for $|n| \ge N/2$, where $N$ is now the full length of the zero-padded data records. Reworking the above derivation in this case we obtain:

$$\left\langle R_{aa}^{(p)} \right\rangle = \frac{N/2 - (p-1)}{N} R_{aa}([p-1]\Delta t) \tag{11.6.14}$$

For $p = 1$ to $N/2$, and

$$\left\langle R_{aa}^{(p)} \right\rangle = \frac{(p-1) - N/2}{N} R_{aa}([p-1-N]\Delta t) \tag{11.6.15}$$

for $p = N/2 + 1$ to $N$. The true correlation $R_{aa}$ can now be recovered from $\langle R_{aa} \rangle$ by straightforward rescaling. Exactly the same procedure can be used to obtain estimates of the cross correlation function from the numerical inverse Fourier transform of cross-spectrum estimates $\langle S_{ab} \rangle$.

## 11.7 Wavenumber spectra and spatial correlations

The discussion of spectral estimates that began in Section 11.4 makes exclusive reference to signals in time and the correlations and frequency spectra they imply. However, the analysis results apply equally well to variations over distances, spatial correlations, and wavenumber spectra with minor modifications due to the conjugate relationship between the space and time Fourier transform definitions adopted in Chapter 3. Specifically, the numerical Fourier transform applied to a variation in space along coordinate $x_1$ sampled at intervals $\Delta x_1$ is

$$\left\langle \tilde{a}_m \right\rangle = \frac{\Delta x_1}{2\pi} \sum_{n=1}^{N} a_n e^{-\frac{2\pi i (n-1)(m-1)}{N}} = \frac{\Delta x_1}{2\pi} \text{DFT}(a_n, m) \tag{11.7.1}$$

where $a_n = a(n\Delta x_1)$, $\left\langle \tilde{a}_m \right\rangle = \left\langle \tilde{a}([m-1]\Delta k_1) \right\rangle$, and $\Delta k_1 = 2\pi/(N\Delta x_1)$. Likewise, the inverse transform is

$$a_n = \frac{2\pi}{N\Delta x_1} \sum_{m=1}^{N} \left\langle \tilde{a}_m \right\rangle e^{\frac{2\pi i (m-1)(n-1)}{N}} = \frac{2\pi}{\Delta x_1} \text{IDFT}\left( \left\langle \tilde{a}_m \right\rangle, n \right) \tag{11.7.2}$$

Wavenumber spectra are very often calculated by numerical Fourier transform of the associated correlation function, at least when the correlation function is obtained in a point-wise fashion using measurements with two or more probes that are traversed to different positions. However, when an array of sensors is used, or measurements are made optically such as in the PIV snapshot of Fig. 11.2, spectra may be calculated by averaging multiple estimates of the wavenumber transform of instantaneous data sampled in space. In this case, analogs of Eqs. (11.6.3), (11.6.4) are used, specifically

$$\left\langle \phi_{aa}^{(m)} \right\rangle = \left\langle \phi_{aa}([m-1]\Delta k_1) \right\rangle$$
$$= \frac{\Delta x_1}{2\pi N N_{rec} \overline{w^2}} \sum_{p=1}^{N_{rec}} \text{DFT}^*\left( a_n^{(p)} w_n, m \right) \text{DFT}\left( a_n^{(p)} w_n, m \right) \tag{11.7.3}$$

where $a_n^{(p)}$ denotes the $p$th record of $N$ samples $a_n = a(n\Delta x_1)$, and

$$\left\langle \phi_{ab}^{(m)} \right\rangle = \left\langle \phi_{ab}([m-1]\Delta k_1) \right\rangle$$
$$= \frac{\Delta x_1}{2\pi N N_{rec} \overline{w^2}} \sum_{p=1}^{N_{rec}} \text{DFT}^*\left( a_n^{(p)} w_n, m \right) \text{DFT}\left( b_n^{(p)} w_n, m \right) \tag{11.7.4}$$

# References

[1] W.N. Alexander, W.J. Devenport, D. Wisda, M.A. Morton, S.A. Glegg, Sound radiated from a rotor and its relation to rotating frame measurements of ingested turbulence, in: 20th AIAA/CEAS Aeroacoustics Conference, Atlanta, GA, 2014.

[2] H.D. Murray, Turbulence and Sound Generated by a Rotor Operating Near a Wall (M.S. thesis), Department of Aerospace and Ocean Engineering, Virginia Tech, Blacksburg, VA, 2016.

[3] Harry Nyquist, Phys. Today 29 (6) (1976) 64.

[4] C. Tropea, A. Yarin, J. Foss, Handbook of Experimental Fluid Mechanics, Springer Verlag, Berlin, 2007.

[5] J.W. Cooley, J.W. Tukey, An algorithm for the machine calculation of complex Fourier series, Math. Comput. 19 (1965) 297–301.

[6] J.S. Bendat, A.G. Piersol, Random Data, fourth ed., Wiley, New York, NY, 2010.

# Phased arrays

**12**

In aeroacoustic measurements we are often faced with the problem of trying to determine the relative levels of two different sources or extracting the sound level of a source in a noisy environment. A classic example is measuring trailing edge noise from an airfoil in a wind tunnel. In this case we need to separate the sources at the trailing edge from the sources at the leading edge and also reject the background noise and reflections from the tunnel walls. To achieve this, phased array technology has been developed over many years and has become a required tool for most aeroacoustic measurements in wind tunnels and on engine test stands. This chapter provides the basic concepts of array technology and the results that can be easily coded to carry out phased array processing. Those students who have already taken a class in acoustics will be familiar with the basic concepts outlined in the first section on the use of line arrays. Section 12.2 and the rest of the chapter extend those concepts to problems that are relevant in aeroacoustics.

## 12.1 Basic delay and sum processing

Phased arrays have been used for many decades in the fields of underwater acoustics, radar, and optics, and the ability of these systems to locate sources that emit propagating waves is very sophisticated. However, the application of phased arrays to aeroacoustics is quite different because the sources of interest are usually continuously distributed, and we are interested in their spatial variation in level. In terms of geometrical optics this represents a near-field problem which is much more difficult to solve than the classic far-field problem where each source can be considered as an isolated point source, regardless of its actual size. In spite of this difference we will start by considering isolated point sources in the geometric far field and show how a simple line array may be used to determine their position and level. This will establish some of the most important characteristics of phased array measurements, such as spatial resolution and aliasing. In the next section we will show how these results can be generalized for arbitrary near- and far-field arrays and how the array design can be evaluated for a particular problem.

### 12.1.1 Basic principles, resolution, and spatial aliasing

We will start with the canonical problem of measuring the location of a simple harmonic source using a line array of $M$ transducers placed far from a source, as illustrated in Fig. 12.1. We can define the acoustic pressure at the array using Eq. (3.4.6) as

$$\hat{p}(\mathbf{x}) = -\frac{i\omega\rho_o Q e^{ikr}}{4\pi r} \tag{12.1.1}$$

Aeroacoustics of Low Mach Number Flows. http://dx.doi.org/10.1016/B978-0-12-809651-2.00012-6

where $r = |\mathbf{x} - \mathbf{y}|$ is the distance from the source at $\mathbf{y}$ to the transducer at $\mathbf{x}$. Each transducer in the array is located at $\mathbf{x}_m = ((m-1)\Delta x - L/2, 0, 0)$, where the subscript $m$ represents the microphone number, as shown in Fig. 12.1. The origin of the coordinate system is at the center of the array, and the total length of the array is $L = (M-1)\Delta x$ (note that this is different from our analyses in previous chapters where we placed the origin in the vicinity of the source or sources). We can then use the far-field approximation (analogous to Eq. 3.6.3) to approximate the propagation distance $r$ when $|\mathbf{y}| \gg |\mathbf{x}|$ to

$$r\left(y_i - x_i^{(m)}\right) \approx r(y_i) - x_i^{(m)}\frac{\partial r(y_i)}{\partial y_i} + \cdots = r_o - x_i^{(m)}y_i/r_o$$

where $r_o = |\mathbf{y}|$ is the distance of the source from the center of the array. Since the transducers are along the $x_1$ axis, this simplifies to $r_o - x_1^{(m)}\sin\theta$, where $y_1/r_o = \sin\theta$ is the angle subtended by the source as shown in Fig. 12.1. This approximation is equivalent to assuming that the array is far enough from the source for the wavefronts it experiences to be planar.

The far-field approximation for the pressure at each transducer is then

$$\hat{p}(\mathbf{x}_m) = \left(\frac{-i\omega\rho_o Q e^{ikr_o}}{4\pi r_o}\right)e^{-ik((m-1)\Delta x - L/2)\sin\theta} \tag{12.1.2}$$

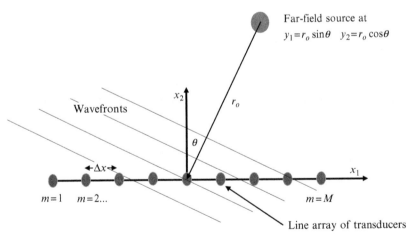

**Fig. 12.1** A line array of $M$ receivers and the waves arriving from a far-field source.

The array output can be obtained by adding the signal from each receiver and dividing the total by $M$, so

$$\hat{p}_t = \frac{1}{M}\sum_{m=1}^{M}\hat{p}(\mathbf{x}_m) = \frac{-i\omega\rho_o Q e^{ikr_o + i\frac{1}{2}kL\sin\theta}}{4\pi r_o}\left(\frac{1}{M}\sum_{m=1}^{M}e^{-ik(m-1)\Delta x\sin\theta}\right) \tag{12.1.3}$$

The summation in the last term is a geometric series and can be summed (see standard mathematical tables) to give

$$\frac{1}{M}\sum_{m=1}^{M}e^{-ik(m-1)\Delta x\sin\theta} = \frac{1-e^{-ikM\Delta x\sin\theta}}{M(1-e^{-ik\Delta x\sin\theta})}$$

$$= \left\{ \frac{\sin\left(\frac{1}{2}kM\Delta x\sin\theta\right)}{M\sin\left(\frac{1}{2}k\Delta x\sin\theta\right)} \right\} e^{-i\frac{1}{2}k(M-1)\Delta x\sin\theta} \qquad (12.1.4)$$

The phase term in Eq. (12.1.4) cancels with the phase term ½$ikL\sin\theta$ in Eq. (12.1.3), so we obtain

$$\hat{p}_t = \frac{-i\omega\rho_o Qe^{ikr_o}}{4\pi r_o}F(kL\sin\theta),$$

$$F(kL\sin\theta) = \left\{ \frac{\sin\left(\frac{1}{2}kM\Delta x\sin\theta\right)}{M\sin\left(\frac{1}{2}k\Delta x\sin\theta\right)} \right\} \qquad (12.1.5)$$

where $F(kL\sin\theta)$ is also dependent on $M$ and is plotted in Fig. 12.2A. This result shows that the array output depends on the angle subtended by the source to the array, as shown in Fig. 12.1. If the source is directly in front of the array, as shown in Fig. 12.3A, then $\theta=0$, all the microphones receive the same in-phase signal, and $F$ has its maximum value ($F=1$). If the angle of the source is increased from zero, then the microphones will start to see out-of-phase signals and $F$ will fall below 1. Eventually, if the wavelength is not too large compared to the microphone spacing, we will reach the situation shown in Fig. 12.3B where the signal received by adjacent microphones is 180 degrees out of phase and $F=0$. Further increase in the source angle will now start to bring the microphone signals back in phase until we reach the situation shown in Fig. 12.3C where $F=1$ once more. We see that there will be a fundamental ambiguity in identifying the source location from the microphone signals. Note that in general if there are angles for which ½$k\Delta x\sin\theta=\pm n\pi$ (where $n$ is an integer) then $F=1$ at that point. Likewise, if there are angles for which ½$kM\Delta x\sin\theta$ is a multiple of $\pi$ then $F=0$, and the array output is zero for these positions.

The overall sensitivity of a linear array for sources as a function of source angle is illustrated in Figs. 12.2B and C for an array with 10 transducers and different values of $kL$. This shows that at low frequencies the sensitivity of the array output to source position is relatively weak (Fig. 12.2B) and the directivity map displaces a single lobe or *beam*. At high frequencies the main lobe of the array sensitivity is much narrower, but we have additional lobes, reflecting the ambiguity noted above where the array output is equally as strong as the main lobe (Fig. 12.2C).

The simplest way to determine the source location is to rotate the array until its output is a maximum. If the microphone spacing is small enough compared to the acoustic wavelength and there is only a single beam in the array sensitivity, as in Fig. 12.2B, then the angle of maximum array output uniquely identifies the direction of the source.

However, if the array sensitivity has multiple beams as shown in Fig. 12.2C then there can be multiple angles where the array output reaches a maximum. The false source positions are called spatial aliases and can be eliminated by ensuring that $|\frac{1}{2}k\Delta x\sin\theta| < \pi/2$ for all angles. Since $k = 2\pi/\lambda$ and $|\sin\theta| < 1$, this requires that the transducer spacing should be less than half an acoustic wavelength, $\Delta x < \lambda/2$.

The resolution of the array is usually given in terms of the width of the straight-ahead lobe, defined by the angle or distance over which the array output drops to 3 dB below its maximum value. This corresponds to $F = 0.707$, which occurs approximately when $kM\Delta x\sin\theta = 2.8$ or $\sin\theta = 1.4\lambda/\pi(L + \Delta x)$, assuming that $M$ is large. Longer arrays (compared to the acoustic wavelength) will therefore have better sensitivity to source position, or resolution. However, to meet the restriction of spatial aliasing $\Delta x < \lambda/2$ and to have a long array for good resolution requires a large number of transducers.

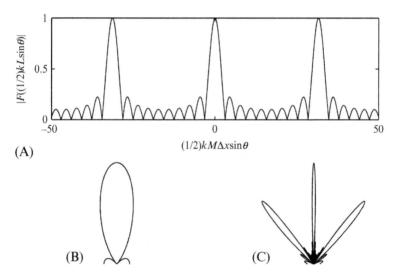

**Fig. 12.2** (A) The array sensitivity $F(kL\sin\theta)$ for waves arriving from different directions. (B) Polar plot of $F(kL\sin\theta)$ for $kL = 8.8$, $M = 10$ and (C) $kL = 88$, $M = 10$.

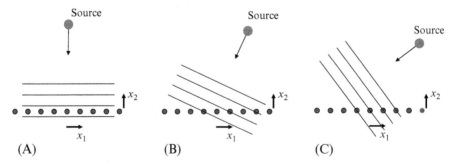

**Fig. 12.3** Wave patterns compared to microphone positions for situations (A) and (C) where the array output is a maximum, and (B) where the output is zero.

## 12.1.2 Beam steering

The array processing described above allows us to determine the direction of wave arrival by physically rotating the array so that it points toward the source. Alternatively, we can steer the array by time shifting the signals, so they add up in phase. From Fig. 12.1 we see that, for a wave arriving from a source at an angle $\theta$ to the array, the time of arrival differs at adjacent transducers by $\Delta\tau = (\Delta x/c_o)\sin\theta$. So, if a phase shift

$$\phi_m = k\left((m-1)\Delta x - \frac{1}{2}L\right)\sin\theta$$

is added to the signal from the $m$th transducer before the summation in Eq. (12.1.3), we obtain

$$\hat{p}_t = \frac{1}{M}\sum_{m=1}^{M}\hat{p}(\mathbf{x}_m)e^{i\phi_m} = \frac{1}{M}\sum_{m=1}^{M}\hat{p}(\mathbf{x}_m)e^{ik\left((m-1)\Delta x - \frac{1}{2}L\right)\sin\theta} = \frac{-i\omega\rho_o Q e^{ikr_o}}{4\pi r_o}$$

which is identical to the signal that is obtained when the array points toward the source. It follows that we can focus the array in different directions by choosing different phase shifts. If we choose

$$\phi_m = k\left((m-1)\Delta x - \frac{1}{2}L\right)\sin\theta_s \qquad\qquad (12.1.6)$$

then the array signal is

$$\hat{p}_t = \frac{i\omega\rho_o Q e^{ikr_o + i\frac{1}{2}kL(\sin\theta - \sin\theta_s)}}{4\pi r_o}\left(\frac{1}{M}\sum_{m=1}^{M}e^{-ik(m-1)\Delta x(\sin\theta - \sin\theta_s)}\right)$$

$$= \frac{-i\omega\rho_o Q e^{ikr_o}}{4\pi r_o}F(kL(\sin\theta - \sin\theta_s)) \qquad\qquad (12.1.7)$$

We see from this result that by choosing different phase shifts $\phi_m$ we have focused the array in different directions without rotating it, and the maximum array output signal will occur when the source is in the direction that the array is focused so $\theta = \theta_s$. One of the important points here is that the phase shift defined Eq. (12.1.6) is equivalent to a time shift $\tau_m = \phi_m/\omega$ applied to each transducer output, so the signal that results from the array processing is

$$p_t(t) = \frac{1}{M}\sum_{m=1}^{M}p(\mathbf{x}_m, t - \tau_m) = \text{Re}\left(\frac{1}{M}\sum_{m=1}^{M}\hat{p}(\mathbf{x}_m)e^{-i\omega t + i\phi_m}\right)$$

As a consequence, this process is often referred to as delay and sum beamforming. In many applications implementing a delay and sum may be computational by more efficient than working in the frequency domain and using the methods described in Section 12.2.

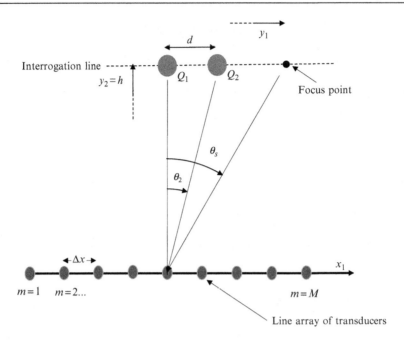

**Fig. 12.4** Imaging of two point sources of equal strength with a linear array.

### 12.1.3 Acoustic images and source levels

The method described above can be used to find the location of a source, but in aeroacoustics the location of the source is usually known, and we need to determine its level relative to the level of other sources that may also be present. To show how this can be achieved using a line array consider the situation shown in Fig. 12.4 in which there are two sources separated by a distance $d$ and a distance $h$ from an array of $M$ transducers. To obtain the array output for different steering angles we can superpose the acoustic fields from each source, and the array output will be

$$
\hat{p}_t = \frac{i\omega\rho_o Q_1 e^{ikh}}{4\pi h} F(kL\sin\theta_s) + \frac{i\omega\rho_o Q_2 e^{ik(d^2+h^2)^{1/2}}}{4\pi(d^2+h^2)^{1/2}} F(kL(\sin\theta_2 - \sin\theta_s))
$$

$$(12.1.8)$$

where $Q_1$ and $Q_2$ are the source strengths of each source, and $\theta_2$ is the angle of source 2 subtended at the array.

To obtain an acoustic image we project the array output onto a line or plane. Since we know the sources lie on the line $y_2 = h$ the obvious location for the source image in this case is on the line $y_{min} < y_1 < y_{max}$ at $y_2 = h$, $y_3 = 0$. Since the sources are in the far field from the array, $H \gg |y_{max}| > |y_{min}| > d$, and we can write

$$
\sin\theta_s = \frac{y_1}{\sqrt{h^2 + y_1^2}} \qquad \sin\theta_2 = \frac{d}{\sqrt{h^2 + d^2}}
$$

Substituting these expressions into Eq. (12.1.8) we can then plot the array output as a function of focus position $y_1$ and obtain the source image $|\hat{p}_t|$ as shown by the solid line in Fig. 12.5. This curve is for the two sources of equal strength separated by 8.2 wavelengths with equal source strength chosen, so $\omega\rho_oQ/4\pi r_o = 1$. We see that this image clearly identifies the two sources in the locations of the two largest peaks. The resolution of the sources is determined by the characteristics of the spatial filter function $F(kL\sin\theta_s)$ given by Eq. (12.1.5).

## 12.1.4 Array shading

One of the problems with the source image is side lobe leakage, which is apparent in Fig. 12.5 in the smaller peaks that occur on either side of the sources. It is usually desirable to minimize the amplitude of the side lobes, even at the expense of broadening the main lobe that identifies the source location. Reducing the side lobes in the image can be achieved in the same way as for the spectrum of a time series (see Chapter 11) by multiplying each array signal by a weighting factor or window function $W_m$ so that the array signal for a single source at $\theta = 0$ is

$$\hat{p}_t = \frac{1}{M}\sum_{m=1}^{M}\hat{p}(\mathbf{x}_m)W_me^{i\phi_m} = \frac{-i\omega\rho_oQe^{ikr_o}}{4\pi r_o}F(kL\sin\theta_s)$$

$$F(kL\sin\theta_s) = \left(\frac{1}{M}\sum_{m=1}^{M}W_me^{i\phi_m}\right)$$

(12.1.9)

Various different options are available, and the most commonly used is a triangular weighting given by $W_m = c_mM/\sum_{m=1}^{M}c_m$ with $c_m = (1 - 2|x_1^{(m)}|/L)$.

Fig. 12.5 compares the source images obtained using a window and no window. The windowed result is smoother than the unwindowed result and has much lower side lobes reducing the chances that one of these would be misidentified as a separate source. An important feature of the image in Fig. 12.5 is that the source image levels exceed one for the unwindowed case because of side lobe leakage. For the windowed image the levels are slightly lower than for the unwindowed image because of the reduced side lobe leakage. The price to be paid for reduced side lobe levels is a loss in resolution, which is apparent in the rounding of the peaks.

Windowing array data is usually desirable and can be optimized when there are only a limited number of point sources in the source distribution. These methods, referred to as adaptive beamforming methods, calculate an optimal window function for which the array output as a zero at source 2 when focused on source 1, and a zero at source 1 when focused on source 2. This process can be extended to several individual sources but fails to work for sources that are continuously distributed. In aeroacoustics most applications involve complicated source environments, and so adaptive beamforming methods have not been as successful as they have been in other areas of array processing.

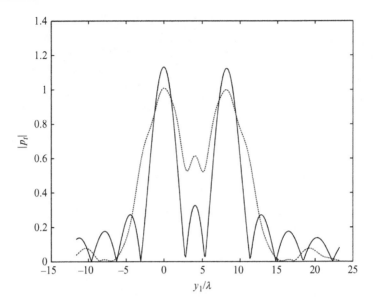

**Fig. 12.5** The source image for two point sources of equal strength separated by $d=0.7$ m at 4000 Hz for an array with $M=15$ transducers and $h=5$ m. The *solid line* is with no window, and the *dotted line* is for a triangular window. The source strengths are chosen, so $\omega\rho_o Q/4\pi r_o=1$.

### 12.1.5 Broadband noise sources

So far we have only considered simple harmonic sources, but the same approach may be used for sources with an arbitrary time dependence. If we take the Fourier transform with respect to time of the signal from each transducer we obtain Eq. (12.1.1) with $Q$ replaced by its Fourier transform, i.e., a function representing the amplitude of the acoustic source at each frequency. We can then follow the same procedures as described above to obtain the array output as a source image. In most cases we are interested in the signals that are broadband in nature, and so we calculate the power spectrum of the array output at each point in the source image, which is given from Eq. (12.1.8), for two sources, for which $d^2 \ll h^2$ by

$$
\begin{aligned}
S_{tt}(\omega) &= \frac{\pi}{T}E\left[|\tilde{p}_t|^2\right] \\
&= \frac{\pi}{T}\left(E\left[|\tilde{Q}_1|^2\right]\left|F(kL\sin\theta_s)\right|^2 + E\left[|\tilde{Q}_2|^2\right]\left|F(kL(\sin\theta_2 - \sin\theta_s))\right|^2\right)(\omega\rho_o/4\pi h)^2 \\
&\quad + \frac{2\pi}{T}\text{Re}\left\{E\left[\tilde{Q}_1^*\tilde{Q}_2\right]F(kL(\sin\theta_2 - \sin\theta_s))F(kL(\sin\theta_s))\right\}(\omega\rho_o/4\pi h)^2
\end{aligned}
$$

(12.1.10)

This result simplifies if the sources are uncorrelated, then the cross spectrum of the signals from each source will be zero. The source image is then the sum of the squares of the source spectra multiplied by the magnitude squared of the array sensitivity function $F$.

This result is of course a function of both frequency and space, and it is often useful to plot these as a contour plot with frequency on the vertical axis and location on the

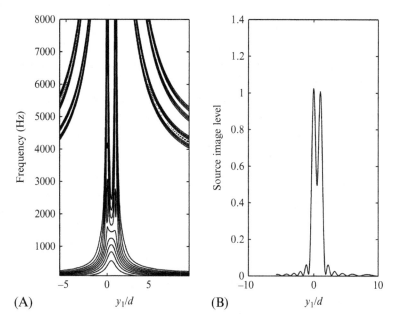

**Fig. 12.6** (A) The contour map of the source image as a function of frequency for two sources of equal strength located at $y_1 = 0$ and $y_1 = d = 0.7$ m obtained using an array of 15 transducers with a separation of 0.1 m. The distance of the array from the sources is 5 m. (B) The source image level as a function of the image location $y_1/d$ at 2000 Hz.

horizontal axis as shown in Fig. 12.6. The figure clearly shows how the ability to resolve the nearby sources and their strengths is frequency dependent and improves at the higher frequencies. It also shows the effects of spatial aliasing of the sources (as illustrated in Fig. 12.3C) in the form of the curved bands that occur at high frequencies and large displacements. We also note that in this case the source at $y_1 = d$ has a peak image level that is lower than the source at $y_1 = 0$. This is a consequence of the proximity of the source to the center of the array.

It is also noteworthy that the sources are clearly identified in Fig. 12.6B at a frequency of 2000 Hz, but the image given for two harmonic sources of the same frequency (as in Fig. 12.5) would not have separated the sources at 2000 Hz. This is a feature of the fact that the sources have been assumed to be uncorrelated in Fig. 12.6B, and so their image depends on $F^2$ not $F$ as in Fig. 12.5, giving an apparent reduction in side lobe level and sharpness of the peak.

# 12.2 General approach to array processing

## 12.2.1 Background

The objective of the initial studies using microphone arrays for acoustic source location in aeroacoustics was to identify the source distributions in jets and on full scale jet engines (Fisher et al. [1], Billingsley and Kinns [2]). The purpose of developing these methods was to separate the noise coming from the engine inlet, the by-pass duct, and the jet on a high by-pass ratio engine operating on a test stand, and to gain more insight into the generation of jet noise from measurements at model scale in anechoic chambers. Since those early studies the applications have been extended to include the measurement of airframe noise by using ground-based acoustic arrays during an aircraft flyover [3], wind tunnel measurements to extract source levels on a model from the background noise [4], and to identify sources on wind turbine blades while they are in operation [5]. Similar array-processing methods are used to extract information about sources in ducts [6], and in particular the modal breakdown of the propagating waves. In other studies, microphone array measurements have been used to determine the unknown parameters in a model of the source distribution under consideration [7,8] thus simplifying the array processing.

Overall the use of microphone arrays for acoustic source measurements can be summarized by the general objective of trying to fit a set of unknown parameters to a specified model of the acoustic sources, with the assumption that the wave propagation from the source to the receiver is known and defined precisely. In the simplest case the source model is a distribution of omnidirectional point sources with specified locations but unknown strengths. In more elaborate models the distribution of relative source strength is defined by a two- or three-parameter model for jet noise [7] or an instability wave [8], and the unknown parameters are estimated from the array measurements using a linear or nonlinear least squares optimization. However, in all cases the process has two steps. The first is to use a source-imaging approach to gain insight into the source distribution, where the dominant sources are located, and the second step is to deconvolve the source image, or model the microphone array data, to obtain the spectra of each acoustic source of interest. In this section we focus on the first step of identifying the dominant sources, and in Section 12.3 we discuss deconvolution methods.

One of the fundamental concerns regarding source location methods is the adequacy or correctness of the source model to represent the actual acoustic source mechanism. For example, jet noise source images are usually modeled by a distribution of stationary omnidirectional uncorrelated monopoles. In contrast Lighthill's model of jet noise represents the jet by quadrupole sources that are convected at a speed close to the speed of sound. It was pointed out by Michel [9] that to correctly interpret jet noise source images they must be compared to a stationary source model. This is an important point because Lighthill's model assumes that the turbulent eddies are acoustically compact in the convected frame of reference, and the directionality is caused by a combination of the source motion and the refraction of sound by the shear layer. If jet noise is modeled in the stationary frame the effective length scale of the source fluctuations is increased, and the sources must be properly phased to give the correct far-field directionality. It follows that the source model is key to understanding the source image obtained by an array of far-field microphones. In many applications such as airframe noise or wind tunnel measurements, the problem is simplified because the source image can be checked by mechanically altering the source distribution. For example, if a side edge flap is identified as an important source of airframe noise during an aircraft flyover it can be retracted and the measurement is repeated. If the source is no longer present in the second test, then the acoustic measurement is verified. Unfortunately, in applications such as jet noise where the source cannot be mechanically removed this option is not available and correctly interpreting the results is more difficult.

In this text we are mainly concerned with low Mach number flows where the acoustic sources are caused by turbulent flow close to impenetrable surfaces. In the rest of this chapter we therefore focus on phased array processing applied to this type of problem. This leaves out the problem of jet noise, and the reader is referred to the chapter by Jordan in Ref. [10] for a more detailed discussion of this topic.

## 12.2.2 The definition of source strength

In Part 1 of this book we showed that for low Mach number flows sound radiation was primarily caused by turbulence in the vicinity of rigid surfaces. The origin of the sound is always the turbulence, but the acoustic levels in the far field are significantly amplified by the scattering of near-field pressure fluctuations by the surface, and these become the dominant sources of sound in most applications.

The objective of array processing, as applied to low Mach number flows, is to identify which parts of a surface are generating the most noise. In flyover noise applications the source is moving, and corrections are needed for both the Doppler shift and the motion of the source. On engine test stands and in wind tunnel applications the sources are stationary and so corrections for source motion are not required, but the environment may be noisy, include reflections from walls and mounting structures, and refraction effects can distort the source image. In addition, the sound being emitted from engine inlets or exits is inherently directional, and this intuitively poses a problem for the geometrical optics approach to array processing that assumes omnidirectional point sources. The source directionality is equivalent to imposing a window function or weighting factor on the signal received by the array, as described by Eq. (12.1.9). If this weighting factor only causes a variation in signal amplitude across the array then its effect will be to smooth out the source image, and it will have a similar, but likely weaker, effect on the source distribution as the triangular weighting shown in Fig. 12.5. However, if the directionality includes a phase variation then the apparent source position can be shifted, and this may lead to erroneous conclusions about the sources if the phase directionality is severe. To avoid these problems, it is necessary to assume that, as seen over the aperture of the array, the sources are approximately omnidirectional. We will return to this issue in Section 12.4 and address the complications of source directionality in that section. However, to establish a general approach to array processing we will start by making this assumption.

In applications where the sources are stationary and the receivers are outside of the flow we can use Curle's equation (4.3.9) to describe the acoustic sources and the signals received by the array. We can reduce the problem to one that can be directly related to a distribution of equivalent sources if we ignore the quadrupole terms in Curle's equation, which describe the sources in the flow, and retain only the linear inviscid surface source terms. Also we will ignore the effect of refraction across the shear layer bounding the mean flow, but, if necessary, wind tunnel corrections can be used to allow for this effect as described in Chapter 10. The net result of these assumptions is that we can specify the sound field by the solution obtained for linear acoustics, Eq. (3.9.12), or its equivalent in the frequency domain, Eq. (3.10.5):

$$\widetilde{p}(\mathbf{x}, \omega) = \int_S \left( \widetilde{p}(\mathbf{y}, \omega) \frac{\partial \widetilde{G}(\mathbf{x}|\mathbf{y})}{\partial y_i} n_i(\mathbf{y}) - \widetilde{G}(\mathbf{x}|\mathbf{y}) \frac{\partial \widetilde{p}(\mathbf{y}, \omega)}{\partial y_i} n_i(\mathbf{y}) \right) dS(\mathbf{y})$$

The surfaces of integration can then be broken down into $N$ sub-surfaces, each one enclosing a different source point. These sub-surfaces can be taken to be arbitrarily small and so are acoustically compact allowing the Green's function to be taken outside of each surface integral giving the signal at the transducer located at $\mathbf{x}^{(m)}$ as

$$\widetilde{p}\left(\mathbf{x}^{(m)}, \omega\right) = \sum_{n=1}^{N} \widetilde{G}\left(\mathbf{x}^{(m)}|\mathbf{y}^{(n)}\right) q_n \tag{12.2.1}$$

where $q_n$ is the effective or equivalent source strength defined as

$$q_n = \int_{S_n} \left( \left( \frac{\widetilde{p}(\mathbf{y}, \omega)}{\widetilde{G}(\mathbf{x}^{(m)}|\mathbf{y}^{(n)})} \right) \frac{\partial \widetilde{G}(\mathbf{x}|\mathbf{y})}{\partial y_i} n_i(\mathbf{y}) - \left( \frac{\widetilde{G}(\mathbf{x}|\mathbf{y})}{\widetilde{G}(\mathbf{x}^{(m)}|\mathbf{y}^{(n)})} \right) \frac{\partial \widetilde{p}(\mathbf{y}, \omega)}{\partial y_i} n_i(\mathbf{y}) \right) dS(\mathbf{y})$$

$$\tag{12.2.2}$$

This shows that, in the geometric far field where $\widetilde{G}\left(\mathbf{x}^{(m)}|\mathbf{y}^{(n)}\right) \approx \left[\widetilde{G}(\mathbf{x}|\mathbf{y})\right]_{S=S_n}$, each source point has two source types, a monopole source that depends on the pressure gradient normal to the surface and a dipole source that depends on the pressure applied to the fluid by the surface. The monopole source is zero for rigid surfaces, but we should not restrict the analysis to that case alone. For example, in an open jet wind tunnel the inlet and the jet catcher may be represented by a Ffowcs Williams and Hawkings surface across the flow, and the parasitic noise from upstream or downstream of the test section will cause equivalent sources that may be of monopole order.

The important part of this model is that the equivalent sources are defined as being omnidirectional, and the actual source directionality must be absorbed into the definition of the source strength, not the Green's function. Substituting the free-field Green's function (Eq. 3.10.8),

$$\widetilde{G}_o(\mathbf{x}|\mathbf{y}) = \frac{e^{ik|\mathbf{x}-\mathbf{y}|}}{4\pi|\mathbf{x}-\mathbf{y}|}$$

we see that in the far field the first term in the integrand of Eq. (12.2.2) depends on

$$\left( \frac{1}{\widetilde{G}_o(\mathbf{x}^{(m)}|\mathbf{y}^{(n)})} \right) \frac{\partial \widetilde{G}_o(\mathbf{x}|\mathbf{y})}{\partial y_i} n_i(\mathbf{y}) \approx \frac{ikx_i n_i\left(\mathbf{y}^{(n)}\right)}{|\mathbf{x}|} \tag{12.2.3}$$

which results in a cosine directionality relative to the normal to the surface $S_o$. If the array aperture is small then this directionality is relatively weak across the array, and the error incurred by assuming an equivalent omnidirectional source is relatively unimportant and, as stated above, will be ignored. If the array processing provides the source strength $q_n$ then we can use Eqs. (12.2.2) and (12.2.3) to interpret the result in terms of the effective source strength as seen by the observer at a particular observer angle.

### 12.2.3 Source images and the point spread function

In order to focus the array onto a specific image point we need to phase shift each microphone signal, multiply by some weighting factor and sum, as in Eq. (12.1.9). This process can be thought of as multiplying the complex amplitudes of the array signals by a steering vector. Using Eq. (12.1.9) as an example, the steering vector in that case was

$$\frac{W_m e^{i\phi_m}}{M}$$

The components of the steering vector are weighting factors generalized to include the phase shift and normalization on the number of microphones $M$. The steering vector of Eq. (12.1.9) assumes sound waves that are planar at the array. A better steering vector uses weighting factors that account for spherical wavefronts from each source point. For example, the steering vector for sources located at $\mathbf{y}^{(j)} = (y_1^{(j)}, 0, 0)$ being focused by a line array of microphones located at $\mathbf{x}^{(m)} = (x_o + (m-1)\Delta x, r_o, 0)$ in the acoustic far field is [11]

$$w_m^{(j)} = W_m \frac{|\mathbf{x}^{(m)} - \mathbf{y}^{(j)}|}{M r_o} e^{-ik\mathbf{x}^{(m)} \cdot \mathbf{y}^{(j)}/r_o} \quad \text{with } W_m = c_m M / \sum_{m=1}^{M} c_m \qquad (12.2.4)$$

where $|\mathbf{x}^{(m)} - \mathbf{y}^{(j)}|$ exactly cancels the effect of spherical spreading as the wave from the source propagates across the array, and $c_m$ is some arbitrary weighting function, such as the triangular weighting used in Eq. (12.1.9). The array output when focused at point $\mathbf{y}^{(j)}$ using this steering vector is then

$$\tilde{p}_t^{(j)}(\omega) = \sum_{m=1}^{M} \left(w_m^{(j)}\right)^* \tilde{p}\left(\mathbf{x}^{(m)}, \omega\right) \qquad (12.2.5)$$

where $\tilde{p}\left(\mathbf{x}^{(m)}, \omega\right)$ is the Fourier transform of the signal from each transducer. We can relate the array output to the source strengths by using Eq. (12.2.1) to give

$$\tilde{p}_t^{(j)}(\omega) = \sum_{m=1}^{M} \sum_{n=1}^{N} \left(w_m^{(j)}\right)^* \tilde{G}\left(\mathbf{x}^{(m)} | \mathbf{y}^{(n)}\right) q_n \qquad (12.2.6)$$

In general, we need to consider broadband sources, and so we evaluate the spectral density of the array output, defined using Eq. (12.2.5) as

$$b_j(\omega) \equiv S_{tt}(\omega) = \frac{\pi}{T} E\left[|\tilde{p}_t^{(j)}(\omega)|^2\right]$$

$$= \sum_{m=1}^{M}\sum_{s=1}^{M} w_s^{(j)}\left(w_m^{(j)}\right)^* \frac{\pi}{T} E\left[\tilde{p}^*\left(\mathbf{x}^{(s)},\omega\right)\tilde{p}\left(\mathbf{x}^{(m)},\omega\right)\right] \qquad (12.2.7)$$

where we have introduced the notation $b_j$ to represent the spectral density of the array output at the $j$th focus point. The last term of the above expression defines the elements of the cross-spectral density (CSD) matrix of the array signals, which is

$$C_{sm} = \frac{\pi}{T} E\left[\tilde{p}^*\left(\mathbf{x}^{(s)},\omega\right)\tilde{p}\left(\mathbf{x}^{(m)},\omega\right)\right] \qquad (12.2.8)$$

To relate the spectrum of the array output $b_j(\omega)$ to the source strengths we first write Eq. (12.2.6) as

$$\tilde{p}_t^{(j)}(\omega) = \sum_{n=1}^{N} F_{jn}q_n, \qquad F_{jn} = \sum_{m=1}^{M}\left(w_m^{(j)}\right)^* \tilde{G}\left(\mathbf{x}^{(m)}|\mathbf{y}^{(n)}\right) \qquad (12.2.9)$$

and so the source image is defined in terms of the source spectrum as

$$b_j(\omega) = \sum_{n=1}^{N}\sum_{p=1}^{N} F_{jn}F_{jp}^* \frac{\pi}{T} E\left[q_p^* q_n\right] \qquad (12.2.10)$$

We now make the assumption that all the sources are uncorrelated. This simplifies the result because only the terms for which $n=p$ will be nonzero. The spectrum of the source strengths can be defined as $Q_{nn}(\omega) = (\pi/T)E[|q_n|^2]$, and so the array output is related to the source strengths by

$$b_j(\omega) = \sum_{n=1}^{N} |F_{jn}|^2 Q_{nn}(\omega) \qquad (12.2.11)$$

where $|F_{jn}|^2$ is referred to as the point spread function for the array focused at $\mathbf{y}^{(j)}$. It is often normalized to be one for $n=j$ and expressed in dB as $20\log_{10}(|F_{nj}|/|F_{nn}|)$. It specifies the source image level at the point $\mathbf{y}^{(j)}$ caused by a source of unit level at the point $\mathbf{y}^{(n)}$. For example, the unweighted point spread function for the linear array described in Section 12.1 is given by Eq. (12.1.9) for a source at $y_1^{(n)}=d$ as

$$|F_{nj}|^2 = \left|\frac{\sin\left(k\left(\sin\theta_n - \sin\theta_j\right)M\Delta x/2\right)}{M\sin\left(k\left(\sin\theta_n - \sin\theta_j\right)\Delta x/2\right)}\right|^2, \qquad \sin\theta_n = \frac{y_1^{(n)}}{\sqrt{r_o^2 + \left(y_1^{(n)}\right)^2}} \qquad (12.2.12)$$

This is shown in Fig. 12.7A for a source located at $y_1^{(n)} = 0$. We see that the point spread function shows a fairly narrow peak at the source location with the two largest sidelobes some 13 dB down to either side. However, Fig. 12.5A also reveals a spurious source image, of equal magnitude as the true image, near $y_1/\lambda = 85$. This results from spatial aliasing (as in Fig. 12.2C). Note that there would also be another spurious image near $y_1/\lambda = -85$.

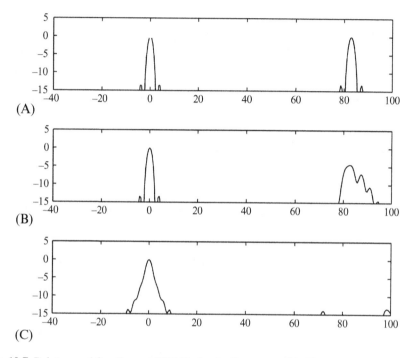

**Fig. 12.7** Point spread functions at 3500 Hz for the line array with 30 transducers and a fixed spacing $\Delta x = 0.375$ m, and for similarly positioned logarithmically spaced array of 30 transducers placed at $x_m = L\log_{10}(m)/\log_{10}(M) - L/2$. Source-array distance $r_o = 30$ m; (A) fixed-spacing array with far-field approximation given by Eq. (12.2.12), (B) fixed-spacing array with steering vectors given by Eq. (12.2.13), and (C) logarithmic array.

The point spread function is important because it determines the array performance for a given array design and a given steering vector. It is not restricted to line arrays and can be used for any type of two- or three-dimensional array or two- or three-dimensional source distribution.

## 12.2.4 Steering vectors

The ability of an array to separate sources and determine their levels depends on the choice of the steering vector, and some general guidelines in making that choice are helpful. The basic rule is that the steering vector should maximize $|F_{jj}|$ and minimize

$|F_{jn}|$ when $n \neq j$. A basic choice for the steering vector is given by Eq. (12.2.4). This relies on the phase shift between the transducers to focus the array and also corrects for the propagation distance from each source to the receiver. The array output is then the spectral level of the signal that would be received from the source at a reference distance $r_o$. However, this form of the steering vector can cause problems when the image includes sidelobes of an aliased source because the weighting factor $|\mathbf{x}^{(m)} - \mathbf{y}^{(j)}|$ that corrects for the spreading loss of far away sources also amplifies the spatial aliases, and this can cause ambiguities at high frequencies. Alternative forms of the steering vector [11] that avoid this problem are

$$
w_m^{(j)} = \frac{\widetilde{G}_o\left(\mathbf{x}_m | \mathbf{y}_j\right)}{\left\| \widetilde{G}_o\left(\mathbf{x}_m | \mathbf{y}_j\right) \right\|} \quad \text{or} \quad w_m^{(j)} = \frac{e^{i\omega \mathbf{x}_m \cdot \mathbf{y}_j / r_c c_\infty}}{M} \tag{12.2.13}
$$

where $\left\| \widetilde{G}_o\left(\mathbf{x}_m | \mathbf{y}_j\right) \right\|$ is the norm of the free-field Green's function for all source and receiver positions. Each steering vector will have its own point spread function and postprocessing algorithm, and so the choice of the weighting in the steering vector is usually determined by the application rather than a general rule, providing the phase shift is correct to focus on a particular source. Fig. 12.7 shows the point spread function for the steering vector given by the first example in Eq. (12.2.13) (Fig. 12.7B) and the far-field approximation given by Eq. (12.2.12) (Fig. 12.7A). It is seen that the point spread function for the source being focused on is identical, but the aliased source image near $y_1/\lambda = 85$ has been altered in Fig. 12.7B because of the weighting factors, reducing its amplitude and shifting its apparent location.

## 12.2.5  Signal-to-noise ratio

Signal-to-noise ratio is also an important consideration. In general, we need to distinguish the sound from the sources that we are interested in from the instrumentation noise that is specific to each receiver. A particular case in point is when the microphones of an array are exposed to a flow and their signals include turbulent pressure fluctuations that are not part of the acoustic field. This noise level can be high but is usually uncorrelated from transducer to transducer, so it will only contribute to the diagonal elements of the measured CSD matrix given by Eq. (12.2.8). In situations where there is significant transducer noise the array processing can be improved by eliminating the diagonal elements from the CSD matrix. This reduces the amount of data that is available to form the source image from $M^2$ to $(M-1)M$ which is not very significant when the number of transducers is large, and the advantage is that the instrumentation noise is eliminated from the processing. However, it can also result in negative source image levels [11] that are unphysical. Therefore, for specific array designs the point spread function always needs to be checked to ensure that diagonal elimination from the cross-spectrum matrix is not detrimental to the expected output of the array.

## 12.2.6 Array design

In Section 12.1 we considered a line array with transducers of equal spacing. It was found that there were competing requirements for resolving sources and preventing spatial aliasing. To reduce aliasing at high frequencies the spacing between the transducers needs to be less than half the acoustic wavelength, but to obtain good resolution at low frequencies the length of the array needs to be as large as possible. For a given number of transducers at equal spacing this limits the operational frequency range to within specific bounds. One approach that can be used to improve array performance is to use an array with unequally spaced transducers so that it includes both closely spaced transducers to eliminate spatial aliasing at high frequencies and a large aperture to maximize source resolution at low frequencies. A common choice is to use logarithmically spaced transducers that are clustered at one end of a line array. An example of the point spread function for a logarithmic array is shown in Fig. 12.7C and can be compared to the point spread function of an array of the same length and with the same number of transducers in Fig. 12.7A. We see that the width of the point spread function near its peak around $y_1 = 0$ is largely unaffected by using the logarithmic spacing, but spatial aliasing that resulted in the spurious image around $y_1 = 85$ is almost completely eliminated. In many applications the logarithmic array can be advantageous because it extends the frequency range that can be considered by a limited number of transducers.

The illustrations so far have only considered line arrays, but in most airframe noise or wind tunnel applications [11,12] there is a requirement to evaluate sources that are located over a two- or three-dimensional region, and so we need an array that can resolve sources in both the $y_1$ and $y_3$ direction. This requires that the transducers in the array are spread over a plane with displacements in the $x_1$ and $x_3$ direction so that the phase of the sound reaching the array from the far field depends on $\exp(-i\omega(x_1y_1 + x_3y_3)/c_\infty r_o)$. It is then possible to construct steering vectors that will focus on a point that depends on $y_1$ and $y_3$ to obtain a source image. Typical array designs that meet this criterion are rectangular or circular arrays in which the transducers are spread over a plane surface. This type of array has been used extensively in underwater acoustics applications. The criteria for spatial aliasing and resolution also apply to planar arrays and depend on both the transducer spacing and maximum dimension of the array in a given direction. The number of transducers in a planar array can rapidly become large, and so the optimization of the array design for a given number of transducers is much more important than for a line array. For this reason, spiral arrays (see Fig. 12.8) are particularly useful in wind tunnel applications [12] because they offer a good compromise between spatial resolution and spatial aliasing for a limited number of receivers. The array includes transducers arranged in a spiral pattern defined by

$$r_m = r_s \exp(h\theta_m)$$

where $r_m$ and $\theta_m$ are the radial position and angle of the $m$th transducer on the spiral ($\theta_m$ can be greater than $2\pi$), and $r_s$ is the radius of the innermost receiver of the spiral pattern. The scaling factor $h$ is determined from the maximum and minimum radius of the array, and the maximum angle as

$$h = \frac{1}{\theta_{max}} \ln\left(\frac{r_{max}}{r_s}\right)$$

For equal spacing along the spiral for $M$ transducers Underbrink [12] gives

$$h\theta_m = \ln\left(1 + \frac{(m-1)(r_{max} - r_s)}{(M-1)r_s}\right)$$

This arrangement is shown in Fig. 12.8A for a spiral array with 40 microphones. The corresponding point spread function is shown in Fig. 12.8B on a plane at a range of 5 m and a frequency of 3000 Hz, for source located at $y_1 = y_3 = 0$. The results show that the array can be focused on a point on a plane and that at these frequencies the side lobes and aliases are not significant. The properties of the array are the same as for a line array and depend on the array diameter to range ratio $2r_{max}/r_o$. Increasing the range reduces the resolution unless the diameter is also increased by the same ratio.

Many other transducer arrangements can be used for specific applications, which are optimized for the requirements of resolution and spatial aliasing in a given environment. In particular, nested spiral arrays [12] can be used to increase the number of transducers in a circular array for a given array diameter. The results are always restricted to a limited frequency range, set by the minimum transducer spacing, the size of the array, and the size and spacing of the sources being measured. The point spread function for a particular design should always be checked before drawing conclusions about array output results. In the next section we will discuss deconvolution methods that allow for improved resolution between sources, but it should be remembered that, as in time series analysis, nothing can be done to eliminate the ambiguity caused by spatial aliasing.

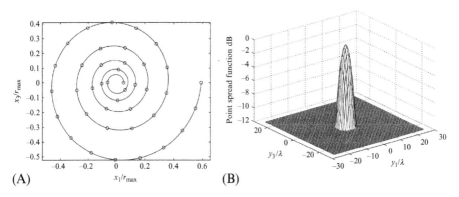

**Fig. 12.8** (A) A spiral array with 40 receivers and (B) its point spread function on a plane at a range of 5 m at 3 kHz. The array diameter is 1.18 m, and the inner diameter of the spiral is 5 cm.

## 12.2.7 Array-processing algorithms

The implementation of the array-processing algorithms described above is particularly simple if they are specified as a series of matrix operations that can be coded directly in Matlab or any computing environment that is optimized for matrix operations. The process is to define the time Fourier transforms of the signal from each transducer at each frequency as a row vector **p** with $M$ elements, defined as

$$\mathbf{p} = \left\{ \tilde{p}\left(\mathbf{x}^{(1)}, \omega\right), \tilde{p}\left(\mathbf{x}^{(2)}, \omega\right), \ldots \right\} \tag{12.2.14}$$

Similarly we can define a row vector with $N$ elements for the source strengths as $\mathbf{q} = \{q_1(\omega), q_2(\omega), \ldots\}$. The relationship between the source strength and the acoustic pressure at each microphone is given by Eq. (12.2.1), and so to implement this we define a rectangular matrix with $M$ rows and $N$ columns for the Green's function, where each element defines the propagation from the source to the receiver, so Eq. (12.2.1) can be written as

$$\mathbf{p}^H = \mathbf{G}\mathbf{q}^H \quad \mathbf{G} \equiv \left[\tilde{G}\left(\mathbf{x}^{(m)}|\mathbf{y}^{(n)}\right)\right] \tag{12.2.15}$$

and the superscript $H$ represents the Hermitian transpose, defined by interchanging the rows and columns of the matrix/vector and taking the complex conjugate. The cross-spectrum matrix is then obtained as in Eq. (12.2.8) as

$$\mathbf{C} = \frac{\pi}{T}E\left[\mathbf{p}^H\mathbf{p}\right] \tag{12.2.16}$$

which has $M$ rows and $M$ columns and can be related to the cross spectrum of the source strengths $\mathbf{Q} = (\pi/T)E[\mathbf{q}^H\mathbf{q}]$ as

$$\mathbf{C} = \mathbf{G}\mathbf{Q}\mathbf{G}^H \tag{12.2.17}$$

To implement array processing we arrange the steering vectors $w_m^{(j)}$ into a rectangular matrix with $M$ rows and $J$ columns, where $J$ is the number of image points so that $\mathbf{W} = \left[w_m^{(j)}\right]$. Then the source image may be calculated using Eq. (12.2.7) as

$$\mathbf{B} = \mathbf{W}^H\mathbf{C}\mathbf{W}, \qquad \mathbf{b} = diag(\mathbf{B}) \tag{12.2.18}$$

where $\mathbf{b}$ is a column vector with elements equal to the spectral density at each point in the image $b_j(\omega)$. These equations can be coded directly and are relatively simple to implement. If the sources are uncorrelated then the source cross-spectrum matrix $\mathbf{Q}$ is diagonal with the nonzero elements equal to $(\pi/T)E[|q_n|^2]$. To obtain the point spread function for the source at $n = j$, the source image should be evaluated by setting $Q_{jj} = 1$ and all other elements of $\mathbf{Q}$ equal to zero. We also note that Eq. (12.2.18) also evaluates the cross spectrum of the source image points, which provides useful additional information about the source distribution. However the source image is defined by the spectrum of the sources at each point, which is given by the diagonal elements of $\mathbf{B}$.

## 12.3 Deconvolution methods

### 12.3.1 Source spectra

The methods given in the previous sections are designed to give a source image that shows the relative strength of each source in a distribution. In most applications we also need to determine the source spectrum for each individual source, and this requires postprocessing of the results. The difficulty is that the sources are not always completely resolved, and so the level at a point in the source image or map may not be an accurate indication of the level of a particular source at that point. To address this, issue a number of different procedures have been developed that deconvolve the source strength from the point spread function and provide a better estimate of the individual source strengths, and their spectra.

Before discussing deconvolution methods, we note that to obtain absolute source levels we only need to know the relative source strengths at each frequency. To show this we evaluate the power spectral density at one microphone in the array, which is designated as the reference microphone, and relate it to the sum of the source strengths assuming no correlation between sources, as

$$C_{mm}(\omega) = \frac{\pi}{T}E\left[\left|\tilde{p}\left(\mathbf{x}^{(m)}, \omega\right)\right|^2\right] = \sum_{n=1}^{N}\left|\tilde{G}\left(\mathbf{x}^{(m)}|\mathbf{y}^{(n)}\right)\right|^2 Q_{nn} \tag{12.3.1}$$

If we then choose a target source, with index $i$, which has nonzero amplitude at the frequency of interest then, we can write this as

$$C_{mm}(\omega) = S_{mm}^{(i)}(\omega)\left(1 + \sum_{n=1,n\neq i}^{N}\frac{\left|\tilde{G}\left(\mathbf{x}^{(m)}|\mathbf{y}^{(n)}\right)\right|^2 Q_{nn}}{\left|\tilde{G}\left(\mathbf{x}^{(m)}|\mathbf{y}^{(i)}\right)\right|^2 Q_{ii}}\right) \tag{12.3.2}$$

$$S_{mm}^{(i)}(\omega) = \left|\tilde{G}\left(\mathbf{x}^{(m)}|\mathbf{y}^{(i)}\right)\right|^2 Q_{ii}$$

where $S_{mm}^{(i)}(\omega)$ is the level that would be measured at the receiver $m$ from the target source $n = i$ in isolation. If we know the relative amplitude of each source $Q_{nn}/Q_{ii}$ then the only unknown in this equation is $S_{mm}^{(i)}(\omega)$, which can be readily evaluated. The spectrum of each of the other sources at the reference receiver can be similarly obtained.

### 12.3.2   The DAMAS method

The task of deconvolution is therefore to accurately define the relative levels of each source in the source map. To place this in context we first define the level at each point in the image using Eq. (12.2.11)

$$b_j(\omega) = \sum_{n=1}^{N}|F_{jn}|^2 Q_{nn}(\omega) \tag{12.3.3}$$

where $j = 1, 2, \ldots J$ specifies the image points. This relationship represents a linear set of $J$ equations with $N$ unknown values $Q_{nn}$. It is often convenient to write these equations in matrix form as

$$\mathbf{b} = \mathbf{Fs} \tag{12.3.4}$$

where $\mathbf{b}$ and $\mathbf{s}$ are column vectors with elements $b_j$ and $Q_{nn}$, respectively, and $\mathbf{F}$ is a rectangular $J \times N$ matrix with elements $|F_{jn}|^2$. In principle, we can solve this equation providing $J > N$ by inverting the matrix $\mathbf{F}$, but this leads to both positive and negative values of the mean square source strength, which is unphysical. Therefore, we impose the constraint that $Q_{nn} > 0$ and solve Eq. (12.3.4) using a nonnegative least squares algorithm. This is the principle of the DAMAS method [11] and yields a stable result with certain limitations. First the number of image points cannot exceed the square of the number of microphones in the array, and second if the image points are placed too close together then the processing is slow with little added information. Brooks et al. [11] suggest that the image point spacing (or "pixel size") $\Delta x$ should be proportional to the beamwidth of the point spread function, defined by the distance between the points that are 3 dB below the peak level. However, at high frequencies this can lead to over resolution, and it is better to fix the pixel size as proportional to $M^2$, using the criteria that $\Delta x$ is proportional to $L/M$, where $L$ is the side length of the source image plane. In Fig. 12.9 we show the deconvolved source image obtained using the DAMAS algorithm for the spiral array of Fig. 12.8. In this case the sources are defined as two line sources to represent the leading and trailing edges of an airfoil. The trailing edge noise sources are placed along the lines $y_1 = -2.1$ and $y_1 = -0.3$ between $-2.1 < y_3 < 1.05$. The level of the sources along the line $y_1 = -2.1$ is 20% of that along $y_1 = 0.3$. From this plot we see that the point spread function (Fig. 12.6A) shows signs of spatial aliasing, and the conventional source image (Fig. 12.9B) does not clearly identify the lower level sources in spite of the fact that they should be resolved. The DAMAS image (Fig. 12.9C), however, reveals the correct location and relative levels of the sources. Similar results can be obtained at lower frequencies providing the pixel size is increased to 40% of the beamwidth when the array fails to resolve the two line sources.

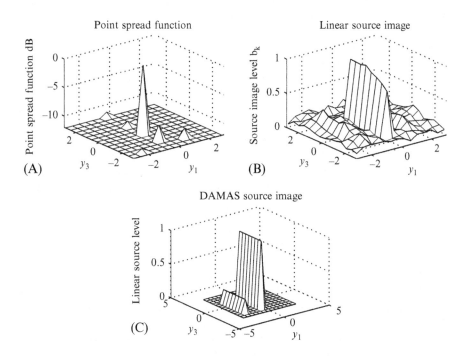

**Fig. 12.9** Examples of the beamformed and deconvolved images using the DAMAS algorithm at 10 kHz for the array used in Fig. 12.8. The image plane lies between $-3 < y_1 < 3$ and $-3 < y_3 < 3$, and the pixel size is $\Delta y \times \Delta y$ with $\Delta y = L/M = 6/40$.

### 12.3.3 The CLEAN algorithm

Another approach that can be used when one source is dominant or can be clearly separated from the others is the CLEAN algorithm [12]. In this approach the source level of the dominant source is assumed to be given by its image level specified by Eq. (12.2.11) and is designated at the location of the maximum in the source image $\mathbf{y}^{(i)}$ with source spectrum given by $Q_{ii} = (4\pi r_o)^2 b_i^{(1)}$ where the superscript indicates the level of the dominant source in the first iteration. The cross spectrum of the array output from this source by itself is then calculated using Eqs. (12.2.1) and (12.2.8) as

$$C_{mj}^{(1)} = \tilde{G}\left(\mathbf{x}^{(m)}|\mathbf{y}^{(i)}\right)\tilde{G}^*\left(\mathbf{x}^{(j)}|\mathbf{y}^{(i)}\right)b_i^{(1)}(4\pi r_o)^2 \tag{12.3.5}$$

This is then subtracted from the measured cross-spectrum matrix. Next the beamformed image is calculated from the corrected cross-spectrum matrix giving a source map that is not contaminated by the side lobes from the dominant source. This process can be repeated a number of times until the noise floor is reached. This has worked well in flyover noise applications [12] where there are distinct isolated source of interest, but its application to distributed sources can be limited.

### 12.3.4 Integrated source maps

In many applications we are interested in the source level of a group of sources that are well resolved in the source map. For example, the source image of trailing edge noise in a wind tunnel, as represented in Fig. 12.9B, is important over a specific area near the blade trailing edge, and we need to extract the noise from that area separately from the noise generated by the blade leading edge or the wind tunnel. To achieve this, we integrate or sum the source image levels over the region of interest. However, the result is influenced by the point spread function for each image point, and so the integrated source level must be corrected for the additional energy caused by the leakage through sidelobes in the point spread function. To specify this we make use of the relationship given in Eq. (12.2.11) and sum the source image levels $b_k$ multiplied by the pixel area $\Delta s_j$ to give the integrated source level over a limited number of image points $j = j_1, j_2, j_3, \dots$ as

$$I = \sum_{j=(j_1,j_2,\dots)} b_j(\omega)\Delta s_j = \sum_{j=(j_1,j_2,\dots)} \sum_{n=1}^{N} |F_{jn}|^2 Q_{nn}(\omega)\Delta s_j$$

To relate the integral of the source image to the actual integrated level we need to define an inner and outer boundary. The outer boundary is the area that the source image has been integrated over and includes all the points $j = j_1, j_2, j_3, \dots$ The inner boundary includes all the sources that contribute to the image within the outer boundary and includes the points $n = n_1, n_2, n_3, \dots$ This is an important approximation and only applies if the source of interest is well resolved and not contaminated by side lobes from other sources in different parts of the system. The integrated source image over the region of interest is then approximated as

$$I \approx \sum_{j=(j_1,j_2,\ldots)} \sum_{n=(n_1,n_2,\ldots)} |F_{jn}|^2 Q_{nn}(\omega) \Delta s_j$$

If the outer boundary is large enough and the pixel size is constant, then we can approximate

$$\sum_{j=(j_1,j_2,\ldots)} |F_{jn}|^2 = F \approx const \tag{12.3.6}$$

so that the integral of the point spread function is independent of the source position. The actual integrated source level over the region of interest is then

$$I_a = \sum_{n=(n_1,n_2,\ldots)} Q_{nn}(\omega) \Delta s_n \approx \frac{I}{F} \tag{12.3.7}$$

This effectively corrects the integrated source image levels for the spreading caused by the finite resolution of the array and allows source spectra to be obtained from source images.

## 12.4 Correlated sources and directionality

The basic assumption applied to the evaluation of the source image was that the sources are omnidirectional and uncorrelated. This assumption is rarely valid for flow noise sources, for example, surface sources are typically of dipole order and have a cosine directionality, edge sources usually have a $\sin(\theta/2)$ directionality, and more extended sources such as rotors have sources that radiate differently in the upstream and downstream directions. Of particular concern are sources that are correlated over an area that is larger than the acoustic wavelength since these sources will have their own distinct acoustic field. An example of this is reflections from rigid surfaces such as wind tunnel walls that cause image sources that can be some distance from the source itself.

First consider the type of error that appears in the source image from the equally spaced line array, used in Fig. 12.7, if the source has a dipole directivity. Two examples are shown in Fig. 12.10, corresponding to a dipole aligned with the $y_1$ axis and a dipole aligned with the $y_2$ axis. For the dipole aligned with the $y_2$ axis the directionality is weak over the array, as shown in Fig. 12.10A, and does not cause a significant variation in level over the array. Consequently, the point spread function is almost the same as it would be for an omnidirectional source. However, for the dipole that has a null in the direction of the array, receivers for $x_1 < 0$ will have signals that are 180 degrees out of phase with the signals for $x_1 > 0$, and so we expect a more noticeable effect on the source image. The worst case scenario is when the array is symmetrical about $x_1 = 0$, and the source image is shown as in Fig. 12.10B. It appears to have two peaks that are symmetric about the actual source point, corresponding to the simple model for a dipole defined by two sources of equal strength and 180 degrees out of phase.

Of particular importance in low Mach number aeroacoustics are trailing edge noise sources that have a $\sin(\theta/2)$ directionality, as shown in Fig. 12.10C. These are typically aligned with the flow, so the null in the directionality points downstream. At 90 degrees to the flow, where measurements are usually made, the directionality is weak, and the source level is retrieved accurately, as shown in Fig. 12.10C.

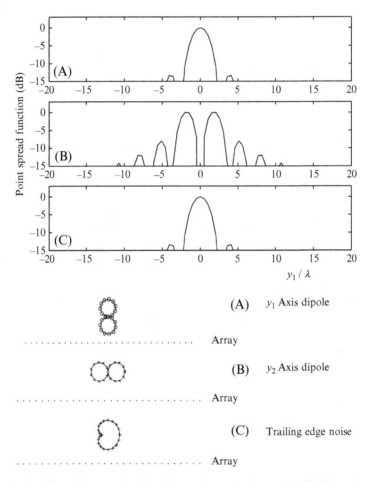

**Fig. 12.10** The effect of source directionality on the source image or effective point spread function. (A) The dipole directionality with a null at 90 degrees to the array, (B) the dipole directionality with a null in the direction of the array shows an image with two sources of equal amplitude, and (C) the $\sin(\theta/2)$ directionality typical of trailing edge noise. The array used is the same as in Fig. 12.7 at 3500 Hz.

In conclusion, the effect of directionality is most important if it includes a null in the far field that falls on the array. A null in the far field is always associated with a 180 degrees change in phase, and so the phase measured across the array is distorted by this effect. If the directionality across the array is relatively weak, as shown by the trailing edge noise or the dipole in the $y_1$ direction in Fig. 12.10, then the assumption of omnidirectional point sources is reasonable.

# References

[1] M.J. Fisher, M. Harper-Bourne, S.A.L. Glegg, Jet engine noise source location: the polar correlation technique, J. Sound Vib. 51 (1977) 23–54.

[2] J. Billingsley, R. Kinns, The acoustic telescope, J. Sound Vib. 48 (1976) 485–510.

[3] U. Michel, J. Helbig, B. Barsikow, M. Hellmig, M. Schuettpelz, Flyover noise measurements on landing aircraft with a microphone array, in: 4th AIAA/CEAS Aeroacoustics Conference, American Institute of Aeronautics and Astronautics, Reston, VA, 1998.

[4] P.T. Soderman, S.J. Noble, Directional microphone array for acoustic studies of wind tunnel models, J. Aircr. 12 (1975) 168–173.

[5] S. Oerlemans, P. Sijtsma, B. Méndez López, Location and quantification of noise sources on a wind turbine, J. Sound Vib. 299 (2007) 869–883.

[6] B. Tester, A. Cargill, B. Barry, Fan noise duct-mode detection in the far-field—simulation, measurement and analysis, in: 5th AIAA Aeroacoustics Conference, Palo Alto, CA, Paper 79-0580, 1979.

[7] B. Tester, M. Fisher, Engine noise source breakdown—theory, simulation and results, in: 7th AIAA Aeroacoustics Conference, Paper 81-2040, 1981. See also P. Strange, G. Podmore, M. Fisher, B. Tester, Coaxial jet noise source distributions, in: 9th AIAA Aeroacoustics Conference, Williamsburg, VA, Paper 84–2361, 1984.

[8] T. Suzuki, T.I.M. Colonius, Instability waves in a subsonic round jet detected using a near-field phased microphone array, J. Fluid Mech. 565 (2006) 197–226.

[9] U. Michel, Characterization of jet noise with phased microphone arrays, in: AARC Engine Noise Phased Array Workshop, Cambridge, MA, 2006.

[10] R. Camussi, Noise Sources in Turbulent Shear Flows: Fundamentals and Applications, Springer, Vienna, 2013.

[11] T.F. Brooks, W.M. Humphreys, A deconvolution approach for the mapping of acoustic sources (DAMAS) determined from phased microphone arrays, J. Sound Vib. 294 (2006) 856–879. See also: T.F. Brooks, W.M. Humphreys, Extension of DAMAS phased array processing for spatial coherence determination (DAMAS-C), in: 12th AIAA/CEAS Aeroacoustics Conference (27th AIAA Aeroacoustics Conference), Paper 2006–2654, 2006.

[12] T. Mueller, Aeroacoustic Measurements, Springer, New York, 2002.

# Part 3

# Edge and boundary layer noise

# The theory of edge scattering

# 13

As we discussed in Chapters 6 and 7 the noise radiated from rotor or fan blades, or a stationary airfoil encountering turbulence, is primarily caused by the interaction of unsteady flow with the leading edge of the blade. In addition, blade boundary layer turbulence can only radiate sound if it encounters a surface discontinuity such as a trailing edge. Both these problems require the detailed analysis of the unsteady flow close to an edge. In this chapter we lay out the details of how edge scattering is calculated so that, in subsequent chapters, we can evaluate leading edge and trailing edge noise.

## 13.1 The importance of edge scattering

We have shown in earlier chapters that sound caused by a turbulent flow in the presence of an airfoil or fan blade is determined by the unsteady pressure on the blade surface. In general, we can model the blade, to first order, by a flat plate of finite chord as shown in Fig. 6.7. The far-field sound was shown in Sections 4.7 and 6.5 to be directly related to the wavenumber transform of the jump in surface pressure across the blade evaluated at the acoustic wavenumber. The physical interpretation of this result is that only waves that propagate across the surface at the speed of sound will couple with the acoustic far field. In principle this is also correct for blades that are acoustically compact in flows of very low Mach number, but in that limit the acoustic wavenumbers are close zero, and the wavenumber spectrum is equal to the net unsteady blade loading for all acoustic wavenumbers of interest.

On a smooth surface, pressure fluctuations associated with turbulence are convected at a speed that is less than (or equal to) the local flow speed, and in most cases of interest this is subsonic. The far-field sound can therefore only be caused by the interaction of the turbulence with an edge, or a discontinuity on the surface, both of which scatter wave energy into acoustic waves. The two most important examples are leading edge noise, where turbulent gust impinges on the leading edge of a blade, and trailing edge noise in which turbulent boundary layer pressure fluctuations are convected downstream across the blade trailing edge. In this chapter both these problems will be considered.

Amiet [1,2] addressed the problems of leading and trailing edge noise by using the solution to the Schwartzschild problem, which was developed for electromagnetic wave scattering in the presence of a semi-infinite half plane. We will derive the solution to this problem using the Weiner Hopf method, which is of general applicability and will be extended to cascades of blades in Chapter 18. The solution to the Schwartzschild problem for electromagnetic waves only applies to a stationary surface in the absence of flow, and so we will also describe how these results can be extended to sound radiation in a uniform flow.

Aeroacoustics of Low Mach Number Flows. http://dx.doi.org/10.1016/B978-0-12-809651-2.00013-8

## 13.2 The Schwartzschild problem and its solution based on the Weiner Hopf method

### 13.2.1 The boundary value problem

The scattering of hydrodynamic pressure waves by a trailing edge can be modeled by considering a surface pressure disturbance traveling downstream over a semi-infinite plate, as shown in Fig. 13.1. The disturbance is expressed in terms of $\Delta p$, the pressure difference between the top and bottom sides of the plate.

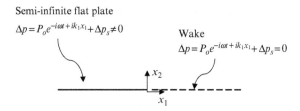

Semi-infinite flat plate
$$\Delta p = P_o e^{-i\omega t + ik_1 x_1} + \Delta p_s \neq 0$$

Wake
$$\Delta p = P_o e^{-i\omega t + ik_1 x_1} + \Delta p_s = 0$$

**Fig. 13.1** A hydrodynamic pressure disturbance traveling downstream over a semi-infinite plate that causes a pressure jump across the plate and a scattered acoustic wave that cancels the pressure jump across the wake.

We assume that the pressure jump $\Delta p$ in the absence of the trailing edge (located at $x_1 = 0$) is given by $P_o \exp(-i\omega t + ik_1 x_1)$ and that the flow Mach number is so small that the mean flow can be ignored in the first instance. Note that we will allow for the possibility that the wavenumber $k_1$ may have a small imaginary part allowing for decay or growth of the disturbance with $x_1$. Both at and downstream of the trailing edge there can be no pressure jump across the surface (the Kutta condition), and so an acoustic wave field must be added that exactly cancels the incident pressure disturbance for $x_1 > 0$, $x_2 = 0$. The scattered acoustic field $p$ must satisfy the acoustic wave equation, and also the boundary condition that $\partial p / \partial x_2 = 0$ on the surface of the plate to match the non-penetration boundary condition. If we initially limit consideration to two dimensions with no mean flow, then the scattered acoustic field is given by the solution of the Schwartzschild problem. This defines a scattered acoustic field with a harmonic pressure fluctuation $\hat{p}(x_1, x_2) \exp(-i\omega t)$ that satisfies the wave equation and the boundary conditions:

$$\frac{\partial^2 \hat{p}}{\partial x_1^2} + \frac{\partial^2 \hat{p}}{\partial x_2^2} + k^2 \hat{p} = 0$$

$$[\Delta \hat{p}_s(x_1)]_{x_1 > 0} = -P_o e^{ik_1 x_1}$$

$$\left[ \frac{\partial \hat{p}(x_1, 0)}{\partial x_2} \right]_{x_1 < 0} = 0$$

(13.2.1)

Schwartzschild showed that this problem can be solved to give the pressure jump over the plate caused by the scattered wave as

$$\Delta \hat{p}_s(x_1) = \frac{-2P_o}{\pi} \int_0^\infty \left( \frac{|x_1|}{\xi} \right)^{1/2} \frac{e^{ik(\xi + |x_1|) + ik_1 \xi}}{\xi + |x_1|} d\xi \qquad x_1 < 0$$

which on evaluation of the integral gives

$$\Delta\hat{p}_s(x_1) = P_o e^{ik_1 x_1}((1-i)E_2((k+k_1)|x_1|) - 1) \qquad x_1 < 0 \tag{13.2.2}$$

where $E_2(x)$ is a modified Fresnel integral defined as

$$E_2(x) = \int_0^x \frac{e^{iq}}{(2\pi q)^{1/2}} dq \tag{13.2.3}$$

which is a function that appears often in the theory of edge scattering in several different forms.

## 13.2.2 Obtaining the Schwartzschild solution using the Weiner Hopf method

The result given by Eq. (13.2.2) is specific to the boundary value problem given by Eq. (13.2.1). However, we can obtain the same result in a way that may be extended to more general cases by considering the scattering in more detail. This will lead to formulations for leading edge noise and far-field radiation that do not follow directly from Eq. (13.2.2).

To obtain Eq. (13.2.2) we start by taking the wavenumber transform of the wave equation with respect to $x_1$, defined such that

$$\tilde{p}(\alpha, x_2) = \frac{1}{2\pi} \int_{-R_\infty}^{R_\infty} \hat{p}(x_1, x_2) e^{-i\alpha x_1} dx_1 \tag{13.2.4}$$

(Note: since this is a wavenumber transform we have chosen the exponential to have a negative sign.) The wave equation then reduces to

$$\frac{\partial^2 \tilde{p}}{\partial x_2^2} + (k^2 - \alpha^2)\tilde{p} = 0$$

and the solution to this equation is

$$\tilde{p}(\alpha, x_2) = A(\alpha)e^{-\gamma x_2} + B(\alpha)e^{\gamma x_2}$$

where

$$\gamma = (\alpha^2 - k^2)^{1/2} \tag{13.2.5}$$

The branch cut for $\gamma$ is chosen so that $\mathrm{Re}(\gamma) > 0$. The presence of the plate implies that the wave field can be discontinuous across the plate where $x_2 = 0$, and so we can define separate solutions for the regions $x_2 > 0$ and $x_2 < 0$. The additional boundary condition that we have is that the acoustic waves must decay at large distances from the plate, and we can only meet that condition if we set $B(\alpha) = 0$ when $x_2 > 0$ and $A(\alpha) = 0$ when $x_2 < 0$. While the plate can support a pressure jump, the normal derivative of the pressure on $x_2 = 0$ must be continuous for both $x_1 > 0$ (where there is no plate) and $x_1 < 0$ (where it is zero). The pressure gradient is therefore

$$\frac{\partial \tilde{p}(\alpha, 0)}{\partial x_2} = [-\gamma A(\alpha)]_{x_2=0^+} = [\gamma B(\alpha)]_{x_2=0^-}$$

and so $A(\alpha) = -B(\alpha)$, and the acoustic field caused by the scattered waves is given by the inverse of Eq. (13.2.4) (i.e., the inverse wavenumber transform)

$$\hat{p}(x_1, x_2) = \int_{-\infty}^{\infty} \text{sgn}(x_2) A(\alpha) e^{i\alpha x_1 - \gamma |x_2|} d\alpha \tag{13.2.6}$$

and allows for a pressure jump across the plate.

In order to solve the unknown function $A(\alpha)$ we use the wavenumber transform of Eq. (13.2.6) on the surface $x_2 = 0^+$ given by

$$A(\alpha) = \frac{1}{2\pi} \int_{-R_\infty}^{R_\infty} \hat{p}(x_1, 0^+) e^{-i\alpha x_1} dx_1$$

However, the boundary condition is different for $x_1 > 0$ and $x_1 < 0$, so we write this as

$$A(\alpha) = A_-(\alpha) + A_+(\alpha) \tag{13.2.7}$$

where

$$A_-(\alpha) = \frac{1}{2\pi} \int_{-R_\infty}^{0} \hat{p}(x_1, 0^+) e^{-i\alpha x_1} dx_1 \tag{13.2.8}$$

and

$$A_+(\alpha) = \frac{1}{2\pi} \int_{0}^{R_\infty} \hat{p}(x_1, 0^+) e^{-i\alpha x_1} dx_1 \tag{13.2.9}$$

The integrand of Eq. (13.2.9) is known from the boundary conditions of the problem, but $A_-(\alpha)$ remains as an unknown. To make use of the non-penetration boundary condition we define the chordwise Fourier transform of the pressure gradient normal to the plate, given by

$$C_+(\alpha) = \frac{1}{2\pi} \int\limits_0^{R_\infty} \frac{\partial \hat{p}(x_1, 0^+)}{\partial x_2} e^{-i\alpha x_1} dx_1 \tag{13.2.10}$$

where the lower limit of the integral is zero because the pressure gradient is zero when $x_1 < 0$. However, by differentiating Eq. (13.2.6) we see that the pressure gradient is given by the inverse Fourier transform

$$\frac{\partial \hat{p}(x_1, x_2)}{\partial x_2} = \int\limits_{-\infty}^{\infty} \left\{ -\gamma A(\alpha) e^{-\gamma|x_2|} \right\} e^{i\alpha x} d\alpha$$

and thus $C_+(\alpha)$ is given by the terms in curly brackets at $x_2 = 0^+$, which using Eq. (13.2.7) is

$$C_+(\alpha) = -\gamma A_-(\alpha) - \gamma A_+(\alpha) \tag{13.2.11}$$

This completely defines the boundary value problem with the conditions set by Eq. (13.2.1). The functions $A_+(\alpha)$ and $A_-(\alpha)$ are required to define the acoustic field using Eq. (13.2.6), but unfortunately only $A_+(\alpha)$ is known, and both $A_-(\alpha)$ and $C_+(\alpha)$ are unknowns, and so it appears that this problem is underdetermined and cannot be solved with these boundary conditions alone.

### 13.2.3 The radiation condition and the Weiner Hopf separation

In solving the wave equation, we used the radiation condition that requires the sound field to decay at large distances from the surface. The same condition applies far upstream or downstream from the edge, and we can use this additional boundary condition to solve for the two unknowns in Eq. (13.2.11). To ensure that waves decay as $|x_1|$ tends to infinity we require that the surface pressure has the asymptotic values

$$\lim_{x_1 \to \infty} \frac{\partial \hat{p}(x_1, 0)}{\partial x_2} < Ce^{-\beta x_1} \qquad \lim_{x_1 \to -\infty} \hat{p}(x_1, 0) < De^{-\beta|x_1|}$$

where $C, D$, and $\beta$ are real positive constants. (There is no need to impose a restriction on the pressure gradient for negative $x_1$ since the gradient is zero here anyway.) The same condition should also be required for the incident pressure disturbance, and this is achieved by requiring that $\text{Im}(k_1) > \beta$. This apparently implies that the incident wave decays from a large value upstream, but this issue can be addressed by assuming that the incident wave is of finite extent in the upstream direction, as would be the case in a real flow.

To show how this additional boundary condition can be used, we make use of some well-known properties of the Laplace transform, which is defined by the transform pairs

$$F(s) = \int_0^\infty f(x)e^{-sx}dx \quad f(x) = \frac{1}{2\pi i}\int_{-i\infty+\beta_o}^{i\infty+\beta_o} F(s)e^{sx}ds \tag{13.2.12}$$

This transform is relevant to the problem being considered because it is essentially the same as the Fourier transform defined in Eq. (13.2.9) over the positive values of $x_1$, so, for example,

$$A_+(\alpha) = F_+(i\alpha)/2\pi \quad \text{where} \quad F_+(s) = \int_0^\infty \hat{p}(x_1, 0^+)e^{-sx_1}dx_1 \tag{13.2.13}$$

Laplace transforms are used extensively in control theory and are often applied to determine if a system is stable or unstable. By definition they represent a function that is zero for $x < 0$, and the system is stable if it does not have exponential growth as $x$ tends to infinity. The criterion to prevent growth at infinity is that the Laplace transform of the function should have no poles, branch cuts, or other singularities on the right side of the complex $s$ plane where $\text{Re}(s) > 0$, as shown in Fig. 13.2.

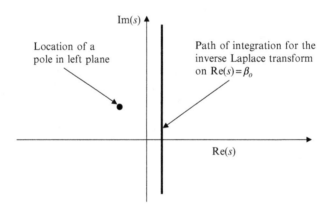

**Fig. 13.2** Laplace transforms in the $s$ plane.

If the function $f(x)$ decays exponentially for large $x > 0$, so $f(x) < C\exp(-\beta x)$, where $C$ and $\beta$ are positive real constants, then the Laplace transform converges when $\text{Re}(s) > -\beta$, and all the singularities of $F(s)$ lie to the left side of the path of integration shown in Fig. 13.2. Since we wish to impose the boundary condition that the scattered field decays for large $|x_1|$, the criterion can be restated as requiring that all the singularities of $F_+(s)$ lie in the region of the $s$ plane where $\text{Re}(s) < -\beta$. We can then impose this restriction on the boundary condition given by Eq. (13.2.12) to eliminate one of the two unknown functions and obtain a complete solution.

However, before we proceed we must define the Laplace transforms of the other terms in Eq. (13.2.11). Since the pressure gradient normal to the plate is zero when $x_1 < 0$ we can define its Laplace transform as in Eq. (13.2.12), by the function

$G_+(s)$, and so $C_+(\alpha)=G_+(i\alpha)/2\pi$. However, for the pressure upstream of the edge, the Laplace transform must be defined by reversing the direction of integration, so (using $\xi=-x_1$)

$$A_-(\alpha)=F_-(-i\alpha)/2\pi \quad \text{where} \quad F_-(s)=\int_0^\infty \hat{p}(-\xi,0^+)e^{-s\xi}d\xi$$

So, the boundary condition may be written as

$$[G_+(s)=-\gamma(s)F_+(s)-\gamma(s)F_-(-s)]_{s=i\alpha} \qquad \gamma(s)=\left(-\left(s^2+k^2\right)\right)^{1/2} \qquad (13.2.14)$$

The functions in this equation are analytic in different parts of the $s$ plane as shown in Fig. 13.3.

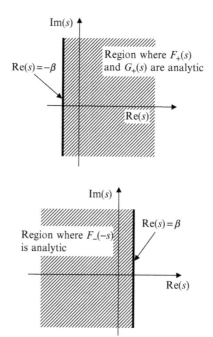

**Fig. 13.3** The regions of the $s$ plane where the functions in Eq. (13.2.14) are analytic.

It follows that the right side of Eq. (13.2.14) only converges in the strip $-\beta<\mathrm{Re}(s)<\beta$. This is important when we evaluate the inverse Laplace transform defined as

$$f(x)=\frac{1}{2\pi i}\int_{\beta_o-i\infty}^{\beta_o+i\infty} F(s)e^{sx}ds \qquad (13.2.15)$$

which is only valid if $F(s)$ converges along the path of integration. The inversion of Eq. (13.2.14) is therefore required to be carried out by choosing $-\beta < \beta_o < \beta$ and is only possible if $\beta > 0$.

We also need to be very specific about the definition of the function $\gamma$. This is a multivalued function and, to meet the radiation condition, is required to have a positive real part. To ensure that this is the case on the path of integration $\text{Re}(s) = \beta_o$ we choose the branch cuts that are shown in Fig. 13.4, and using the criteria given in Appendix B, define

$$\gamma(s) = J_+(s)J_-(s) \quad J_+(s) = e^{i\pi/4}(s - ik)^{1/2} \quad J_-(s) = e^{i\pi/4}(s + ik)^{1/2} \quad (13.2.16)$$

and require that $k$ has a small positive imaginary part that moves the branch points away from the path of integration as shown.

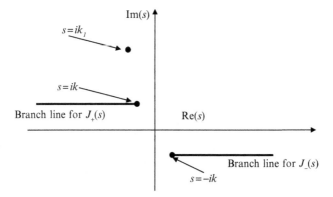

**Fig. 13.4** The branch cuts and poles of Eq. (13.2.12) in the $s$ plane.

All the terms in Eq. (13.2.14) are now defined along the path of integration required for the inverse Laplace transform. However, on the left side of this equation $G_+(s)$ represents a function that is zero upstream of the edge and decays at large $x_1$, so it can have no singularities for $\text{Re}(s) > -\beta$. The same is therefore true for the right-hand side of this equation. This gives the additional restriction that is needed to solve Eq. (13.2.14) for the two unknowns, and we can manipulate the terms so that this criterion is met, which is the basis of the Weiner Hopf method. First divide by $J_+(s)$ so that

$$\left[\frac{G_+(s)}{J_+(s)} = -J_-(s)F_+(s) - J_-(s)F_-(-s)\right]_{s=i\alpha} \quad (13.2.17)$$

The first term on the right side of this equation is a mixture of functions that have non-analytic features for both positive and negative values of real $s$. However, it can be split into two parts, one which exactly cancels the second term on the right (which only has singularities for $\mathrm{Re}(s) > \beta$) and the second that has no singularities on the left side of the $s$ plane. This separation is dependent on the form of $F_+(s)$ which is given by the boundary condition for $x_1 > 0$ in Eq. (13.2.1) with the pressure at $x_2 = 0^+$ being taken as half of the pressure difference $\Delta p_s$, and thus

$$F_+(s) = -\frac{P_o}{2} \int_0^\infty e^{-(s-ik_1)x_1} dx_1 = -\frac{P_o}{2(s-ik_1)}$$

This has a simple pole at $s = ik_1$ that is shown in Fig. 13.4. By removing the residue at the pole we can then write Eq. (13.2.15) as

$$\left[\frac{G_+(s)}{J_+(s)} - P_o \frac{J_-(ik_1)}{2(s-ik_1)} = P_o\left(\frac{J_-(s) - J_-(ik_1)}{2(s-ik_1)}\right) - J_-(s)F_-(-s)\right]_{s=i\alpha}$$

If we take the inverse transform of this equation along the imaginary axis $s = i\alpha$ the right side will give a function that is zero for $x_1 > 0$ because all the terms are analytic for $\mathrm{Re}(s) < 0$ as shown in Fig. 13.3. Similarly, the left side will give an equation that is zero for $x_1 < 0$ (providing that $J_+(s)$ is not zero when $\mathrm{Re}(s) > 0$). The only possible solution therefore is that the two sides are independently equal to zero (or equal to a function whose inverse transform is zero). We then obtain the solution for the unknown function that gives the pressure on the upstream surface as

$$\left[F_-(-s) = P_o\left(\frac{J_-(s) - J_-(ik_1)}{2J_-(s)(s-ik_1)}\right)\right]_{s=i\alpha} \tag{13.2.18}$$

Using the definitions in Eq. (13.2.16) the pressure on the upstream part of the surface can be obtained from the inverse Laplace transform of

$$\left[F_-(s) = \frac{P_o}{2}\left(\frac{(ik+ik_1)^{1/2}}{(-s+ik)^{1/2}(s+ik_1)} - \frac{1}{(s+ik_1)}\right)\right]_{s=i\alpha} \tag{13.2.19}$$

as a function of $\xi = -x_1$. Both terms in this equation have a pole at $s = -ik_1$ in the right half plane that individually would cause a growing wave, but since the residues of the terms cancel we can move the path of integration to the right of the pole and substitute $s = s_1 + ik$, so the inverse transform (designated by $L^{-1}\{\}$) is

$$f(\xi) = \frac{P_o e^{ik\xi}}{2} L^{-1}\left\{\frac{(-i(k+k_1))^{1/2}}{(s_1 + i(k+k_1))\sqrt{s_1}} - \frac{1}{(s_1 + i(k+k_1))}\right\}$$

Standard tables give the inverse Laplace transform of the functions $1/(s+a)$ and $1/s^{1/2}$, which should be combined as a convolution integral when multiplied together. We then obtain the pressure jump across the surface as $2f(\xi)$ where

$$\Delta \hat{p}_s(-x_1) = P_o \left( e^{-i\pi/4}(k_1 + k)^{1/2} \int_0^{|x_1|} \frac{e^{i(k_1+k)\tau}}{\sqrt{\pi\tau}} d\tau - 1 \right) e^{ik_1 x_1} \quad x_1 < 0$$

which is identical to the result given by Eq. (13.2.2).

### 13.2.4 Generalized Fourier transforms and Laplace transforms

The analysis given in the previous section was based on Laplace transforms to obtain the solution to the Schwartzschild problem. In much of the literature (Noble [3], Morse and Feshbach [4]) the Weiner Hopf method is carried out using Generalized Fourier transforms of the type

$$F_+(\alpha) = \frac{1}{2\pi} \int_0^\infty f(x) e^{\pm i\alpha x} dx \qquad F_-(\alpha) = \frac{1}{2\pi} \int_{-\infty}^0 f(x) e^{\pm i\alpha x} dx$$

These are equivalent to the approach given above with $\pm i\alpha$ replacing the Laplace transform variable $s$. The advantage of the current approach is that there are well-documented tables of Laplace transforms and their inverses that are readily available and can be used to simplify the results, and so this is the reason that this approach has been used here. Generalized Fourier transforms can always be obtained from Laplace transforms, providing the correct substitutions are made. However, care must be exercised in the location of branch cuts in the $s$ plane as a function of different variables. If the wrong branch cut is chosen, then the incorrect result will be obtained.

The other benefit of the present approach is that, while different authors use different conventions for Fourier transforms, the convention for Laplace transforms is universal, and so there is little ambiguity in the final result.

## 13.3 The effect of uniform flow

In Chapter 6 we discussed thin airfoil theory and showed how the unsteady loading of an airfoil in a uniform flow could be modeled by a flat plate at zero angle of attack that satisfied the convective wave equation. In trailing edge noise applications, we need to account for the mean flow over the surface, and so the Schwartzschild problem described in the previous section must be modified to account for the flow. In addition, the pressure perturbation over the surface is usually three dimensional, and so the two-dimensional analysis used in Section 13.2 must be extended to the three-dimensional case.

To account for these effects, the perturbations in pressure and velocity potential must satisfy the convective wave equation (6.4.3) so that

$$\frac{1}{c_\infty^2} \frac{D_\infty^2 \phi}{Dt^2} - \nabla^2 \phi = 0 \quad \text{and} \quad \frac{1}{c_\infty^2} \frac{D_\infty^2 p'}{Dt^2} - \nabla^2 p' = 0 \quad \text{where} \quad p' = -\rho_o \frac{D_\infty \phi}{Dt} \qquad (13.3.1)$$

These equations account for the convection of the sound waves by the uniform free stream and are thus no different than the sound propagation equations for a stationary medium derived in Chapter 3, except that the time derivative is expressed in the frame of reference of an observer who is moving relative to the medium rather than one who is fixed with respect to it. The pressure perturbation on the surface can still be considered harmonic, but is convected downstream in the direction of the flow, and can include a harmonic spanwise variation so that the boundary condition given by Eq. (13.2.1) is modified to

$$[\Delta\hat{p}(x_1,x_3)]_{x_1>0} = -P_o e^{ik_1x_1 + ik_3x_3} \tag{13.3.2}$$

where $k_1 = \omega/U_c$ with $U_c$ being the convection speed of the incident pressure perturbation. Since the incident disturbance is harmonic in time and the surface geometry is independent of the spanwise direction, we can specify the scattered acoustic field as

$$[\hat{p}(x_1,x_2,x_3)]_{3D} = [\hat{p}(x_1,x_2)]_{2D} e^{ik_3x_3} \tag{13.3.3}$$

so that the scattered pressure has the form $\hat{p}(x_1,x_2)\exp(-i\omega t + ik_3x_3)$. Substituting this form into the convective wave equation we obtain

$$\beta^2 \frac{\partial^2 \hat{p}}{\partial x_1^2} + 2ikM\frac{\partial \hat{p}}{\partial x_1} + \frac{\partial^2 \hat{p}}{\partial x_2^2} + (k^2 - k_3^3)\hat{p} = 0 \tag{13.3.4}$$

with $M = U/c_\infty$, $\beta = (1 - M^2)^{1/2}$, and $k = \omega/c_\infty$.

To solve this equation, we proceed as in Section 13.2 and take its wavenumber Fourier transform with respect to $x_1$, giving

$$\frac{\partial^2 \widetilde{\hat{p}}}{\partial x_2^2} - (\alpha^2\beta^2 + 2\alpha kM - k^2 + k_3^2)\widetilde{\hat{p}} = 0$$

We solve this equation as in Section 13.2 subject to the same boundary conditions for the pressure jump across the surface and the continuity of the pressure gradient and obtain the solution as before, so the three-dimensional pressure is given in the same form as Eq. (13.2.6) by

$$\hat{p}(x_1,x_2,x_3) = \int_{-\infty}^{\infty} \text{sgn}(x_2)A(\alpha)e^{i\alpha x_1 - \gamma|x_2| + ik_3x_3} d\alpha \tag{13.3.5}$$

where in this case

$$\gamma = \sqrt{\alpha^2\beta^2 + 2\alpha kM - k^2 + k_3^2}$$

The Schwartzschild problem can be solved in exactly the same way as was done in Section 13.2 the only difference being in the definition of $\gamma$, which can be taken into account by factorizing $\gamma$ and specifying, for $s = i\alpha$

$$\gamma(s) = J_+(s)J_-(s) \qquad J_+(s) = e^{i\pi/4}\beta(s + ik_oM - i\kappa)^{1/2}$$
$$J_-(s) = e^{i\pi/4}(s + ik_oM + i\kappa)^{1/2} \tag{13.3.6}$$

$$\kappa = \sqrt{k_o^2 - k_3^2/\beta^2} \qquad k_o = \omega/\beta^2 c_\infty \tag{13.3.7}$$

The unsteady surface pressure in the no flow case was shown to be determined by the function $J_-(s) = (s + ik)^{1/2}$, and so the effect of flow is to modify the result by replacing $k$ with $\kappa + k_oM$ in the final result, so from Eq. (13.2.3) we obtain

$$\Delta\hat{p}_s(x_1, x_3) = P_o e^{ik_1x_1 + ik_3x_3}((1-i)E_2((\kappa + k_oM + k_1)|x_1|) - 1) \quad x_1 < 0 \tag{13.3.8}$$

This extends the two-dimensional analysis to the three-dimensional case with flow and gives a procedure for simplifying the rather complex problem of a convected flow to a tractable problem.

The blade response to an incident pressure disturbance is important in the evaluation of trailing edge noise, as will be discussed in detail in Chapter 15. In this case the nondimensional response $g_{te}(x_1, k_1, k_3)$ is defined so that

$$\Delta\hat{p}_s(x_1, x_3) = P_o g_{te}(x_1, k_1, k_3)e^{ik_3x_3} \quad x_1 < 0$$
$$g_{te}(x_1, k_1, k_3) = e^{ik_1x_1}((1-i)E_2(k_1A|x_1|) - 1) \tag{13.3.9}$$

where $A = 1 + (\kappa + k_oM)/k_1$. Some care needs to be used when evaluating $\kappa$ because this can be imaginary for large values of $k_3$. However, for surface pressure fluctuations that couple to the acoustic far field $\kappa$ must always be real valued as will be discussed in Chapter 14. For the case when $k_3 = 0$ we find that $A = 1 + U_c/c_\infty(1 - M)$ which is always greater than 1. The effect of the mean flow on the surface pressure only appears as the factor of $(1 - M)$ in the definition of $A$ and tends to make the response more rapid in the vicinity of $x_1 = 0$.

## 13.4 The leading edge scattering problem

The unsteady loading on a flat plate of finite chord caused by an unsteady upwash gust can also be evaluated in a compressible flow by using a modification to the Schwartzschild problem described in the previous sections. However, the effect of finite chord is important in this problem, and so we need to include the effect of both the leading edge and the trailing edge of the blade in the solution. This cannot be achieved in closed form, but an iterative method based on the successive approximations of the blade by semi-infinite flat plates converges quite quickly and is a good approximation to the complete response. The approach is to first solve the problem of an upwash gust incident on a semi-infinite flat plate and then add a correction based on the Schwartzschild solution described in Section 13.2 to ensure that there is no pressure jump across the wake downstream of the blade trailing edge. This correction however introduces a small jump in potential upstream of the leading edge, and so additional terms need to be added to correct this. However, as we will show below this is a relatively small correction that can usually be ignored.

### 13.4.1  The leading edge response

To solve the first-order problem of an upwash gust encountering a semi-infinite flat plate, as shown in Fig. 13.5, we will use the same approach as was used in Section 13.2 but with the velocity potential as the dependent variable in the wave equation rather than the pressure.

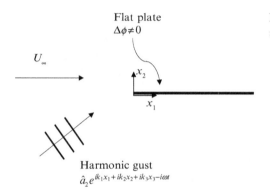

Flat plate
$\Delta\phi\neq0$

$U_\infty$

Harmonic gust
$\hat{a}_2 e^{ik_1x_1+ik_2x_2+ik_3x_3-i\omega t}$

**Fig. 13.5** An upwash gust incident on a semi-infinite flat plate.

The wave equation in this case is defined by Eq. (13.3.1); the potential of the scattered acoustic field has the form $\hat{\phi}(x_1,x_2,x_3)\exp(-i\omega t)=\hat{\phi}(x_1,x_2)\exp(-i\omega t+ik_3x_3)$, and the boundary conditions are

$$\left[\Delta\hat{\phi}(x_1x_3)\right]_{x_1<0}=0 \qquad \left[\frac{\partial\hat{\phi}(x_1,0,x_3)}{\partial x_2}+\hat{a}_2 e^{ik_1x_1+ik_3x_3}\right]_{x_1>0}=0 \qquad (13.4.1)$$

for a harmonic upwash gust convected by the mean flow, so $k_1=\omega/U_\infty$ in this case. The solution can be obtained as before by taking the wavenumber transform of the wave equation. The final result will be of the same form as Eq. (13.3.5) with the velocity potential replacing the unsteady pressure and the requirement that the gradient of the velocity potential is continuous across the plate. The boundary conditions can then be defined using the half-range transforms given by Eqs. (13.2.7) through (13.2.10), with the subscripts referring to the upstream or downstream part of the $x_1$ axis, so

$$C_-(\alpha)+C_+(\alpha)=-\gamma A_-(\alpha)-\gamma A_+(\alpha)$$

In this case $A_-(\alpha)=0$ and

$$C_+(\alpha)=\frac{-\hat{a}_2}{2\pi i(\alpha-k_1)}$$

As before this equation has two unknowns $A_+(\alpha)$ and $C_-(\alpha)$, so we need to use the radiation condition as $|x_1|$ tends to infinity to find a solution. The procedure used in Section 13.2 is followed again by writing the boundary condition in terms of the Laplace transforms defined in Eq. (13.2.14) with the velocity potential as the dependent variable and $s=i\alpha$, so

$$\left[\frac{-\hat{a}_2}{(s-ik_1)}+G_-(-s)=-\gamma(s)F_+(s)\right]_{s=i\alpha}$$

We need to separate the terms in this equation that contribute to the wave field upstream or downstream of the leading edge, and this is achieved by factorizing $\gamma(s)$, as in Eq. (13.3.6) identifying those terms that have singularities in the right or left side of the $s$ plane. From Fig. 13.4 we see that the factored equation takes the form

$$\left[\frac{-\hat{a}_2}{(s-ik_1)J_-(s)}-\frac{-\hat{a}_2}{(s-ik_1)J_-(ik_1)}+\frac{G_-(-s)}{J_-(s)}=-J_+(s)F_+(s)-\frac{-\hat{a}_2}{(s-ik_1)J_-(ik_1)}\right]_{s=i\alpha}$$

As before the inverse Laplace transform of the left side of this equation gives a function that is zero for $x_1>0$, while the right side gives a function that is zero for $x_1<0$ providing that the imaginary part of $k_1$ is greater than zero. We then obtain the Laplace transform of the jump in potential across the surface downstream of the leading edge as

$$F_+(s)=\frac{\hat{a}_2}{(s-ik_1)J_-(ik_1)J_+(s)} \tag{13.4.2}$$

or, in terms of the wavenumber spectrum of the potential jump,

$$A(\alpha)=A_+(\alpha)=\frac{\hat{a}_2}{2\pi i(\alpha-k_1)J_-(ik_1)J_+(i\alpha)} \tag{13.4.3}$$

The inverse Laplace transform of Eq. (13.4.2) gives the potential jump, but in this case it is useful to evaluate the unsteady pressure jump

$$\Delta\hat{p}^{(1)}(x_1,x_3)e^{-i\omega t}=-\rho_o\frac{D_\infty}{Dt}\left(\Delta\hat{\phi}(x_1,x_3)e^{-i\omega t}\right)$$

and since $(\omega-\alpha U_\infty)=(k_1-\alpha)U_\infty$ and the pressure jump is the inverse transform of $2A(\alpha)$, we obtain

$$\Delta \hat{p}^{(1)}(x_1 x_3) = \frac{-\rho_o U_\infty \hat{a}_2 e^{ik_3 x_3}}{\pi \beta (k_1 + k_o M + \kappa)^{1/2}} \int_{-\infty}^{\infty} \frac{e^{i\alpha x_1}}{(\alpha + k_o M - \kappa)^{1/2}} d\alpha \qquad (13.4.4)$$

The inverse transform can be evaluated from tables, and we find that

$$\Delta \hat{p}^{(1)}(x_1, x_3) = \pi \rho_o \hat{a}_2 U_\infty g^{(1)}(x_1, k_3, \omega, M) e^{ik_3 x_3}$$

$$g^{(1)}(x_1, k_3, \omega, M) = \frac{-2 e^{i(\kappa - Mk_o)x_1 + i\pi/4}}{\pi (k_1 + \kappa \beta^2)^{1/2} \sqrt{\pi x_1}} \qquad x_1 > 0 \qquad (13.4.5)$$

(where we have used $\beta^2(k_1 + k_o M) = k_1$). The important conclusion from this result is that the pressure jump at the sharp edge has a square root singularity and tends to zero at large distances downstream of the leading edge. The implication is that both the pressure jump and the particle velocity at the leading edge are infinite, but in reality the velocity at the edge will be limited by viscous effects and the details of the geometry. The modeling that has been used here is based on thin airfoil theory, which ignores that rounding of the leading edge of a blade. When the details of the rounding are included in the analysis the leading edge pressure jump will be quite different as discussed in Chapter 7 for incompressible flow.

### 13.4.2   The trailing edge correction

For a blade with a finite chord these results need to be corrected for the effect of the trailing edge and the Kutta condition. We can calculate a correction that satisfies the trailing edge boundary condition by using the trailing edge scattering theory described in Section 13.2. However, this result is for a semi-infinite flat plate, and so there will be a residual error that occurs at the blade leading edge. Successive corrections can be calculated using the methods described here, but first we will consider the first-order trailing edge correction that is obtained by solving the trailing edge boundary value problem described with the incident pressure jump given by Eq. (13.4.5). Since the origin in the leading edge problem is at a distance $c$ upstream of the trailing edge the result given by Eq. (13.4.5) must be redefined with the origin at the trailing edge, giving

$$\Delta \hat{p}^{(1)}(x_1, x_3) = \pi \rho_o \hat{a}_2 U_\infty \left\{ \frac{-2 e^{i(\kappa - Mk_o)(x_1 + c) + i\pi/4}}{\pi (k_1 + \kappa \beta^2)^{1/2} \sqrt{\pi (x_1 + c)}} \right\} e^{ik_3 x_3}$$

Carrying out the full Weiner Hopf procedure on this function is difficult, but a simplified result is obtained in the high-frequency limit if the amplitude variation with $x_1$ is ignored, and we approximate the pressure near the trailing edge as

$$\Delta \hat{p}(x_1, x_3) \approx \pi \rho_o \hat{a}_2 U_\infty \left\{ \frac{-2e^{i(\kappa - Mk_o)c + i\pi/4}}{\pi (k_1 + \kappa \beta^2)^{1/2} \sqrt{\pi c}} \right\} e^{i(\kappa - Mk_o)x_1 + ik_3 x_3}$$

We can then use the analysis given in Section 13.2 to find the trailing edge correction as the additional pressure jump

$$\Delta \hat{p}^{(2)}(x_1, x_3) = \frac{-2\pi \rho_o \hat{a}_2 U_\infty e^{i(\kappa - Mk_o)(x_1 + c) + i\pi/4}}{\pi (k_1 + \kappa \beta^2)^{1/2} \sqrt{\pi c}} \{ (1 - i)E_2(2\kappa |x_1|) - 1 \} e^{ik_3 x_3}$$

$$x_1 < 0$$

If the origin of the coordinate system is relocated back to the leading edge, we obtain the corrected nondimensional surface pressure function as

$$g^{(1+2)}(x_1, k_3, \omega, M) = \frac{-2e^{i(\kappa - Mk_o)x_1 + i\pi/4}}{\pi (k_1 + \kappa \beta^2)^{1/2} \sqrt{\pi c}} \left\{ \sqrt{\frac{c}{x_1}} + (1 - i)E_2(2\kappa |x_1 - c|) - 1 \right\}$$

$$0 < x_1 < c \tag{13.4.6}$$

The corrected solution given by Eq. (13.4.6) is only an approximate solution, but includes the major effects caused by the trailing edge, and ensures that the Kutta condition is satisfied. The pressure jump includes a discontinuity upstream of the leading edge because we have used a trailing edge correction that assumes it is the same as the response of a semi-infinite flat plate upstream of the trailing edge. However, for large arguments we find that

$$(1 - i)E_2(2\kappa c) \approx 1 + O\left( (2\kappa c)^{-1/2} \right)$$

so this correction is small when the blade chord is large compared to the acoustic wavelength.

In conclusion we have developed expressions for the response of a flat plate airfoil to a harmonic gust and obtained the unsteady surface pressure distribution, which can be used to calculate the unsteady loading and the far-field sound. These results are crucial to the calculation of sound radiation from thin airfoils and, although the derivation is complex, the results are important to our understanding of the problem. We will discuss the features and characteristics of the results in the following chapters.

# References

[1] R.K. Amiet, Noise due to turbulent flow past a trailing edge, J. Sound Vib. 47 (1976) 387–393. See also, R.K. Amiet, Effect of incident surface pressure field on noise due to turbulent flow past a trailing edge, J. Sound Vib. 57 (1978) 305–306.

[2] R.K. Amiet, High frequency thin airfoil theory for subsonic flow, AIAA J. 14 (1976) 1076–1082.

[3] B. Noble, Methods Based on the Wiener-Hopf Technique for the Solution of Partial Differential Equations, Chelsea Publishing Company, New York, NY, 1958.

[4] P.M.C. Morse, H. Feshbach, Methods of Theoretical Physics, McGraw-Hill, London, 1953.

# Leading edge noise

# 14

One of the most important noise sources in low Mach number flows is a blade moving through a region of unsteady flow or turbulence. The sound that radiates to the acoustic far field is referred to as leading edge noise. There are many examples where this sound source occurs including fans of all types, helicopter rotors and propellers. This chapter discusses the mechanisms of leading edge noise and shows how it can be calculated for a stationary blade in a uniform flow.

## 14.1 The compressible flow blade response function

In Chapter 6 we introduced thin airfoil theory and showed how the unsteady flow over a thin blade could be approximated to first order by considering the same flow over an infinitely thin flat plate at zero angle of attack, as shown in Fig. 6.7. This approximation is valid providing that the mean flow speed around the blade does not differ from the free stream flow speed $U_\infty$ by more than $\pm \varepsilon U_\infty$ and the amplitude of the unsteady gust is also of order $\varepsilon U_\infty$, where $\varepsilon \ll 1$ is a small parameter. We then considered the case of an acoustically compact stationary blade in an incompressible mean flow and showed that the source of sound was equivalent to an acoustic dipole with its axis normal to the flow, and strength equal to the net unsteady force produced by the blade. The relationship between the unsteady force and the amplitude of a harmonic incident gust was given by Sears function and was critically dependent on the application of the Kutta condition at the blade trailing edge since this controls the circulation about the blade. However, in aeroacoustic applications we are also concerned with compressible flows, and the modeling of the unsteady loading by an incompressible flow approximation is a severe limitation. We must therefore investigate in detail the effects of compressibility on the blade response function so that the limitations of incompressible flow theory are well defined.

### 14.1.1 The compressible and incompressible flow blade response to a step gust

If a step upwash gust is swept past a stationary blade, then there will be a sudden change in angle of attack that convects across the blade surface. In an incompressible flow the whole flow reacts instantaneously to any change in boundary conditions, and so when the gust strikes the leading edge of the blade vorticity is shed into the blade wake at the same instant in order to satisfy the Kutta condition at the trailing edge. In a compressible fluid the physical mechanism is quite different. In general, information propagates through the medium at the speed of sound, and so the trailing edge boundary condition will be unaltered until the acoustic wave generated at the leading edge

Aeroacoustics of Low Mach Number Flows. http://dx.doi.org/10.1016/B978-0-12-809651-2.00014-X

by the gust interaction reaches the trailing edge. Since the wave propagation speed is enhanced by the mean flow convection the time taken for the acoustic wave to travel from the leading edge to the trailing edge is $c/(c_\infty + U_\infty)$, where $c$ is the blade chord, $U_\infty$ is the mean flow speed, and $c_\infty$ is the speed of sound. To determine if this delay is important we must consider the phase shift between the leading edge pressure fluctuations and the trailing edge pressure. If the frequency of interest is $\omega$ then the phase shift will be $\omega c/(c_\infty + U_\infty)$. For incompressible flow $c_\infty$ is infinite, and so the phase shift is zero. A step gust however excites all frequencies, and so an incompressible flow calculation of the unsteady loading from a step gust will only be valid for the low frequency part of the loading spectrum.

We can write this criterion in terms of the acoustic wavenumber $k = \omega/c_\infty$ and the Mach number $M = U_\infty/c_\infty$ and require that

$$\frac{kc}{1+M} \ll i \tag{14.1.1}$$

for incompressible flow theory to be valid. This is the same as requiring that the blade chord be acoustically compact, which was the assumption used in Section 6.4. However, in many applications this requirement is not met. For example, a frequency of 1000 Hz and a blade chord of 30 cm imply $kc = 5.5$ in air, and so the incompressible flow assumption is far from valid. In underwater applications with the same dimensions and at the same frequency $kc = 1.25$, and so compressibility effects are also important even though the medium is almost incompressible.

An alternative perspective is obtained if the incoming gust has a streamwise scale $L_w$, so the peak frequency excited by the gust is $\omega = 2\pi U_\infty/L_w$. The criterion given by Eq. (14.1.1) then becomes

$$\frac{2\pi cM}{(1+M)L_w} \ll 1 \tag{14.1.2}$$

Since the lengthscale of the gust is likely to be of the same magnitude as the blade chord, the criterion is primarily determined by the flow Mach number $M$. The Mach number for a $\pi/2$ phase shift is then $M = 0.333$, and at flow speeds substantially less than this the incompressible flow approximation is valid. However, some caution needs to be used in extending this concept to frequencies that are well above the peak frequency of the incoming gust and lie in the range where $kc$ is of order one, and this can occur if the lengthscale of the incident turbulence is much smaller than the blade chord.

## 14.1.2 Leading and trailing edge solutions

In most aeroacoustics applications the Mach number is usually 0.3 or greater, and the frequencies of interest are 500 Hz and above, so the effect of compressible flow on the blade response function cannot be ignored. Unfortunately, there is no complete analytical solution to the flat plate blade response function in a compressible flow, and numerical methods have to be used for the full calculation. However, there are close approximations to the solution based on an iterative approach proposed by Landahl [1]. In the first iteration the blade is assumed to be a semi-infinite flat plate, as shown in Fig. 13.5. This will give a solution that satisfies the boundary conditions on the plate and in the upstream flow, but it also causes a pressure jump across the wake that does not satisfy the Kutta condition. To ensure a zero pressure jump across the wake a second solution is added to the first solution that exactly cancels the pressure jump across the wake but induces no additional upwash upstream of the trailing edge. This second solution is found by solving a semi-infinite plate trailing edge problem as shown in Fig. 13.1. The sum of the two solutions satisfies the blade boundary conditions and the Kutta condition but induces a pressure discontinuity upstream of the leading edge. To correct this a third solution is added which eliminates the upstream pressure jump and satisfies the nonpenetration boundary condition on the blade surface but induces a pressure jump over the wake which needs to be corrected. This process can be repeated indefinitely and converges to a solution if the residual pressure jumps in the wake become smaller and smaller. Fortunately, the convergence is quite quick at high frequencies, and usually only the first two terms in the series are needed to achieve acceptable results.

### 14.1.3    The first-order solution for the surface pressure

The first-order solution can be obtained using the Weiner–Hopf method as described in Section 13.4. The unsteady pressure jump across a blade, modeled by a semi-infinite flat plate, caused by a harmonic upwash gust of amplitude $\hat{a}_2 \exp(ik_1 y_1 + ik_2 y_2 + ik_3 y_3 - i\omega t)$, as shown in Fig. 13.5, was given by Eq. (13.4.5) and is repeated here as

$$\Delta \hat{p}^{(1)}(y_1 y_3) = \pi \rho_o U_\infty \hat{a}_2 g^{(1)}(y_1, k_3, \omega, M) e^{ik_3 y_3}$$

$$g^{(1)}(y_1, k_3, \omega, M) = \frac{-2e^{i(\kappa - Mk_o)y_1 + i\pi/4}}{\pi^{3/2} (k_1 + \kappa \beta^2)^{1/2} y_1^{1/2}} \quad y_1 > 0 \tag{13.4.5}$$

where positions have been written in terms of coordinate $y$, and we have explicitly included the time dependency. The origin of $y$ is at the leading edge. For a gust convected by the mean flow the wavenumbers are given by

$$k_1 = \frac{\omega}{U_\infty} \qquad \kappa = \left(k_o^2 - \frac{k_3^2}{\beta^2}\right)^{1/2} \qquad k_o = \frac{k_1 M}{\beta^2} \qquad \beta = (1 - M^2)^{1/2} \tag{14.1.3}$$

The first point to note from this result is that the surface pressure tends to infinity as $y_1^{-1/2}$ at the leading edge of the blade. This behavior is expected from the steady flow around an airfoil at a small angle of attack and is the result of the flat plate approximation. If the airfoil had a rounded leading edge then the pressure would remain finite as $y_1$ tends to zero. Next consider the phase which depends on $(\kappa - Mk_o)y_1 + k_3 y_3$ and indicates a wave propagating in the positive $y_1$ and $y_3$ directions, assuming that $k_1 > 0$ and $k_3 > 0$. If $k_3 = 0$ then the phase reduces to

$$k_o(1 - M)y_1 = \frac{\omega y_1}{c_\infty + U_\infty}$$

which represents a wave propagating downstream at the speed of sound plus the free stream speed, as discussed earlier. The surface pressure is therefore controlled by an acoustic wave propagating over the surface downstream from the leading edge.

If $k_3$ is given a value in the range $0 < k_3 < \beta k_o$ then we can write $k_3 = \beta k_o \sin\varphi$ and $\kappa = k_o \cos\varphi$. The phase variation in Eq. (13.4.5) then represents an acoustic wave propagating at an angle to the $y_1$ direction defined by

$$\varphi_e = \tan^{-1}\left(\frac{k_3}{\kappa - Mk_o}\right) = \tan^{-1}\left(\frac{\beta\sin\varphi}{\cos\varphi - M}\right)$$

as shown in Fig. 14.1A, for example, for $M = 0.3$, $\varphi = \pi/8$ radians, giving $\varphi_e = 30.3$ degrees.

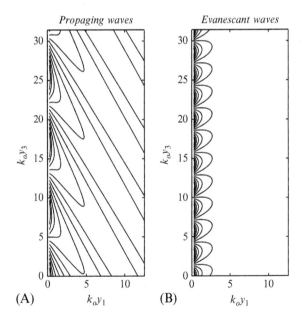

**Fig. 14.1** The surface pressure for different values of the spanwise wavenumber for $M = 0.3$. (A) $k_3 = \beta k_o \sin\varphi$ with $\varphi = \pi/8$ and (B) $k_3 = 1.1k_o$.

Because this is an acoustic wave it will couple strongly with the acoustic far field, and there will be radiation in the direction of $\varphi_e$. This is referred to as a super-critical wave [2]. In contrast for the case when $|k_3| > \beta k_o$ we find from Eq. (14.1.3) that $\kappa$ is a positive imaginary number (because of the choice of the branch cut), and the wave decays exponentially in the $y_1$ direction. This is referred to as an evanescent or subcritical wave and is illustrated in Fig. 14.1B. Because this wave does not propagate it will not radiate to the acoustic far field.

Turbulent flows that are dominated by eddies with small lengthscales in the spanwise direction will contain significant energy at wavenumbers where $k_3$ is sub-critical. The fluctuations at these wavenumbers will have little impact on the acoustic far field because they do not couple with the acoustic wavenumbers that radiate sound. However, evanescent waves can dominate the pressure fluctuations measured on the surface close to the leading edge. Consequently, trying to relate measured pressure fluctuations on the blade surface near the leading edge to the far-field sound is not straightforward. Eq. (13.4.5) shows that at high spanwise wavenumbers $|k_3| \gg \beta k_o$ and the surface pressure fluctuations caused by subcritical waves will decay with distance from the leading edge as $\exp\left(-|k_3||y_1|\right)/y_1^{1/2}$. So, the unsteady pressure caused by the gust may be quite local to the leading edge, and its initial decay may be much faster than would be expected if the evanescent wave was not considered. However, because of this rapid decay, the surface pressure should asymptote to the super-critical case for surface pressures measured more than an acoustic wavelength downstream of the leading edge.

An alternative interpretation of this effect is obtained if we consider the speed with which the intersection between the gust wavefront and the leading edge propagates along the blade in the spanwise direction. Since the gust wavelength in the spanwise direction is $2\pi/k_3$ and the gust period experienced at the leading edge is $2\pi/\omega$, this speed is $c_i = \omega/k_3$. For this wave to be supercritical we require that $|k_3| < \beta k_o = \omega/\beta c_\infty$ or $\omega/\beta|k_3|c_\infty > 1$, and it follows that we require $|c_i|/\beta c_\infty > 1$. The wavefront must therefore propagate across the surface in the spanwise direction at a speed that is greater than the speed of sound.

The first-order solution for the surface pressure given above is for a semi-infinite flat plate and can be corrected to allow for finite chord using the approach given in Section 13.6. The correction is obtained by adding a solution to the boundary value problem that ensures that there is no pressure jump across the wake of the blade that extends downstream from $y_1 = c$ while maintaining zero velocity normal to the blade surface for $y_1 < c$. The correction discussed in Section 13.4 is only a first-order approximation to the complete solution because the surface pressure induced by the leading edge is assumed to be a simply convected wave with no decay with distance downstream. Given this approximation the blade response function with the trailing edge correction is given by Eq. (13.4.6) as

$$g^{(1+2)}(y_1, k_3, \omega, M) = \frac{-2e^{i(\kappa - Mk_o)y_1 + i\pi/4}}{\pi(k_1 + \kappa\beta^2)^{1/2}\sqrt{\pi c}} \left\{ \sqrt{\frac{c}{y_1}} + (1 - i)E_2(2\kappa|y_1 - c|) - 1 \right\}$$

$$0 < y_1 < c \qquad\qquad\qquad (13.4.6)$$

where, as above, positions have been written in terms of the coordinate $y_1$. The additional correction includes the complex Fresnel integral defined as

$$E_2(x) = \int_0^x \frac{e^{iq}}{(2\pi q)^{1/2}} dq$$

which is shown in Fig. 14.2.

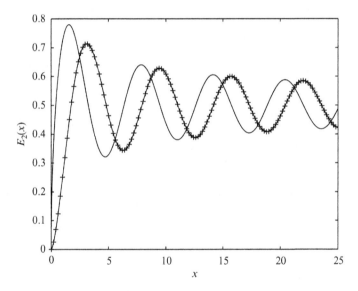

**Fig. 14.2** The complex Fresnel integral $E_2(x) = C_2(x) + iS_2(x)$. Solid line is the real part $C_2(x)$, and the marked line $-+-$ is the imaginary part $S_2(x)$.

For large arguments the complex Fresnel integral has the asymptotic value of $(1 + i)/2$, so the last two terms of Eq. (13.4.6) cancel as $2\kappa|y_1 - c|$ tends to infinity. It follows therefore that at high frequencies the trailing edge correction causes a small oscillation of the pressure about first-order solution for positions significantly upstream of the trailing edge and ensures the pressure is zero at the trailing edge as illustrated in Fig. 14.3. Most of the physics of the problem can therefore be obtained from the first-order solution, and at high frequencies, the combined first- and second-order solution is well approximated by the first-order solution if it is truncated at the trailing edge of the blade.

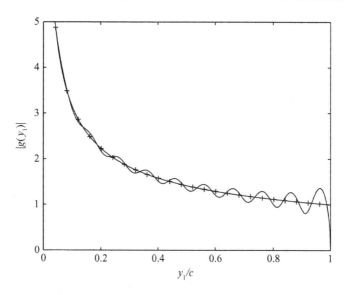

**Fig. 14.3** The distribution of surface pressure on the blade surface for $k_o c = 10\pi$ and $k_3 = 0$. Solid line shows $|g^{(1+2)}|$, and the marked line $-+-$ shows $|g^{(1)}|$.

## 14.1.4 The unsteady lift in compressible flow

The blade response function also gives the total lift on the blade surface caused by the upwash gust, and this can be obtained from Eq. (13.4.7) by integrating the pressure over both the blade span and the blade chord. For blades with large span the spanwise integral will be zero if $k_3 \neq 0$ because of the oscillatory nature of $\Delta p$ with $y_3$, and so the loading depends only on the amplitude of the gust with zero spanwise wavenumber. Integrating Eq. (13.4.6) is complicated by the trailing edge correction, but, as discussed earlier, this will have a small effect on the result at high frequencies. A good approximation is obtained by integrating the first-order solution given by Eq. (13.4.5). We then obtain the nondimensional unsteady lift per unit span, normalized by $\pi \rho_o U_\infty \hat{a}_2 c$ as in Eq. (6.4.4), as

$$S^{(1)}(\sigma, M) = -\frac{1}{c} \int_0^c g^{(1)}(y_1, 0, \omega, M) dy_1 \qquad (14.1.4)$$

where the reduced frequency $\sigma = \omega a / U_\infty$ and $a = c/2$ is the semi-chord. The negative sign is because $\Delta p$ represents the pressure on the top of the airfoil less that on the bottom. This integral can be evaluated directly to give

$$S^{(1)}(\sigma, M) = \frac{2\sqrt{2} E_2(k_o(1-M)c)e^{i\pi/4}}{\pi(k_1 + k_o\beta^2)^{1/2}(k_o(1-M))^{1/2}c}$$

This is often written in terms of the nondimensional variables

$$\mu = k_o(1-M)c \qquad (k_1 + k_o\beta^2)c = \mu(1+M)^2/M$$

which gives (after using $(1+i) = \sqrt{2}\exp(i\pi/4)$)

$$S^{(1)}(\sigma, M) = \frac{2(1+i)M^{1/2}}{\pi\mu(1+M)}E_2(\mu) \qquad (14.1.5)$$

In the high frequency limit the function $E_2(x)$ will asymptote to $(1+i)/2$, and so we can approximate

$$S^{(1)}(\sigma, M) \approx \frac{2iM^{1/2}}{\pi\mu(1+M)} = \frac{i}{\pi\sigma M^{1/2}} \qquad (14.1.6)$$

where we have used the relationship $\mu = 2\sigma M/(1+M)$. This shows that the first-order approximation to the compressible blade response function is inversely proportional to the nondimensional frequency at high frequencies. This scaling is quite different from the incompressible results given by Sears function (Eqs. 6.4.4, 6.4.5) that scaled inversely with the square root of the nondimensional frequency. The compressibility effect is therefore to reduce the high-frequency response of the blade, and this has an increasingly important effect in high speed flows. Fig. 14.4 illustrates this difference for a flow Mach number of $M = 0.3$. At low frequencies the magnitude of the compressible solution is slightly larger than the incompressible solution, but this is caused by the approximate nature of the compressible solution which only includes the first-order approximation without the trailing edge or additional leading edge corrections. However, at high frequencies, where the first-order compressible blade response function is a more accurate approximation, the incompressible and compressible solutions are very different. Amiet [3] notes that the incompressible solution is appropriate to use when $\sigma M/\beta^2 < \pi/4$, which corresponds to the frequency where the blade chord (divided by $\beta^2$) is less than a quarter of the acoustic wavelength. For the example shown in Fig. 14.4 the compressible blade response function should be used at nondimensional frequencies $\sigma > 2.38$ according to Amiet's criterion, as shown by the vertical line in the figure. This corresponds to the lowest frequency where the compressible blade response is less than the incompressible response.

### 14.1.5  An arbitrary gust

The results given so far in this section have been for a harmonic gust with space dependence $\hat{a}_2 \exp(ik_1(y_1 - U_\infty t) + ik_3y_3)$ in the plane of the blade, $y_2 = 0$. To extend these results to an arbitrary gust we consider the upwash velocity on the blade surface $y_2 = 0$ to be the superposition of gusts with a spectrum of wavenumbers, so

$$u_2(y_1 - U_\infty t, 0, y_3) = \int\limits_{-\infty}^{\infty} \int\limits_{-\infty}^{\infty} \int\limits_{-\infty}^{\infty} \tilde{u}_2(\mathbf{k}) e^{ik_1(y_1 - U_\infty t) + ik_3 y_3} d\mathbf{k} \qquad (14.1.7)$$

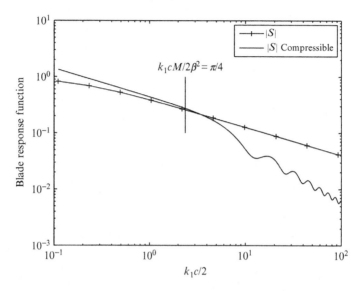

**Fig. 14.4** The compressible blade response function given by Eq. (14.1.5) compared to Sears Function as a function of the nondimensional frequency $\sigma = k_1 c/2$, for a flow Mach number of $M = 0.3$.

The pressure jump across the blade is then obtained by superimposing the contributions from each wavenumber component of the gust, so using Eq. (13.4.5) with $g^{(1+2)}$ from Eq. (13.4.6), and substituting $\tilde{u}_2(\mathbf{k})$ for $\hat{a}_2$, including the time dependence that is suppressed in Eq. (13.4.5), we obtain

$$\Delta p(y_1, y_3, t) = \int\limits_{-\infty}^{\infty} \int\limits_{-\infty}^{\infty} \int\limits_{-\infty}^{\infty} \pi \rho_o U_\infty \tilde{u}_2(\mathbf{k}) g^{(1+2)}(y_1, k_3, k_1 U_\infty, M) e^{-ik_1 U_\infty t + ik_3 y_3} d\mathbf{k}$$

$$(14.1.8)$$

The Fourier transform of this signal with respect to time is obtained by replacing $k_1$ by $\omega/U_\infty$, and so the integrand over $k_1$ in Eq. (14.1.8) is simply an inverse Fourier transform with respect to time, as defined in Eq. (3.10.2), and so

$$\Delta \tilde{p}(y_1, y_3, \omega)$$
$$= \int\limits_{-\infty}^{\infty} \int\limits_{-\infty}^{\infty} \pi \rho_o \tilde{u}_2(\omega/U_\infty, k_2, k_3) g^{(1+2)}(y_1, k_3, \omega, M) e^{ik_3 y_3} dk_2 dk_3 \qquad (14.1.9)$$

Further simplification is possible if we note that the gust response is independent of the $k_2$ wavenumber, so we have

$$\Delta \tilde{p}(y_1, y_3, \omega) = \int_{-\infty}^{\infty} \pi \rho_o \tilde{w}_2(\omega/U_\infty, k_3) g^{(1+2)}(y_1, k_3, \omega, M) e^{ik_3 y_3} dk_3 \qquad (14.1.10)$$

where $\tilde{w}_2(k_1, k_3)$ is the planar wavenumber transform of $u_2(y_1, 0, y_3)$ in the plane $y_2 = 0$, so

$$\tilde{w}_2(k_1, k_3) = \int_{-\infty}^{\infty} \tilde{u}_2(k_1, k_2, k_3) dk_2$$

$$= \frac{1}{(2\pi)^2} \int_{-R_\infty}^{R_\infty} \int_{-R_\infty}^{R_\infty} u_2(y_1, 0, y_3) e^{-k_1 y_1 - ik_3 y_3} dy_1 dy_3 \qquad (14.1.11)$$

These results give the unsteady loading on a blade encountering an arbitrary gust. It is important to note that there are some subtle differences between the results for a harmonic gust given by Eq. (13.4.5) and the frequency domain result given by Eq. (14.1.10). The spanwise dependence of the general gust is completely determined by the wavenumber integral defined in Eq. (14.1.11), and so any spanwise dependence can be included. For example, if the gust is of finite spanwise extent then this characteristic will appear as part of the wavenumber spectrum as a function of $k_3$.

## 14.2 The acoustic far field

### 14.2.1 The acoustic far field from the leading edge interaction

The acoustic field that results from unsteady pressure fluctuations on a thin plate in uniform flow is given by Eq. (6.5.5), as

$$\left(\tilde{\rho}(\mathbf{x}, \omega) c_\infty^2\right)_{\text{dipole}} \approx \frac{-i\omega x_2 e^{ik_o r_e}}{4\pi c_\infty r_e^2} \int_{-d}^{d} \int_{0}^{c} \Delta \tilde{p}(\mathbf{y}, \omega) e^{-ik_o x_1 y_1/r_e - ik_o x_3 y_3 \beta^2/r_e - ik_o M(x_1 - y_1)} dy_1 dy_3$$

$$k_o = \frac{\omega}{\beta^2 c_\infty} \qquad r_e = \sqrt{x_1^2 + \beta^2 (x_2^2 + x_3^2)} \qquad (6.5.5)$$

In obtaining this result we made use of Prantl-Glauert coordinates and the far-field approximation, so the blade chord $c$ and span $b = 2d$ are required to be small compared to the propagation distance $r_e$.

It will be convenient, when we come to rotor noise calculations in Chapter 16, to have this result expressed in terms of the acoustic pressure perturbation $p' = \rho' c_\infty^2$ and the wavenumber transform of the surface pressure, as given by Eqs. (4.7.12) and (6.5.6),

$$\tilde{p}\left(\mathbf{x}, \omega\right) \approx \left(-\frac{i\pi\omega x_2 e^{ik_or_e - ik_oMx_1}}{c_\infty r_e^2}\right) \Delta\tilde{p}\left(k_1^{(o)}, k_3^{(o)}, \omega\right) \text{ where } k_1^{(o)}$$

$$= k_o\left(\frac{x_1}{r_e - M}\right) \qquad \text{and} \qquad k_3^{(o)} = \frac{k_o x_3 \beta^2}{r_e} \tag{14.2.1}$$

and

$$\Delta\tilde{p}\left(k_1^{(o)}, k_3^{(o)}, \omega\right) = \frac{1}{(2\pi)^2} \int_{-d}^{d} \int_0^c \Delta\tilde{p}(\mathbf{y}, \omega) e^{-ik_1^{(o)}y_1 - ik_3^{(o)}y_3} dy_1 dy_3 \tag{4.7.12}$$

The important feature of this result is that $\Delta\tilde{p}\left(k_1^{(o)}, k_3^{(o)}, \omega\right)$ has dimensions of a force (per unit frequency, per unit wavenumber squared) and so, in the limit that $k_o$ is very small, $\Delta\tilde{p}\left(k_1^{(o)}, k_3^{(o)}, \omega\right) \approx \Delta\tilde{p}(0, 0, \omega)$, and Eq. (14.2.1) reduces to a dipole source that depends on the force produced by the blade. However, when the acoustic wavelength is comparable to the blade span or chord then the unsteady loading on the blade surface couples to the acoustic field at the wavenumbers $k_1^{(o)}$ and $k_3^{(o)}$. To obtain the far-field sound, $\Delta\tilde{p}\left(k_1^{(o)}, k_3^{(o)}, \omega\right)$ must therefore be evaluated at the appropriate wavenumbers which will depend on the observer location. A special case is for semi-infinite blade chord because in that case the blade chord is always large compared to the wavelength, and so the directionality will be distinctly different from the case of a blade with an acoustically compact chord.

In order to obtain the wavenumber transform of the surface pressure we use Eq. (14.1.10) in Eq. (4.7.12). First consider the integral over the span. In Eq. (14.1.10) the dependence on span is given by an inverse Fourier transform over $k_3$, whereas Eq. (4.7.12) evaluates a forward transform as a function of $y_3$. It follows that the integrand in Eq. (14.1.10) matches the forward transform in Eq. (4.7.12). Given this observation we obtain

$$\Delta\tilde{p}\left(k_1^{(o)}, k_3^{(o)}, \omega\right) = \frac{1}{2}\rho_o c\tilde{w}_2\left(\omega/U_\infty, k_3^{(o)}\right)\Lambda\left(k_1^{(o)}, k_3^{(o)}, \omega, M\right) \tag{14.2.2}$$

where the integral over the chord is specified by

$$\Lambda\left(k_1^{(o)}, k_3^{(o)}, \omega, M\right) = \frac{1}{c}\int_0^c g^{(1+2)}\left(y_1, k_3^{(o)}, \omega, M\right) e^{-ik_1^{(o)}y_1} dy_1 \tag{14.2.3}$$

The function $\Lambda$ defined in Eq. (14.2.3) is closely related to the nondimensional lift on the airfoil surface. Since this is accurately represented by the first-order approximation to the unsteady surface pressure response we can simplify the evaluation of Eq. (14.2.3) by using $g^{(1)}$ instead of $g^{(1+2)}$. The function $\Lambda$ can then be interpreted as the nondimensional blade response function as observed in the acoustic far field and can be evaluated directly from Eq. (13.4.5) as

$$\Lambda\left(k_1^{(o)}, k_3^{(o)}, \omega, M\right) = \frac{-2(1+i)E_2\left(\left(\kappa - Mk_o - k_1^{(o)}\right)c\right)}{\pi\left(\omega/U_\infty + \kappa\beta^2\right)^{1/2}\left(\kappa - Mk_o - k_1^{(o)}\right)^{1/2}c} \tag{14.2.4}$$

providing that $\kappa$ (evaluated here using $k_3^{(o)}$) is real and $\kappa > k_o x_1/r_e$. This condition is always met when the spanwise wavenumber of the incident gust is zero, but some care must be taken when evaluating this result for larger spanwise wavenumbers particularly where the observer is near the $x_1$ axis.

## 14.2.2  The far-field directionality and scaling

To provide some insight into the scaling in the acoustic far field we consider the case when the observer is in the plane $x_3 = 0$, and so spanwise gust wavenumber is zero $k_3^{(o)} = 0$. In this case $\kappa = k_o = \omega M/U_\infty\beta^2$, and we obtain

$$\Lambda\left(k_1^{(o)}, 0, \omega, M\right) = \frac{-(1-M)^{1/2}(1+i)E_2(k_o c(1 - x_1/r_e))}{\pi\sigma M^{1/2}(1 - x_1/r_e)^{1/2}} \tag{14.2.5}$$

In the high-frequency limit the Fresnel function tends to $(1+i)/2$, and we obtain

$$\Lambda\left(k_1^{(o)}, 0, \omega, M\right) = \frac{-i(1-M)^{1/2}}{\pi\sigma M^{1/2}(1 - x_1/r_e)^{1/2}} \tag{14.2.6}$$

An important feature of this result is that it provides the scaling of the far-field sound on the mean flow Mach number.

The directionality in the far field is also quite different from the dipole directionality discussed in Chapter 4 for acoustically compact surfaces. Combining Eqs. (14.2.1), (14.2.2), and (14.2.6) shows that the far-field sound depends on

$$\left(\frac{x_2\beta}{r_e}\right)\frac{1}{(1 - x_1/r_e)^{1/2}} = \frac{\sin\theta_e}{(1 - \cos\theta_e)^{1/2}} = \sqrt{2}\cos(\theta_e/2) \qquad \theta_e = \tan^{-1}(x_2\beta/x_1) \tag{14.2.7}$$

which gives a cardioid-shaped directionality as shown in Fig. 14.5A and is quite different from the dipole directivity given by $\sin\theta_e$ shown in Fig. 14.5B. However, if the effect of finite chord is included and the directionality is calculated by using Eq. (14.2.5), the directionality has multiple lobes and a null in the upstream and downstream directions similar to that of a dipole (see Fig. 14.5C and D).

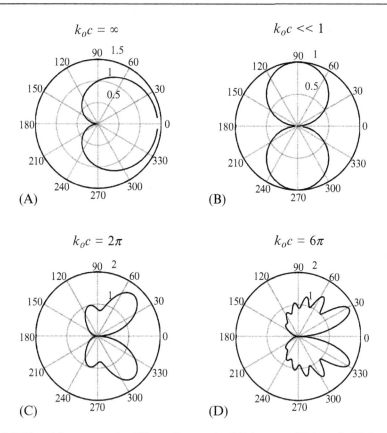

**Fig. 14.5** Directivity patterns for different frequencies (A) $k_o c = \infty$, (B) $k_o c = 0$, (C) $k_o c = 2\pi$, and (D) $k_o c = 6\pi$.

### 14.2.3 Impulsive gusts of finite span

Further insight is obtained by considering an impulsive gust of very short duration that is constant over a finite span $b$. The impulsive gust can be characterized by a Dirac delta function scaled on the blade semi-chord in the direction of the flow, and so the upwash is

$$u_2(y_1, 0, y_3) = \frac{w_o c}{2} \delta(y_1) H(b/2 - |y_3|)$$

In this case we find that the wavenumber transform in Eq. (14.1.11) reduces to

$$\tilde{\tilde{w}}_2\left(k_1^{(o)}, k_3^{(o)}\right) = \frac{w_o c b}{2(2\pi)^2} \left\{ \frac{\sin\left(k_3^{(o)} b/2\right)}{k_3^{(o)} b/2} \right\} \tag{14.2.8}$$

When this result is used in Eq. (14.2.2) to calculate the acoustic far field it is seen that the directionality in the $y_3$ direction is determined by the terms in $\{\ \}$. If the spanwise extent of the gust is large then the directionality will have a clearly defined beam of sound that is strongest in the direction where $k_3^{(o)}b = k_ox_3b\beta^2/r_e = 0$. The sound radiation is therefore primarily in the direction normal to the plane of the blade where $x_3 = 0$. However, when the spanwise extent of the gust is much smaller than the acoustic wavelength then $k_ob \ll 1$, and the terms in $\{\ \}$ are approximately unity for all observer angles, so the directionality in both the $x_1$ and $x_3$ direction is determined by $x_2\Lambda/r_e$ and is independent of the gust. This is important in cases where the spanwise correlation lengthscale of a gust is small because in that case the directionality is determined entirely by the acoustics and the blade response function.

### 14.2.4  A step gust

A similar important example is given by a step gust for which

$$u_2(y_1, 0, y_3) = w_oH(-y_1)H(b/2 - |y_3|)$$

Ahead of the gust the upwash velocity is zero, and behind the gust the upwash is $w_o$. Providing that spanwise extent of the gust is much smaller than the acoustic wavelength we obtain using Eq. (14.1.11)

$$\tilde{w}_2\left(k_1^{(o)}, k_3^{(o)}\right) = \lim_{\varepsilon \to 0} \frac{iw_ob}{(2\pi)^2\left(k_1^{(o)} + i\varepsilon\right)}$$

where a small imaginary part has been added to the wavenumber to ensure the wavenumber transform converges. The difference between the step upwash gust and the impulsive gust is that the step gust is inversely proportional to the wavenumber $k_1^{(o)} = k_o(x_1/r_e - M)$, and this will alter the far-field directionality and the spectral shape. At high frequencies and for an observer at $x_3 = 0$ we can use Eqns. (14.2.1), (14.2.2), and (14.2.7) to give the directionality for a step gust as

$$\left(\frac{x_2\beta}{r_e}\right)\frac{1}{(1 - x_1/r_e)^{1/2}(x_1/r_e - M + i\varepsilon)} = \frac{\sqrt{2}\cos(\theta_e/2)}{(x_1/r_e - M + i\varepsilon)}$$

The interesting characteristic of this result is that for a step gust there will be a strong beam of radiation at the angle where $x_1/r_e = M$, which is caused by the sudden change in the unsteady lift. This is a characteristic of the step gust, specifically its permanent change in the angle of attack of the mean flow.

## 14.3   An airfoil in a turbulent stream

One of the most important applications of leading edge noise is to blades embedded in a turbulent flow. While these flows are usually inhomogeneous, they can often be modeled by a homogeneous turbulent flow with the same characteristics, such as turbulence intensity and lengthscale. To determine the characteristics of the radiated sound we will combine the results obtained above with the models of homogeneous turbulence discussed in Section 9.1, to obtain an estimate of the power spectrum of the far-field noise.

The power spectrum of a signal can be obtained from its Fourier transform using Eq. (8.4.13), and so if we consider a blade of large span we can use Eqs. (14.2.1) and (14.2.2) to define the far-field spectral density of the acoustic pressure as

$$S_{pp}(\mathbf{x}, \omega) \approx \left(\frac{\pi \omega x_2}{c_\infty r_e^2}\right)^2 \frac{S_{FF}\left(\omega, k_1^{(o)}, k_3^{(o)}\right)}{(2\pi)^4}$$

where

$$
\frac{S_{FF}\left(\omega, k_1^{(o)}, k_3^{(o)}\right)}{(2\pi)^4} = \frac{\pi}{T} E\left[\left|\Delta \tilde{p}\left(k_1^{(o)}, k_3^{(o)}, \omega\right)\right|^2\right]
$$

$$
= \left|\frac{\rho_o c}{2} \Lambda\left(k_1^{(o)}, k_3^{(o)}, \omega, M\right)\right|^2 \frac{\pi}{T} E\left[\left|\tilde{w}_2\left(\omega/U_\infty, k_3^{(o)}\right)\right|^2\right]
$$

(14.3.1)

and $T$ is half the averaging time used to obtain the spectral estimate. The term $S_{FF}(\omega, k_1^{(o)}, k_3^{(o)})$ represents the spectrum of the unsteady force produced by the airfoil that couples with the acoustic field that radiates to an observer at $\mathbf{x}$. At low frequencies when the blade is acoustically compact it is simply the unsteady loading spectrum on the blade. However, at high frequencies it will depend on the observer location $\mathbf{x}$ as well as the flow Mach number.

This result is still quite general and applies to both a homogeneous and an inhomogeneous flow providing that we can define the expected value of the turbulence spectrum. We can use Eq. (8.4.38) written as

$$
\phi_{22}(k_1, k_3) = \frac{\pi^2}{R_\infty^2} E\left[\left|\tilde{w}_2(k_1, k_3)\right|^2\right]
$$

(14.3.2)

and thus,

$$
\frac{\pi}{T} E\left[\left|\tilde{w}_2(k_1, k_3)\right|^2\right] = \frac{R_\infty^2}{\pi T} \phi_{22}(k_1, k_3)
$$

(14.3.3)

where $\phi_{22}$ is the planar wavenumber spectrum and can be defined using Eqs. (9.1.14) or (9.1.23), using different empirical models for the turbulent spectrum. The averaging time $T$ is defined by the time it takes the volume of turbulence to pass over the blade, and so $R_\infty/T = U_\infty$. In this result the size of the turbulent region also determines the wetted span of the blade, and so $R_\infty = b/2$. We then obtain the effective loading spectrum as

$$\frac{S_{FF}\left(\omega, k_1^{(o)}, k_3^{(o)}\right)}{(2\pi)^4} \approx \left|\frac{\rho_o c}{2}\Lambda\left(k_1^{(o)}, k_3^{(o)}, \omega, M\right)\right|^2 \left(\frac{U_\infty b}{2\pi}\right)\phi_{22}\left(\omega/U_\infty, k_3^{(o)}\right)$$

$$(14.3.4)$$

A suitable model for the planar wavenumber spectrum is given by the von Kármán turbulence model and is defined by Eq. (9.1.14) as

$$\phi_{22}(k_1, k_3) = \frac{4}{9\pi}\frac{\overline{u^2}}{k_e^2}\frac{(k_1^2 + k_3^2)/k_e^2}{\left[1 + (k_1^2 + k_3^2)/k_e^2\right]^{7/3}} \qquad k_e = \frac{\sqrt{\pi}}{L_f}\frac{\Gamma(5/6)}{\Gamma(1/3)} \qquad (9.1.14)$$

The scaling of the far-field sound is revealed by using the approximation given by Eq. (14.2.6) for an observer in the $y_3 = 0$ plane. This gives the far-field pressure as

$$S_{pp}(\mathbf{x}, \omega) \approx \frac{4}{9\pi}\left(\frac{\rho_o^2 U_\infty \overline{u^2} bM}{\pi k_e^2 r_e^2 (1 + M)}\right)\left(\frac{(\omega/k_e U_\infty)^2}{\left[1 + (\omega/k_e U_\infty)^2\right]^{7/3}}\right)\cos^2\left(\frac{\theta_e}{2}\right) \qquad (14.3.5)$$

This reveals that the spectral density scales with the fourth power of the flow speed, since $\overline{u^2}$ will scale with $U_\infty^2$, which is a direct consequence of the leading edge scattering mechanism and the fact that the spectral density has been chosen to describe the far field. If a *spectral level* is measured then it will depend on the bandwidth of the measurement and will be given by $\Delta\omega S_{pp}(\omega)$. The right side of the equation is then adjusted by multiplying by $\Delta\omega = (k_e U_\infty)(\Delta\omega/k_e U_\infty)$, and the scaling of the far-field sound will depend on the fifth power of the mean flow speed. This is an important difference from the dipole scaling laws discussed in Chapter 4 that suggest the sound radiation scales on the sixth power of the flow speed and is a direct consequence of the compressibility effects that are controlling the unsteady loading at the leading edge of the blade. Another important feature of this result is that the spectral level $\Delta\omega S_{pp}(\omega)$ scales as $b/k_e \sim bL_f$, so it depends on the blade span multiplied by the integral lengthscale of the turbulence. Reducing the integral lengthscale of the turbulence therefore reduces the overall level of the spectrum. However, it also alters the spectral shape. To illustrate this point Fig. 14.6 shows a typical spectrum plotted as a function of the nondimensional frequency for different values of $L_f/c$. It is seen that the spectrum shifts to the right as the integral length scale is reduced, increasing the high frequency content of the spectrum, but the low-frequency levels are reduced.

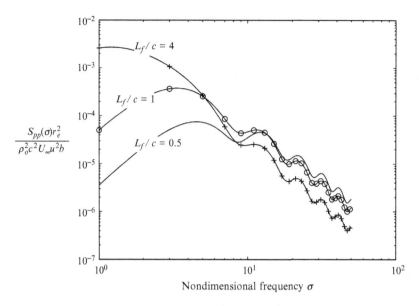

**Fig. 14.6** Normalized far-field spectrum at $\mathbf{x} = (0, x_2, 0)$ for different turbulence lengthscales for a flat plate airfoil at $M = 0.3$, as a function of the nondimensional frequency $\sigma$. The spectrum is normalized as $S_{pp}(\sigma)r_e^2/\rho_o^2c^2U_\infty\overline{u^2}b$.

# 14.4 Blade vortex interactions in compressible flow

Blade vortex interaction is an important source of sound in situations where a blade tip vortex on a rotor washes over a structure or is re-ingested into the rotor. An example is a helicopter executing a maneuver in which the blade tip vortices are washed back into the rotor plane and are cut by the blades. As a result, a loud thumping sound is heard, and the far-field sound levels can be very high. This topic will be discussed in more detail in Chapter 16, and in this section we describe the approach to calculating the sound from a blade vortex interaction developed by Amiet [4].

In Section 7.5 we discussed the unsteady loading on a blade caused by a passing vortex in incompressible flow. For rotor blades moving with tip speeds (such as a helicopter rotor) that approach the speed of sound the incompressible flow approximation is no longer valid, and so in this section we use the compressible blade response function to calculate the radiated sound. However, the approach is not limited to compressible flows and can also be used at low Mach numbers with the correct adjustment to the blade response function.

## 14.4.1 The upwash velocity spectrum from a blade vortex interaction

In this section we consider a three-dimensional vortex incident on a stationary blade in a uniform flow, Fig. 14.7. The vortex axis makes an angle $\phi_v$ to the leading edge of the blade, as shown. The vortex is convected with the speed $U_\infty$ in the $y_1$ direction and, as in Section 7.5, it is a height $h$ above the blade.

We can use the general gust description given by Eq. (14.1.11) to calculate the upwash spectrum that is needed to obtain the acoustic field using Eqs. (14.2.1) and (14.2.2). This is convenient for a flow that is dominated by a line vortex because we can use Eq. (6.3.10) to relate the wavenumber spectrum of the velocity perturbation to the wavenumber spectrum of the vorticity as

$$\frac{\tilde{\tilde{\mathbf{u}}}(\mathbf{k}) = i\mathbf{k} \times \tilde{\tilde{\boldsymbol{\omega}}}(\mathbf{k})}{|\mathbf{k}|^2} \quad \text{where} \quad \tilde{\tilde{\boldsymbol{\omega}}}(\mathbf{k}) = \frac{1}{(2\pi)^3} \int_V \boldsymbol{\omega}(\mathbf{y}) e^{-i\mathbf{k}\cdot\mathbf{y}} dV \qquad (14.4.1)$$

For a line vortex the vorticity can be easily expressed using the coordinates $(\xi_1, \xi_2, \xi_3)$, where $\xi_3$ is aligned with the vortex core, that is, in the direction of the unit vector $\mathbf{z}$ as shown in Fig. 14.7. Coordinate $\xi_2$ is defined normal to the blade surface. The vorticity is then

$$\boldsymbol{\omega}(\mathbf{y}) = \Gamma \mathbf{z} \delta(\xi_1) \delta(\xi_2 - h)$$

In blade-based coordinates $y_i$ we have that $\mathbf{z} = (\sin\phi_v, 0, \cos\phi_v)$, where $\phi_v$ is the angle that the vortex core makes with the blade leading edge, $\xi_2 = y_2$ and

$$y_1 = \xi_1 \cos\phi_v + \xi_3 \sin\phi_v \qquad\qquad y_3 = \xi_3 \cos\phi_v - \xi_1 \sin\phi_v$$

The wavenumber spectrum of the vorticity is then obtained by integrating over $\xi_1$ and $\xi_2$ and noting that the arguments of the Dirac delta functions are zero when $y_2 = h$ and $\xi_1 = 0$, so

$$\tilde{\tilde{\boldsymbol{\omega}}}(\mathbf{k}) = \frac{\Gamma \mathbf{z}}{(2\pi)^3} \int_{-R_\infty}^{R_\infty} e^{-ik_2 h - i(k_1 \sin\phi_v + k_3 \cos\phi_v)\xi_3} d\xi_3$$

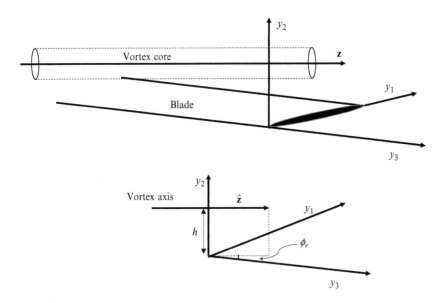

**Fig. 14.7** A blade vortex interaction showing a vortex cutting a blade and the coordinate system used for the analysis.

If the vortex is of finite length $L$ then we can take $R_\infty = L/2$ and

$$\tilde{\omega}(\mathbf{k}) = \frac{\Gamma L z e^{-ik_2 h}}{(2\pi)^3}\left(\frac{\sin\left((k_1\sin\phi_v + k_3\cos\phi_v)L/2\right)}{(k_1\sin\phi_v + k_3\cos\phi_v)L/2}\right)$$

To obtain the upwash spectrum on the blade surface we use Eqs. (14.1.10) and (14.4.1), noting that the component of $\mathbf{k}\times\mathbf{z}$ in the $y_2$ direction is $k_3\sin\phi_v - k_1\cos\phi_v$, so

$$\tilde{w}_2(k_1,k_3) = \frac{i\Gamma L(k_3\sin\phi_v - k_1\cos\phi_v)}{(2\pi)^3}\left(\frac{\sin\left(k_1\sin\phi_v + k_3\cos\phi_v L/2\right)}{k_1\sin\phi_v + k_3\cos\phi_v L/2}\right)$$

$$\int_{-\infty}^{\infty}\frac{e^{-ik_2 h}}{k_1^2 + k_2^2 + k_3^2}dk_2$$

The integral can be completed analytically using tables of Fourier transforms, and we obtain

$$\tilde{w}_2(k_1,k_3) = \frac{i\Gamma L(k_3\sin\phi_v - k_1\cos\phi_v)e^{-k_{13}h}}{(2\pi)^2 2k_{13}}\left(\frac{\sin\left(k_1\sin\phi_v + k_3\cos\phi_v L/2\right)}{k_1\sin\phi_v + k_3\cos\phi_v L/2}\right)$$

$$(14.4.2)$$

where $k_{13}^2 = k_1^2 + k_3^2$.

Based on this result we can calculate the acoustic far field using Eqs. (14.2.1) and (14.2.2), by setting $k_3 = k_o x_3\beta^2/r_e$ and $k_1 = \omega/U_\infty$. In this case, the argument of the sinc function above becomes

$$(k_1\sin\phi_v + k_3\cos\phi_v)L/2 = \frac{\omega L\cos\phi_v}{2U_\infty}\left(\tan\phi_v + \frac{M\beta^2 x_3}{r_e}\right)$$

This implies that there will be a strong beam of radiation in the direction where

$$\beta x_3/r_e = -\tan\phi_v/\beta M$$

Note that if the angle of the vortex to the leading edge of the blade $\phi_v$ is large enough then the right-hand side of this expression will be greater than 1, this beam will not occur, and the peak level of the acoustic field will be greatly reduced. Blade vortex interaction noise is therefore only significant when

$$\tan\phi_v < \beta M$$

which gives a design criterion for this type of source. If the aerodynamics of a rotor can be altered so that the interaction angle meets this criterion, then the loud thumping sound associated with a BVI is avoided.

We also note that the strength of the acoustic field depends on

$$\exp\left(-k_{13}h\right) = \exp\left(-|\omega h/U_\infty|\sqrt{1 + \left(Mx_3\beta^2/r_e\right)^2}\right)$$

This differs from the two-dimensional result by a factor dependent on $M^2$ and can be ignored for most Mach numbers of interest. The scaling of a blade vortex interaction on frequency will be strongly affected by the displacement of the vortex above the blade, given by $h$. To illustrate this, consider a parallel interaction at the observer location $x_3 = 0$ (so $\phi_v = 0$ and $k_3^{(o)} = 0$) and use Eq. (14.4.2) in Eqs. (14.2.1) and (14.2.2) to give

$$\tilde{p}\left(\mathbf{x}, \omega\right) \approx \left(\frac{-i\pi\omega x_2 e^{ik_o r_e - ik_o Mx_1}}{c_\infty r_e^2}\right)\left(\frac{-i\rho_o c \Gamma L e^{-|\omega|h/U_\infty}}{4(2\pi)^2}\right)\Lambda\left(k_1^{(o)}, 0, \omega, M\right)$$

$$(14.4.3)$$

In the high-frequency limit we can use Eq. (14.2.6) to approximate $\Lambda$, and so

$$\tilde{p}\left(\mathbf{x}, \omega\right) \approx \left(\frac{x_2 e^{ik_o r_e - ik_o Mx_1}}{r_e^2(1 - x_1/r_e)^{1/2}}\right)\left(\frac{i\rho_o \Gamma UL(1 - M)^{1/2}M^{1/2}e^{-|\omega|h/U_\infty}}{2(2\pi)^2}\right)$$

This shows the cardioid directivity expected from a leading edge interaction for blades with noncompact chords and a scaling with flow speed that is much stronger than a traditional unsteady loading source. (Since $\Gamma$ would be expected to scale on $M$ the Mach number scaling for the mean-square acoustic pressure is approximately $M^3$.) However, the exponential decay of the spectrum with frequency is the dominant effect.

## References

[1] M.T. Landhahl, Unsteady Transonic Flow, Cambridge University Press, Cambridge, 1961.
[2] J.M.R. Graham, Similarity rules for thin aerofoils in non-stationary subsonic flows, J. Fluid Mech. 43 (1970) 753–766.
[3] R.K. Amiet, Acoustic radiation from an airfoil in a turbulent stream, J. Sound Vib. 41 (1975) 407–420.
[4] R.K. Amiet, Airfoil gust response and the sound produced by airfoil-vortex interaction, J. Sound Vib. 107 (1986) 487–506.

# Trailing edge and roughness noise

# 15

In the last chapter we examined leading edge noise—noise generated by an airfoil subjected to free stream unsteadiness. Airfoils and other fluid dynamic devices also generate noise as a result of turbulence and unsteadiness in the boundary layers and other viscous flow regions that grow from their surfaces. This is called *self-noise*. In this chapter we focus most of our discussion on self-noise generated by turbulent boundary layers, specifically the flow of a turbulent boundary layer over a sharp trailing edge and the flow of the boundary layer over a rough wall.

## 15.1  The origin and scaling of trailing edge noise

Trailing edge noise is fundamentally a consequence of the interaction between unsteadiness in the flow and the sharp corner formed by the trailing edge of a lifting surface. Consider this situation in its most idealized form, illustrated in Fig. 15.1. The trailing edge is modeled as a semi-infinite flat plate of zero thickness immersed in a uniform flow with a sweep angle $\Lambda_o$. This approximation, in which the leading edge is ignored, is realistic as long as the airfoil chord remains large compared to the acoustic wavelength of the sound produced.

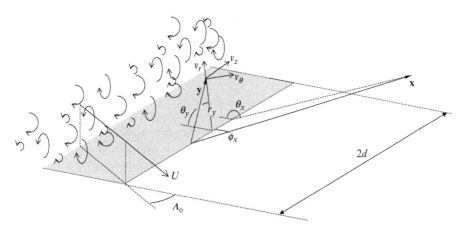

**Fig. 15.1** Nomenclature for the problem of a swept trailing edge behind a semi-infinite flat plate.

Turbulence generated by the boundary layers formed on both sides of the plate is convected over the trailing edge. Sound generated in this scenario can be characterized in terms of a solution to Lighthill's equation. One strategy, first considered by Ffowcs

Aeroacoustics of Low Mach Number Flows. http://dx.doi.org/10.1016/B978-0-12-809651-2.00015-1

Williams and Hall in 1970 [1], is to employ a tailored Green's function chosen to satisfy the rigid wall boundary condition on the plate. The sound produced is then given by Eq. (4.5.6) purely in terms of the Lighthill stress tensor,

$$\widetilde{p}(\mathbf{x}, \omega) = \int_V \left( \frac{\partial^2 \widetilde{G}_T(\mathbf{x}|\mathbf{y})}{\partial y_i \partial y_j} \right) \widetilde{T}_{ij}(\mathbf{y}, \omega) dV(\mathbf{y}) \tag{15.1.1}$$

where we have expressed the sound field in terms of the pressure fluctuations it produces. Ffowcs-Williams and Hall solved this equation using an exact expression for the Green's function and by ignoring the viscous and nonisentropic contributions to the Lighthill stress tensor, approximations consistent with high Reynolds number and low Mach number flow. In this case the sound source is only a function of the Reynolds stress fluctuations $T_{ij} = \rho v_i v_j$. Simplifying the result for turbulence close to the edge and for a far-field observer, they developed an expression for the radiated sound

$$\widetilde{p}(\mathbf{x}, \omega) \approx -\frac{\sqrt{i}}{2} k^2 \sqrt{\sin\phi_x} \cos\left(\tfrac{1}{2}\theta_x\right) \int \left\{ \begin{array}{c} \left[ \widetilde{\rho v_r^2}(\mathbf{y}, \omega) - \widetilde{\rho v_\theta^2}(\mathbf{y}, \omega) \right] \cos\left(\tfrac{1}{2}\theta_y\right) \\ -2\widetilde{\rho v_r v_\theta}(\mathbf{y}, \omega) \sin\left(\tfrac{1}{2}\theta_y\right) \end{array} \right\}$$

$$\times \frac{e^{ik|\mathbf{x}| - ik\mathbf{y} \cdot \mathbf{x}/|\mathbf{x}|}}{(2\pi k r_y)^{3/2} |\mathbf{x}|} dV(\mathbf{y})$$

$$\tag{15.1.2}$$

As illustrated in Fig. 15.1, $\theta_x$ and $\theta_y$ are the angles of the observer and source points, $\mathbf{x}$ and $\mathbf{y}$, measured from the plate in a plane perpendicular to its edge; $r_y$ is the source distance from the edge, in that plane; and $\phi_x$ is the angle of the observer measured from the edge. The result is expressed in terms of the polar velocity components about the trailing edge $v_r$ and $v_\theta$ and the acoustic wavenumber $k$.

Eq. (15.1.2) reveals immediately that there is no significant sound generated by velocity fluctuations parallel to the trailing edge. This is consistent with expectations of the theory of vortex sound (Section 7.1, discussed for trailing edge applications by Howe [2]) from which we would expect the streamwise component of the vorticity to have no impact on the sound generated.

In order to examine the scaling of the sound implied by Eq. (15.1.2) we note that since the mean velocity of the flow must be parallel to the flat plate, then, for example,

$$v_r^2 = \left( -U \cos\Lambda_o \cos\theta_y + u_r \right)^2 = \left( U \cos\Lambda_o \cos\theta_y \right)^2 - 2Uu_r \cos\Lambda_o \cos\theta_y + u_r^2$$

We can ignore the first term since it is steady and does not contribute to the sound, and the last term on the basis that it is negligible if the turbulent fluctuations are small compared to the mean velocity. If we restrict ourselves to low Mach number flows then we also ignore the density fluctuation contribution to the Lighthill stresses and replace $\rho$ with $\rho_o$. Applying similar considerations to $v_\theta^2$ and $v_r v_\theta$, Eq. (15.1.2) becomes

$$\widetilde{p}(\mathbf{x}, \omega) \approx -\frac{\sqrt{i}}{2} k^2 \sqrt{\sin\phi_x} \cos\left(\frac{1}{2}\theta_x\right) \rho_o U \cos\Lambda_o \int \frac{\widetilde{v_i f_i}(\theta_y) e^{ik|\mathbf{x}| - ik\mathbf{y}\cdot\mathbf{x}/|\mathbf{x}|}}{\left(2\pi k r_y\right)^{3/2} |\mathbf{x}|} dV(\mathbf{y})$$

$$(15.1.3)$$

where subscript $i = 1, 2$ for the cylindrical components $v_r$ and $v_\theta$, and $f_1$ and $f_2$ involve only trigonometric functions of $\theta_y$. Note that the form of Eq. (15.1.2) is unchanged if we choose to use Cartesian velocity components inside the integral. We can now estimate the spectrum of the far-field sound as

$$S_{pp}(\mathbf{x}, \omega) = \frac{\pi}{T} E[\widetilde{p}^*(\mathbf{x}, \omega)\widetilde{p}(\mathbf{x}, \omega)]$$

$$\approx \frac{k \sin\phi_x \cos^2\left(\frac{1}{2}\theta_x\right) \rho_o^2 U^2 u^2 \cos^2\Lambda_o}{4(2\pi)^3 |\mathbf{x}|^2} \int\int \frac{S_{ij}(\mathbf{y}, \mathbf{y}', \omega)}{u^2} f_i(\theta_y) f_j\left(\theta'_y\right)$$

$$\frac{e^{ik(\mathbf{y}-\mathbf{y}')\cdot\mathbf{x}/|\mathbf{x}|}}{\left(r_y r'_y\right)^{3/2}} dV(\mathbf{y}) dV(\mathbf{y}') \qquad (15.1.4)$$

where we have introduced the two-point velocity cross spectrum,

$$S_{ij}(\mathbf{y}, \mathbf{y}', \omega) = \frac{\pi}{T} E\left[\widetilde{v}_i^*(\mathbf{y}, \omega)\widetilde{v}_j(\mathbf{y}', \omega)\right]$$

and also the constant $u$ representing the velocity scale of the turbulent fluctuations. Now, the frequency of the sound will be controlled by the frequency with which eddies pass the trailing edge, $U/L$, and thus we expect $k = U/Lc_\infty$, where $L$ is the lengthscale of the turbulence. We also expect $L$ to be the scale of the distance between the sources and the trailing edge, $r_y$ and $r_y'$. Incorporating these observations, Eq. (15.1.4) becomes

$$S_{pp}(\mathbf{x}, \omega) \approx \frac{\sin\phi_x \cos^2\left(\frac{1}{2}\theta_x\right) \rho_o^2 U^3 u^2 \cos^2\Lambda_o}{4(2\pi)^3 |\mathbf{x}|^2 c_\infty L^4}$$

$$\times \int\int \frac{S_{ij}(\mathbf{y}, \mathbf{y}', \omega)}{u^2} f_i(\theta_y) f_j\left(\theta'_y\right) e^{ik(\mathbf{y}-\mathbf{y}')\cdot\mathbf{x}/|\mathbf{x}|} \frac{dV(\mathbf{y})}{\left(r_y/L\right)^{3/2}} \frac{dV(\mathbf{y}')}{\left(r'_y/L\right)^{3/2}}$$

$$(15.1.5)$$

The inner integral in this expression is the volume under a weighted correlation coefficient function and is expected to scale with the cube of $L$. The outer integral will multiply this result by the span of the trailing edge $b$, and a distance proportional to $L$

in each of the other two directions where the weighting in $r_y'$ is effective. In conclusion, we have that the far-field sound spectrum will scale approximately as

$$S_{pp}(\mathbf{x}, \omega) \sim \frac{\sin\phi_x \cos^2\left(\frac{1}{2}\theta_x\right)\rho_o^2 U^3 u^2 L b \cos^2\Lambda_o}{|\mathbf{x}|^2 c_\infty} S(\omega) \qquad (15.1.6)$$

where $S(\omega)$ provides the shape of the normalized pressure spectrum and has units of seconds.

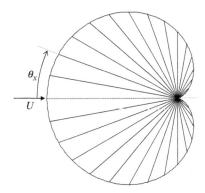

**Fig. 15.2** Directivity of trailing edge noise for an observer in a plane normal to the trailing edge.

This expression reveals a number of characteristics typical of trailing edge noise. Most importantly trailing edge noise is seen to scale approximately on the fifth power of the flow velocity, or more accurately, as the fourth power of the velocity times the Mach number. This clearly demonstrates that at low Mach number the presence of the trailing edge greatly amplifies the direct sound radiation from the turbulence which, in the absence of the trailing edge, would scale as $U^4 M^4$ according to Eq. (4.4.11). Interestingly, trailing edge noise is also slightly more efficient than the classical dipole scaling given by Eq. (4.4.8), much like high-frequency leading edge noise. Trailing edge noise also has a cardioid directivity, $\cos^2(\frac{1}{2}\theta_x)$ (Fig. 15.2), in which the loudest sound is directed upstream and no sound propagates directly downstream. Sound is emitted above the airfoil in antiphase with that emitted below. Eq. (15.1.6) reveals that trailing edge noise can be reduced by sweeping the trailing edge, relative to the flow it experiences (though this effect may be partly compensated for if accompanied by a corresponding increase in the trailing edge length), or by decreasing the scale of the turbulence $L$. Note that Eq. (15.1.5) suggests that trailing edge noise can also be controlled if the turbulent sources can be moved away from the edge, increasing $r_y$.

The most important results from Ffowcs-Williams and Hall's analysis are these scaling and directivity observations. More detailed analysis requires estimation of the complicated two-point velocity spectrum function that forms the source term.

# 15.2  Amiet's trailing edge noise theory

A different strategy for the analysis of trailing edge noise is to seek a solution through Curle's equation (4.3.12). In this case we will have a quadrupole contribution from the sound generated directly by the turbulence and a dipole contribution from the sound produced by the unsteady loading on the airfoil which, of course, is an integration of the pressure fluctuations experienced on the airfoil surface. Based on the scaling arguments outlined in Section 4.4, we expect the dipole contributions to dominate in low Mach number flows.

Amiet [3,4] exploited this observation to develop an analytical approach to the quantitative prediction of trailing edge noise. Like Ffowcs-Williams and Hall he used the thin-airfoil theory approximation of a flat plate (Fig. 15.3) in a uniform flow in the $x_1$ direction. Statistically identical turbulent boundary layers are assumed to form on either side of the airfoil.

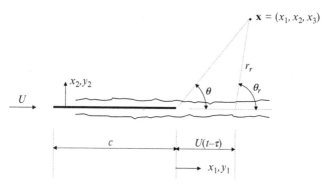

**Fig. 15.3** Nomenclature for Amiet's trailing edge noise theory. Coordinates $x_3$ and $y_3$ are perpendicular to the plane of the page in which direction the trailing edge has a span of $b$.

Amiet's theory takes advantage of the fact that the spectral form of the wall pressure fluctuations imposed by a turbulent boundary layer on a flat plate, in the absence of the trailing edge, is well known. At least in the idealized circumstances pictured in Fig. 15.3, the convected vorticity that defines the boundary-layer turbulence will not be modified as it flows over the trailing edge since no flow distortion occurs. The changes in the pressure field that occur in the vicinity of the trailing edge, including the radiation of sound, are therefore an irrotational response of the flow to the removal of the non-penetration condition and the imposition of the Kutta condition at the trailing edge. Amiet modeled this situation by treating the pressure fluctuation as having two components, one representing the undisturbed boundary layer and the other resulting from the response of the trailing edge to this disturbance.

We begin with Eq. (6.5.5) for the dipole sound radiated by a flat plate airfoil in terms of the pressure difference it experiences. Integrating this expression gives the sound expressed in terms of the wavenumber transform of the acoustic pressure

$$\widetilde{p}(\mathbf{x}, \omega) \approx \frac{-i\pi\omega x_2 \Delta\widetilde{p}\left(k_1^{(o)}, k_3^{(o)}, \omega\right)e^{ik_o r_e - ik_o M x_1}}{c_\infty r_e^2} \tag{15.2.1}$$

where the wavenumber arguments to $\Delta p$ are given by Eq. (6.5.6)

$$k_1^{(o)} = k_o\left(\frac{x_1}{r_e} - M\right) \quad k_3^{(o)} = k_o\beta^2\left(\frac{x_3}{r_e}\right) \tag{6.5.6}$$

and thus account for the convection of the sound waves in the uniform stream surrounding the airfoil. Note that here $k_o = \omega/(\beta^2 c_\infty)$, $r_e = \sqrt{x_1^2 + \beta^2\left(x_2^2 + x_3^2\right)}$, and $\beta^2 = 1 - M^2$. Using the definition of the wavenumber transform of the pressure, adapted from Eq. (4.7.12), the full wavenumber transform of the surface pressure jump is related to the spanwise wavenumber transform as

$$\Delta\widetilde{p}(k_1, k_3, \omega) = \frac{1}{2\pi}\int_{-c}^{0} \Delta\widetilde{p}(y_1, k_3, \omega)e^{-ik_1 y_1}dy_1 \tag{15.2.2}$$

We can therefore rewrite Eq. (15.2.1) to give the far-field sound in terms of the pressure differences at a given chordwise location on the airfoil

$$\widetilde{p}(\mathbf{x}, \omega) \approx \frac{-i\omega x_2 e^{ik_o r_e - ik_o M x_1}}{2c_\infty r_e^2}\int_{-c}^{0} \Delta\widetilde{p}\left(y_1, k_3^{(o)}, \omega\right)e^{-ik_1^{(o)} y_1}dy_1 \tag{15.2.3}$$

Note that in Eqs. (15.2.2), (15.2.3) we have prescribed the limit so as to recognize the finite size of the airfoil, which is taken as extending from $-c$ to $0$ in the chordwise direction $y_1$. We are interested in the spectrum of the far-field sound, rather than the acoustic pressure at any particular instant, and this we can obtain using Eq. (8.4.14) as

$$S_{pp}(\mathbf{x}, \omega) = \frac{\pi}{T}E[\widetilde{p}^*(\mathbf{x}, \omega)\widetilde{p}(\mathbf{x}, \omega)]$$

$$\approx \left(\frac{\omega x_2}{2c_\infty r_e^2}\right)^2 \int_{-c}^{0}\int_{-c}^{0}\frac{\pi}{T}E\left[\Delta\widetilde{p}^*\left(y_1, k_3^{(o)}, \omega\right)\Delta\widetilde{p}\left(y_1', k_3^{(o)}, \omega\right)\right]e^{-ik_1^{(o)}\left(y_1' - y_1\right)}dy_1 dy_1'$$

$$\approx \left(\frac{\omega x_2}{2c_\infty r_e^2}\right)^2 \frac{b}{2\pi}\int_{-c}^{0}\int_{-c}^{0}\phi_{qq}\left(y_1, y_1', k_3^{(o)}, \omega\right)e^{-ik_1^{(o)}\left(y_1' - y_1\right)}dy_1 dy_1'$$

where $\phi_{qq}$ is the spanwise wavenumber transform of the cross spectrum of pressure difference fluctuations between any two points on the airfoil surface. Note that we have taken the range $\pm R_\infty$ of the spanwise Fourier transform to be the airfoil span, of $\pm b/2$, effectively assuming that this is very large compared to the spanwise scales of the turbulence encapsulated in $\phi_{qq}$. We will restrict ourselves to an overhead observer for whom $x_3 = k_3^{(o)} = 0$, and so

$$S_{pp}(x_1, 0, x_3, \omega) \approx \left(\frac{\omega x_2}{2 c_\infty r_e^2}\right)^2 \frac{b}{2\pi} \int\limits_{-c}^{0} \int\limits_{-c}^{0} \phi_{qq}(y_1, y_1', 0, \omega) e^{-i k_1^{(o)}(y_1' - y_1)} dy_1 dy_1'$$

(15.2.4)

In order to specify $\phi_{qq}$ we prescribe the boundary-layer pressure fluctuations in two parts. The first part is the pressure fluctuations that would be produced by the boundary layer in the absence of the trailing edge $p_{bl}$. In terms of the wavenumber frequency transform of $p_{bl}$ this is

$$p_{bl}(\mathbf{y}, t) = \int\limits_{-\infty}^{\infty} \int\limits_{-\infty}^{\infty} \int\limits_{-\infty}^{\infty} \widetilde{\widetilde{p}}_{bl}(k_1, k_3, \omega) e^{-i(\omega t - k_1 y_1 - k_3 y_3)} dk_1 dk_3 d\omega$$

(15.2.5)

Using Taylor's frozen flow hypothesis we assume that pressure fluctuations do not evolve and are convected downstream along the airfoil surface at a uniform speed $U_c$ so that $k_1 = K_1 \equiv \omega/U_c$. Following the discussion at the end of Chapter 8, we expect this to be 60–80% of the boundary-layer edge velocity. This means that the wavenumber transform of the boundary-layer pressure fluctuations can be written as

$$\widetilde{\widetilde{p}}_{bl}(k_1, k_3, \omega) = \widetilde{\widetilde{p}}_{bl}(k_3, \omega) \delta(k_1 - \omega/U_c)$$

Thus Eq. (15.2.5) can be reduced to

$$\begin{aligned} p_{bl}(\mathbf{y}, t) &= \int\limits_{-\infty}^{\infty} \int\limits_{-\infty}^{\infty} \int\limits_{-\infty}^{\infty} \widetilde{\widetilde{p}}_{bl}(k_3, \omega) \delta(k_1 - \omega/U_c) e^{-i(\omega t - k_1 y_1 - k_3 y_3)} dk_1 dk_3 d\omega \\ &= \int\limits_{-\infty}^{\infty} \int\limits_{-\infty}^{\infty} \widetilde{\widetilde{p}}_{bl}(k_3, \omega) e^{-i(K_1(U_c t - y_1) - k_3 y_3)} dk_3 d\omega \end{aligned}$$

Note that because of the overhead observer simplification we will only need to consider contributions to $p_{bl}$ at zero-spanwise wavenumber, $k_3 = 0$. The second part of $\phi_{qq}$ is the pressure fluctuation that results from the response of the trailing edge to this disturbance, which is given by the Schwarzschild solution detailed in Chapter 13. For a pressure perturbation,

$$P_o e^{-i(K_1(U_c t - y_1) - k_3 y_3)}$$

the Schwarzschild solution Eq. (13.3.9) gives the trailing edge pressure response as

$$P_o g_{te}(y_1, K_1, k_3) e^{-i(K_1 U_c t - k_3 y_3)}$$

So, for all wavenumber components, the total pressure fluctuation on the airfoil surface will be

$$p(\mathbf{y}, t) = \int\limits_{-\infty}^{\infty} \int\limits_{-\infty}^{\infty} \widetilde{\widetilde{p}}_{bl}(k_3, \omega) \left[ e^{iK_1 y_1} + g_{te}(y_1, K_1, k_3) \right] e^{-i(K_1 U_c t - k_3 y_3)} \, dk_3 \, d\omega$$

Taking the Fourier transform in $y_3$ and time, we obtain

$$\widetilde{\widetilde{p}}(y_1, k_3, \omega) = \widetilde{\widetilde{p}}_{bl}(k_3, \omega) \left[ e^{iK_1 y_1} + g_{te}(y_1, k_1, k_3) \right] \tag{15.2.6}$$

For the response function $g_{te}$ we ignore the leading edge of the airfoil and use the result for the trailing edge response of a semi-infinite airfoil, from Eq. (13.3.9)

$$g_{te}(y_1, K_1, 0) = \{(1-i)E_2(K_1 A|y_1|) - 1\} e^{iy_1 K_1}$$
$$\text{where} \quad A = \frac{U_c / c_\infty}{1 - M} + 1 \quad \text{and} \quad E_2(x) = \int\limits_0^x \frac{e^{iq}}{(2\pi q)^{1/2}} \, dq \tag{15.2.7}$$

We expect this to be a good approximation as long as the airfoil chord remains large compared to the acoustic wavelength and thus $K_1 c \gg 1$. Substituting into Eq. (15.2.6) we obtain

$$\widetilde{\widetilde{p}}(y_1, 0, \omega) = (1-i)\widetilde{\widetilde{p}}_{bl}(0, \omega) e^{iK_1 y_1} E_2(K_1 A|y_1|)$$

Now, the spectrum of the pressure difference across the airfoil will be twice the spectrum of the pressure itself, as we do not expect the boundary-layer pressure fluctuations to correlate between the two sides. So,

$$\phi_{qq}(y_1, y_1', 0, \omega) = \frac{2\pi^2}{TR_\infty} E\left[ \widetilde{\widetilde{p}}^*(y_1, 0, \omega) \widetilde{\widetilde{p}}(y_1', 0, \omega) \right]$$
$$= \frac{2\pi^2}{TR_\infty} E\left[ \widetilde{\widetilde{p}}_{bl}^*(0, \omega) \widetilde{\widetilde{p}}_{bl}(0, \omega) \right] (1+i)(1-i) E_2^*(K_1 A|y_1|) E_2(K_1 A|y_1'|) e^{iK_1(y_1' - y_1)}$$
$$= 2\phi_{pp}(0, \omega)(1+i)(1-i) E_2^*(K_1 A|y_1|) E_2(K_1 A|y_1'|) e^{iK_1(y_1' - y_1)} \tag{15.2.8}$$

where $\phi_{pp}$ is the spanwise wavenumber frequency spectrum of the undisturbed boundary-layer pressure fluctuations. That is, in terms of the full wavenumber frequency spectrum $\Phi_{pp}$,

$$\phi_{pp}(k_3, \omega) = \int\limits_{-\infty}^{\infty} \Phi_{pp}(k_1, k_3, \omega) dk_1$$

Substituting from Eq. (15.2.8) into Eq. (15.2.4), we recover

$$S_{pp}(\mathbf{x}, \omega) \approx \left(\frac{\omega x_2}{2c_\infty r_e^2}\right)^2 \frac{b}{\pi} \phi_{pp}(0, \omega) \int\limits_{-c}^{0} \int\limits_{-c}^{0} (1+i)(1-i) E_2^*(K_1 A |y_1|) E_2(K_1 A |y_1'|)$$

$$e^{iK_1(y_1' - y_1)} e^{-ik_1^{(o)}(y_1' - y_1)} dy_1 dy_1'$$

(15.2.9)

The integrals inside over $y_1$ and $y_1'$ can be separated and are conjugates of each other, and so our final result can be written as

$$S_{pp}(x_1, x_2, 0, \omega) \approx \left(\frac{\omega x_2}{2c_\infty r_e^2}\right)^2 \frac{b}{\pi} \phi_{pp}(0, \omega) \left| \frac{2}{c} \int\limits_{-c}^{0} (1-i) E_2(K_1 A |y_1|) e^{i(K_1 - k_1^{(o)}) y_1} dy_1 \right|^2$$

or,

$$S_{pp}(x_1, x_2, 0, \omega) \approx \pi b \left(\frac{\omega c x_2}{4\pi c_\infty r_e^2}\right)^2 \phi_{pp}(0, \omega) |L|^2 \qquad (15.2.10)$$

where $\mathcal{L}$, the integral inside the absolute value, is Amiet's [3,4] generalized lift function. Note that the sound predicted by Eq. (15.2.10) is twice that given by Amiet since we are accounting for both airfoil boundary layers. The integration defining $\mathcal{L}$ can be done analytically, but Amiet [4] argues that the integrand should be modified to soften the effects of the leading edge limit (at $-c$) which otherwise implies unsteady loading generated by the plane-wall boundary-layer pressure fluctuations $p_{bl}$ acting at the sharp leading edge of the flat plate. These do not exist for a real airfoil where the boundary layer originates at a rounded leading edge. To avoid this problem, the complex exponential in Eq. (15.2.6) is modified to include a term that ensures decay with distance from the trailing edge, i.e., the exponential becomes $\exp(K_1 y_1 [i + \varepsilon])$ and the undisturbed boundary-layer contribution to $\mathcal{L}$ is then assessed in the high-frequency limit as $\varepsilon$ tends to zero. With this adjustment Amiet [4] gives the magnitude of $\mathcal{L}$ as

$$|\mathcal{L}| = \frac{1}{\Theta} \left| (1-i) \left\{ \sqrt{\frac{A/B}{1+x_1/r_e}} E_2(BK_1 c(1+x_1/r_e)) e^{2i\Theta} - E_2(AK_1 c) \right\} + 1 \right|$$

(15.2.11)

where

$$\Theta = K_1 c (1 + B(M - x_1/r_e))/2 \quad \text{and} \quad B = \frac{U_c/c_\infty}{1 - M^2} = \frac{A-1}{1+M}$$

Note that the source term $\phi_{pp}$ in Eq. (15.2.10) is related by Fourier transform to the spanwise cross spectrum of surface pressure fluctuations

$$\phi_{pp}(0, \omega) = \frac{1}{2\pi} \int_{-R_\infty}^{R_\infty} S_{pp}(\Delta y_3, \omega) d\Delta y_3$$

Thus we can write

$$\phi_{pp}(0, \omega) = \frac{1}{\pi} l_p(\omega) S_{pp}(\omega) \quad \text{where} \quad l_p(\omega) = \int_{0}^{R_\infty} \frac{S_{pp}(\Delta y_3, \omega)}{S_{pp}(\omega)} d\Delta y_3 \qquad (15.2.12)$$

and express the source term as the product of a frequency-dependent spanwise pressure lengthscale $l_p(\omega)$ and the autospectrum of the surface pressure fluctuations $S_{pp}(\omega)$. At any given frequency we expect the scale of the turbulence in the streamwise and spanwise directions to be the comparable and therefore anticipate that $l_p(\omega)$ will vary as $U/\omega$.

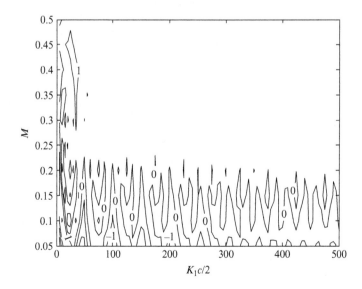

**Fig. 15.4** Plot of $(K_1 c/2)^2 |\mathcal{L}|^2 M$ vs. Mach number and wavenumber, evaluated for $U_c/U = 0.8$ and $x_1/r_e = 0.5$, in decibels.

To determine the scaling and directivity implied by Amiet's result we note, as before, that the frequency of the sound $\omega$ will be determined by the flow speed divided by the turbulence scale $U/L$. Roughly speaking we also expect the pressure spectrum to vary as $\rho_o^2 U^4 S(\omega)$. The scaling of $|\mathcal{L}|^2$ can be determined from the fact that $\text{Lim}_{x \to \infty}[E_2(x)] = \frac{1}{2}(1+i)$. Thus for large $K_1 c$ (implying, not unreasonably, that the eddies are small compared to the chord length)

$$|\mathcal{L}|^2 \approx \frac{1}{(K_1 c/2)^2} \left( \frac{A/B}{1+x_1/r_e} \right) \sim \frac{4L^2}{c^2} \frac{1}{M} \frac{1}{1+x_1/r_e}$$

since at low Mach number $A \approx 1$ and $B \approx M$. The realism of this characterization can be seen in Fig. 15.4 which shows that $(K_1 c/2)^2 |\mathcal{L}|^2 M$ is equal to 1 over a substantial range of frequencies and Mach numbers with an error of less than $\pm 1$ dB. Indeed estimating $|\mathcal{L}|^2$ as $(K_1 c/2)^{-2} M^{-1}$ may be sufficiently accurate for some applications. Finally, we note that ignoring convective effects, $r_e = |\mathbf{x}|$, $x_2/r_e = x_2/|\mathbf{x}| = \sin\theta$, and $x_1/r_e = \cos\theta$, where $\theta$ is the directivity angle from the downstream direction defined in Fig. 15.3. Combining these observations, we conclude that in the absence of convective effects

$$S_{pp}(x_1, x_2, 0, \omega) \sim \frac{\sin^2\left(\frac{1}{2}\theta\right)\rho_o^2 U^5 Lb}{c_\infty |\mathbf{x}|^2} S(\omega) \tag{15.2.13}$$

in complete agreement with Ffowcs-Williams and Hall's result (Eq. 15.1.6). Note that accounting for convective effects [5] produces a more complex directivity in terms of the angle $\theta_r$, the observer location measured from the retarded source position (Fig. 15.3), giving

$$S_{pp}(x_1, x_2, 0, \omega) \sim \frac{\sin^2\left(\frac{1}{2}\theta_r\right)\rho_o^2 U^5 Lb}{(1 + M\cos\theta_r)[1 + (M - M_c)\cos\theta_r]^2 c_\infty r_r^2} S(\omega) \tag{15.2.14}$$

where $r_r$ is the distance from the retarded source position to the observer.

Amiet's major contribution is in formulating a method in which well-defined and established empirical information can be used to predict the form and level of a trailing edge noise spectrum from first principles. As an example, we consider sound predictions for a NACA 0012 airfoil at zero angle of attack. The data [6], shown in Fig. 15.5, reflect a chordlength of 0.23 m and a span of 1.22 m, with an observer 3 m directly overhead over the mid-span of the trailing edge. The flow speed is 55.5 m/s giving a chord Reynolds number close to 660,000. The airfoil boundary layer is heavily tripped, and the trailing edge displacement thickness $\delta^*$ is estimated as 2.4 mm. From the discussion following Eq. (9.2.12) we take $u_\tau/U_e = 4\%$. We would expect, approximately, $\delta/\delta^* = 8$ and $U_c/U_e = 0.8$ (see Section 9.2.2). With these parameters we can use Eqs. (9.2.29) or (9.2.37) to estimate the autospectrum of the boundary-layer pressure fluctuations $S_{pp}(\omega)$. For $l_p(\omega)$ we use Amiet's suggested relation

$$l_p(\omega) = 2.1 U_c / \omega \qquad\qquad (15.2.15)$$

Note that we could alternatively get an expression for $l_p(\omega)$, or for $\phi_{pp}(0,\omega)$ as a whole, by integrating one of the wavenumber spectrum models for the wall pressure presented at the end of Chapter 9.

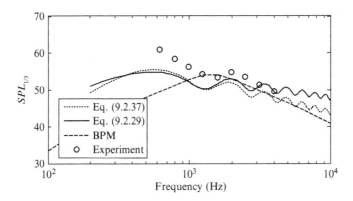

**Fig. 15.5** Comparison of trailing edge noise spectra computed using Eq. (15.2.10), with different boundary-layer pressure spectrum models, compared with measurements for a NACA 0012 airfoil at zero angle of attack [6].

The comparison of predictions performed using the boundary-layer spectrum models and the data are presented in Fig. 15.5. The predictions have been scaled to one-third octave band SPL to make this comparison, by multiplying by the frequency and by $(2^{1/6} - 2^{-1/6})$. Overall the spectrum is predicted quite well, including some features of the shape. The predictions appear some 5 dB below the measurements at lower frequencies. This type of disagreement is not unexpected. The adverse pressure gradient experienced toward the rear of the airfoil would be expected to increase the energy of larger boundary-layer eddies, and the amplitude of the pressure fluctuations they produce, and increase the spanwise lengthscale over which the pressure fluctuations correlate $l_p$ when compared to a flat plate boundary layer. Using pressure spectra and scales measured on the airfoil itself has been shown to improve agreement between measurements and predictions [5,7].

## 15.3   The method of Brooks, Pope, and Marcolini [8]

While Amiet's theory establishes the fundamental relationship between trailing edge noise and the boundary-layer pressure fluctuations on the airfoil generating the sound, it does not in most cases serve as a very practical prediction tool. While the flat plate serves as an adequate aeroacoustic model of a real airfoil, it is a poor model for determining the detailed aerodynamics of the boundary layer. Airfoil geometry and angle of attack have a strong influence on the boundary-layer turbulence, particularly at the trailing edge. One might therefore propose an experimental campaign to measure the zero-spanwise wavenumber spectrum $\phi_{pp}(0,\omega)$ over a large test matrix of practical conditions and fit the resulting data to correlations to be used in trailing edge noise prediction. Measurements of pressure fluctuations and their scale near a trailing edge are difficult, however, and this approach requires that we choose the "right" measurement location in an evolving flow and also separate out the boundary-layer pressure fluctuations and the trailing edge response. A much better plan is to run the same test matrix but instead measure the far-field trailing edge noise at each condition, since this comparatively simple measurement contains all the needed information about the source.

Brooks, Pope, and Marcolini (BPM) [8] took precisely this approach, measuring the self-noise of airfoils over a large range of conditions and forming these data into empirical correlations that then serve as a prediction tool. Since its publication in 1989, the BPM method has become the tool of first (and often last) choice for airfoil self-noise predictions. It has also become the comparison baseline for more sophisticated noise prediction schemes, such as those using large eddy simulation, or measurements on configurations or at conditions that lie outside of the range of tests that underpinned the original correlations.

BPM did not use their sound measurements to invert Amiet's equation (15.2.10) and determine the source component of the surface pressure spectrum at each condition. Though perfectly possible, this is an unnecessary step. The empirical information can more directly be incorporated by scaling the measured sound spectra according to the form derived by Amiet and Ffowcs-Williams and Hall (Eq. 15.2.14) and then establishing normalized spectral forms and correlations for the boundary-layer scaling variables as functions of conditions.

In their study of turbulent boundary-layer trailing edge noise, BPM placed a series of two-dimensional NACA 0012 airfoils, of chordlength from 1 in. to 1 ft, in the test section of the Quiet Flow Facility at NASA Langley (Fig. 10.4). They positioned a pair of far-field microphones on either side of the chord line of each airfoil in order to separate the trailing edge noise from facility noise by using its antiphase relationship. They also used a second pair of microphones placed fore and aft of each airfoil to eliminate parasitic leading edge noise originating from the airfoil model mounting. The airfoils were tested over a range of angles of attack in clean configuration and with a heavy boundary-layer trip, to ensure transition, and at flow speeds that provided chord Reynolds numbers up to 1.5 million, and Mach numbers up to 0.2.

They extracted from this database all the cases where the far-field sound appeared to be produced by turbulent boundary-layer trailing edge noise. They chose to model the contributions to the noise spectrum using the form

$$
S_i(\mathbf{x}, \bar{f}_i; \text{Re}, \alpha, M) = \frac{2\sin^2\left(\frac{1}{2}\theta_r\right)\sin^2(\phi_r)}{(1 + M\cos\theta_r)[1 + (M - M_c)\cos\theta_r]^2}
$$
$$
\times \frac{2d\delta_i^* M^5}{r_r^2} \times S_i(\bar{f}_i; \text{Re}, \alpha, M)
$$
(15.3.1)

where the normalized frequency is defined as

$$\bar{f}_i \equiv \frac{2\pi\omega\delta_i^*}{U}$$

**Fig. 15.6** Definition of directivity angles for the BPM method.

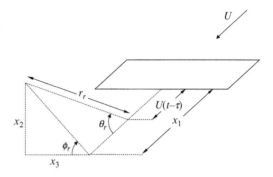

and where subscript $i$ distinguishes the contributions from different components of the boundary-layer turbulence, and we have explicitly indicated the airfoil chord Reynolds number Re, the absolute value of the angle of attack $\alpha$, and Mach number $M$ as parameters controlling the spectrum. Comparing with Eq. (15.2.14) we can see that they chose the boundary-layer displacement thickness (dependent on $\alpha$ and Re) as the measure of the turbulence scale, and they replaced the $U^5$ scaling with $M^5$. The residual factors of $\rho_o^2$ and $c_\infty^4$ are absorbed into the dimensional model spectral form $S_i$. The directivity includes the term $\sin^2\phi_r$ which prescribes a standard dipole out-of-plane directivity to allow results to be used for all observer locations. The definitions of $\theta_r$ and $\phi_r$ in three-dimensional space are shown in Fig. 15.6. As a practical matter, BPM chose to express their spectral curve fits in terms of one-third octave band SPL. Thus we write

$$SPL_i(\mathbf{x}, \bar{f}_i; \text{Re}, \alpha, M) = 10\log_{10}\left(\frac{2\sin^2\left(\frac{1}{2}\theta_r\right)\sin^2(\phi_r)}{(1+M\cos\theta_r)[1+(M-M_c)\cos\theta_r]^2} \times \frac{2d\delta_i^* M^5}{r_r^2}\right)$$
$$+ F_i(\bar{f}_i; \text{Re}, \alpha, M)$$

where $F_i(\bar{f}; \text{Re}, \alpha, M) = 10\log_{10}\left(\frac{S_i(\bar{f}_i; \text{Re}, \alpha, M)}{4\times 10^{-10}\text{Pa}^2}\right)$

Two of the contributions to the spectrum are from the pressure- and suction-side boundary layers, $SPL_p$ and $SPL_s$, and use the corresponding displacement thicknesses. The third contribution $SPL_\alpha$, which also uses the suction-side boundary-layer displacement thickness, was hypothesized by BPM to be associated with a thin region of separated flow formed on the suction side. They introduced this when they found that the attached boundary-layer correlations could not account for the frequency shifts seen in measured sound spectra with angle of attack. Since the contributions add linearly in power, the total trailing edge noise is

$$SPL_{tot} = 10\log_{10}\left(10^{SPL_p/10} + 10^{SPL_s/10} + 10^{SPL_a/10}\right)$$

To evaluate this expression we need the displacement thicknesses $\delta_s^*$ and $\delta_p^*$ and the spectral functions $F_s$, $F_p$, and $F_a$. The displacement thicknesses can be supplied from measurements, calculations (e.g., using a software tool such as Xfoil), or curve fits given by BPM based on integrating hot-wire velocity profiles measured just downstream of the NACA 0012 airfoil trailing edges. The spectral functions are defined as follows

$$F_s(\bar{f}_s; \text{Re}, \alpha, M) = F_A(\bar{f}_s; \text{Re}, M) + F_{K1}(\text{Re}) - 3$$
$$F_p(\bar{f}_p; \text{Re}, \alpha, M) = F_A(\bar{f}_p; \text{Re}, M) + F_{K1}(\text{Re}) + F_{\Delta K}(\text{Re}, \alpha) - 3$$
$$F_a(\bar{f}_s; \text{Re}, \alpha, M) = F_B(\bar{f}_s; \text{Re}, \alpha, M) + F_{K2}(\alpha, M)$$

The function $F_{K1}$ sets the level of the model spectrum $F_A$ and has the form

$$F_{K1}(\text{Re}) = \begin{cases} -4.31\log_{10}\text{Re} + 156.3 & \text{Re} < 247,000 \\ -9.0\log_{10}\text{Re} + 181.6 & 247,000 \le \text{Re} \le 800,000 \\ 128.5 & 800,000 < \text{Re} \end{cases}$$

$F_{\Delta K}$ provides an additional adjustment to the level of the model spectrum for the pressure side contribution

$$F_{\Delta K}(\text{Re}, \alpha) = \begin{cases} \alpha\left(1.43\log_{10}\left(\text{Re}\,\delta_p^*/c\right) - 5.29\right) & \text{Re}\,\delta_p^*/c \le 5000 \\ 0 & \text{Re}\,\delta_p^*/c > 5000 \end{cases}$$

where the group $\text{Re}\delta_p^*/c$ is the Reynolds number based on the pressure side boundary-layer displacement thickness. For brevity we have not explicitly indicated functional dependence on $\delta_p^*/c$. Function $F_{K2}$ sets the level of the model spectrum due to separation, $F_B$, and is

$$F_{K2}(\alpha, M) = F_{K1}(\text{Re}) + \begin{cases} -1000 & \alpha < \gamma_o - \gamma \\ \sqrt{\beta^2 - (\beta/\gamma)^2(\alpha - \gamma_o)^2} + \beta_o & \gamma_o - \gamma \le \alpha \le \gamma_o + \gamma \\ -12 & \gamma_o + \gamma < \alpha \end{cases}$$

where

$$\gamma = 27.094M + 3.31 \quad \gamma_o = 23.43M + 4.651$$
$$\beta = 72.65M + 10.74 \quad \beta_o = -34.19M - 13.82$$

The function $F_A$ defining the spectral shape and frequency scaling for attached flow is

$$F_A(\bar{f}_i; \text{Re}, M) = F_A^{(\min)}(a) + F_A^{(R)}(a_o)\left[F_A^{(\max)}(a) - F_A^{(\min)}(a)\right]$$

where

$$a = \left|\log_{10}\left(50\bar{f}_i M^{0.6}\right)\right|$$

$$a_0 = \begin{cases} 0.57 & \text{Re} < 95,200 \\ -9.57 \times 10^{-13}(\text{Re} - 857,000)^2 + 1.13 & 95,200 < \text{Re} < 857,000 \\ 1.13 & 857,000 < \text{Re} \end{cases}$$

$$F_A^{(\min)}(a) = \begin{cases} \sqrt{67.552 - 886.788a^2} - 8.219 & a < 0.204 \\ -32.665a + 3.981 & 0.204 \leq a \leq 0.244 \\ -142.795a^3 + 103.656a^2 - 57.757a + 6.006 & 0.244 < a \end{cases}$$

$$F_A^{(\max)}(a) = \begin{cases} \sqrt{67.552 - 886.788a^2} - 8.219 & a < 0.130 \\ -15.901a + 1.098 & 0.130 \leq a \leq 0.321 \\ -4.669a^3 + 3.491a^2 - 16.699a + 1.149 & 0.321 < a \end{cases}$$

$$F_A^{(R)}(a_o) = \frac{-20 - F_A^{(\min)}(a_o)}{F_A^{(\max)}(a_o) - F_A^{(\min)}(a_o)}$$

For the separated flow portion the spectral shape and frequency scaling function $F_B$ is similarly defined as

$$F_B(\bar{f}_i; \text{Re}, \alpha, M) = F_B^{(\min)}(b) + F_B^{(R)}(b_o)\left[F_B^{(\max)}(b) - F_B^{(\min)}(b)\right]$$

where

$$b = \begin{cases} \left|\log_{10}\left(50\bar{f}_s M^{0.6}\right)\right| & \alpha < 1.33° \\ 10^{0.0054(\alpha-1.33)^2}\left|\log_{10}\left(50\bar{f}_s M^{0.6}\right)\right| & 1.33° \leq \alpha \leq 12.5° \\ 4.72\left|\log_{10}\left(50\bar{f}_s M^{0.6}\right)\right| & 12.5° < \alpha \end{cases}$$

$$b_0 = \begin{cases} 0.3 & \text{Re} < 95,200 \\ -4.48 \times 10^{-13}(\text{Re} - 857,000)^2 + 0.56 & 95,200 < \text{Re} < 857,000 \\ 0.56 & 857,000 < \text{Re} \end{cases}$$

$$F_B^{(\min)}(b) = \begin{cases} \sqrt{16.888 - 886.788b^2} - 4.109 & b < 0.130 \\ -83.607b + 8.138 & 0.130 \leq b \leq 0.145 \\ -817.810b^3 + 355.210b^2 - 135.024b + 10.619 & 0.145 < b \end{cases}$$

$$F_B^{(\max)}(b) = \begin{cases} \sqrt{16.888 - 886.788b^2} - 4.109 & b < 0.100 \\ -31.330b + 1.854 & 0.100 \leq b \leq 0.187 \\ -80.541b^3 + 44.174b^2 - 39.381b + 2.344 & 0.187 < b \end{cases}$$

$$F_B^{(R)}(b_o) = \frac{-20 - F_B^{(\min)}(b_o)}{F_B^{(\max)}(b_o) - F_B^{(\min)}(b_o)}$$

Note that all angles are in degrees, and that these relations are valid for angles of attack from 0 to 12.5 degrees or $\gamma_o$, whichever is the smallest.

The above equations constitute a prediction method that accounts not only for Reynolds number and boundary-layer thickness effects but also for the most important aerodynamic parameter—the angle of attack. At the conditions of the BPM measurements, these formulae constitute a curve fit to their experimental data and can be used to illustrate some basic variations, as we have done in Figs. 15.7 and 15.8.

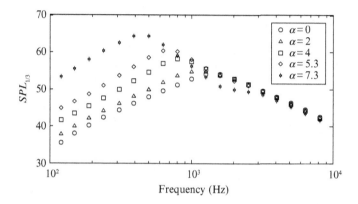

**Fig. 15.7** Trailing edge noise spectra for a tripped NACA 0012 airfoil as a function of angle of attack, in degrees, evaluated using the BPM method [8]. Conditions are $U = 55.5$ m/s, $c = 22.86$ cm, $b = 1.22$ m, $\theta_r = \phi_r = 90$ degrees, $r_r = 3$ m, and standard atmosphere. Boundary layer displacement thicknesses calculated using the BPM correlations.

Fig. 15.7 shows the effect of angle of attack on the trailing edge noise radiated by a tripped 22.86 cm chord blade. Angle of attack has surprisingly little effect on the high-frequency portion of the spectrum. At low frequencies the sound level increases with angle of attack at an increasing rate, and the frequency at which the sound level reaches its maximum decreases by roughly a factor of 3 between 0 and 7.3 degrees. This change presumably reflects the effects of adverse pressure gradient on the suction-side boundary layer, increasing the scale and intensity of turbulent motions.

Fig. 15.8 shows the effect of chordlength on the sound at zero angle of attack. The obvious effect here is that as the chordlength is increased, the trailing edge boundary layer grows, and so the frequency of the sound drops, these effects being roughly in proportion. In Fig. 15.8 we have factored this variation out by plotting the sound against frequency normalized on chord and flow speed, as $fc/U$, where $f$ is frequency in Hz. In this form we can see that the simple chord scaling just described works well at frequencies below $fc/U \approx 5$ (i.e., turbulent scales larger than about 20% of the chord). At higher frequencies the sound level increases with the chordlength because the greater the chordlength, the higher the Reynolds number of the boundary layer, and the greater the range and energy of the turbulence it produces at small scales.

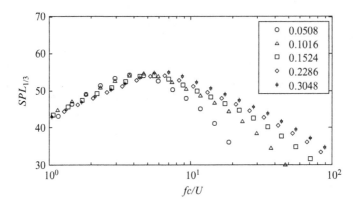

**Fig. 15.8** Trailing edge noise spectra for a tripped NACA 0012 airfoil as a function of chordlength, in meters, evaluated using the BPM method [8]. Conditions are $\alpha = 0$, $U = 55.5$ m/s, $c = 22.86$ cm, $b = 1.22$ m, $\theta_r = \phi_r = 90$ degrees, $r_r = 3$ m, and standard atmosphere. Boundary layer displacement thicknesses calculated using the BPM correlations.

Fig. 15.5 includes a prediction using the BPM method of the same NACA 0012 airfoil results compared with Amiet's method. The BPM prediction is also quite good, though curiously shows a larger difference at low frequencies. Trailing edge noise measurement at low frequencies is notoriously difficult, and the differences may reflect the fact that BPM include only noise contributions in antiphase across the chord line, whereas the measurements of [6] were made using a phased microphone array to distinguish trailing edge noise sources from one side of the airfoil.

A particular advantage of the BPM scheme is that it is not restricted to self-noise generated by turbulent boundary-layer flow over a sharp trailing edge. Self-noise generated by vortex shedding from a trailing edge (due to trailing edge bluntness or a laminar trailing edge boundary layer), stalled flow over the airfoil, and interaction of boundary-layer turbulence with the tip of a blade during the formation of a trailing vortex can be handled by essentially the same approach. BPM reviewed experimental data and present further curve fits that allow prediction of these sources and, in the case of laminar shedding noise and stall noise, the conditions at which these become the dominant self-noise mechanisms.

The BPM method is often used for airfoils that do not have a NACA 0012 section. This may be done by entering trailing edge boundary-layer thicknesses computed or measured on the airfoil in question and entering an angle of attack referenced to zero lift. It is also quite common to see predictions at conditions that exceed the range of the experiments on which the method is based. (BPM did so themselves on a model scale rotor in the DNW-LLF, as discussed in Appendix C of Ref. [8], which actually caused them to refine some of their prediction methods.) These approaches are fine for rough estimates, but it is important to remember that such extrapolations can lead to substantial error.

## 15.4   Roughness noise

We have seen above that the intense surface pressure fluctuations generated by a turbulent boundary layer are a significant source of sound at low Mach numbers when they are scattered from a trailing edge. Given the form of Curle's equation (4.3.9) one might be forgiven for thinking that turbulent boundary-layer flow over a smooth flat surface would be a similarly important source of noise. This is not the case. To see why, imagine the application of Eq. (4.3.9) to flow over an infinite rigid wall defined by the plane $x_2 = y_2 = 0$, as shown in Fig. 15.9.

We apply this equation by taking the volume of integration $V$ as existing only above the surface so that its enclosing surface $S$ is formed by the wall and hemispherical surface of much greater radius than the extent of the flow and the turbulent boundary layer. Given that the velocities on the surface will be zero, because of the non-penetration and no-slip conditions, then Eq. (4.3.9) becomes

$$p'(\mathbf{x}, t) = \int_{-T}^{T} \int_{S} p_{ij} \frac{\partial G}{\partial y_i} n_j dS(\mathbf{y}) d\tau + \int_{-T}^{T} \int_{V} \left( \frac{\partial^2 G}{\partial y_i \partial y_j} \right) T_{ij}(\mathbf{y}, \tau) dV(\mathbf{y}) d\tau \qquad (15.4.1)$$

where the sound field has been written in terms of its pressure fluctuations, and $p_{ij}$ is the sum of the pressure and viscous stresses $p\delta_{ij} - \sigma_{ij}$. At low Mach number, the only significant viscous stresses acting on the wall are those associated with viscous shear parallel to the wall, $\sigma_{12} = \sigma_{21} = \mu \partial v_1 / \partial y_2$ and $\sigma_{23} = \sigma_{32} = \mu \partial v_3 / \partial y_2$. Thus, since $n_j = (0,1,0)$ the first term expands to give

$$\begin{aligned}
p'(\mathbf{x}, t) = {} & \int_{-T}^{T} \int_{S} \left[ p\delta_{i2} - \mu \left( \delta_{i1} \frac{\partial v_1}{\partial y_2} + \delta_{i3} \frac{\partial v_3}{\partial y_2} \right) \right] \frac{\partial G}{\partial y_i} dS(\mathbf{y}) d\tau \\
& + \int_{-T}^{T} \int_{V} \left( \frac{\partial^2 G}{\partial y_i \partial y_j} \right) T_{ij}(\mathbf{y}, \tau) dV(\mathbf{y}) d\tau
\end{aligned} \qquad (15.4.2)$$

This equation can be evaluated in two ways, first using the free-field Green's function, $G_o$, i.e.,

$$\begin{aligned}
p'(\mathbf{x}, t) = {} & \int_{-T}^{T} \int_{S} \left[ p\delta_{i2} - \mu \left( \delta_{i1} \frac{\partial v_1}{\partial y_2} + \delta_{i3} \frac{\partial v_3}{\partial y_2} \right) \right] \frac{\partial G_o}{\partial y_i} dS(\mathbf{y}) d\tau \\
& + \int_{-T}^{T} \int_{V} \left( \frac{\partial^2 G_o}{\partial y_i \partial y_j} \right) T_{ij}(\mathbf{y}, \tau) dV(\mathbf{y}) d\tau
\end{aligned} \qquad (15.4.3)$$

Alternatively, we can evaluate Eq. (15.4.2) using the tailored Green's function of Eq. (4.5.3) that accounts for sound reflection from a flat surface

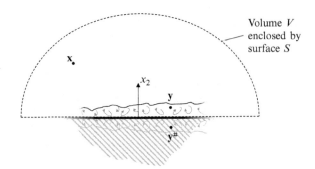

**Fig. 15.9** Control volume and surface for the case of a turbulent boundary layer growing on an infinite plane wall.

$$G_T = \frac{\delta(t - \tau - |\mathbf{x} - \mathbf{y}|/c_\infty)}{4\pi|\mathbf{x} - \mathbf{y}|} + \frac{\delta(t - \tau - |\mathbf{x} - \mathbf{y}^{\#}|/c_\infty)}{4\pi|\mathbf{x} - \mathbf{y}^{\#}|} = G_o + G_o^{\#} \qquad (15.4.4)$$

where $\mathbf{y}^{\#} = (y_1, -y_2, y_3)$ identifies locations of the image sources in the wall. As indicated, this function is merely the sum of the free-field Green's function for the source and image points. Differentiation of Eq. (15.4.4) shows that on the wall at $y_2 = 0$, $\partial G_T/\partial y_2 = 0$, $\partial G_T/\partial y_1 = 2\partial G_o/\partial y_1$, and $\partial G_T/\partial y_3 = 2\partial G_o/\partial y_3$. So, substituting these relations into Eq. (15.4.2) we lose the pressure term and obtain

$$p'(\mathbf{x}, t) = 2 \int_{-T}^{T} \int_{S} \mu \left( \delta_{i1} \frac{\partial v_1}{\partial y_2} + \delta_{i3} \frac{\partial v_3}{\partial y_2} \right) \frac{\partial G_o}{\partial y_i} dS(\mathbf{y}) d\tau$$
$$+ \int_{-T}^{T} \int_{V} \left( \frac{\partial^2 G_o}{\partial y_i \partial y_j} \right) T_{ij}(\mathbf{y}, \tau) dV(\mathbf{y}) d\tau + \int_{-T}^{T} \int_{V} \left( \frac{\partial^2 G_o^{\#}}{\partial y_i \partial y_j} \right) T_{ij}(\mathbf{y}, \tau) dV(\mathbf{y}) d\tau$$

$$(15.4.5)$$

By comparing Eqs. (15.4.3), (15.4.5) we see that effect of the surface pressure term is equivalent to a doubling of the viscous shear stress dipole and the addition of a quadrupole term representing the noise made by the image of the boundary-layer turbulence in the wall. The shear dipole is expected to be negligibly weak except perhaps at low Reynolds numbers, and even then it has been argued [9] that this term is not a true source but acts as a modifier to the propagation of acoustic waves adjacent to the surface. The quadrupole term is, of course, small in any low Mach number flow.

The relative silence of turbulent boundary layers is not maintained when the wall is rough (Fig. 15.10). In this case the unsteady boundary-layer pressure field is experienced on an uneven surface so that individual features of the roughness (which we will refer to as *roughness elements*) are subjected to an unsteady pressure force tangential to the wall that acts as a dipole source, radiating sound back out into the flow. This source is enhanced by the effect that the roughness has on the boundary layer itself, intensifying the turbulence and the associated surface pressure fluctuations. To analyze this situation, we begin once more with Eq. (15.4.1) but without the viscous and Lighthill stress tensor terms

$$p'(\mathbf{x}, t) = \int\limits_{-T}^{T} \int\limits_{S} p \frac{\partial G}{\partial y_i} n_i dS(\mathbf{y}) d\tau \tag{15.4.6}$$

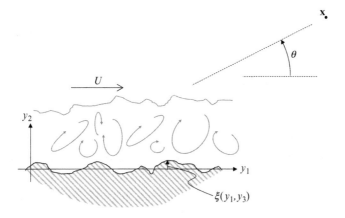

**Fig. 15.10** Nomenclature for noise radiated from flow over a rough wall.

As shown in Fig. 15.10, we define our coordinates so that the mean plane of the wall is at $y_2 = 0$ and with the surface shape defined by the function $\xi$ so that on the surface $y_2 = \xi(y_1, y_3)$. Note that we are assuming that $\xi$ is single valued function, i.e., the surface has no overhanging elements. The apparently complex task of evaluating the derivative of the Green's function on the rough surface is simplified by using once more the tailored Green's function of Eq. (15.4.4). Since $\partial G / \partial y_2 = 0$ at $y_2 = 0$ we can write the derivatives of this Green's function at the rough surface in terms of the Taylor series expansions

$$\begin{aligned}
\left[\frac{\partial G_T}{\partial y_1}\right]_{y_2=\xi} &= \left[\frac{\partial G_T}{\partial y_1}\right]_{y_2=0} + \frac{\xi^2}{2}\left[\frac{\partial^3 G_T}{\partial y_1 \partial y_2^2}\right]_{y_2=0} + O(\xi^3) \\
\left[\frac{\partial G_T}{\partial y_2}\right]_{y_2=\xi} &= \xi\left[\frac{\partial^2 G_T}{\partial y_2^2}\right]_{y_2=0} + O(\xi^3) \\
\left[\frac{\partial G_T}{\partial y_3}\right]_{y_2=\xi} &= \left[\frac{\partial G_T}{\partial y_3}\right]_{y_2=0} + \frac{\xi^2}{2}\left[\frac{\partial^3 G_T}{\partial y_3 \partial y_2^2}\right]_{y_2=0} + O(\xi^3)
\end{aligned} \tag{15.4.7}$$

Substituting Eq. (15.4.4) into Eq. (15.4.7), and ignoring all but the leading order terms, we obtain

$$\left[\frac{\partial G_T}{\partial y_1}\right]_{y_2=\xi} \approx \frac{x_1}{2\pi|\mathbf{x}|^2 c_\infty}\delta'\left(t-\tau-\frac{|\mathbf{x}|}{c_\infty}+\frac{x_1 y_1+x_3 y_3}{|\mathbf{x}|c_\infty}\right)$$

$$\left[\frac{\partial G_T}{\partial y_2}\right]_{y_2=\xi} \approx \frac{-x_2^2 \xi}{2\pi|\mathbf{x}|^3 c_\infty^2}\delta''\left(t-\tau-\frac{|\mathbf{x}|}{c_\infty}+\frac{x_1 y_1+x_3 y_3}{|\mathbf{x}|c_\infty}\right) \qquad (15.4.8)$$

$$\left[\frac{\partial G_T}{\partial y_3}\right]_{y_2=\xi} \approx \frac{x_1}{2\pi|\mathbf{x}|^2 c_\infty}\delta'\left(t-\tau-\frac{|\mathbf{x}|}{c_\infty}+\frac{x_1 y_1+x_3 y_3}{|\mathbf{x}|c_\infty}\right)$$

Where these results have been simplified using far-field approximations. For example, the result for $\partial G_T/\partial y_1$ without simplification is

$$\left[\frac{\partial G_T}{\partial y_1}\right]_{y_2=\xi} = \frac{x_1-y_1}{2\pi|\mathbf{x}-\mathbf{y}|^2 c_\infty}\delta'\left(t-\tau-\frac{|\mathbf{x}-\mathbf{y}|}{c_\infty}\right)+\frac{x_1-y_1}{2\pi|\mathbf{x}-\mathbf{y}|^3}\delta\left(t-\tau-\frac{|\mathbf{x}-\mathbf{y}|}{c_\infty}\right)$$

$$(15.4.9)$$

In the far-field $\mathbf{x}$ is large (using an origin close to the source), so we can neglect the second term compared to the first, ignore $y_1$ relative to $x_1$ in the remaining numerator, and replace $|\mathbf{x}-\mathbf{y}|^2$ in the remaining denominator by $|\mathbf{x}|^2$ and $|\mathbf{x}-\mathbf{y}|$ in the argument of $\delta'$ by $|\mathbf{x}|-x_i y_i/|\mathbf{x}|$, thereby giving the result in Eq. (15.4.8). Substituting Eq. (15.4.8) into Eq. (15.4.6) we obtain

$$p'(\mathbf{x},t) = \frac{1}{2\pi|\mathbf{x}|^2 c_\infty}\int_S\int_{-T}^{T} x_1\delta'(\tau^*-\tau)pn_1 - \frac{x_2^2 \xi}{|\mathbf{x}|c_\infty}\delta''(\tau^*-\tau)pn_2 + x_3\delta'(\tau^*-\tau)pn_3 d\tau dS(\mathbf{y})$$

$$(15.4.10)$$

where

$$\tau^* = t-\frac{|\mathbf{x}-\mathbf{y}|}{c_\infty}\approx t-\frac{|\mathbf{x}|}{c_\infty}+\frac{x_1 y_1+x_3 y_3}{|\mathbf{x}|c_\infty}$$

Performing the time integration then yields

$$p'(\mathbf{x},t) = \frac{1}{2\pi|\mathbf{x}|^2 c_\infty}\int_S (x_1 n_1+x_3 n_3)\left[\frac{\partial p}{\partial \tau}\right]_{\tau=\tau^*} - \frac{x_2^2 \xi n_2}{|\mathbf{x}|c_\infty}\left[\frac{\partial^2 p}{\partial \tau^2}\right]_{\tau=\tau^*} dS(\mathbf{y}) \qquad (15.4.11)$$

where $dS$ and $n_i$ represent the surface area and unit outward normal on the uneven rough wall. As shown in Fig. 15.11 these can be expressed in terms of the roughness height function $\xi$ as

$$n_i dS(\mathbf{y}) = \left( -\frac{\partial \xi}{\partial y_1}, 1, -\frac{\partial \xi}{\partial y_3} \right) dy_1 dy_3 \tag{15.4.12}$$

and so,

$$p'(\mathbf{x}, t) = \frac{-1}{2\pi |\mathbf{x}|^2 c_\infty} \int_S \left( x_1 \frac{\partial \xi}{\partial y_1} + x_3 \frac{\partial \xi}{\partial y_3} \right) \left[ \frac{\partial p}{\partial \tau} \right]_{\tau = \tau^*} + \frac{x_2^2 \xi}{|\mathbf{x}| c_\infty} \left[ \frac{\partial^2 p}{\partial \tau^2} \right]_{\tau = \tau^*} dy_1 dy_3$$

$$\tag{15.4.13}$$

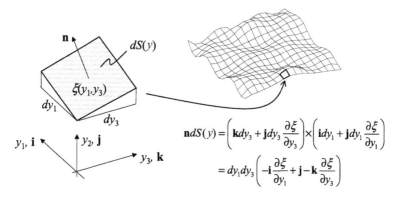

**Fig. 15.11** Expression of the rough surface area element $ndS(y)$ in terms of the roughness height function $\xi$.

The two terms inside the integral are not of the same order. If we characterize the time-scale of the flow in terms of the turbulence lengthscale $L$ divided by the flow velocity $U$ and denote the characteristic height of the roughness as $h$, then the second term is smaller than the first by a factor of $ML_\xi/L$, where $L_\xi$ is the streamwise or spanwise scale of the roughness. The second derivative term thus scales as if it were a quadrupole, and we neglect it as we have already neglected the quadrupole turbulence sources. Taking the Fourier transform of the result with respect to observer time we obtain

$$\widetilde{p}(\mathbf{x}, \omega) = \frac{-1}{4\pi^2 |\mathbf{x}|^2 c_\infty} \int_S \left( x_1 \frac{\partial \xi}{\partial y_1} + x_3 \frac{\partial \xi}{\partial y_3} \right) \int_{-T}^{T} \left[ \frac{\partial p}{\partial \tau} \right]_{\tau = \tau^*} e^{i\omega t} dt dy_1 dy_3 \tag{15.4.14}$$

Changing the variable of time integration to $\tau^*$, and remembering that taking the time derivative of a variable is equivalent to multiplying its Fourier transform by $-i\omega$, we obtain

$$\tilde{p}(\mathbf{x}, \omega) = \frac{ike^{ik|\mathbf{x}|}}{2\pi|\mathbf{x}|^2} \int_S \left( x_1 \frac{\partial \xi}{\partial y_1} + x_3 \frac{\partial \xi}{\partial y_3} \right) \tilde{p}(\mathbf{y}, \omega) e^{-ik(x_1y_1 + x_3y_3)/|\mathbf{x}|} dy_1 dy_3 \qquad (15.4.15)$$

where $k$ is the acoustic wavenumber $\omega/c_\infty$. From the definition of the inverse spatial Fourier transform, we can express $\tilde{p}(\mathbf{y}, \omega)$ in terms of the wavenumber frequency transform of the wall pressure $\tilde{\tilde{p}}(k_1, k_3, \omega)$ as

$$\tilde{p}(\mathbf{y}, \omega) = \int_{-\infty}^{\infty} \int_{-\infty}^{\infty} \tilde{\tilde{p}}(k_1, k_3, \omega) e^{i(k_1y_1 + k_3y_3)} dk_1 dk_3 \qquad (15.4.16)$$

giving

$$\tilde{p}(\mathbf{x}, \omega) = \frac{ike^{ik|\mathbf{x}|}}{2\pi|\mathbf{x}|^2} \int_{-\infty}^{\infty} \int_{-\infty}^{\infty} \tilde{\tilde{p}}(k_1, k_3, \omega) \int_S \left( x_1 \frac{\partial \xi}{\partial y_1} + x_3 \frac{\partial \xi}{\partial y_3} \right) e^{-i\mathbf{\kappa} \cdot \mathbf{y}} dy_1 dy_3 dk_1 dk_3$$

$$(15.4.17)$$

where

$$\mathbf{\kappa} = \left( -k_1 + \frac{kx_1}{|\mathbf{x}|}, 0, -k_3 + \frac{kx_3}{|\mathbf{x}|} \right)$$

The inner integral of Eq. (15.4.17) has the form of a wavenumber transform. Indeed, if we introduce the wavenumber transform of the surface height,

$$\tilde{\tilde{\xi}}(\kappa_1, \kappa_3) = \frac{1}{(2\pi)^2} \int_S \xi(y_1, y_3) e^{-i\mathbf{\kappa} \cdot \mathbf{y}} dy_1 dy_3 \qquad (15.4.18)$$

and note that $i\kappa_1 \tilde{\tilde{\xi}}$ and $i\kappa_3 \tilde{\tilde{\xi}}$ are the wavenumber transforms of the surface slopes in the $y_1$ and $y_3$ directions, then Eq. (15.4.17) can be reduced to

$$\tilde{p}(\mathbf{x}, \omega) = \frac{2\pi ike^{ik|\mathbf{x}|}}{|\mathbf{x}|^2} \int_{-\infty}^{\infty} \int_{-\infty}^{\infty} \tilde{\tilde{p}}(k_1, k_3, \omega)(i\kappa_1 x_1 + i\kappa_3 x_3) \tilde{\tilde{\xi}}(\kappa_1, \kappa_3) dk_1 dk_3 \qquad (15.4.19)$$

This is the equation for the roughness noise. If the features of the roughness are acoustically compact in all directions so that $|k_1| \ll |k|$ and $|k_3| \ll |k|$, then $\kappa_1$ and $\kappa_1$ are indistinguishable from the negative of $k_1$ and $k_3$.

For Eq. (15.4.19) to be practically useful, we need to express the far-field sound in terms of its frequency spectrum

$$S_{pp}(\mathbf{x}, \omega) = \frac{\pi}{T} E[\widetilde{p}^*(\mathbf{x}, \omega)\widetilde{p}(\mathbf{x}, \omega)]$$

$$= \frac{4\pi^2 k^2}{|\mathbf{x}|^4} \int\limits_{-\infty}^{\infty} \int\limits_{-\infty}^{\infty} \frac{\pi}{T} E\left[\widetilde{\widetilde{p}}^*(k_1, k_3, \omega)\widetilde{\widetilde{p}}(k_1', k_3', \omega)\right]$$

$$\times (\kappa_1 x_1 + \kappa_3 x_3)(\kappa_1' x_1 + \kappa_3' x_3)\widetilde{\widetilde{\xi}}^*(\kappa_1, \kappa_3)\widetilde{\widetilde{\xi}}(\kappa_1', \kappa_3')dk_1 dk_3 dk_1' dk_3' \quad (15.4.20)$$

Now, if we assume that the portion of the fluctuating surface pressure field that is responsible for the roughness noise is homogeneous then we can write the expected value in the integral in terms of the wavenumber spectrum of the surface pressure $\Phi_{pp}$. Note that this assumption excludes pressure fluctuations directly associated with the individual flows around roughness elements, such as due to the local effect of eddy shedding from an element or the interstitial flows between adjacent elements. From Eq. (8.4.37), we have

$$\frac{\pi}{T} E\left[\widetilde{\widetilde{p}}^*(k_1, k_3, \omega)\widetilde{\widetilde{p}}(k_1', k_3', \omega)\right] = \Phi_{pp}(k_1, k_3, \omega)\delta(k_1 - k_1')\delta(k_3 - k_3')$$

and so,

$$S_{pp}(\mathbf{x}, \omega) = \frac{4\pi^2 k^2}{|\mathbf{x}|^2} \int\limits_{-\infty}^{\infty} \int\limits_{-\infty}^{\infty} \Phi_{pp}(k_1, k_3, \omega)\left(\frac{\kappa_1 x_1 + \kappa_3 x_3}{|\mathbf{x}|}\right)^2 \left|\widetilde{\widetilde{\xi}}(\kappa_1, \kappa_3)\right|^2 dk_1 dk_3$$

$$(15.4.21)$$

Eq. (15.4.21) is an expression for the sound field that assumes that we know the precise roughness geometry. In many situations it is more likely that we will only have statistical knowledge of the roughness, and for this situation the expected value operator on the right-hand side of Eq. (15.4.20) needs to encompass the surface coordinate terms as well as the pressure terms. Assuming there is direct correlation between the details of the pressure field and those of the rough surface (which is inevitable if the homogeneity assumption is valid) then the result of this derivation becomes

$$S_{pp}(\mathbf{x}, \omega) = \frac{k^2 h^2 \Sigma}{|\mathbf{x}|^2} \int\limits_{-\infty}^{\infty} \int\limits_{-\infty}^{\infty} \Phi_{pp}(k_1, k_3, \omega)\Gamma(\mathbf{\kappa}, \mathbf{x})dk_1 dk_3 \quad (15.4.22)$$

where $\Sigma$ is the area of the surface, projected onto the $y_2 = 0$ plane, $h$ is the RMS roughness, and

$$\Gamma(\mathbf{\kappa}, \mathbf{x}) = \left(\frac{\kappa_1 x_1 + \kappa_3 x_3}{|\mathbf{x}|}\right)^2 \left\{\frac{4\pi^2}{\Sigma h^2} E\left[\widetilde{\widetilde{\xi}}^*(\kappa_1, \kappa_3)\widetilde{\widetilde{\xi}}(\kappa_1, \kappa_3)\right]\right\} \quad (15.4.23)$$

Note that the term inside the curly braces is the wavenumber spectrum of the surface height, normalized on its mean square. We see that both Eqs. (15.4.21), (15.4.22) cleanly separate the flow-dependent acoustic source term, represented by the surface pressure wavenumber spectrum, from the surface geometry term representing the spectrum of the surface slope in the direction of the observer.

These equations provide perfectly practical expressions for roughness noise prediction if we know the surface geometry, or its typical statistical form, and the boundary-layer parameters needed to define one of the wavenumber frequency spectrum models of the surface pressure, such as Eq. (9.2.34) or (9.2.38). Given the fact that we have neglected sound generated by the individual flows around roughness elements, we might expect these equations only to be accurate when the roughness comparable to the viscous scales of the boundary layer and such flows are suppressed. However, this is not the case and accurate predictions have been made in flows where the roughness elements are at least as large as the displacement thickness [10]. The implication is that the scattering of the homogeneous components of the surface pressure spectrum is a much greater contributor to roughness noise than the self-noise of individual elements. For a broadband rough surface, the predicted far-field sound has been found to be only weakly dependent on the form of the wavenumber frequency spectrum [10], so the exact choice or accuracy of this model is not likely to be critical in most circumstances.

As regards the scaling implicit in Eq. (15.4.22), we expect that $k \sim M/L$, that the wavenumber spectrum of surface pressure will vary approximately as $\rho_o^2 U^4 S(\omega) L^2$, where $L$ represents the scale of the turbulence, and that the wavenumbers $k_1$ and $k_3$ (and $\kappa_1$ and $\kappa_3$) representing flow scales will vary as $1/L$. We thus have, for an observer above the mid-span of the surface ($x_3 = 0$),

$$S_{pp}(x_1, x_2, 0, \omega) \sim \frac{\cos^2\theta \rho_o^2 U^4 M^2 h^2 \Sigma}{|\mathbf{x}|^2 L^2} S(\omega) \tag{15.4.24}$$

where we are taking $\kappa_i \approx -k_i$ and introducing $\cos\theta = x_1/|\mathbf{x}|$ and $S(\omega)$, with units of seconds, which gives the normalized spectral shape. This shows that roughness noise has classical dipole scaling and directivity, the dipole being oriented parallel to the wall. The sound increases in proportion to the area of the surface and the square of the height of the roughness elements. Dividing Eq. (15.4.24) by Eq. (15.2.13) for trailing edge noise and ignoring the differences in spectral shape and directivity gives

$$\frac{S_{pp}\big|_{\text{roughness}}}{S_{pp}\big|_{\text{trailing edge}}} \sim \frac{M h^2 \Sigma}{L^3 L_{TE}} \tag{15.4.25}$$

Since $h/L$ is likely to be small, we see that at low Mach number roughness noise will be the dominant self-noise source only when the surface area of the roughness is very large compared to the product of the turbulence scale and the total edge length. This happens with undersea vehicles, for example, which can have large surface area but small lifting surfaces. Roughness noise can also be important in determining the acoustic noise floor of wind tunnels [11], where an observer in a hard-wall test section finds themselves far from any edge sources, but in the direct line of streamwise roughness noise dipoles associated with the walls of the diffuser and contraction.

The above scaling argument ignores factors controlling the shape of the noise spectrum. From Eq. (15.4.22) we see that the observer hears a sound spectrum whose shape is determined by wavenumber filtering of the surface pressure fluctuation field, the filter being the surface slope spectrum represented by $\Gamma$. For rough surfaces with random roughness elements defined by near-vertical sides it can be shown [12] that the slope spectrum is wavenumber white, i.e., $\Gamma$ is simply a constant. In this case, Eq. (15.4.22) becomes dependent on the unweighted wavenumber spectrum integrated with respect to $k_1$ and $k_3$, which is simply equal to the wall pressure frequency spectrum, i.e.,

$$S_{pp}(\mathbf{x}, \omega) = \frac{k^2 h^2 \Sigma \Gamma}{|\mathbf{x}|^2} \int\limits_{-\infty}^{\infty} \int\limits_{-\infty}^{\infty} \Phi_{pp}(k_1, k_3, \omega) dk_1 dk_3 = \frac{k^2 h^2 \Sigma \Gamma(\mathbf{x}) G_{pp}(\omega)}{2|\mathbf{x}|^2} \quad (15.4.26)$$

Thus the roughness noise spectrum normalized on the surface pressure frequency spectrum should simply be proportional to the frequency squared, the mean square roughness height, and the area of the rough surface, i.e.,

$$\frac{S_{pp}(\mathbf{x}, \omega)}{G_{pp}(\omega)} = \frac{\omega^2 h^2 \Sigma \Gamma(\mathbf{x})}{2c_\infty^2 |\mathbf{x}|^2} \quad (15.4.27)$$

Fig. 15.12 shows this type of behavior on roughness noise measurements made over a series of sandpaper surfaces of different grit size [13]. The smallest grit size of 100 implies roughness elements with heights comparable to the viscous sublayer thickness, whereas the largest, of 20, indicates roughness elements that penetrate into the log layer and are comparable to the displacement thickness. The sound spectra are normalized on the surface pressure frequency spectrum and the square of the roughness height and convincingly follow a line proportional to $\omega^2$, except at high frequencies for the largest roughness sizes.

Sandpaper and other stochastic surfaces do not strictly have wavenumber white surface slope spectra, but their slope spectra are expected to have a broad maximum. In the vicinity of this maximum not only $\Gamma$ is approximately constant, but the surface pressure wavenumber spectrum is most heavily weighted by $\Gamma$, making it probable that this portion of the integral in Eq. (15.4.22) will dominate what is heard in the far field. Thus Eq. (15.4.27) provides a simple means of estimating the roughness noise spectrum generated by many random surfaces.

The fact that the rough surface acts as a filter on the surface pressure field in determining the far-field sound raises the intriguing possibility of using such a surface as a

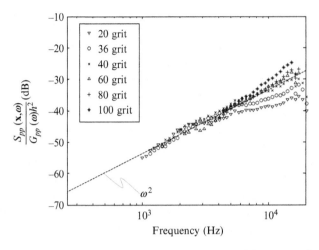

**Fig. 15.12** Narrowband spectra of the roughness noise generated by rectangular sandpaper surfaces exposed to a wall-jet boundary layer in air. Boundary layer edge velocity and thickness are approximately 14 m/s and 20 mm, respectively.
Data from Ref. [13].

probe. That is, designing a deterministic surface with a shape chosen to reveal a specific part of the surface pressure wavenumber frequency spectrum in the noise it radiates. For example, consider a wall with sinusoidal ribs of wavenumber $\mathbf{k}^{(w)} = \left( k_1^{(w)}, 0, k_3^{(w)} \right)$. The shape of the wall is

$$\xi(y_1, y_3) = h_o \cos\left( k_1^{(w)} y_1 + k_3^{(w)} y_3 \right) \tag{15.4.28}$$

which has the wavenumber transform

$$\widetilde{\widetilde{\xi}}(k_1, k_3) = \frac{h_o}{2} \left( \delta\left( k_1^{(w)} - k_1 \right) \delta\left( k_3^{(w)} - k_3 \right) + \delta\left( k_1^{(w)} + k_1 \right) \delta\left( k_3^{(w)} + k_3 \right) \right) \tag{15.4.29}$$

Substituting this into Eq. (15.4.19) and assuming that the surface is acoustitally compact, so that $\boldsymbol{\kappa} = -\mathbf{k}$, we obtain

$$\widetilde{p}(\mathbf{x}, \omega) = \frac{h_o \pi i k e^{ik|\mathbf{x}|}}{|\mathbf{x}|^2} \int\limits_{-\infty}^{\infty} \int\limits_{-\infty}^{\infty} \widetilde{\widetilde{p}}(k_1, k_3, \omega)(ik_1 x_1 + ik_3 x_3)$$

$$\left( \delta\left( k_1^{(w)} - k_1 \right) \delta\left( k_3^{(w)} - k_3 \right) + \delta\left( k_1^{(w)} + k_1 \right) \delta\left( k_3^{(w)} + k_3 \right) \right) dk_1 dk_3$$

$$= + \frac{h_o \pi k e^{ik|\mathbf{x}|}}{|\mathbf{x}|^2} \mathbf{k}^{(w)} \cdot \mathbf{x} \left[ \widetilde{\widetilde{p}}\left( k_1^{(w)}, k_3^{(w)}, \omega \right) + \widetilde{\widetilde{p}}\left( -k_1^{(w)}, -k_3^{(w)}, \omega \right) \right] \tag{15.4.30}$$

So that the far-field sound spectrum is

$$S_{pp}(\mathbf{x}, \omega) = \frac{\pi}{T} E[\tilde{p}^*(\mathbf{x}, \omega)\tilde{p}(\mathbf{x}, \omega)]$$

$$= \left(\frac{h_o \pi k}{|\mathbf{x}|^2}\right)^2 \left(\mathbf{k}^{(w)} \cdot \mathbf{x}\right)^2 \left[\Phi_{pp}\left(k_1^{(w)}, k_3^{(w)}, \omega\right) + \Phi_{pp}\left(-k_1^{(w)}, -k_3^{(w)}, \omega\right)\right]$$

(15.4.31)

Note that if our coordinate system is oriented so that $k_1$ is measured in the flow direction, then the wavenumber frequency spectrum at negative $k_1^{(w)}$ will be negligible since this implies upstream traveling turbulence, and we can truncate this result to

$$S_{pp}(\mathbf{x}, \omega) = \left(\frac{h_o \pi k}{|\mathbf{x}|^2}\right)^2 \left(\mathbf{k}^{(w)} \cdot \mathbf{x}\right)^2 \Phi_{pp}\left(k_1^{(w)}, k_3^{(w)}, \omega\right)$$

(15.4.32)

So, a sinusoidal surface of wavenumber $\mathbf{k}^{(w)}$ radiates only the portion of the wavenumber frequency spectrum of the wall pressure at that wavenumber. The shape of the far-field sound spectrum gives the frequency dependence or the surface pressure spectrum at that wavenumber, multiplied by $\omega^2$. By turning a sinusoidal surface to different angles, different parts of the wavenumber frequency spectrum can be revealed. Fig. 15.13 illustrates just how this works. The wavenumber frequency spectrum of the wall pressure is represented in terms of contour surfaces drawn using the Chase model of Eq. (9.2.34). We are considering the sound radiated by a circular patch of sinusoidal ridges, with ridges oriented perpendicular to the direction $\alpha_w$ where $\cos \alpha_w = k_1^{(w)}/|\mathbf{k}^{(w)}|$. This surface will radiate sound that reveals the shape of the

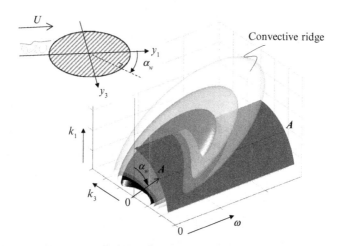

**Fig. 15.13** Schematic showing the wavenumber frequency spectrum of the wall pressure, and the cut $AA$ radiated by a sinusoidal surfaces with ridges oriented at angle $\alpha_w$. Adapted from Ref. [14].

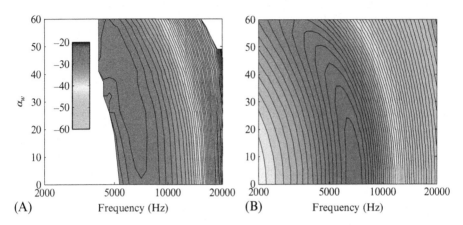

**Fig. 15.14** Contours of the wavenumber frequency spectrum of surface pressure fluctuations for a wall-jet boundary layer in air along the cylindrical cut illustrated in Fig. 15.13. Boundary layer edge velocity and thickness are 22 m/s and 17 mm, respectively. (A) Measurements. (B) Prediction using the Chase model of Eq. (9.2.34).
Adapted from Ref. [14].

wavenumber frequency spectrum along the line *AA*. By rotating the surface to vary $\alpha_w$ we can reveal an entire cylindrical cut through the wavenumber frequency spectrum, as shown. Fig. 15.14 shows results from just such a measurement used to reveal the wavenumber spectrum of a wall-jet boundary layer using a ridged surface with a 1.26-mm wavelength [14]. This is a far smaller wavelength (and thus higher wavenumber) that could be probed using an array of conventional surface microphones and is a much easier measurement, requiring only a single far-field microphone. The results (Fig. 15.14A) are plotted as a map, which reveals the convective ridge forming a near-vertical arc in the plot. This appears to be in fair agreement with the prediction of the map using the Chase model (Fig. 15.14B).

# References

[1] J.E. Ffowcs Williams, L.H. Hall, Aerodynamic sound generation by turbulent flow in the vicinity of a scattering half plane, J. Fluid Mech. 40 (1970) 657–670.

[2] M.S. Howe, Acoustics of Fluid–structure Interactions, Cambridge University Press, Cambridge, 1998.

[3] R.K. Amiet, Noise due to turbulent flow past a trailing edge, J. Sound Vib. 47 (1976) 387–393.

[4] R.K. Amiet, Effect of incident surface pressure field on noise due to turbulent flow past a trailing edge, J. Sound Vib. 57 (1978) 305–306.

[5] R.H. Schlinker, R.K. Amiet, Helicopter rotor trailing edge noise, NASA CR 3470, 1981.

[6] W.J. Devenport, R.A. Burdisso, A. Borgoltz, P.A. Ravetta, M.F. Barone, K.A. Brown, M. A. Morton, The Kevlar-walled anechoic wind tunnel, J. Sound Vib. 332 (2013) 3971–3991.

[7] T.F. Brooks, T.H. Hodgson, Trailing edge noise prediction from measured surface pressures, J. Sound Vib. 78 (1981) 69–117.

[8] T.F. Brooks, D.S. Pope, M.A. Marcolini, Airfoil self-noise and prediction, NASA RP 1218, 1989.

[9] M.S. Howe, The role of surface shear stress fluctuations in the generation of boundary layer noise, J. Sound Vib. 65 (1979) 159–164.

[10] W.N. Alexander, W.J. Devenport, S.A.L. Glegg, Predictions of sound from rough wall boundary layers. AIAA J. 51 (2013) 465–475, http://dx.doi.org/10.2514/1.j051840.

[11] E.G. Duell, J. Yen, J. Walter, S. Arnette, Boundary layer noise in aeroacoustics wind tunnels, in: 42nd AIAA Aerospace Sciences Meeting and Exhibit, Reno, Nevada, 2004.

[12] S.A.L. Glegg, W. Devenport, The far-field sound from rough wall boundary layers. Proc. R. Soc. A 465 (2009) 1717–1734, http://dx.doi.org/10.1098/rspa.2008.0318.

[13] W. Devenport, D. Grissom, W. Alexander, B. Smith, S. Glegg, Measurements of roughness noise. J. Sound Vib. 330 (2011) 4250–4273, http://dx.doi.org/10.1016/j.jsv.2011.03.017.

[14] W. Devenport, E. Wahl, S. Glegg, W. Alexander, D. Grissom, Measuring surface pressure with far field acoustics. J. Sound Vib. 329 (2010) 3958–3971, http://dx.doi.org/10.1016/j.jsv.2010.03.012.

# Part 4

# Rotating blades and duct acoustics

# Open rotor noise

## 16

Up to this point in the text we have derived the basic equations of aero and hydroacoustics and have presented the analytical methods needed to solve them. We have also considered idealized problems such as leading and trailing edge noise. We will now turn our attention to specific problems that can be addressed using these theories. The first problem we will consider is the noise from propellers and rotors.

## 16.1  Tone and broadband noise

There are many applications in which rotor noise is a serious problem and a cause for concern. The commercial usage of propeller-driven aircraft is limited by high levels of cabin noise. Ship propellers are a major cause of both shipboard noise and sound radiation to the far field. Airboats are propeller driven and generate high noise levels when running at full speed. On a larger scale, wind turbines can be very noisy if designed incorrectly. Helicopter rotors have many of the same characteristics as propellers, and there are both military and civil applications in which the reduction of helicopter noise is important. In all these examples the same source mechanisms are found, but the dominant processes depend on the application. In this section we discuss the different mechanisms that cause noise from propellers and rotors, and then, in the following sections, we derive prediction methods for each source type.

Rotating blades emit two distinctly different types of acoustic signatures. The first is referred to as tone or harmonic noise and is caused by sources that repeat themselves exactly during each rotation. The second is broadband noise which is a random, nonperiodic signal caused by turbulent flow over the blades. Fig. 16.1 illustrates these signals and shows how they combine. In Fig. 16.1A the signature from a single blade is shown during the period of one revolution $T_p$. If the rotor has three blades, then this signature is repeated at the blade-passing frequency (BPF), and the sum (see Fig. 16.1B) is a signature that repeats itself with a period of $T_p/3$. A typical broadband signal is shown in Fig. 16.1C, and this is seen to have a quite different character, with no associated periodicity, but an envelope that varies periodically. The sum of the signal types is shown in Fig. 16.1D. Note that how the sum tends to hide the details of the broadband component.

The best way to determine the relative importance of tone and broadband noise is to consider the narrow-band frequency spectrum of the signal. The *Spectrum Level* is defined as the root mean square of the signal, which has passed through a frequency filter of bandwidth $\Delta f$ and centered on the frequency $f$. In rotor noise applications we always deal with harmonic signals, and it is important to use the Spectrum Level defined in this way rather than the Power Spectrum of the signal, which gives the

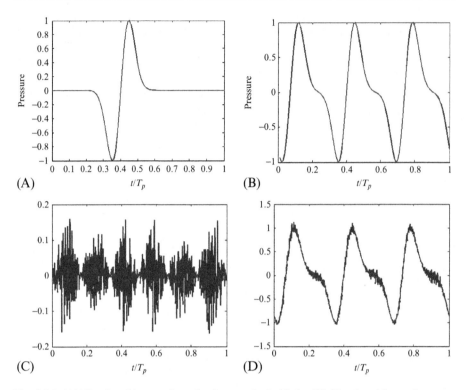

**Fig. 16.1** (A) The time history of a pulse from a single blade. (B) The time history from a three-bladed rotor. (C) The time history of broadband rotor noise. (D) The time history of rotor noise over one period.

signal energy per unit Hz. Fig. 16.2 shows a typical example of a rotor noise spectrum with a 3 Hz bandwidth for a three-bladed rotor at 600 rpm. The peaks define the tone noise and occur at the blade-passing frequencies, which in this case are multiples of 30 Hz. At higher frequencies the broadband random noise dominates the spectrum. This type of analysis is vital to the evaluation of rotor noise because it enables us to unambiguously distinguish between the tone and broadband noise, hence allows us to determine the most important noise source mechanisms.[1]

---

[1] If we increase the bandwidth of the analysis from the 3 Hz specified above, the Spectrum Level of the broadband noise will also increase because more broadband energy is passing through the filter, but the tone level will remain the same. In contrast, if we had used a power spectrum analysis the broadband level would be unaltered by changing the analysis bandwidth because the energy per Hz remains the same. However, the tone levels in a power spectrum are reduced by increasing the analysis bandwidth, and this is why it is important to evaluate rotor noise using the spectrum level, and to specify the analysis bandwidth when reporting results.

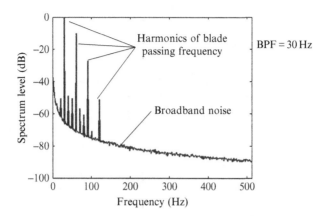

**Fig. 16.2** The spectrum of rotor noise showing harmonics at blade-passing frequency and broadband noise.

The primary sources of tone noise depend on the rotor tip speed, and the flow conditions in which the rotor is operating. Our understanding of rotor noise is based on the Ffowcs-Williams and Hawkings equation given by Eq. (5.2.13):

$$\rho'(\mathbf{x}, t)c_{\infty}^2 = \frac{\partial^2}{\partial x_i \partial x_j} \int_{V_o} \left[ \frac{T_{ij}}{4\pi r |1 - M_r|} \right]_{\tau=\tau^*} dV(\mathbf{z})$$

$$- \frac{\partial}{\partial x_i} \int_{S_o} \left[ \frac{p_{ij} n_j}{4\pi r |1 - M_r|} \right]_{\tau=\tau^*} dS(\mathbf{z}) + \frac{\partial}{\partial t} \int_{S_o} \left[ \frac{\rho_{\infty} V_j n_j}{4\pi r |1 - M_r|} \right]_{\tau=\tau^*} dS(\mathbf{z})$$

$$(5.2.13)$$

At low speeds the loading noise, given by the second term in Eq. (5.2.13), is usually the dominant source of sound. This indicates that the steady and unsteady pressure on the blade surface is the basis for the radiated sound. There are many effects that can influence the blade loading. If the propeller is operating with a completely clean inflow (uniform flow with no turbulence), which is rarely the case, then the blade loading is steady in blade-based coordinates, but the component of the force in the direction of the observer varies as the blade rotates. For example, consider the sound radiated out to the sides of the propeller in the plane of the rotor, Fig. 16.3. A far-field observer close to that plane "sees" a blade drag force that continuously changes direction, and so its component in the direction of the observer varies with time and a sound wave is generated. The same is true, but to a lesser extent, for the thrust force. The amount of load variation that is "seen" by the observer is obviously very dependent on the observer location, and the sound field is therefore very directional.

**Fig. 16.3** The lift and drag forces on a propeller.

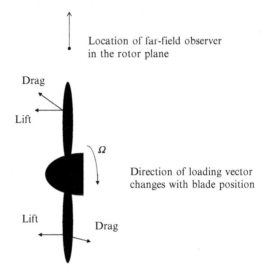

The sound caused by the time variation of the steady loading applies to all propellers, but it is a relatively weak source of sound compared to the unsteady loading. Most propellers operate in a nonuniform distorted inflow, and so their angle of attack varies continuously as they rotate as shown in Fig. 16.4. (For helicopter rotors this is a necessary design feature for level flight.) Smoothly varying changes of angle of attack are usually not very important, but when the blade encounters a sudden velocity deficit in the flow then the angle of attack change causes a rapid change in blade loading. The discussion of Eq. (5.2.13) showed that in the acoustic far field we can approximate $\partial/\partial x_i \sim (-x_i/|\mathbf{x}|c_\infty)\partial/\partial t =\sim (-x_i/(1-M_r)|\mathbf{x}|c_\infty)\partial/\partial\tau$, which shows that it is the time variation of the loading that generates sound. Consequently, a blade encountering a velocity deficit that causes a rapid change in loading can be a very efficient source of sound. A classic example of this is a wind turbine which can be designed so that the blades operate either upwind or downwind of the tower (Figs. 16.5 and 16.6). In the downwind design the tower causes a significant velocity deficit that the blade moves though, and so a strong acoustic pulse is generated (Fig. 16.5). In contrast if the wind turbine is designed so that the tower is downstream of the blades then the blades never pass through the velocity deficit, and they only encounter a small velocity perturbation as they pass the tower (Fig. 16.6). The upwind design of wind turbine is therefore significantly quieter than the downwind design. This principle applies to any propeller and, wherever possible, mounting of the rotor so that inflow distortions are minimized and slowly varying will reduce noise.

A special case of unsteady loading noise is caused by blade vortex interactions (BVIs) in helicopter rotors (Fig. 16.7). During forward flight the tip vortices of a helicopter can be ingested into the rotor and, given the right conditions, the helicopter blades can pass near the core of the vortex, and this causes a local, very rapid, change in angle of attack and a sudden change in blade load. This interaction emits a loud "thumping" sound and is often the dominant cause for complaints about helicopter noise. If the operational conditions are changed so that the wake is ingested in a different way then this noise source is eliminated, but unfortunately this is not always possible for some maneuvers.

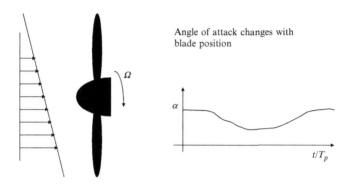

**Fig. 16.4** A propeller operating in a distorted inflow.

In addition to unsteady loading noise the third term of Eq. (5.2.13) shows that there is a contribution from the blade surface motion that is referred to as thickness noise. We will show later that this is only important at tip speeds with Mach numbers in excess of 0.7. However, for high-speed helicopters and transonic propellers this source can be important. The mechanism for this source is the time varying displacement of fluid by the blade volume as it rotates. To the fixed observer in the acoustic far field it is as if the blade volume changes as it rotates, and this apparent variation in volume causes a sound wave in the far field. The simplest way to reduce thickness noise is to reduce the blade volume near the blade tip. If the blade thickness is halved in the tip region then the thickness noise is reduced by 6 dB, which is not insignificant and can be an effective way to reduce the noise from high-speed rotors.

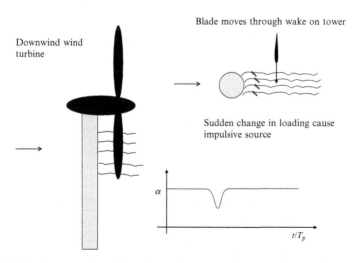

**Fig. 16.5** Illustration of a downwind wind turbine.

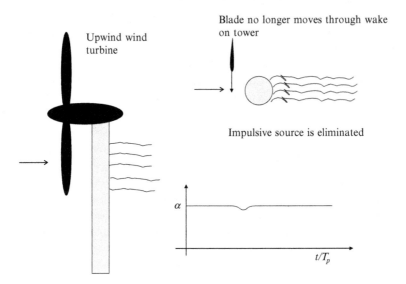

**Fig. 16.6** Illustration of an upwind wind turbine.

When the blade tip speed is transonic or supersonic then shock discontinuities can occur both on the blade surface and in the fluid surrounding the blade tips (Fig. 16.8). This is considered quadrupole noise because it is a source in the fluid volume as distinct from on the blade surface. From the observer's perspective, the shocks apparently change as the blade rotates, and so they generate sound. This mechanism can be just as important as thickness noise in some rotor designs. In general, the shocks are weaker if the blades are thinner, and so thinning of the blade tips is always advantageous for the reduction of transonic and supersonic rotor noise.

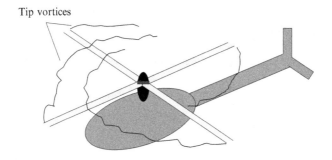

**Fig. 16.7** Illustration of blade vortex interactions.

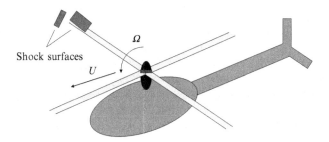

**Fig. 16.8** The shock surfaces which can produce quadrupole noise from rotating blades.

Broadband rotor noise is always caused by random variations in blade loading resulting from the interaction of the blades with turbulence. The turbulence is often generated upstream of the propeller and ingested into the rotor, but it can also be self-generated in the blade boundary layer or at the blade tips. An example where inflow turbulence is important is on ships where the propellers operate in a very disturbed flow underneath the hull. On helicopters, the trailing tip vortices that cause BVI noise can be surrounded by high levels of turbulence that generate broadband noise. This is referred to as blade wake interaction noise. The turbulence in the blade boundary layer does not generate much sound by itself, but when it passes the blade trailing edge the local boundary conditions change rapidly, and significant sound generation can occur (Fig. 16.9). As discussed in Chapter 15, this is trailing edge noise and is often considered as the most important mechanism of broadband noise generation in fans and propellers. Broadband noise from the turbulence at blade tips is not well understood at this time but may be important on low aspect ratio blades and should not be discounted as a possible noise source mechanism.

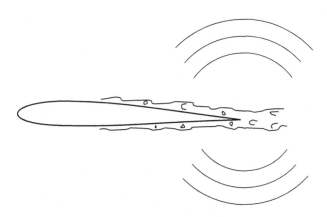

**Fig. 16.9** Trailing-edge noise from the blade boundary layer interacting with the trailing edge of a blade.

In the above we have summarized all the important source mechanisms for propeller and rotor noise. It is clear from this discussion that there are a number of competing mechanisms that are all important. In any particular application or set of operational conditions there may be several equally significant mechanisms or one may completely dominate. To determine the correct approach to sound reduction it is very important to be able to predict the noise levels from each of the sources described above. In the following sections we will discuss the prediction methodology for propeller and rotor noise and then, in Chapter 18, we will extend these ideas to ducted fans that have many of the same problems.

## 16.2   Time domain prediction methods for tone noise

### 16.2.1   Loading noise

Loading noise is caused by the time variation of the compressive stress tensor $p_{ij}$ on a blade surface as it rotates and may be predicted by the second term in the Ffowcs-Williams and Hawkings equation (5.2.13), which gives the radiated acoustic pressure as

$$(p'(\mathbf{x}, t))_{loading} = -\frac{\partial}{\partial x_i} \int_{S_o} \left[ \frac{p_{ij}n_j}{4\pi r |1 - M_r|} \right]_{\tau = \tau^*} dS(\mathbf{z}) \tag{16.2.1}$$

The surface integral in Eq. (16.2.1) should be carried out over the complete surface of the rotor blade, but in many instances it can be simplified to an integral over the blade planform. The planform is the projection of the rotor geometry into the rotor disc plane, as shown in Fig. 16.10. This approximation is valid if the blades are thin and the acoustic wavelength at the maximum frequency of interest is much larger than the blade thickness. We can then ignore the displacement of the upper surface from the lower surface and define the force per unit area applied to the fluid by the difference between the values of $p_{ij}n_j$ evaluated on the upper and lower surfaces for a given point on the blade planform (see Fig. 16.10). We define

$$f_i d\Sigma = \left( [p_{ij}n_j]_{upper} - [p_{ij}n_j]_{lower} \right) dS \quad \text{and} \quad |n_1| dS = d\Sigma$$

where $d\Sigma$ is the planform area in the disc plane (see Fig. 16.10) of the surface element $dS$. Substituting into Eq. (16.2.1) we obtain the thin airfoil approximation

$$(p'(\mathbf{x}, t))_{loading} = -\frac{\partial}{\partial x_i} \int_{\Sigma_o} \left[ \frac{f_i(\mathbf{z}, \tau)}{4\pi r |1 - M_r|} \right]_{\tau = \tau^*} d\Sigma(\mathbf{z}) \tag{16.2.2}$$

For blades with lean and sweep there may also be a considerable displacement of the mean camber line of the blade from the rotor disc plane. This displacement should be included in the retarded time calculation that takes place inside the integral in Eq. (16.2.2).

If the variation of the surface loading $f_i$ repeats itself during each blade rotation, then tone noise results. To calculate the acoustic field, we need to know the precise variation of $f_i$ over the complete blade surface, as a function of emission time $\tau$. However, there are some inherent difficulties with this requirement because the surface integral must be evaluated at a fixed observer time, and so the emission time $\tau$ will vary over the surface of the blade.

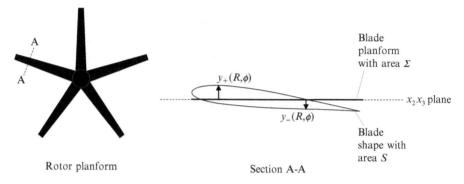

**Fig. 16.10** Rotor blade planform relative to the upper and lower surface of the blade.

To address these difficulties, we use the approach pioneered by Farassat [1] and shift the spatial derivatives to source time, as explained in Section 5.2. The result was given by Eq. (5.5.2) for a far-field observer. If we project the blade loading onto the surface planform as described above, we obtain

$$(p'(\mathbf{x},t))_{\text{loading}} \approx \frac{1}{c_\infty} \int_{\Sigma_o} \left[ \frac{x_i}{4\pi|\mathbf{x}|^2(1-M_r)^2} \left\{ \frac{\partial f_i}{\partial \tau} + \frac{f_i}{(1-M_r)} \frac{\partial M_r}{\partial \tau} \right\} \right]_{\tau=\tau^*} d\Sigma(\mathbf{z})$$

$$(16.2.3)$$

To evaluate Eq. (16.2.3) we need to calculate the location of each surface element defined in the integrand at a given observer time.

To proceed we need to define the coordinate system of the blade and the observer. There are several different choices for the coordinates, and care must be used in specifying the convention that will be used. In aircraft and marine propeller applications the convention [2] is to define coordinates $(x,y,z)$ with $x$ pointing in the direction of thrust. In helicopter applications the $z$ coordinate is chosen in the direction of thrust. In actuator disc theory the $x$ coordinate is chosen in the direction of flow through the propeller. In this chapter we choose the aircraft propeller convention with $x$ or $x_1$ in the direction of thrust as shown in Fig. 16.11. Furthermore, we need to choose the direction of blade rotation. A propeller rotating anticlockwise when viewed from upstream is defined as having "right-hand rotation." If it rotates in a clockwise direction it is said to have "left-hand rotation." We will choose right-hand rotation as shown in Fig. 16.11.

To evaluate the integrand in Eq. (16.2.3) we solve the equation $t - \tau - r(\tau)/c_\infty = 0$, so we can specify the azimuthal location $\phi = \phi_1 + \Omega\tau$ of each surface element as a function of observer time $t$, radius $R$, and azimuthal location $\phi_1$ in blade-based coordinates. For example, consider a rotor which is in the $y_2y_3$ plane at $\tau = 0$ and is moving with linear velocity $(U_o, 0, 0)$ as shown in Fig. 16.11. In blade-based coordinates the location of each surface element is $\mathbf{z} = (0, R\cos\phi_1, R\sin\phi_1)$, and in fixed coordinates the surface element is located at $\mathbf{y} = (U_o\tau,\ R\cos(\phi_1 + \Omega\tau),\ R\sin(\phi_1 + \Omega\tau))$. The retarded time equation is given by $t - \tau - |\mathbf{x} - \mathbf{y}(\tau)|/c_\infty = 0$, and to solve this equation we use an interpolation method. First we evaluate $t$ using a uniformly spaced set of points $\tau_m = m\Delta\tau$ in source time and then interpolate the results to obtain values of $\tau$ at fixed intervals in observer time $t_j = j\Delta t$. Fig. 16.12 shows a plot of source location vs observer time for a stationary propeller rotating with a tip Mach number of 0.8 for source located at 30%, 60%, and 90% of the tip radius $R_{tip}$ and an observer in the acoustic far field at $\mathbf{x} = (40R_{tip}, 30R_{tip}, 0)$. Notice how the curves cross at different observer times because the source at the outer radius is sometimes closer, and sometimes further from the observer as the blade rotates.

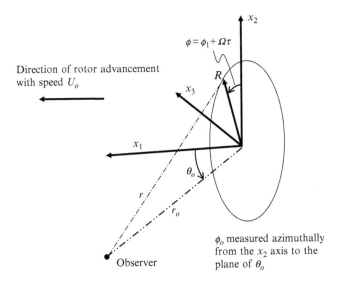

**Fig. 16.11** Coordinate system used for propeller and rotor noise calculations.

For blades with a subsonic tip speeds it is a relatively simple task to use this curve to compute the location of each blade element for uniformly spaced steps in observer time. However, for supersonic propellers this process becomes more difficult because the curve is no longer monotonically increasing, and we will return to this issue in Section 16.2.3.

The noise from steady loading is readily computed by defining the blade thrust $F_L$ and the blade drag $F_D$ for blade element of span $\Delta R$ located at $R$. (Note the sign convention requires that $f_i$ is the force per unit area applied to the fluid and is equal and opposite to the force per unit area applied to the blade.) For steady loading $f_1 d\Sigma = -F_L$ is constant, but the direction of the drag force varies with blade location, so $f_2 d\Sigma = F_D \sin(\phi_1 + \Omega\tau)$ and $f_3 d\Sigma = -F_D \cos(\phi_1 + \Omega\tau)$. Typically the drag force is about 10% of the thrust. The acoustic far field for each surface element can be computed using Eq. (16.2.3).

To illustrate a typical calculation, consider an observer at $\mathbf{x} = (40R_{tip}, 30R_{tip}, 0)$ for surface elements at 30%, 60%, and 90% of the blade radius. To evaluate Eq. (16.2.3) we need to define both the blade loading and the relative Mach number $M_r$, which is given, for $x_3 = 0$ and $U_o = 0$, by

$$M_r = -\frac{x_2 \Omega R \sin(\phi_1 + \Omega\tau)}{|\mathbf{x}| c_\infty} \tag{16.2.4}$$

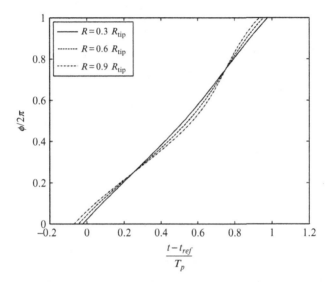

**Fig. 16.12** The blade location $\phi$ at radii of $0.3R_{tip}$, $0.6R_{tip}$, and $0.9R_{tip}$ as a function of observer time normalized by the period of rotation $T_p$ (to normalize this plot $t_{ref}$ was chosen as $|\mathbf{x}|/c_o$). The blade tip Mach number is $M_{tip} = 0.8$, and the observer location is $\mathbf{x} = (40R_{tip}, 30R_{tip}, 0)$.

We then obtain a nondimensional acoustic signal defined as

$$
\begin{aligned}
C_p(t) &= \frac{4\pi |\mathbf{x}| c_\infty (p'(\mathbf{x}, t))_{\text{loading}}}{\Omega F_L} \approx -\left[ \frac{x_1}{|\mathbf{x}|(1 - M_r)^3} \frac{\partial M_r}{\partial(\Omega\tau)} \right]_{\tau = \tau^*} \\
&+ \left[ \frac{x_2 (D/L)}{|\mathbf{x}|(1 - M_r)^2} \left\{ \cos(\phi_1 + \Omega\tau) + \frac{\sin(\phi_1 + \Omega\tau)}{(1 - M_r)} \frac{\partial M_r}{\partial(\Omega\tau)} \right\} \right]_{\tau = \tau^*}
\end{aligned} \tag{16.2.5}
$$

with

$$\frac{\partial M_r}{\partial(\Omega\tau)} = -\left\{ \frac{x_2 \Omega R}{|\mathbf{x}| c_\infty} \cos(\phi_1 + \Omega\tau) \right\}$$

There are several important features to this result. First notice how the loading closest to the tip generates the most significant part of the signature because the rate of change of $M_r$ is largest at this radius. Second note that if drag to lift ratio is small then the lift force dominates the calculation. In this case $F_D/F_L = 0.1$ and the peak of the pressure pulse is only marginally increased by including the drag term (Fig. 16.13).

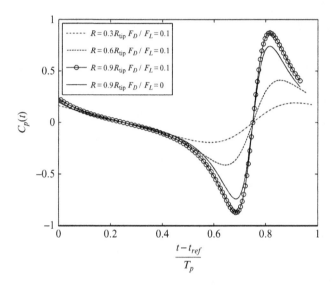

**Fig. 16.13** The acoustic signature from a rotating steady loading at radii of $0.3R_{tip}$, $0.6R_{tip}$, and $0.9R_{tip}$ as a function of observer time normalized by the period of rotation $T_p$ (to normalize this plot $t_{ref}$ was chosen as $|\mathbf{x}|/c_o$). The blade tip Mach number is $M_{tip} = 0.8$, and the observer location is $\mathbf{x} = (40R_{tip}, 30R_{tip}, 0)$.

The magnitude of the signature in this example is small because the thrust force is constant during the rotation of the blade. However, if it has a harmonic time dependence, so $f_1 d\Sigma = -F_L f(\tau)$, where $f(\tau) = \cos(m\Omega\tau)$, then Eq. (16.2.5) becomes

$$C_p(t) \approx -\left[\frac{x_1}{|\mathbf{x}|(1-M_r)^2}\left\{\frac{\partial f(\tau)}{\partial(\Omega\tau)} + \frac{f(\tau)}{(1-M_r)}\frac{\partial M_r}{\partial(\Omega\tau)}\right\}\right]_{\tau=\tau^*} \tag{16.2.6}$$

For large values of $m$ the rate of change of $f(\tau)$ is much greater than the rate of change of the relative Mach number $M_r$, and so the radiated levels increase dramatically as shown in Fig. 16.14 for the case when $m = 10$.

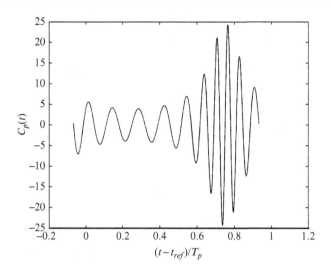

**Fig. 16.14** The time history of the acoustic signature from a propeller with a harmonic unsteady loading with a source frequency of 10 times the rotational frequency. Source located at $R = 0.9R_{tip}$ and $M_{tip} = 0.8$, and the observer location is $\mathbf{x} = (40R_{tip}, 30R_{tip}, 0)$.

The peak signal level in Fig. 16.14 is approximately 30 times larger than in Fig. 16.13, showing the importance of the unsteady loading in these calculations. Also note how the frequency of the signature varies during the blade rotation. During the first part of the cycle the blade is moving away from the observer, and the frequency is lower than the source frequency because of a Doppler frequency shift. In the second part of the cycle the blade is moving toward the observer, and the Doppler shift increases the observed frequency so that it is higher than the source frequency. This effect is dependent on the observer position and is more dramatic in the plane of the rotor than along the rotor axis. The importance of unsteady loading is even more significant when the blade encounters a sudden change in angle of attack caused by an inflow distortion, an encounter with a wake, or a vortex in a BVI. To illustrate this, consider a fluctuation in thrust on the blade segment of span $\Delta R$, located at $R$, given by

$$f_1 d\Sigma = F_L((\tau - \tau_o)/T_v) \exp\left(-(\tau - \tau_o)^2/T_v^2\right) \tag{16.2.7}$$

as illustrated in Fig. 16.15. The choice of $T_v = 0.02T_p$ gives a pulse which lasts for approximately one-tenth of the period of rotation in source time. The observed signature from this pulse at the observer location is given in Fig. 16.16 and is much shorter than the source signature as shown in Fig. 16.15. The source is at 90% of the blade span, and so the Doppler frequency shift can be large, and the observed signature becomes very impulsive, with a peak level that is much greater than for a harmonic variation in loading as shown in Fig. 16.14.

**Fig. 16.15** The time history of an impulsive load in source time.

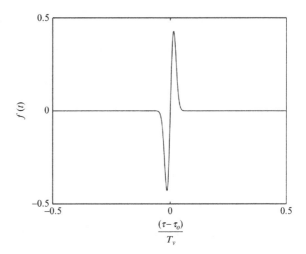

In the examples given above the signature from individual blade elements has been shown. To obtain a complete prediction of rotor noise the surface integral must be evaluated, and this can be achieved by numerical integration or summing the calculations for each blade element across the span at the correct retarded time. This is a relatively straightforward extension of the procedures described above providing that the blade surface loading is known as both a function of blade radius and azimuth. In low-frequency applications it is often reasonable to ignore the distribution of loading in the chordwise direction and replace $f_i$ with the sectional blade lift and drag per unit span for each blade radius. This assumes that the blade chord is small compared to the acoustic wavelength and is not always a good assumption, but it simplifies the calculations considerably and is a valid approach to obtain first estimates of blade noise signatures at low frequencies.

**Fig. 16.16** The observed acoustic signature from the impulsive source signature shown in Fig. 16.15. The location of the impulse occurs at 90% of the blade span. The blade tip Mach number is $M_{tip} = 0.8$, and the observer location is $\mathbf{x} = (40R_{tip}, 30R_{tip}, 0)$.

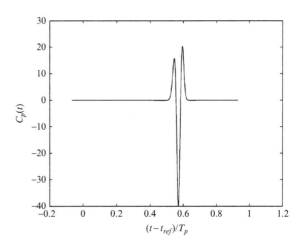

## 16.2.2   Thickness noise

Thickness noise is described by the third term in the Ffowcs-Williams and Hawkings equation (5.2.13) which gives the acoustic field as

$$(p'(\mathbf{x},t))_{\text{thickness}} = \frac{\partial}{\partial t} \int\limits_{S_o} \left[ \frac{\rho_\infty V_j n_j}{4\pi r |1 - M_r|} \right]_{\tau=\tau^*} dS(\mathbf{z}) \tag{16.2.8}$$

The evaluation of this source term for a propeller can be achieved using the method that was described in Section 16.2.1. The same procedure for evaluating the retarded time solution can be employed with the same limitations on numerical accuracy. The main difference is the appropriate evaluation of the source strength on the blade surface, which is easier in this case because it is defined exactly by the blade geometry.

As before we can replace the surface integral for thin blades by an integral over the blade planform in the rotor disk plane. We denote the upper and lower surface coordinates, displaced axially (in the $y_1$ direction from the rotor disk plane) as $y_+(R,\phi)$ and $y_-(R,\phi)$. As shown in Fig. 16.10 we define the blade volume in terms of the upper surface function $f_+(R,\phi) = y_1 - y_+(R,\phi) = 0$ and the lower surface function $f_-(R,\phi) = y_-(R,\phi) - y_1 = 0$. The blade thickness is then defined by $h = y_+ - y_-$. The normal to the blade surface is then given by $\mathbf{n} = \nabla f_+/|\nabla f_+|$ on the upper surface and $\mathbf{n} = \nabla f_-/|\nabla f_-|$ on the lower surface. It follows that on the upper surface

$$V_j n_j = \frac{1}{|\nabla f_+|} \left( V_1 - V_2 \frac{\partial y_+}{\partial y_2} - V_3 \frac{\partial y_+}{\partial y_3} \right) \tag{16.2.9}$$

Similarly, on the lower surface

$$V_j n_j = \frac{1}{|\nabla f_-|} \left( -V_1 + V_2 \frac{\partial y_-}{\partial y_2} + V_3 \frac{\partial y_-}{\partial y_3} \right) \tag{16.2.10}$$

Since $dS = d\Sigma/|n_1|$, $\nabla h = \nabla y_+ - \nabla y_-$, and $|n_1| = 1/|\nabla f_+|$ or $1/|\nabla f_-|$, we find that the contributions from the upper and lower surfaces can be combined to give

$$V_j n_j dS = -\left( V_2 \frac{\partial h}{\partial y_2} + V_3 \frac{\partial h}{\partial y_3} \right) d\Sigma = -\mathbf{V} \cdot \nabla h \, d\Sigma \tag{16.2.11}$$

Thickness noise can then be defined as an integral over the blade planform as

$$(p'(\mathbf{x},t))_{\text{thickness}} = -\frac{\partial}{\partial t} \int\limits_{\Sigma_o} \left[ \frac{\rho_\infty \mathbf{V} \cdot \nabla h}{4\pi r |1 - M_r|} \right]_{\tau=\tau^*} d\Sigma(\mathbf{z}) \tag{16.2.12}$$

This shows how the blade thickness contributes to the source strength, and the result is independent of the blade angle of attack or camber. Consequently, this source is defined as "thickness" noise.

Evaluation of the source term in this equation at the correct retarded time can be challenging numerically, but an elegant alternative [1] simplifies this considerably. To derive this alternate form, we return to the Ffowcs-Williams and Hawkings equation (5.2.4) and only retain the thickness noise terms on the right-hand side. For an impermeable surface for which $V_j n_j = v_j n_j$ we obtain

$$\left[ \frac{1}{c_\infty^2} \frac{\partial^2 (H_s p')}{\partial t^2} - \frac{\partial^2 (H_s p')}{\partial x_i^2} \right]_{\text{thickness}} = \frac{\partial}{\partial t} \left[ \rho_\infty V_j n_j \delta(f) |\nabla f| \right] \tag{16.2.13}$$

However, for a body that does not deform as it moves we have, using Eqs. (5.1.12), (5.1.3),

$$\frac{\partial}{\partial t}(1 - H_s(f)) = -\frac{\partial f}{\partial t} \delta(f) = V_j n_j \delta(f) |\nabla f| \tag{16.2.14}$$

so the wave equation for thickness noise sources can be given in an alternate form as

$$\left[ \frac{1}{c_\infty^2} \frac{\partial^2 (H_s p')}{\partial t^2} - \frac{\partial^2 (H_s p')}{\partial x_i^2} \right]_{\text{thickness}} = \frac{\partial^2}{\partial t^2} \left[ \rho_\infty (1 - H_s(f)) \right] \tag{16.2.15}$$

The term $(1 - H_s(f))$ represents the volume inside the moving surface, and so thickness noise can be considered as being completely equivalent to the sound from a displaced mass of fluid moving through a stationary medium, and this is referred to as Isom's result [1]. The solution to Eq. (16.2.15) is obtained following the procedures given in Section 5.2, with the double derivative with respect to time requiring integration by parts twice, followed by the integration (5.2.12), giving

$$[p'(\mathbf{x}, t)]_{\text{thickness}} = \frac{\partial^2}{\partial t^2} \int_{V_\infty} \left[ \frac{\rho_\infty (1 - H_s(f))}{4\pi r |1 - M_r|} \right]_{\tau = \tau^*} dV(\mathbf{z}) \tag{16.2.16}$$

This result is relatively straightforward to evaluate numerically by breaking the volume of the blade down into acoustically compact volume elements of volume $\Delta V_k$, centered on $\mathbf{y}^{(k)}$, and using $\partial/\partial t = (1 - M_r) \partial/\partial \tau$, so

$$[p'(\mathbf{x}, t)]_{\text{thickness}} = \sum_k \rho_\infty \Delta V_k \left[ \frac{1}{(1 - M_r)} \frac{\partial}{\partial \tau} \left( \frac{1}{(1 - M_r)} \frac{\partial}{\partial \tau} \left( \frac{1}{4\pi r |1 - M_r|} \right) \right) \right]_{\substack{\mathbf{y} = \mathbf{y}^{(k)} \\ \tau = \tau^*}} \tag{16.2.17}$$

This result shows that thickness noise can be computed directly from volume elements within the blade, with the correct allowance for the propagation distance from the source to the observer. The signature from different parts of the blade planform will arrive at the observer at the same time, so the calculation is nontrivial, but it is tractable using numerical methods.

For an observer in the far field terms of order $r^{-2}$ can be dropped, and this result simplifies to

$$[p'(\mathbf{x}, t)]_{\text{thickness}} = \sum_k \rho_\infty \Delta V_k \left[ \frac{1}{4\pi r} \left( \frac{1}{(1-M_r)^4} \frac{\partial^2 M_r}{\partial \tau^2} + \frac{3}{(1-M_r)^5} \left( \frac{\partial M_r}{\partial \tau} \right)^2 \right) \right]_{\substack{\mathbf{y}=\mathbf{y}^{(k)} \\ \tau=\tau^*}}$$

(16.2.18)

We can normalize this result to obtain a nondimensional thickness noise pulse for each volume element as

$$C_q(t) = \frac{4\pi |\mathbf{x}| c_\infty \, (p'(\mathbf{x}, t))_{\text{thickness}}}{\Omega \left( \Omega^2 R \rho_\infty \Delta V_k \right)}$$

$$= \left[ \frac{c_\infty}{\Omega R} \left( \frac{1}{(1-M_r)^4} \frac{\partial^2 M_r}{\partial (\Omega \tau)^2} + \frac{3}{(1-M_r)^5} \left( \frac{\partial M_r}{\partial (\Omega \tau)} \right)^2 \right) \right]_{\substack{\mathbf{y}=\mathbf{y}^{(k)} \\ \tau=\tau^*}} \quad (16.2.19)$$

Using this normalization, we see that the thickness noise is scaled on an equivalent force equal to $\Omega^2 R \rho_o \Delta V_k$, and so will depend on the blade volume. As an example the normalized signature for an observer at $(40R_{\text{tip}}, 30R_{\text{tip}}, 0)$ and a source at 90% of the blade span, with tip Mach number $M_{\text{tip}} = 0.8$, is shown in Fig. 16.17. The thickness noise gives a clearly defined pulse as the blade moves toward the observer, and its level exceeds that of a steady loading signature for high tip Mach numbers for the same equivalent blade loading. It is important to appreciate, however, that this

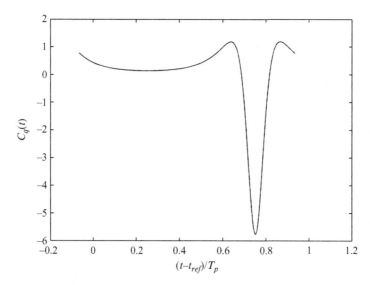

**Fig. 16.17** The time history of the thickness noise signature from a propeller. Source located at $R = 0.9R_{\text{tip}}$ and $M_{\text{tip}} = 0.8$, and the observer location is $\mathbf{x} = (40R_{\text{tip}}, 30R_{\text{tip}}, 0)$.

calculation does not include all the effects of blade shape and retarded time across the blade chord, and more detailed calculations are required to define the exact thickness noise pulse.

### 16.2.3 Supersonic tip speeds

The calculations that have been described above are for blades with subsonic tip speed. When the blade tip Mach number exceeds one then a number of effects occur. These include the presence of shock surfaces in the flow and a singularity in the integrals (16.2.1), (16.2.2), and (16.2.14) when $M_r = 1$. Another important issue for supersonic rotors is that there is no longer a one-to-one relationship between the emission time and the observer time. Fig. 16.18 shows the calculation of blade position for a given observer time for a supersonic rotor when the observer is in the plane of the rotor. Close to the blade tip it is seen that there are in fact three solutions for $\phi$ for an observer time of $t/T_p = 0.75$. Consequently, the source strength from many parts of the rotor disc is concentrated at this instant of observer time, and a large impulsive peak in the sound signature is expected. Numerical calculations in and around this instant are clearly complex, and the reader is referred to the papers by Farassat [1] on the details of how to address this problem. Providing the correct asymptotic numerical coding is used, the time domain methods described above can be extended to supersonic rotors. However, as will be shown in the next section, many of these numerical difficulties with time domain methods are overcome if frequency domain methods are employed, at the expense, unfortunately, of other numerical issues.

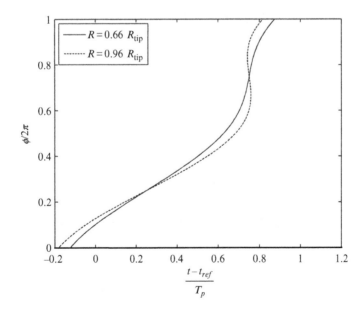

**Fig. 16.18** The blade locations $\phi$ at a given observer time for a rotor with a tip Mach number of 1.2 for an observer in the plane of the rotor.

## 16.3    Frequency domain prediction methods for tone noise

In this section we discuss the prediction of tone noise from a propeller or rotor using frequency domain methods [2]. These complement the time domain methods described in the previous section and also provide greater insight into the important physics of rotor noise.

### 16.3.1    Harmonic analysis of loading and thickness noise

In Section 16.2 we showed that the loading and thickness noise from a rotor could be defined as an integral over the blade planform, which was given by the combination of Eqs. (16.2.2), (16.2.14) as

$$p'(\mathbf{x}, t) = -\frac{\partial}{\partial x_i} \int_{\Sigma_o} \left[ \frac{f_i}{4\pi r |1 - M_r|} \right]_{\tau = \tau^*} d\Sigma(\mathbf{z}) - \frac{\partial}{\partial t} \int_{\Sigma_o} \left[ \frac{\rho_\infty \mathbf{V}.\nabla h}{4\pi r |1 - M_r|} \right]_{\tau = \tau^*} d\Sigma(\mathbf{z})$$

(16.3.1)

For tone noise the resulting acoustic signature will be periodic, and the contribution from each blade will be repeated after each revolution. In this section we make use of this characteristic to evaluate the integrals in Eq. (16.3.1). We limit consideration to stationary propellers, but note that the theory can be readily extended to propellers and rotors in flight.

The signature from each blade is identical, and so if $p_1(\mathbf{x}, t)$ is the signature from one blade in isolation, the signature from a rotor with $B$ blades is

$$p'(\mathbf{x}, t) = \sum_{n=1}^{B} p_1\left(\mathbf{x}, t + nT_p/B\right)$$

(16.3.2)

where $T_p$ is the period of one rotation. Because the acoustic signature is periodic we can expand the signature from a single blade as a Fourier series of the type

$$p_1(\mathbf{x}, t) = \sum_{j=-\infty}^{\infty} c_j e^{-2\pi i j t / T_p} = \sum_{j=-\infty}^{\infty} c_j e^{-ij\Omega t}$$

(16.3.3)

where $\Omega = 2\pi/T_p$ is the rotational frequency. Combining Eqs. (16.3.2), (16.3.3) then gives

$$p'(\mathbf{x}, t) = \sum_{j=-\infty}^{\infty} c_j \sum_{n=1}^{B} e^{-ij(\Omega t - 2\pi n/B)} = B \sum_{m=-\infty}^{\infty} c_{mB} e^{-imB\Omega t}$$

(16.3.4)

where we have made use of the fact that the sum over $n$ is only nonzero when $j = mB$. The important feature of this result is that we can define the time history from a rotor with $B$ blades from the Fourier coefficients of the time history from a single blade, but only the coefficients of order $mB$ will contribute.

To determine the Fourier coefficients required in Eq. (16.3.3) we evaluate the integral

$$c_n(\mathbf{x}) = \frac{\Omega}{2\pi} \int_0^{T_p} p_1(\mathbf{x}, t) e^{in\Omega t} dt \tag{16.3.5}$$

Combining this with Eq. (16.3.1) then gives

$$c_n(\mathbf{x}) = -\frac{\Omega}{2\pi} \int_0^{T_p} \left\{ \frac{\partial}{\partial x_i} \int_{\Sigma_o} \left[ \frac{f_i}{4\pi r |1 - M_r|} \right]_{\tau=\tau^*} d\Sigma(\mathbf{z}) \right\} e^{in\Omega t} dt$$

$$- \frac{\Omega}{2\pi} \int_0^{T_p} \left\{ \frac{\partial}{\partial t} \int_{\Sigma_o} \left[ \frac{\rho_\infty \mathbf{V} \cdot \nabla h}{4\pi r |1 - M_r|} \right]_{\tau=\tau^*} d\Sigma(\mathbf{z}) \right\} e^{in\Omega t} dt \tag{16.3.6}$$

In the far field the differentiation of the first integral can be changed to a differential over time (see Section 5.2) using $\partial/\partial x_i \sim (-x_i/r_o c_o)\partial/\partial t$, where $r_o = |\mathbf{x}|$ so that by using the properties of Fourier series we obtain

$$c_n(\mathbf{x}) = \frac{in\Omega^2}{2\pi} \left( \int_{\Sigma_o} \int_0^{T_p} \left[ \frac{-x_i f_i}{4\pi r_o^2 c_o |1 - M_r|} \right]_{\tau=\tau^*} e^{in\Omega t} dt d\Sigma(\mathbf{z}) \right.$$

$$+ \left. \int_{\Sigma_o} \int_0^{T_p} \left[ \frac{\rho_\infty \mathbf{V} \cdot \nabla h}{4\pi r_o |1 - M_r|} \right]_{\tau=\tau^*} e^{in\Omega t} dt d\Sigma(\mathbf{z}) \right) \tag{16.3.7}$$

The integral over observer time can be changed to an integral over emission time because $t = \tau + r(\tau)/c_o$, and so the time differentials are related by $dt = |1 - M_r| d\tau$. It follows that since the source terms are also periodic with the same time scale, Eq. (16.3.7) reduces to

$$c_n(\mathbf{x}) = \frac{in\Omega^2}{2\pi} \left( \int_{\Sigma_o} \int_0^{T_p} \frac{-x_i f_i(\mathbf{z}, \tau)}{4\pi r_o^2 c_o} e^{in\Omega(\tau + r(\tau)/c_o)} d\tau d\Sigma(\mathbf{z}) \right.$$

$$+ \left. \int_{\Sigma_o} \int_0^{T_p} \frac{\rho_\infty \mathbf{V}(\mathbf{z}, \tau) \cdot \nabla h(\mathbf{z})}{4\pi r_o} e^{in\Omega(\tau + r(\tau)/c_o)} d\tau d\Sigma(\mathbf{z}) \right) \tag{16.3.8}$$

One of the most significant aspects of this transformation is that it eliminates the singularity that occurs when $M_r = 1$, so the integrals are harmonic and well behaved for all blade speeds.

For an observer in the acoustic far field the integral over time in Eq. (16.3.8) can be evaluated directly. Consider the blade element at radius $R$ and azimuthal location $\phi_1$ in blade-fixed coordinates, which is at $\mathbf{y} = (0, R\cos(\phi_1 + \Omega\tau), R\sin(\phi_1 + \Omega\tau))$. Using the approximation $r \sim r_o - \mathbf{x} \cdot \mathbf{y}/|\mathbf{x}|$ the propagation distance from this element to the far-field observer is given by (see Fig. 16.11).

$$r(\tau) \approx r_o - x_2 R \cos(\phi_1 + \Omega\tau)/|\mathbf{x}| - x_3 R \sin(\phi_1 + \Omega\tau)/|\mathbf{x}| \tag{16.3.9}$$

It is advantageous to specify the observer location in spherical coordinates so that

$$x_1 = r_o \cos\theta_o \quad x_2 = r_o \sin\theta_o \cos\phi_o \quad x_3 = r_o \sin\theta_o \sin\phi_o \tag{16.3.10}$$

then

$$r(\tau) \approx r_o - R \sin\theta_o \cos(\phi_1 - \phi_o + \Omega\tau) \tag{16.3.11}$$

The final step in the analysis is to make use of the Fourier series expansion

$$e^{-i\alpha\cos\theta} = \sum_{m=-\infty}^{\infty} J_m(\alpha) e^{-im(\theta + \pi/2)} \tag{16.3.12}$$

where $J_m(\alpha)$ is a Bessel function of the first kind of order $m$. This function has well-known properties and is readily available on most computational systems, so making use of it to simplify the analysis is advantageous. However, it is difficult to compute accurately for large orders, and so asymptotic formulae for the Bessel function are sometimes required. We can now write

$$e^{in\Omega r(\tau)/c_\infty} = e^{in\Omega r_o/c_\infty} \sum_{m=-\infty}^{\infty} J_m\left(\frac{n\Omega R \sin\theta_o}{c_o}\right) e^{-im(\phi_1 - \phi_o + \Omega\tau + \pi/2)} \tag{16.3.13}$$

Combining all these results into Eq. (16.3.8) gives the Fourier series coefficients in a convenient form as

$$c_n(\mathbf{x}) = \frac{in\Omega e^{in\Omega r_o/c_o}}{4\pi r_o c_o} \sum_{m=-\infty}^{\infty} \int_{R_i}^{R_{\text{tip}}} Q_{m,n}(R, \mathbf{x}) J_m\left(\frac{n\Omega R \sin\theta_o}{c_o}\right) dR \tag{16.3.14}$$

with the blade planform surface element defined as $d\Sigma = R d\phi_1 dR$ and the blade surface given by $\phi_{TE}(R) < \phi_1 < \phi_{LE}(R)$ and $R_i < R < R_{\text{tip}}$. The source term is given as

$$Q_{m,n}(R, \mathbf{x}) = \frac{\Omega}{2\pi} \int_{\phi_{TE}}^{\phi_{LE}} \int_0^{T_p} \left\{ \frac{-x_i f_i(\mathbf{z}, \tau)}{r_o} + \rho_\infty c_o \mathbf{V}(\mathbf{z}, \tau) \cdot \nabla h(\mathbf{z}) \right\} e^{i(n-m)\Omega\tau - im(\phi_1 - \phi_o + i\pi/2)} R d\phi_1 d\tau$$

$$\tag{16.3.15}$$

We have now reduced the problem to two relatively straightforward integrals. The first integral in Eq. (16.3.14) is over the blade span, and the second, in Eq. (16.3.15), gives the source term and requires integrals over the blade chord and the period of rotation. We will start by considering the case in which we can ignore the unsteady loading terms and limit consideration to only the steady loading and thickness noise terms. If the thrust per unit area on the blade surface is $f_L(R, \phi_1)$ and the drag is $f_D(R, \phi_1)$ then we have $f_i = (-f_L, -f_D \sin(\Omega \tau + \phi_1), f_D \cos(\Omega \tau + \phi_1))$, and we can define

$$\frac{x_i f_i}{r_o} = -f_L \cos \theta_o - f_D \sin \theta_o \sin (\Omega \tau + \phi_1 - \phi_o) \tag{16.3.16}$$

Similarly, the thickness noise term may be simplified as in Section 16.2.2, as

$$\rho_\infty \mathbf{V}(\mathbf{z}, \tau) \cdot \nabla h(\mathbf{z}) = \rho_\infty (\Omega R) \frac{1}{R} \frac{\partial h}{\partial \phi_1} \tag{16.3.17}$$

which is independent of time. Using these results in Eq. (16.3.15) then gives

$$Q_{m,n}(R, \mathbf{x}) = \frac{\Omega}{2\pi} \int_{\phi_{TE}}^{\phi_{LE}} \int_0^{T_p} \left\{ f_L(R, \phi_1) \cos \theta_o + f_D(R, \phi_1) \sin \theta_o \sin (\Omega \tau + \phi_1 - \phi_o) \right.$$
$$\left. + \rho_\infty c_o \Omega \frac{\partial h(R, \phi_1)}{\partial \phi_1} \right\} e^{i(n-m)\Omega\tau - im(\phi_1 - \phi_o + \pi/2)} R d\phi_1 d\tau \tag{16.3.18}$$

We have now completely specified the time dependence of the sources, and so the integral over time may be evaluated analytically. We note that both the thrust term and the thickness noise term are independent of time, and so their integral is only nonzero when $n = m$. Similarly, for the drag term the sine function may be split into the sum of two exponentials, and so the integral over time will only be nonzero when $n = m \pm 1$. Finally, we note that the integral of the thickness noise term over $\phi_1$ can be carried out by parts, and providing $h = 0$ at the leading and trailing edges we obtain

$$Q_{m,n}(R, \mathbf{x}) = \int_{\phi_{TE}}^{\phi_{LE}} \left\{ f_L(R, \phi_1) \cos \theta_o \delta_{mn} + f_D(R, \phi_1) \sin \theta_o \left( \frac{\delta_{n,m-1} - \delta_{n,m+1}}{2i} \right) \right.$$
$$\left. + im\Omega \rho_\infty c_o h(R, \phi_1) \delta_{mn} \right\} e^{-im(\phi_1 - \phi_o + \pi/2)} R d\phi_1 \tag{16.3.19}$$

This provides a simple formula for the source term that can be readily evaluated. We still need to know the distribution of the thrust, drag, and thickness on the blade surface, but this can be obtained from the blade design characteristics. As a first approximation we can use a point-loading approximation, but this is inaccurate for the higher harmonics. One of the most important features of this result is that the integral over time has reduced the summation in Eq. (16.3.14) to only three values of $m$ for a given harmonic $n$. The summation in Eq. (16.3.14) is therefore limited to only three terms (and can be further simplified by using properties of Bessel functions if required). This would not have been the case for unsteady loading because $f_L$ would then be a function of time. To show how unsteady loading can be incorporated into the calculation, we can assume that the loading is periodic in source time and can therefore be expanded as a Fourier series, so

$$f_L(R, \phi_1, \tau) = \sum_{k=-\infty}^{\infty} f_L^{(k)}(R, \phi_1) e^{-ik\Omega\tau} \tag{16.3.20}$$

and similarly for the drag term. Using this result in Eq. (16.3.18) and integrating over time shows that the integral will only be nonzero when $n - m = k$ (or $n - m = k \pm 1$ for the drag term), so we can rewrite Eq. (16.3.19) to include unsteady loading as

$$Q_{m,n}(R, \mathbf{x}) = \int_{\phi_{TE}}^{\phi_{LE}} \left\{ f_L^{(n-m)}(R, \phi_1) \cos\theta_o + f_D^{(k)}(R, \phi_1) \sin\theta_o \left( \frac{\delta_{n,m+k+1} - \delta_{n,m+k-1}}{2i} \right) \right.$$
$$\left. + im\Omega\rho_\infty c_o h(R, \phi_1) \delta_{mn} \right\} \times e^{-im(\phi_1 - \phi_o + \pi/2)} R d\phi_1 \tag{16.3.21}$$

In this result the number of terms required in the summation over $m$ in Eq. (16.3.14) has been significantly increased, and the expansion Eq. (16.3.20) may converge only slowly. So, the evaluation of Eq. (16.3.21) may be time-consuming, but the computational procedure is relatively straightforward. One of the advantages of this approach is that it allows the distribution of blade loading and thickness over the chord to be readily included in the calculation. It should also be noted that the radial integral in Eq. (16.3.14) includes a dependence on the Bessel function $J_m(n\Omega R \sin\theta_o / c_o)$ which has a strong impact on the directionality of the acoustic field. These functions can be cumbersome to compute, but fortunately there are a number of asymptotic approximations, which simplify the task of computing the radial integral.

As an illustrative example we will calculate the Fourier series coefficients $c_n(\mathbf{x})$ for the Gaussian derivative impulsive blade loading given by Eq. (16.2.7). The loading coefficients in Eq. (16.3.20) can be calculated analytically in the limit that $T_v \ll T_p$ and are

$$f_L^{(k)}(R, \phi_1) = \frac{ikF_L(\Omega T_v)^2 \sqrt{\pi}}{4\pi} \exp\left( ik\Omega\tau_o - (k\Omega T_v/2)^2 \right) \left\{ \frac{\delta(\phi_1)\delta(R)}{R} \right\} \tag{16.3.22}$$

where the terms in { } are required so that the force acts at a point. It follows then that

$$Q_{m,n}(R, \mathbf{x}) = \frac{i(n-m)F_L(\Omega T_v)^2 \sqrt{\pi}}{4\pi} e^{i(n-m)\Omega\tau_o - ((n-m)\Omega T_v/2)^2} \delta(R) \cos\theta_o e^{im(\phi_o - \pi/2)} \tag{16.3.23}$$

A contour plot of these coefficients is shown in Fig. 16.19 and clearly shows the importance of the terms with increasing $m$ and $n$.

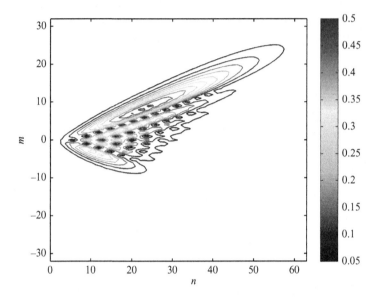

**Fig. 16.19** The contours of the function $Q_{m,n}$ specified in Eq. (16.3.23).

The nondimensional Fourier series coefficients of the acoustic signature are then obtained as

$$C_p^{(n)} = \frac{4\pi r_o c_\infty c_n(\mathbf{x})}{\Omega F_L} = in\Omega e^{in\Omega r_o/c_o} \sum_{m=-\infty}^{\infty} J_m\left(\frac{n\Omega R \sin\theta_o}{c_o}\right) \cos\theta_o e^{im(\phi_o - \pi/2)}$$

$$\times \left\{ \frac{i(n-m)(\Omega T_v)^2}{4\sqrt{\pi}} e^{i(n-m)\Omega\tau_o - ((n-m)\Omega T_v/2)^2} \right\} \qquad (16.3.24)$$

and the amplitude of these are plotted on a dB scale in Fig. 16.20. Note that the scale has a very large range, and the harmonics that contribute significantly are only those with indices $<32$. Evaluating the pulse observed in the far field from these harmonics reproduces the signature calculated directly using the time series approach, as shown in Fig. 16.16.

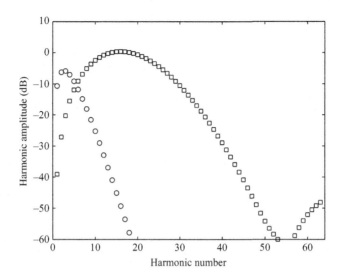

**Fig. 16.20** The amplitude of the harmonics given in Eq. (16.3.24) for the radiated sound from a rotor with a loading pulse *(squares)* as shown in Fig. 16.15, and the thickness noise harmonics *(circles)* given by Eq. (16.3.25). The observer is at $\mathbf{x} = (40 \, R_{\text{tip}}, 30R_{\text{tip}}, 0)$.

We can also evaluate the harmonics of the thickness noise pulse that was given in Fig. 16.17. The nondimensional form of these harmonics is given by

$$C_q^{(n)} = \frac{4\pi r_o c_\infty c_n(\mathbf{x})}{\Omega\left(\rho_o \Omega^2 R \Delta V_k\right)} = -\left(\frac{c_\infty}{\Omega R}\right) n^2 J_n\left(\frac{n\Omega R \sin\theta_o}{c_\infty}\right) e^{\text{im}(\phi_o - \pi/2) + in\Omega r_o/c_\infty}$$

$$(16.3.25)$$

and these are also shown in Fig. 16.20. From this result we see that thickness noise tends to dominate for the lower harmonics, and unsteady loading noise dominates for the higher harmonics. Note also that for a rotor or propeller with $B$ blades only the harmonics $n = mB$ will contribute, so the contribution of thickness noise may be limited to the first two blade-passing frequencies.

## 16.4 Broadband noise from open rotors

In the previous sections we have discussed the harmonic content of the sound from a rotor, referred to as tone noise. This assumes that the acoustic sources are periodic, and the same signature is generated by each blade. In addition to tone noise, rotors also generate broadband noise that is not periodic with source fluctuations that are typically uncorrelated from blade to blade.

In Chapters 14 and 15 broadband noise sources on blades in a uniform flow were discussed assuming that the blade and observer were fixed and embedded in a uniform flow, as in a wind tunnel. To use these results for a blade that is moving relative to a fixed observer we can use *Amiet's approximation*, which applies when the time scale of the source fluctuations on a rotating blade is very much less than the time it takes for one rotor revolution. This will be the case for trailing edge noise sources that depend on the blade boundary layer, as described in Chapter 15, and for leading-edge noise sources when the lengthscale of the incoming turbulence is of the order of a blade chord or less. It also applies to a BVI and leading-edge noise when small-scale turbulence is stretched in the direction of the rotor inflow. In this limit Amiet [3] argued that the rotor noise signal could be accurately estimated by breaking down each revolution of the rotor into the suitable number of time segments and 10–15 radial segments as shown schematically in Fig. 16.21 and that (1) the blade could be assumed to be in rectilinear motion in each time segment and (2) the sources in each segment were uncorrelated, so the far-field sound was the incoherent sum of the mean square levels in each segment. The advantage of this approach is that the results from wind tunnel testing of fixed blades can be incorporated directly into the source terms for rotor noise, and this greatly simplifies the scaling of the results from model scale to full scale.

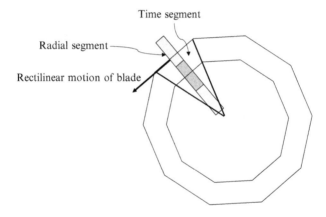

**Fig. 16.21** Segmenting of a rotor noise calculation.

To apply Amiet's approximation the approach described in Section 16.3 will be used, but in this case we will evaluate the Fourier transform of the time history rather than its Fourier series coefficients. We start by evaluating the dipole source term in the Ffowcs-Williams and Hawkings equation for a single blade, which is

$$p'(\mathbf{x}, t) = -\frac{\partial}{\partial x_i} \int_{\Sigma_o} \left[ \frac{f_i(\mathbf{z}, \tau)}{4\pi r |1 - M_r|} \right]_{\tau=\tau^*} d\Sigma(\mathbf{z}) \qquad (16.2.2)$$

In general, we are interested in obtaining the spectrum of the far-field sound, and so we need to evaluate the Fourier transform of this equation with respect to time. As before we place the observer in the geometric far field and make the far-field approximation that $\partial/\partial x_i \sim (x_i/r_o c_\infty)\partial/\partial t$, so the Fourier transform of Eq. (16.2.2) with respect to time is

$$\widetilde{p}(\mathbf{x}, \omega) = -\frac{1}{2\pi} \int_{-T}^{T} \frac{x_i}{4\pi r_o^2} \left\{ \frac{\partial}{\partial t} \int_{\Sigma_o} \left[ \frac{f_i(\mathbf{z}, \tau)}{|1 - M_r|} \right]_{\tau=\tau^*} d\Sigma(\mathbf{z}) \right\} e^{i\omega t} dt \tag{16.4.1}$$

where $r_o = |\mathbf{x}|$. The differential with respect to observer time is equivalent to multiplying the Fourier transform by $-i\omega$, and as in Chapters 14 and 15 we specify the unsteady loading per unit area on the blade in terms of the pressure jump in the direction of the blade normal (which will be different for each segment), so

$$f_i(\mathbf{z}, \tau) = n_i(\tau)\Delta p(\mathbf{z}, \tau) \tag{16.4.2}$$

and we obtain

$$\widetilde{p}(\mathbf{x}, \omega) = \frac{1}{2\pi} \int_{-T}^{T} \frac{i\omega x_i}{4\pi r_o^2} \left\{ \int_{\Sigma_o} \left[ \frac{n_i(\tau)\Delta p(\mathbf{z}, \tau)}{|1 - M_r|} \right]_{\tau=\tau^*} d\Sigma(\mathbf{z}) \right\} e^{i\omega t} dt$$

The integral over observer time can be shifted to an integral over source time by noting that as before $dt = |1 - M_r| d\tau$ and $t = \tau + r(\tau)/c_\infty$, so we have

$$\widetilde{p}(\mathbf{x}, \omega) = \left( \frac{i\omega x_i}{4\pi r_o^2} \right) \frac{1}{2\pi} \int_{T_1}^{T_2} \left\{ \int_{\Sigma_o} n_i(\tau)\Delta p(\mathbf{z}, \tau) d\Sigma(\mathbf{z}) \right\} e^{i\omega\tau + i\omega r(\tau)/c_\infty} d\tau \tag{16.4.3}$$

where $T_1$ and $T_2$ are the source times that correspond to the observer times $\pm T$ and tend to infinity.

So far only the far-field approximation has been used, and so the result given by Eq. (16.4.3) can be used as a starting point for an exact analysis. However, if we make Amiet's approximation and segment the source time history into discrete intervals of length $\Delta\tau$, the spanwise integral into finite parts Eq. (16.4.3) takes the form

$$\widetilde{p}(\mathbf{x}, \omega) = \left( \frac{i\omega x_i}{4\pi r_o^2} \right) \frac{1}{2\pi} \sum_{m=1}^{M} \sum_{n=1}^{N} \int_{\tau_m - \Delta\tau/2}^{\tau_m + \Delta\tau/2} \left\{ \int_{R_n}^{R_{n+1}} \int_0^c n_i(\tau)\Delta p(\mathbf{z}, \tau) d\xi_1 dR \right\} e^{i\omega\tau + i\omega r(\tau)/c_\infty} d\tau$$

$$\tag{16.4.4}$$

where $\tau_m = m\Delta\tau$, $\tau_1 - \Delta\tau/2 = T_1$, and $\tau_M + \Delta\tau/2 = T_2$. Similarly, $R_1$ is the inner radius of the blade, and $R_{N+1}$ is the outer radius of the blade, and the segmentation in the radial direction does not have to be uniform. Also we have defined the distance from the blade leading edge in the chordwise direction as $\xi_1$ and the spanwise direction as $R$.

In the geometric far field from the rotor we can approximate

$$r(\tau) \approx r_o - x_i y_i(\tau)/r_o$$

where $y_i(\tau)$ defines the location of the blade in fixed coordinates and can be evaluated from the blade location shown in Fig. 16.11 as

$$y_1 = U_o\tau + \xi_1 \sin\beta_o \quad y_2 = R\cos(\phi_1 + \Omega\tau) \quad y_3 = R\sin(\phi_1 + \Omega\tau)$$

$$\phi_1 = -\tan^{-1}\left(\frac{\xi_1 \cos\beta_o}{R}\right) \tag{16.4.5}$$

where the pitch angle $\beta_o$ is shown in Fig. 16.22. In Amiet's approximation both $\Omega\Delta\tau$ and $\phi_1$ are limited to small angles, and so we can approximate

$$x_i y_i(\tau) \approx x_i \left( y_i(\tau_m) + (\tau - \tau_m)\frac{\partial y_i(\tau_m)}{\partial\tau} + \xi_1 \frac{\partial y_i(\tau_m)}{\partial\xi_1} + (R - R_n)\frac{\partial y_i(\tau_m)}{\partial R} \cdots \right)$$

$$= x_i y_i(\tau_m) - \xi_1 \cos\beta_o\{x_3 \cos(\Omega\tau_m) - x_2 \sin(\Omega\tau_m)\}$$

$$+ (R - R_n)\{x_2 \cos(\Omega\tau_m) + x_3 \sin(\Omega\tau_m)\} + (\tau - \tau_m)U_r$$

where $U_r$ is the velocity of the blade in the direction of the observer.

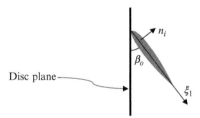

**Fig. 16.22** Blade pitch angle.

Making the geometric far-field approximations the Fourier transform Eq. (16.4.4) then becomes

$$\widetilde{p}(\mathbf{x}, \omega) = \left(\frac{i\omega x_i}{4\pi r_o^2}\right)\frac{1}{2\pi}\sum_{m=1}^{M}\sum_{n=1}^{N} n_i^{(m,n)} e^{i\omega\tau_m + i\omega r(\tau_m)/c_\infty}$$

$$\times \int_{-\Delta\tau/2}^{\Delta\tau/2}\left\{\int_0^{c}\int_0^{R_{n+1}-R_n} \Delta p(\xi_1, R_n + R', \tau_m + \tau')e^{i\omega\tau'\left(1 - M_r^{(m,n)}\right) - ika_m\xi_1 - ikb_m R'}\,d\xi_1\,dR'\right\}d\tau'$$

$$\tag{16.4.6}$$

where $k = \omega/c_\infty$, the displacement variables $\tau' = \tau - \tau_m$ and $R' = R - R_n$ have been introduced, and the blade normal is assumed constant over the surface and defined for each segment, Fig. 16.21. Also we have defined the Mach number in the direction of the observer for each segment, and

$$a_m = -\cos\beta_o\{x_3\cos(\Omega\tau_m) - x_2\sin(\Omega\tau_m)\}/r_o$$
$$b_m = \{x_2\cos(\Omega\tau_m) + x_3\sin(\Omega\tau_m)\}/r_o \qquad (16.4.7)$$

Since the pressure jump is zero everywhere, but on the blade the integral over the chord in Eq. (16.4.6) can be extended to infinity in the $\xi_1$ direction and becomes a wavenumber transform. However, the spanwise integral in each segment needs to be truncated at the edge of the segment, as discussed in Chapter 14, and some care is needed when applying this approach to sources that have significant spanwise correlation. By choosing the spanwise segments to be acoustically compact the phase variation can be eliminated, and the integral over the span is then well approximated by simply integrating the pressure jump over the span. Evaluating the integrals inside the curly braces then gives a result in terms of the wavenumber transform of the pressure jump, so

$$\left\{\int_0^c \int_0^{R_{n+1}-R_n} \Delta p(\xi_1, R_n + R', \tau_m + \tau')e^{i\omega\tau'\left(1 - M_r^{(m,n)}\right) - ika_m\xi_1 - ikb_mR'}d\xi_1 dR'\right\}$$
$$= (2\pi)^2 \int_{-\infty}^{\infty} \Delta\widetilde{p}^{(m,n)}(ka_m, kb_m, \omega_o)e^{-i\omega_o(\tau_m + \tau') + i\omega\tau'\left(1 - M_r^{(m,n)}\right)}d\omega_o \qquad (16.4.8)$$

The integral over $\tau'$ can then be carried out explicitly, and we obtain

$$\widetilde{p}(\mathbf{x}, \omega) = \left(\frac{i\omega x_i}{4\pi r_o^2}\right)\frac{1}{2\pi}\sum_{m=1}^{M}\sum_{n=1}^{N} n_i^{(m,n)} e^{i\omega\tau_m + i\omega r(\tau_m)/c_\infty}$$
$$\times (2\pi)^2\Delta\tau \int_{-\infty}^{\infty} \Delta\widetilde{p}^{(m,n)}(ka_m, kb_m, \omega_o)\left(\frac{\sin\left(\left(\omega_o - \omega\left(1 - M_r^{(m,n)}\right)\right)\Delta\tau/2\right)}{\left(\omega_o - \omega\left(1 - M_r^{(m,n)}\right)\right)\Delta\tau/2}\right)e^{-i\omega_o\tau_m}d\omega_o$$
$$(16.4.9)$$

This is the basis for Amiet's approximation and shows how the far-field sound can be directly related to the wavenumber transform of the surface pressure jump on the blades, evaluated at the wavenumbers corresponding to the waves in the acoustic far field. When $k$ tends to zero this is the unsteady loading on the blade segment, and so scales as a dipole source as described in Chapter 4. However, at higher frequencies the results given in Chapters 14 and 15 must be used and can be inserted directly into this equation.

The size of the segment directly impacts the frequency scaling of the result. The integral over $\omega_o$ represents a frequency filter that depends on the size of the segment used to separate the time steps in the process. The filtering process peaks when the source frequency is equal to the observer frequency reduced by a Doppler factor to account for source motion. The wavenumbers used in the evaluation of $\Delta p$ are different from those used in Chapters 14 and 15 because in the present case the source is moving and the observer is stationary, whereas in the earlier examples the source and observer were stationary and the flow was moving. The consequence of this shift is that the source frequency is Doppler shifted to account for the source motion, but the spatial scales remain the same in both the observer-based coordinates and the moving source coordinates, so the wavenumbers are defined in the observer frame of reference.

The far-field spectral density from each segment can be defined by using the definition given in Eq. (8.4.13). If we assume that the segmentation is sufficient to resolve the frequency content of the signal so that the filter has no effect, we can specify the power spectral density from each segment as

$$
S_{pp}^{(m,\,n)}(\mathbf{x},\,\omega) = \left( \frac{\pi \omega x_i n_i^{(m,\,n)}}{c_\infty r_o^2} \right)^2 \frac{\pi}{T} E\left[ \left| \Delta \widetilde{p}\left( k a_m, k b_m, \omega \left( 1 - M_r^{(m,\,n)} \right) \right) \right|^2 \right]
$$

$$(16.4.10)$$

and, if the fluctuations are statistically independent for each value of $m$ and $n$, then the total far-field sound is obtained by summing the spectrum generated by each segment and multiplying by the number of blades. Amiet also pointed out that the averaging time in the fixed frame of reference would be different from the frame of reference of a moving blade by a factor of $1 - M_r$, and so Eq. (16.4.10) should also be corrected by this factor when using stationary blade data or models as inputs. This is expected to be a good approximation for trailing edge noise that depends on very small scale turbulence and is often suitable for leading-edge noise as well. Note that the expected value used in Eq. (16.4.10) is used in its most general sense in that the averaging is done on a blade-by-blade basis while it passes through each segment multiple times. This implies that the spectrum is obtained by averaging over many rotor revolutions and is a key part of this approach. It is important to appreciate that this result is equivalent to Eq. (14.3.1) and can be used with an inflow turbulence spectrum such as Eq. (14.3.4) to give the far-field sound from inflow turbulence. The overall characteristics of the far-field sound are then very similar to the noise from a blade that is not rotating, since the Doppler frequency shift in Eq (16.4.10) will tend to average out when the summation is applied across all the blade segments. The spectra shown in Fig. 14.6 are then expected to be typical of the spectra from a rotor when inflow turbulence dominates the far-field noise signature. For trailing edge noise, a similar conclusion can be drawn and modeling functions such as Eq. (15.2.10) can be readily adapted to provide the input required for Eq. (16.4.10). We will discuss how this is modified in situations where the segment signals are not statistically independent in the next section.

The valuable part of Eq. (16.4.10) is that it shows how measurements made in a wind tunnel on a stationary blade in a uniform flow can be applied directly to the sources on a rotating blade. The wavenumber spectrum of the blade surface pressure was defined for a blade in a uniform flow in Sections 14.3, and from Eq. (14.3.1) we have that

$$\frac{\pi}{T} E\left[\left|\Delta\widetilde{\widetilde{p}}\left(k_1^{(o)}, k_3^{(o)}, \omega\right)\right|^2\right] = \left[\frac{S_{pp}(\mathbf{x}, \omega)}{\left(\pi\omega x_2/c_\infty r_e^2\right)^2}\right]_{\text{WindTunnel}}$$

where $k_1^{(o)} = k_o(x_1/r_e - M)$ $k_3^{(o)} = k_o x_3 \beta^2/r_e$ (16.4.11)

It follows that measurements made in the wind tunnel at the locations

$$\frac{x_1}{r_e} = \left[\frac{a_m}{1 - M_r^{(m,n)}}\right]_{\text{rotor}} \beta^2 + M \quad \frac{x_3}{r_e} = \left[\frac{b_m}{1 - M_r^{(m,n)}}\right]_{\text{rotor}}$$ (16.4.12)

and normalized as in Eq. (16.4.11) can be used a direct input into Eq. (16.4.10) for both leading-edge and trailing edge noise. This of course also applies to the Brooks, Pope, and Marcolini model for trailing edge noise (Section 15.3) and provides a suitable prediction method for fan noise.

# 16.5 Haystacking of broadband noise

In the previous section the broadband noise from a rotor was considered assuming that the flow scales that caused the sound were of sufficiently small that the pressure fluctuations on each blade segment, and each blade, were statistically independent and had the same statistics at all blade positions. This is a significant approximation, and we must also consider those situations where the blade pressure fluctuations are not uniform but vary at different points in the rotor plane, or are correlated from blade to blade. The first of these effects is referred to as amplitude modulation and is illustrated in Fig. 16.1C. The second is caused by the stretching of turbulent eddies as they enter the rotor and is sometimes referred to as inflow distortion noise.

## 16.5.1 Amplitude modulation

Amplitude modulation occurs when the source level on a rotating blade varies significantly with position. An example is a rotor operating in a very uneven inflow, such as the rotor operating near a wall that was discussed in Chapter 10. In this case the rotor blades pass through a turbulent boundary layer that extends over about one-fourth of the rotor disc plane. The noise levels from a particular blade are low when it is in the free stream flow outside the boundary layer, and all the leading-edge noise is generated when the blade passes through the high levels of turbulence in the boundary layer. The signal from each blade is therefore strongly modulated, but the modulation is partially mitigated by the number of blades in the rotor. If the blade count is low, then the effect of modulation is significant because there are times when no blade is in the region of high-level turbulence. However, if the blade count is high then there are always a number of blades in the region where noise is produced and the signal has very little variation with time. Another example is the effect of a nonuniform mean flow on trailing edge noise. If the nonuniform mean flow causes a significant change in blade angle of attack, then the trailing edge noise will be locally increased (or decreased as the case may be) and the far-field sound will have a signal that is modulated. In each case we will assume that the signal from each blade is uncorrelated, which will be the case for trailing edge noise but not always the case for leading-edge noise, and we will discuss the impact of blade-to-blade correlation in the next section.

To illustrate the effect that amplitude modulation has on the measured spectrum consider a source signal from each blade that can be modeled as $f_s(t)$, where $s$ is the blade number and $f_s(t)$ is uncorrelated for different blades. If the signal is modulated as the blade rotates by a mean flow effect, then the modulation can be represented by the function $g(t - sT_p/B)$ for blade number $s$, where $T_p$ is the time for one blade rotation and $B$ is the blade count. The total signal is then

$$a(t) = \sum_{s=1}^{B} f_s(t) g(t - sT_p/B)$$

An example of the signal from one blade is shown in Fig. 16.23. This signal is only nonzero for one-fifth of the period of blade rotation, and the signals from successive blades should be added to this. Fig. 16.23B shows the signal for seven blades summed with the correct time delay. The signal appears to be continuous, and no periodic character is apparent from the overall signature, which is misleading.

The modulating function is periodic and so can be expressed as a Fourier series

$$g(t - sT_p/B) = \sum_{n=-\infty}^{\infty} g_n e^{2\pi i n (t - sT_p/B)/T_p}$$

and so the Fourier transform of the signal is given by

$$\tilde{a}(\omega) = \sum_{s=1}^{B} \sum_{n=-\infty}^{\infty} \tilde{f}_s(\omega - 2\pi n/T_p) g_n e^{2\pi i n s/B}$$

The power spectrum of the signal is given by

$$S_{aa}(\omega) = \sum_{s=1}^{B} \sum_{r=1}^{B} \sum_{n=-\infty}^{\infty} \sum_{m=-\infty}^{\infty} \frac{\pi}{T} Ex\left[\tilde{f}_s(\omega - 2\pi n/T_p)\tilde{f}_r^*(\omega - 2\pi m/T_p)\right] g_n g_m^* e^{2\pi i (ns - mr)/B}$$

If we assume that the loading on each blade is statistically independent the double summation is reduced to a single summation, and if the base signal is statistically stationary and independent of blade number, then only those terms for which $n = m$ will be nonzero, so

$$S_{aa}(\omega) = B \sum_{n=-\infty}^{\infty} S_{ff}(\omega - 2\pi n/T_p)|g_n|^2 \tag{16.5.1}$$

As an example we can consider a signal that is modulated by a periodic rectangular window of length $T_g$, where $T_g = T_p/5$ as shown in Fig. 16.23A, and assume that the source spectrum is given by

$$S_{ff}(\omega) = \frac{(\omega T_s)^2}{1 + (\omega T_s)^4}$$

as shown in Fig. 16.23C. The signal has a time scale $T_s = 0.3T_p$ and the spectrum peaks at the frequency $\omega = \Omega/2$. When the signal is modulated and summed as in Eq. (16.5.1) the resulting spectrum is quite different and shows a series of peaks and an oscillatory level sometimes referred to as scalloping. The peaks do not occur at exact multiples of the rotation frequency or blade passage frequency because $S_{ff}$ peaks at a frequency that is nonzero. The spectrum, however, is quite different from that of a single blade as a direct result of the periodic modulation of the signal.

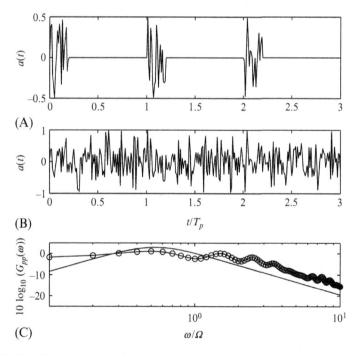

**Fig. 16.23** The effect of modulation on the broadband signature from a seven-bladed rotor: (A) The signature from one blade that is only nonzero for one-fifth of a rotor revolution; (B) the sum of the signals from $B = 7$ blades; and (C) the spectrum of the signals, *solid line* is the spectrum $G_{ff}(\omega) = 2S_{ff}(\omega)$, and -o- is $G_{pp}(\omega) = 2S_{pp}(\omega)$ given by Eq. (16.5.1).

### 16.5.2   Blade-to-blade correlation

In Amiet's method it was assumed that the signals generated by each blade are statistically independent, and so there is no blade-to-blade correlation. This is a reasonable approximation for trailing edge noise that depends on the individual blade boundary layers and for situations where the length scale of the incoming turbulence is much smaller than the distance between the blades. However, it is known that when turbulence enters a rotor it can be stretched in the direction of the flow, and this extends its axial length scale. This stretching can be substantial and alone can result in a single turbulent structure being cut multiple times by successive blades so that the blade loading is correlated between the blades. This results in a quasiperiodic signal in the acoustic far field that includes bursts of pulses at the BPF and a sound spectrum with broadband peaks at near multiples of the BPF, referred to as haystacks. The criterion for this effect to occur is that the BPF should be significantly higher than the axial inflow speed $U_o$ divided by the axial turbulence length scale $B\Omega L/U_o \gg 1$. If this is a large parameter then blade-to-blade correlation needs to be considered, if not then Amiet's approximation can be applied.

This feature was originally identified by Sevik [4] for a propeller operating in grid-generated isotropic homogeneous turbulence. It was expected that the spectrum from this interaction would be a smooth function of frequency determined by the wavenumber content of the inflow turbulence. However, Sevik found that the spectrum included a series of humps that peaked at frequencies slightly above the BPF. It was later shown by Martinez [5] that the shifting of the humps from the BPF was caused by the pitch angle of the blades to the axial flow.

To illustrate the characteristics of blade-to-blade correlation, consider a series of pulses that persist for a limited period of time such as would be generated by a rotor with $B$ blades cutting through an eddy at the blade-passing interval. The signature from each blade passage is the same, but the amplitude is modulated by the variation in the strength of the eddy as it passes though the rotor. A model for the time history of the sound produced by one eddy is

$$p(t) = \sum_{n=-\infty}^{\infty} f(t - 2\pi n/B\Omega)e^{-(U_o t/L)^2} \tag{16.5.2}$$

Here the envelope $\exp(-(U_o t/L)^2)$ defines the time variation in the strength of the eddy as it is convected through the rotor with axial velocity $U_o$. We model the observed signature of a single-blade passage through the eddy as $f(t) = (Ut/L_o)\exp(-(Ut/L_o)^2)$ which is the same shape as the signal shown in Fig. 16.15. In this model $L$ represents the axial lengthscale of the turbulence, $L_o$ is the transverse lengthscale, and $U$ is the flow speed relative to the blade.

Fig. 16.24A shows the time history modeled by Eq. (16.5.2). The spectrum of these pulses is shown in Fig. 16.24B for $B\Omega L/U_o = B\pi/4$ and $B\Omega L_o/U = 2$. The time scale of the interaction is relatively small in this case indicating that the axial flow speed $U_o$ is large and the axial lengthscale $L$ is small, so the blades only chop the eddy once, and the spectrum of the pressure time history is comparatively smooth. However, when the axial flow speed is reduced, and the lengthscale $L$ is increased, the criterion for haystacking $B\Omega L/U_o$ is increased. Fig. 16.25A gives the time history for the case when $B\Omega L/U_o = B\pi$. Because several blades interact with a single eddy the time history has multiple

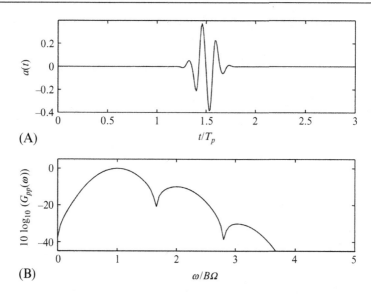

**Fig. 16.24** Signal from interactions of rotor blades with a single small eddy modulated by the time variation of the eddy strength at the rotor position. (A) Time history modeled by Eq. (16.5.2) and (B) spectrum of the time histories in figure (A). $B=7$, $B\Omega L/U_o=B\pi/4$, and $B\Omega L_o/U=2$.

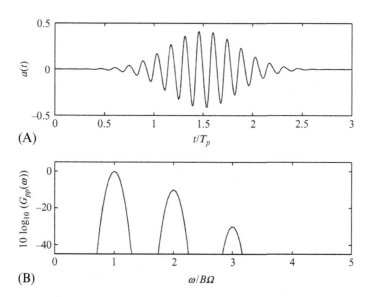

**Fig. 16.25** Signal from interactions of rotor blades with a single large eddy modulated by the time variation of the eddy strength at the rotor position. (A) Time history modeled by Eq. (16.5.2) and (B) the spectra of the time histories in figure (A). $B=7$, $B\Omega L/U_o=B\pi$, and $B\Omega L_o/U=2$.

pulses at the blade-passing interval and the spectrum (Fig. 16.25B) shows clearly defined peaks at the blade-passing frequencies. This is the haystacking phenomenon.

Details of a time domain approach to this problem can be found in Glegg et al. [6].

## 16.6   Blade vortex interactions

When a helicopter undertakes a maneuver the trailing tip vortices shown in Fig. 16.7 can be ingested into the rotor and a BVI can occur. In certain flight regimes these interactions can occur when the axis of the vortex is parallel to the blade leading edge as shown in Fig. 16.26 for an advancing rotor.

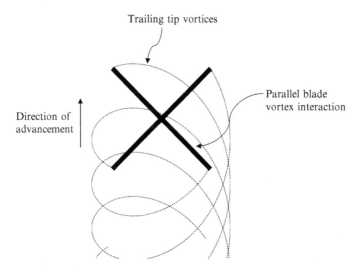

**Fig. 16.26** The trailing tip vortices from a helicopter in flight. The *direction* of advancement shows the flight direction, and the *dashed lines* show the trailing tip vortices from each blade.

In Section 14.4 we showed that BVI depended on the angle that the axis of the vortex makes with the blade leading edge and the distance of the vortex core from the blade. Parallel BVIs are therefore the most important sources of sound. The characteristic sound of a BVI is a loud thumping sound that is caused by the impulsive unsteady loading described in both Sections 14.4 and 7.5 and, because the time scale of the pulse is usually short compared to the blade passage interval, the pulses are heard as individual events and can be analyzed as such. The spectrum from multiple pulses will then be given by the spectral harmonics of a single pulse at the blade-passing frequencies.

To calculate the acoustic field from a BVI we can use Amiet's method and consider the pulse to be short enough that the blade is in linear motion during the BVI. The BVI takes place at a specific point in the rotor disc plane, and so only one or two segments of the plane, as shown in Fig. 16.21, need to be included in the analysis. We can therefore use Eq. (16.4.9) to predict the far-field sound by only considering the specific values of $m,n$ that define the location of the BVI. The pulses are of short duration, and so the segment time interval required in Amiet's method can be taken as being long compared to the pulse so that the integral in Eq. (16.4.9) is dominated by the sinc function and Eq. (16.4.9) is well approximated by

$$\widetilde{p}(\mathbf{x}, \omega) = \left( \frac{i\pi\omega x_i n_i^{(m,\,n)} e^{i\omega r(\tau_m)/c_\infty}}{r_o^2} \right) \Delta \widetilde{p}^{\,(m,\,n)} \left( ka_m, kb_m, \omega \left( 1 - M_r^{(m,\,n)} \right) \right) \quad (16.6.1)$$

for the segment where the BVI occurs. To evaluate this expression, we require the wavenumber spectrum of the pressure jump across the blade, and this is obtained from Eq. (14.2.2) and takes the form

$$\Delta \widetilde{p}(ka_m, kb_m, \omega) = \frac{1}{2} \rho_o c \widetilde{w}_2(\omega/U, kb_m) \Lambda(ka_m, kb_m, \omega, M) \quad (16.6.2)$$

where the upwash velocity spectrum is given by Eq. (14.4.2) as

$$\widetilde{w}_2(k_1, k_3) = \frac{i\Gamma L(k_3 \sin\phi_v - k_1 \cos\phi_v) e^{-k_{13}h}}{(2\pi)^2 2k_{13}} \left( \frac{\sin\left((k_1 \sin\phi_v + k_3 \cos\phi_v)L/2\right)}{(k_1 \sin\phi_v + k_3 \cos\phi_v)L/2} \right)$$

$$(16.6.3)$$

with $k_{13} = \left(k_1^2 + k_3^2\right)^{1/2}$.

The key features of this result are the angle of the BVI $\phi_v$ and the vortex miss distance $h$. As discussed in Section 14.4 if the interaction angle does not meet the criterion that

$$\tan\phi_v < \beta M$$

then no significant sound will be radiated because the vortex interaction with the blade leading edge moves subsonically (see Section 14.4). However, if this criterion is met then a BVI that radiates sound will occur and will depend on the miss distance of the vortex from the blade. This introduces a factor $\exp(-k_{13}h)$ in Eq. (16.6.3) which is well approximated by $\exp(-|\omega h/U|)$ and suppresses the high-frequency content of the gust when the miss distance $h$ is large. It is seen from these results that both increasing the vortex miss distance $h$ and the interaction angle $\phi_v$ reduce the sound level, and for noise control purposes they can be optimized to be an effective noise reduction tool.

# References

[1] F. Farassat, Linear acoustic formulas for calculating rotor blade noise, AIAA J. 19 (1981) 1122–1130.
[2] D. Hanson, Helicoidal surface theory for harmonic noise of propellers in the far field, AIAA J. 18 (1980) 1213–1220.

[3] R.K. Amiet, Noise produced by turbulent flow into a propeller or helicopter rotor, AIAA J. 15 (1977) 307–308.

[4] M. Sevik, Sound radiation from subsonic rotor subjected to turbulence, in: International Symposium on Fluid Mechanics and Design of Turbomachinery, University Park, PA, 1973.

[5] R. Martinez, Asymptotic theory of broadband rotor thrust, part II: analysis of the right frequency shift of the maximum response, J. Appl. Mech. 63 (1996) 143–148.

[6] S.A.L. Glegg, W. Devenport, N. Alexander, Broadband rotor noise predictions using a time domain approach, J. Sound Vib. 335 (2015) 115–124.

# Duct acoustics

**17**

In many applications aeroacoustic sources occur in ducted environments. A most important application is, of course, the internal sources of noise on a high bypass-ratio turbofan engine that is commonly used in commercial aircraft transportation. The duct has a large impact on both the flow through the engine and the acoustic source efficiency. In this chapter the important issues of duct propagation will be discussed, including the effect of acoustic absorption by material that can be placed on the duct walls to attenuate the sound before it is radiated from the duct exits to the acoustic far field.

## 17.1  Introduction

In the early days of commercial air transportation the noise from aircraft was dominated by jet noise, which scales with the sixth or eighth power of the jet velocity depending on the temperature of the jet. However, in the 1970s high bypass-ratio turbofan engines were introduced, and this enabled the same thrust to be obtained with a lower jet exit velocity relative to the surrounding flow and a corresponding reduction in jet noise. Aircraft noise levels were significantly reduced as a consequence, and the fan noise sources became comparable in level to the noise from the jet. To further reduce aircraft noise, the fan noise sources had to be minimized as well as the jet noise, and consequently ducted fan noise has become an important consideration in low noise aircraft engine design.

The design of a typical high bypass-ratio aircraft engine is shown in Fig. 17.1. The outer duct extends from the engine inlet to the bypass duct exit and is supported by the stator vanes and struts that are downstream of the fan. Just aft of the fan is the compressor inlet, which leads to the combustion chamber and the turbine, and the high speed flow generated by combustion exhausts through the turbine exit to form the jet core. The fan generates thrust, and the turbulent flow that results from the loaded fan blades impinges on the downstream stator vanes. The wake flow is highly turbulent and has a swirling motion, and the stator vanes are designed to reduce the swirl, recovering the energy lost to angular momentum downstream of the fan. The primary source of noise in the engine has been found to be the impingement of the wake from the fan onto the stator vanes.

In general, the outer duct of the engine is circular, but in some applications the inlet is modified so that it interferes less with the aerodynamic performance of the aircraft or enables additional ground clearance when the engine is mounted below the wing. The duct shape also has a varying cross section, and this can impact the propagation of acoustic waves along the duct. However, a great deal can be learned from studying the

Aeroacoustics of Low Mach Number Flows. http://dx.doi.org/10.1016/B978-0-12-809651-2.00017-5

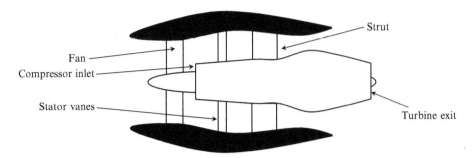

**Fig. 17.1** Schematic of a high bypass-ratio turbofan engine.

acoustic propagation in circular ducts and treating the variations in the duct cross section as a second-order effect. However, there are instances when the variation in duct cross section is vital to the understanding of sound propagation, and we will discuss these effects in more detail later.

In the analysis given in this chapter we will assume that sound propagation along the duct is linear, which implies the use of Goldstein's equation given in Chapter 6, and excludes the nonlinear propagation of sound associated with buzz-saw noise (noise caused by the rotating shock structure produced when the rotor is operated supersonically). We will limit consideration to circular ducts with a steady mean flow, which may be a function of radius and may include a swirling flow. We will also evaluate the effect of liners on the duct walls and radiation from the duct exits upstream or downstream of the fan. In general, we will consider these effects as idealized with simple boundary conditions so that we can identify the physical effects that are taking place. For an accurate calculation for aeroacoustic sources in a duct, numerical methods must be used. However, these are beyond the scope of the current treatment, and so references will be provided when appropriate.

## 17.2   The sound in a cylindrical duct

### 17.2.1   General formulation

We start by considering the linear acoustics problem of sound propagation in a cylindrical duct that has a mean flow. It will be assumed that the mean flow is uniform in the axial direction and that there are no vortical or turbulent perturbations introduced upstream of the region of interest. The duct will initially be taken to be infinitely long, but in later sections we will consider ducts of finite length.

Given these assumptions we can use Goldstein's equation to describe the acoustic waves in the duct in terms of the velocity potential $\phi$ as a function of the cylindrical coordinates $(R, \varphi, x)$ shown in Fig. 17.2A. We will also consider cases when the duct has a center body as shown in Fig. 17.2B. Expressing Goldstein's equation (in the form of Eq. (6.3.2) with no source term) in cylindrical coordinates with an axisymmetric mean flow and a uniform sound speed and mean density, yields

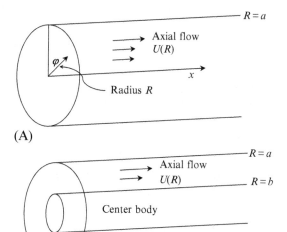

**Fig. 17.2** (A) A cylindrical duct of radius $a$ with no center body with an axial mean flow $U$. Cylindrical coordinates are shown as $(R, \varphi, x)$, and the outer duct radius is $R = a$. (B) A cylindrical duct with a center body. The outer duct radius is $a$, and the center body radius is $b$.

$$\frac{1}{c_\infty^2}\frac{D_\infty^2\phi}{Dt^2} - \frac{\partial^2\phi}{\partial x^2} - \frac{1}{R^2}\frac{\partial^2\phi}{\partial\varphi^2} - \frac{1}{R}\frac{\partial}{\partial R}\left(R\frac{\partial\phi}{\partial R}\right) = 0 \tag{17.2.1}$$

We will solve this equation for a harmonic time dependence $\exp(-i\omega t)$ and a mean flow in the axial direction. We make use of the fact that the sound field is periodic in the azimuthal direction, so the potential can be expanded as a Fourier series giving

$$\phi(R, \varphi, x, t) = \sum_{m=-\infty}^{\infty} \hat{\phi}_m(R, x)e^{-i\omega t - im\varphi} \tag{17.2.2}$$

where $\hat{\phi}_m(R, x)$ are the complex Fourier series coefficients of order $m$ that define the sound field. This simplifies Eq. (17.2.1) to

$$\left(k^2 - \frac{m^2}{R^2}\right)\hat{\phi}_m + 2ikM\frac{\partial\hat{\phi}_m}{\partial x} + (1 - M^2)\frac{\partial^2\hat{\phi}_m}{\partial x^2} + \frac{1}{R}\frac{\partial}{\partial R}\left(R\frac{\partial\hat{\phi}_m}{\partial R}\right) = 0 \tag{17.2.3}$$

where $M = U/c_\infty$ is the Mach number of the axial flow. Eq. (17.2.3) is the basic equation that describes the wave propagation in a circular duct.

## 17.2.2 Hard-walled ducts

The solution of this equation will depend on the boundary conditions that are imposed. We will start by requiring that the duct walls are hard and velocity perturbations in the direction normal to the wall are zero. If the mean flow is independent of the radius, then we can use the method of separation of variables to find a solution to Eq. (17.2.3). In this approach we specify

$$\hat{\phi}_m(R, x) = X_m(x)\psi_m(R) \tag{17.2.4}$$

then Eq. (17.2.3) can be simplified by substituting for $\hat{\phi}_m$ and dividing by $X_m\psi_m$, so

$$\left(k^2 - \frac{m^2}{R^2}\right) + \frac{2ikM}{X_m}\frac{\partial X_m}{\partial x} + \frac{(1-M^2)}{X_m}\frac{\partial^2 X_m}{\partial x^2} + \frac{1}{R\psi_m}\frac{\partial}{\partial R}\left(R\frac{\partial \psi_m}{\partial R}\right) = 0 \tag{17.2.5}$$

The second and third terms are independent of $R$ and so must be equal to a constant if this equation is to apply for all values of $x$. We will choose this constant to be $-\kappa^2$, and so we obtain

$$\frac{2ikM}{X_m}\frac{\partial X_m}{\partial x} + \frac{(1-M^2)}{X_m}\frac{\partial^2 X_m}{\partial x^2} = -\kappa^2$$
$$\left(k^2 - \frac{m^2}{R^2} - \kappa^2\right) + \frac{1}{R\psi_m}\frac{\partial}{\partial R}\left(R\frac{\partial \psi_m}{\partial R}\right) = 0 \tag{17.2.6}$$

To solve the first of these two equations we can seek a solution in the form $X_m(x) = C_m\exp(i\mu x)$, where $C_m$ is a constant. The value of $\mu$ is then obtained from the dispersion relationship, by substituting this expansion for $X_m(x)$ into the first expression in Eq. (17.2.6)

$$\mu^2 + \frac{2kM\mu}{\beta^2} = \frac{\kappa^2}{\beta^2}$$

which has two possible solutions given by

$$\mu = \frac{kM}{\beta^2} \pm \sqrt{\left(\frac{kM}{\beta^2}\right)^2 + \frac{\kappa^2}{\beta^2}} \tag{17.2.7}$$

Similarly, we can solve the second of the two equations in Eq. (17.2.6) by multiplying through by $\psi_m$ and expanding the differential, so

$$\frac{\partial^2 \psi_m}{\partial R^2} + \frac{1}{R}\frac{\partial \psi_m}{\partial R} + \left(\alpha^2 - \frac{m^2}{R^2}\right)\psi_m = 0, \qquad \alpha^2 = k^2 - \kappa^2 \tag{17.2.8}$$

This is Bessel's equation which has the well-known solution

$$\psi_m(\alpha R) = AJ_m(\alpha R) + BY_m(\alpha R)$$

where $J_m(\alpha R)$ and $Y_m(\alpha R)$ are Bessel functions of the first and second kind of order $m$ and are illustrated in Fig. 17.3. We see that the Bessel function of the first kind is finite for all values of $\alpha R$ and is zero at $\alpha R = 0$ for all orders $m \neq 0$. In contrast the

Bessel functions of the second kind are infinite at $\alpha R = 0$ for all orders. Both functions are oscillatory for large values of $\alpha R$ and decay to zero as $(\alpha R)^{-1/2}$ for large arguments.

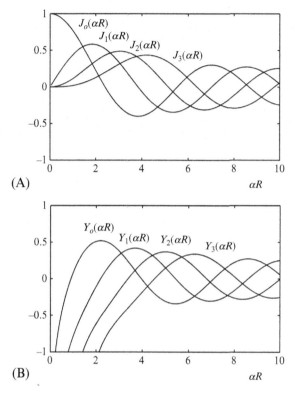

**Fig. 17.3** Examples of Bessel functions (A) of the first kind $J_m(\alpha R)$ and (B) of the second kind $Y_m(\alpha R)$.

For the special case when the duct has no center body the sound field must remain finite at $R = 0$, and this eliminates $Y_m(\alpha R)$ as a possible solution. The sound field in the duct then only depends on $J_m(\alpha R)$. However, the solution must also match the non-penetration boundary condition on the outer duct wall, and so we require that the derivative of $J_m(\alpha R)$ with respect to $R$ is zero at the wall where $R = a$. This is only possible for values of $\alpha$ for which $\partial J_m(\alpha R)/\partial R = 0$ when $R = a$. There will be an infinite number of values of $\alpha$ that meet this condition, and they will be defined as $\alpha_{mn}$. For example, Fig. 17.4A and B shows the functions $\psi_m(\alpha_{mn}R) = J_m(\alpha_{mn}R)$ for different values of $m$ and $n$. Of particular note is the case when $m = 0$, $n = 0$ for which $\alpha_{mn} = 0$. This represents the plane wave mode that has no radial variation, so the sound field is constant across the duct.

If the duct has a center body with radius $b$ then the boundary condition at $R = b$ is satisfied if the derivative of $\psi_m(\alpha R)$ with respect to $R$ is zero at both walls. The boundary condition at $R = b$ is met if

$$\psi_m(\alpha R) = J_m(\alpha R) - \frac{J'_m(\alpha b)Y_m(\alpha R)}{Y'_m(\alpha b)} \tag{17.2.9}$$

where the prime represents a differentiation with respect to the argument of the function. To satisfy the boundary condition on the outer wall $\alpha$ must take on the values for which

$$\psi'_m(\alpha a) = J'_m(\alpha a) - \frac{J'_m(\alpha b)Y'_m(\alpha a)}{Y'_m(\alpha b)} = 0$$

As with the duct without a center body there are an infinite number of solutions to this equation, and each represents a characteristic function or mode of propagation defined by the functions $\psi_m(\alpha_{mn}R)$. Examples of these modes for a duct with a center body are shown in Fig. 17.4C and D. Note that the $m=0$, $n=0$ case yields a plane wave for which $\alpha_{mn}=0$ as was the case for the duct without a center body.

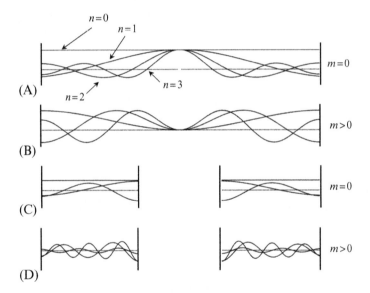

**Fig. 17.4** Mode shapes for hard-walled ducts (A) no center body, $m=0$; (B) no center body, $m>0$; (C) with center body, $m=0$; and (D) with center body, $m>0$. The *dashed line* gives $\psi_m(\alpha_{mn}R)=0$.

Eq. (17.2.8) is also identifiable as a Sturm Liouville equation with boundary conditions $\psi'_m(\alpha a) = \psi'_m(\alpha b) = 0$. The theory of differential equations shows that the solution of this equation is given by the sum of a set of eigenfunctions of the form

$$\sum_{n=0}^{\infty} A_{mn}\psi_m(\alpha_{mn}R)$$

We commonly refer to the eigenfunctions as modes of propagation, and each mode individually satisfies the boundary conditions. Furthermore, Sturm Liouville theory shows that the modes are orthogonal which means that

$$\int_a^b \psi_m(\alpha_{mn}R)\psi_m(\alpha_{ms}R)RdR = \delta_{ns}C_{mn}$$

where $C_{mn}$ is a constant. Using the properties of the Fourier series that was used to expand the solution as a function of azimuthal angle $\varphi$ (Eq. 17.2.2) we then find that

$$\int_a^b \int_0^{2\pi} \psi_m(\alpha_{mn}R)\psi_j(\alpha_{js}R)e^{i(m-j)\varphi}RdRd\varphi = \delta_{ns}\delta_{mj}\Lambda_{mn} \tag{17.2.10}$$

This is a valuable property that we will use to identify modes and evaluate how they are coupled to acoustic sources in the duct. Note that the integral is now over the duct cross-sectional area, and we can define the constant $\Lambda_{mn}$ as

$$\Lambda_{mn} = \int_S |\psi_m(\alpha_{mn}R)|^2 dS \tag{17.2.11}$$

where $S$ is the duct cross sectional area.

The key element required to define the modes is the wavenumbers $\alpha_{mn}$ that are solutions to $\psi_m'(\alpha_{mn}a) = 0$. For the duct without a center body for which $\psi_m(\alpha_{mn}a) = J_m(\alpha_{mn}a)$ these values are tabulated and readily available. For large values of $n$ there is also an asymptotic solution given by

$$\alpha_{mn}a \approx f_{mn} - \frac{(4m^2 + 3)}{f_{mn}} + \cdots O(1/f_{mn}^3),$$

$$f_{mn} = (n + m/2 - 3/4)\pi \gg 1 \tag{17.2.12}$$

For a duct with a center body an approximate high-frequency solution can be obtained for the mode shapes given by Eq. (17.2.9) by using the large argument approximations of the Bessel functions

$$J_m(\alpha R) \approx \sqrt{\frac{2}{\pi \alpha R}} \left\{ \cos\left(\alpha R - \frac{\pi}{4} - \frac{m\pi}{4}\right) + O\left(\frac{1}{\alpha R}\right) \right\}$$

$$Y_m(\alpha R) \approx \sqrt{\frac{2}{\pi \alpha R}} \left\{ \sin\left(\alpha R - \frac{\pi}{4} - \frac{m\pi}{4}\right) + O\left(\frac{1}{\alpha R}\right) \right\}$$

The approximate mode shapes of a hard-walled duct with a center body are then

$$\psi_m(\alpha R) = \frac{\cos(\alpha(R - b))}{\sqrt{R/(a + b)}} \tag{17.2.13}$$

where the normalization of the amplitude of the mode has been chosen for convenience (see below), and the wavenumbers are given by the solutions to $\tan(\alpha(a-b))=1/2\alpha a$. For large arguments this gives, to a first approximation, $\alpha_{mn}=n\pi/(a-b)$. The important aspect of this result is that for ducts with center bodies the mode shapes are characterized by a relatively simple function that closely resembles a cosine wave with maxima or minima at the duct walls and $n$ zero crossing points, as illustrated in Fig. 17.4C and D.

Finally, we note that the normalization parameter for a duct without a center body is given by

$$\Lambda_{mn} = \pi\left(a^2 - \frac{m^2}{\alpha_{mn}^2}\right)J_m^2(\alpha_{mn}a) \tag{17.2.14}$$

and for the duct modes with a center body $\Lambda_{mn}=(a^2-b^2)\pi$ which is the cross-sectional area of the duct and justifies the choice of normalization used in Eq. (17.2.13).

### 17.2.3  Modal propagation

We can now summarize these results and combine Eqs. (17.2.2), (17.2.4) to give the modal description of sound propagation in a duct as

$$\phi(R,\varphi,x,t) = \sum_{m=-\infty}^{\infty}\sum_{n=0}^{\infty}\tilde{A}_{mn}\Psi_m(\alpha_{mn}R)e^{-i\omega t - im\varphi + i\mu_{mn}^{\pm}x} \tag{17.2.15}$$

where from the dispersion relationship (17.2.7) and (17.2.8) we define

$$\mu_{mn}^{\pm} = -\frac{kM}{\beta^2} \pm \sqrt{\left(\frac{kM}{\beta^2}\right)^2 + \frac{k^2 - \alpha_{mn}^2}{\beta^2}}$$

as the wavenumber that specifies the propagation in the axial direction. This is more conveniently written by rearranging terms as

$$\mu_{mn}^{\pm} = \frac{-kM \pm k_{mn}}{\beta^2}, \qquad k_{mn} = \sqrt{k^2 - \beta^2\alpha_{mn}^2}, \qquad \beta^2 = 1 - M^2 \tag{17.2.16}$$

This wavenumber tells a great deal about the wave propagation in the duct. First we note that the $\pm$ sign is chosen to represent waves propagating in the positive or negative $x$ direction when the value of the square root is real and taken to be positive. When the argument of the square root is negative (which occurs when $\beta\alpha_{mn} > k$) then it must have a positive imaginary part to ensure that the wave decays in the direction of propagation. It follows that waves will either propagate as waves or decay with distance in either the upstream or downstream direction as defined by the $\pm$ sign in Eq. (17.2.16). When the waves decay they are classified as being *cutoff*, and when they propagate they are defined as being *cut on*. The rate of decay depends on the "cutoff" ratio

$$\chi_{mn} = \beta \alpha_{mn}/k \tag{17.2.17}$$

If the cutoff ratio is large $\chi_{mn} \gg 1$ then the value of $\mu_{mn}^{\pm}$ has a large positive or negative imaginary part, and the duct mode of order $m,n$ decays rapidly with distance along the duct. On the other hand, when the cutoff ratio is very small then the value of $k_{mn}$ is real, and the duct mode of order $m,n$ propagates along the duct without attenuation with the wavenumber

$$\frac{k}{\beta^2}(-M \pm 1) = \frac{\pm k}{1 \pm M}$$

which is consistent with upstream or downstream propagation of a plane wave in a uniform flow with Mach number $M$. When the cutoff ratio $\chi_{mn}$ is of order 1 then the modes are said to be close to cutoff and $k_{mn}$ tends to zero.

The decay of cutoff modes is an important feature of duct acoustics because it limits the number of acoustic modes that will propagate from a source to a duct exit, where they can radiate to the acoustic far field. If the source is a large distance from the duct exit then the cutoff modes play no role in the far-field radiation, but if the source is close to the duct exit then cutoff modes cannot be neglected. The rate of decay of a cutoff mode depends on

$$\exp\left(-\left(k|x|/\beta^2\right)\sqrt{\chi_{mn}^2 - 1}\right)$$

where $|x|$ is the distance from the source. When $\chi_{mn} \gg 1$ the amplitude of the mode decays to zero over a distance that is a fraction of an acoustic wavelength. However, when the mode is close to cut off then the decay is relatively slow. This is important because the fan design can be tailored so that the acoustic modes are cutoff, and this can result in significant far-field noise reductions.

An important property of Bessel functions is that the first zero of $J'_m(\alpha a)$ will occur when $\alpha a > m$. Consequently, for a duct without a center body there will only be propagating duct modes when

$$ka > \beta\alpha_{m1}a > \beta m$$

and, as a consequence, modes will only propagate when $2\pi a/\lambda > \beta m$. This is the ratio of the duct circumference to the acoustic wavelength and must be greater than $m$ in order for the mode of order $m$ to be cut on.

Another feature of the modal expansion of the sound field (Eq. 17.2.15) is that the modes are spinning as they propagate along the duct. The axial propagation speed is given by $\omega/\mu_{mn}^{\pm}$, and the speed of angular rotation is given by $\omega/m$. Since $ka/\beta > m$ it follows that the angular speed at the outer duct wall is $\omega a/m > \omega\beta/k = c_{\infty}\beta$. The angular speed of a propagating mode at the duct wall is therefore supersonic. This is an important characteristic of duct propagation because it implies that rotating sources, such as fans with subsonic tip speeds, will not couple directly with propagating acoustic modes. This issue will be discussed in more depth in the next chapter where it will be shown that subsonic fan noise sources only couple with duct modes if their rate of rotation is "stepped up" or increased by source interactions that effectively increase their rate of rotation.

## 17.3   Duct liners

In the previous section it was assumed that the duct had a hard wall so that the acoustic perturbation velocity normal to the wall was zero. In a lined duct the relationship is more complicated and defined by the liner impedance which is a complex quantity given by the ratio of the pressure imposed on the liner to the acoustic velocity normal to the surface $z = \hat{p}/\hat{v}_s$ where both the pressure and particle velocity have a harmonic time dependence $\exp(-i\omega t)$. The wall is assumed to be locally reacting, which implies that acoustic waves do not propagate within the liner. To match the motion of the fluid with the motion at the liner we must ensure that the displacement normal to the surface $\xi$ is equal to the fluid particle displacement at the same point. The particle velocity of the fluid normal to the surface is given by the rate of change of the particle displacement in a frame of reference moving with the mean flow, so the relationship between the acoustic velocity potential $\phi$ and the acoustic particle displacement is

$$\mathbf{n} \cdot \nabla \phi = \frac{D\xi}{Dt} \tag{17.3.1}$$

where $\mathbf{n}$ is the unit normal to the surface. To obtain a boundary condition for the acoustic velocity potential with a harmonic time dependence we use the relationship between the pressure and the potential given by Eq. (6.1.7) so that

$$\hat{p}e^{-i\omega t} = -\rho_o \frac{D_o}{Dt}\left(\hat{\phi}e^{-i\omega t}\right)$$

For the liner, however, the relationship between its displacement and velocity normal to the surface is $v_s = \partial\xi/\partial t$, and so

$$-\rho_o \frac{D_o}{Dt}\left(\hat{\phi}e^{-i\omega t}\right) = -i\omega z\hat{\xi}e^{-i\omega t} \tag{17.3.2}$$

We take the substantial derivative of Eq. (17.3.2) and linearize Eq. (17.3.1) about the mean flow to obtain the boundary condition

$$-\rho_o \frac{D_o^2}{Dt^2}\left(\hat{\phi}e^{-i\omega t}\right) = -i\omega z \frac{D_o}{Dt}\left(\hat{\xi}e^{-i\omega t}\right) = -i\omega z\mathbf{n} \cdot \nabla\hat{\phi}e^{-i\omega t} \tag{17.3.3}$$

The boundary conditions for the Sturm Liouville problem discussed in the previous section need to be modified to account for the liner. Using Eq. (17.3.3) we find that boundary condition for the radial modes in Eq. (17.2.15) becomes

$$\psi'_m(\alpha_{mn}a) = \frac{\rho_o(\omega - \mu^\pm_{mn}U)^2 \psi_m(\alpha_{mn}a)}{-i\omega\alpha_{mn}z} \tag{17.3.4}$$

where the wavenumber of propagation along the duct is defined as a function of the radial wavenumber as given by Eq. (17.2.16). This is often written in terms of the nondimensional admittance $\beta_a = \rho_o c_\infty/z$ and takes the form

$$\frac{\alpha_{mn}\psi'_m(\alpha_{mn}a)}{\psi_m(\alpha_{mn}a)} = ik\beta_a\left(1 - \mu^\pm_{mn}M/k\right)^2 \tag{17.3.5}$$

This is a far more complicated boundary condition than for a hard-walled duct (which has an admittance of zero) and needs to be computed numerically. However, some insight can be obtained by considering approximate solutions for small admittances and no axial flow. Considering a Taylor series expansion of Eq. (17.3.5) gives

$$\psi'_m(\alpha_{mn}a) \approx \psi'_m\left(\alpha^{(o)}_{mn}a\right) + \left(\alpha_{mn} - \alpha^{(o)}_{mn}\right)a\psi''_m\left(\alpha^{(o)}_{mn}a\right) + \cdots$$

where $\alpha^{(o)}_{mn}a$ are the solutions for the hard-walled duct, and so $\psi'_m\left(\alpha^{(o)}_{mn}a\right) = 0$. From Eq. (17.2.8) we find that

$$\psi''_m\left(\alpha^{(o)}_{mn}a\right) = -\left(1 - \left(\frac{m}{\alpha^{(o)}_{mn}a}\right)^2\right)\psi_m\left(\alpha^{(o)}_{mn}a\right)$$

and so we can approximate Eq. (17.3.5) when $M=0$ as

$$-\left(\alpha_{mn} - \alpha^{(o)}_{mn}\right)\alpha^{(o)}_{mn}a\left(1 - \left(\frac{m}{\alpha^{(o)}_{mn}a}\right)^2\right) = ik\beta_a \tag{17.3.6}$$

This gives a relatively simple approximation for the radial wavenumber.

To determine the effect on mode attenuation along the duct we need to consider the axial wavenumber, which, for zero flow, can be approximated from Eq. (17.2.16) as

$$k_{mn} = k^{(o)}_{mn} - \left(\alpha_{mn} - \alpha^{(o)}_{mn}\right)\frac{\alpha^{(o)}_{mn}}{k^{(o)}_{mn}} + \cdots$$

and combining this with Eq. (17.3.6) we obtain

$$k_{mn} = k_{mn}^{(o)} + \frac{ik\beta_a}{k_{mn}^{(o)}a\left(1 - (m/\alpha^{(o)}{}_{mn}a)^2\right)} + \cdots \tag{17.3.7}$$

This shows that the effect of the liner impedance is to give the axial wavenumber an imaginary part, which causes the modes to attenuate as they propagate along the duct. The amount of attenuation is determined by not only the admittance but also the wavenumbers of propagation. For the lowest order mode, we have noted in Section 17.2 that $m/\alpha_{mo}^{(o)} a$ is approximately 1, and so this mode will be highly attenuated by the liner because the imaginary part of the axial wavenumber will be large. However, this effect is reduced for the higher-order radial modes for which $m/\alpha_{mo}^{(o)}a \ll 1$. Similarly, close to the cutoff frequency $k_{mn}^{(o)}$ will be small, and so the effect of the liner will be large. It follows that liners are important at reducing levels close to cutoff where typically duct mode power levels are high.

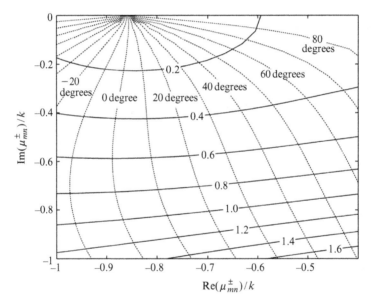

**Fig. 17.5** The admittance for different axial wavenumbers for a cylindrical duct without a center body for mode order $m=1$ with a flow Mach number $M=0.3$ at a frequency $ka=2$. The *solid lines* (horizontal) show the magnitude of the admittance, and the *dashed lines* (vertical) show lines of constant phase.

The more general result, with flow, requires a solution to Eqs. (17.3.5), (17.2.16), and can be achieved by evaluating these functions for different values of $\alpha_{mn}/k$ (both real and imaginary) and determining $\beta_a$ and $\mu_{mn}^{\pm}/k$. For example, in Fig. 17.5 we show contours of the admittance $|\beta_a|$, and its phase plotted against the real and imaginary parts of $\mu_{mn}^{\pm}/k$. In this case the flow Mach number is 0.3 and $ka=2$, and there is no center body in the duct. Note that for a hard wall the admittance is zero, and the curves lie on the real axis at $\mu_{mn}^{\pm}/k = -0.855$, which is the phase speed for the $m = 1$ mode propagating upstream at this frequency. As the magnitude of the admittance is increased the imaginary part of $\mu_{mn}^{\pm}/k$ is no longer zero and takes on significant negative values, which attenuates the mode as it propagates. For the wave to decay as it propagates upstream the imaginary part of $\mu_{mn}^{\pm}/k$ must be negative as shown, and the attenuation of this mode in decibels is given by $20\log_{10}\left(\exp\left(-|\mathrm{Im}(\mu_{mn}^{\pm})x|\right)\right)$ and gives an attenuation of about 10 dB per wavelength when the magnitude of the admittance is $> 0.2$ and the phase lies in the range $\pm 30$ degrees.

## 17.4 The Green's function for a source in a cylindrical duct

To analyze aeroacoustic sources in a duct we can make use of either Curle's theorem or the Ffowcs-Williams and Hawkings equation written in terms of a suitable Green's function. In general, a Green's function in a duct with a nonuniform mean flow and liners is not readily available in an analytical form, but we can obtain useful results by limiting consideration to a uniform flow in a hard-walled duct. The more complex situation of sources in a nonuniform swirling flow will be considered in Section 17.6. For the case of uniform flow in a hard-walled duct the Green's function is chosen so that its derivative normal to the duct wall is zero. The surface integrals over the duct walls in the Ffowcs-Williams and Hawkings equation (5.2.9) are thus eliminated, and only the surface integrals over the fan blades and stator vanes need to be considered. It should also be noted that Lighthill's equation only applies in a uniform mean flow. If the flow in the duct is nonuniform, then this has to be accounted for separately in the evaluation of the source terms. In principle this can be achieved by coupling the wave field in the duct to the source terms on a Ffowcs-Williams and Hawkings surface surrounding the sources in the acoustic near field and using a separate solution for the wave propagation in the duct. However, the key part of the calculation remains as the evaluation of the Ffowcs-Williams and Hawkings equation using a Green's function with a uniform mean flow, that is the solution to

$$\frac{1}{c_{\infty}^2}\frac{D_{\infty}^2 G_e}{D\tau} - \nabla^2 G_e = \delta(\mathbf{x} - \mathbf{y})\delta(t - \tau)$$

As in Section 3.10 we will first consider the Green's function defined in terms of its Fourier transform with respect to time. We can then use Eq. (3.10.7) to obtain the equivalent result in the time domain, which can be used in Eq. (5.4.4) for a source in a duct with flow. Substituting Eq. (3.10.7) into the equation above and evaluating all the integrals gives the Green's function in the frequency domain as the solution to

$$\nabla^2 \widetilde{G}_e + k^2 \widetilde{G}_e - 2ikM\frac{\partial \widetilde{G}_e}{\partial x_o} - M^2\frac{\partial^2 \widetilde{G}_e}{\partial x_o^2} = -\delta(R - R_o)\delta(x - x_o)\delta(\varphi - \varphi_o)/R$$

$$(17.4.1)$$

where the right-hand side of this equation is defined in cylindrical coordinates and meets the requirements of Eq. (3.9.4). The source location is given by $(R_o, \varphi_o, x_o)$ and the observer location by $(R, \varphi, x)$, and the flow is defined as being uniform in the axial direction with speed $U$ so that $M = U/c_\infty$. Note also that the sign of the exponent in the definitions in Eq. (3.10.7) determines the signs of the terms on the left of this equation because the differential is carried out with respect to $\tau$.

In the region $x < x_o$ or $x > x_o$ this equation is identical to the homogeneous equation for sound propagation in a duct with flow $-U$ in the axial direction (see Eq. 17.2.3), so we expect the solution to have the same form as Eq. (17.2.15), and we can expand the Greens function as a set of modes given by

$$\tilde{G}_e(R,\varphi,x|R_o,\varphi_o,x_o) = \sum_{m=-\infty}^{\infty}\sum_{n=0}^{\infty} \tilde{A}_{mn}(R,\varphi,x)\psi_m(\alpha_{mn}R_o)e^{-im\varphi_o + ikx_o M/\beta^2 \pm ik_{mn}x_o/\beta^2}$$

where the sign in the exponent is chosen as positive when $x_o > x$ and as negative when $x_o < x$ (as will be justified below). As it stands this solution is discontinuous at $x = x_o$ and if it is used in Eq. (17.4.1) we will have to account for the discontinuity in the evaluation of the derivatives with respect to $x_o$. If the mode amplitudes are continuous at the source point, then it follows that

$$\tilde{G}_e(R,\varphi,x|R_o,\varphi_o,x_o) = \sum_{m=-\infty}^{\infty}\sum_{n=0}^{\infty} \tilde{A}_{mn}(R,\varphi)\psi_m(\alpha_{mn}R_o)e^{-im\varphi_o + ik(x_o-x)M/\beta^2 + ik_{mn}|x_o-x|/\beta^2}$$

When this result is used in Eq. (17.4.1) and the derivatives are evaluated using

$$\frac{\partial^2}{\partial x_o^2}\left(e^{ik(x_o-x)M/\beta^2 + ik_{mn}|x_o-x|/\beta^2}\right)$$

$$= \left(\frac{2ik_{mn}}{\beta^2}\right)\delta(x_o-x) - \left(\frac{kM + \text{sgn}(x_o-x)k_{mn}}{\beta^2}\right)^2 e^{ik(x_o-x)M/\beta^2 + ik_{mn}|x_o-x|/\beta^2}$$

we find that the differential equation for the Green's function reduces to

$$\sum_{m=-\infty}^{\infty}\sum_{n=0}^{\infty} 2ik_{mn}\tilde{A}_{mn}(R,\varphi)\psi_m(\alpha_{mn}R_o)e^{-im\varphi_o} = -\delta(R-R_o)\delta(\varphi-\varphi_o)/R$$

Since the modes form an orthogonal set we can use Eq. (17.2.10) to extract the mode amplitudes from this equation by multiplying each side by $\psi_m(\alpha_{mn}R)e^{im\varphi}$ and integrating over the duct cross section, giving

$$\tilde{A}_{mn}(R,\varphi) = \frac{-1}{2ik_{mn}\Lambda_{mn}}\psi_m(\alpha_{mn}R)e^{im\varphi}$$

and so the Green's function is given by

$$
\widetilde{G}_e(R,\varphi,x|R_o,\varphi_o,x_o) =
$$
$$
\sum_{m=-\infty}^{\infty} \sum_{n=0}^{\infty} \frac{i\psi_m(\alpha_{mn}R)\psi_m(\alpha_{mn}R_o)e^{-im(\varphi_o-\varphi)+ik(x_o-x)M/\beta^2+ik_{mn}|x_o-x|/\beta^2}}{2k_{mn}\Lambda_{mn}} \tag{17.4.2}
$$

and the solution in the time domain is obtained from the inverse transform specified in Eq. (3.10.7).

This result is fundamental to the understanding of sound propagation in a duct because it shows how sound will propagate from a source at $(R_o,\varphi_o,x_o)$ to an observer at $(R,\varphi,x)$. When $x>x_o$ then the waves propagate downstream with the same phase dependence as given by the modal expansion (17.2.15) and (17.2.16). Similarly, when $x<x_o$ the waves are propagating upstream with the correct phase dependence. We also note that this result is not reciprocal. If the source and observer positions are interchanged then the phase factor that depends on $ik(x_o-x)M/\beta^2$ will be reversed, and so the wave fields are not identical unless $M=0$. Reciprocity is only achieved if the direction of the flow is also reversed, and this is known as the reverse flow theorem, which applies for all sources in a steady flow not only when they are in a duct.

The Green's function represents the sound from a point monopole or volume displacement source, and Eq. (17.4.2) shows how this depends on the radial location of the source. If the source is located at a null of the radial mode shape, then it does not couple with the acoustic field in the duct and that mode is not excited. In contrast if the source is located at a maximum of the radial mode shape then that mode is strongly excited, and if that mode is close to cutoff ($k_{mn}\approx 0$) then it could dominate the sound field in the duct. Similar results apply to dipole and quadrupole sources since their acoustic efficiency is given by the derivative of the Green's function along the axis of the dipole. This can impact the efficiency of each mode because the derivatives will introduce a factor that depends on the mode wavenumber and/or order. This will be discussed in more detail in Chapter 18.

## 17.5  Sound power in ducts

In Section 2.6 we introduced the concept of sound power. An important application of this concept is to duct acoustics. Since waves propagate along the duct and out of the duct exit to the far field via a complicated path (see Fig. 17.1) the concept that sound power is conserved allows us to relate the in duct sound levels directly to the far field, assuming that there is no absorption at the duct walls and no sound power is reflected back toward the source by the duct exits or internal features. This of course is an important assumption that only applies for ducts of large diameter compared to the acoustic wavelength. For small diameter ducts such as car exhausts, the acoustic wavelength is large compared to the duct diameter and the reflections at the exhaust exit and at changes of cross section, such as the muffler, completely control the sound power radiated to the far field. In contrast on an aero engine the duct diameter is so large that waves propagate freely out of the inlet or exhaust, and reflections back toward the source are of secondary importance. This characteristic is very important for engine design purposes because it means that the sound power provides a measure of the effective noise source level, which corresponds to the expected level of the far-field sound.

In Section 2.6 we defined the sound power from a source in a volume bounded by the surface $S$ as

$$W = \int_S \mathbf{I} \cdot \mathbf{n} dS$$

where $\mathbf{I}$ is the acoustic intensity vector and $\mathbf{n}$ is the unit normal vector pointing out of the volume containing the sources. In a hard-walled duct, the intensity is zero normal to the duct walls, and so the integral is carried out over the duct cross sections upstream or downstream of the source. We can therefore split the sound power into its upstream and downstream components, which radiate from the engine inlet or exit, respectively.

The acoustic intensity in a moving fluid is given by Eq. (2.6.17) in terms of the acoustic particle velocity and acoustic pressure perturbations as

$$\mathbf{I} = E[(\rho_o \mathbf{u} + \rho' \mathbf{U})(p'/\rho_o + \mathbf{U} \cdot \mathbf{u})]$$

In the absence of vortical waves $\mathbf{u} = \nabla \phi$ and the pressure perturbation is given by $p' = \rho' c_\infty^2 = -\rho_o D_o \phi / Dt$, so

$$(p'/\rho_o + \mathbf{U} \cdot \mathbf{u}) = -\frac{D_o \phi}{Dt} + \mathbf{U} \cdot \nabla \phi = -\frac{\partial \phi}{\partial t}$$

The intensity can then be written in terms of the velocity potential as

$$\mathbf{I} = E\left[ -\rho_o \frac{\partial \phi}{\partial t} \left( \nabla \phi - \frac{\mathbf{U}}{c_\infty^2} \frac{D_o \phi}{Dt} \right) \right] \tag{17.5.1}$$

For waves with harmonic time dependence we can reduce this result using the approach given in Section 3.8 so that

$$\mathbf{I} = \frac{-\rho_o}{2} \operatorname{Re} \left( -i\omega \hat{\phi}^* \left( \nabla \hat{\phi} - \frac{\mathbf{U}}{c_\infty^2} (\mathbf{U} \cdot \nabla \hat{\phi} - i\omega \hat{\phi}) \right) \right) \tag{17.5.2}$$

For upstream or downstream propagating waves we need to consider the component of the intensity in the positive or negative axial direction and integrate over the cross section of the duct. For a uniform flow we can define the acoustic velocity potential using the modal expansion given by Eq. (17.2.15) so that

$$\left( \nabla \hat{\phi} - \frac{\mathbf{U}}{c_\infty^2} \left( \mathbf{U} \cdot \nabla \hat{\phi} - i\omega\hat{\phi} \right) \right) \cdot \mathbf{n}$$

$$= \pm i \sum_{m=-\infty}^{\infty} \sum_{n=0}^{\infty} \tilde{A}_{mn} \psi_m(\alpha_{mn} R) \left\{ \mu_{mn}^{\pm} - M^2 \left( \mu_{mn}^{\pm} - \omega/U \right) \right\} e^{-im\varphi + i\mu_{mn}^{\pm} x}$$

where the normal vector points away from the source and the $\pm$ refers to downstream or upstream propagation, respectively. The terms in the curly braces simplify to give $\mu_{mn}^{\pm}\beta^2 + kM = \pm k_{mn}$. When these results are used in Eq. (17.5.2) and the surface integral is carried out over the duct cross section, then we can make use of the orthogonality of the duct modes, given by Eq. (17.2.10), to obtain the sound power in either the upstream or downstream directions as

$$W_{\pm} = \frac{\omega \rho_o}{2} \sum_{m=-\infty}^{\infty} \sum_{n=0}^{\infty} |\tilde{A}_{mn}|^2 \operatorname{Re}(k_{mn}) \Lambda_{mn} \qquad (17.5.3)$$

This remarkably simple result has some important implications. First we note that the power for each mode is uncoupled, so we can treat a noise control problem mode by mode. Furthermore, the level is not simply a function of the mode amplitude but is also determined by the normalization factor $\Lambda_{mn}$ and the real part of the wavenumber $k_{mn}$. If the mode is cutoff, then $k_{mn}$ is imaginary, and there is no sound power transmitted in that mode. Consequently, only propagating modes contribute to the far-field sound power levels. An important result is that for a source in a duct the duct mode amplitudes are given by the Green's function specified in Eq. (17.4.2). This shows that the duct mode amplitudes are inversely proportional to $k_{mn}\Lambda_{mn}$, and so the sound power will tend to infinity at the cut on frequency where $k_{mn}$ is zero unless the source strength also tends to zero. This issue will be discussed in more detail in Chapter 18.

The consequence of this result is that if we use the modal sound power to evaluate sources in the duct only the amplitude of propagating modes needs be considered. Since these are limited to a finite number of modes the infinite summations in Eq. (17.5.3) are no longer required making their evaluation tractable.

## 17.6 Nonuniform mean flow

In Section 17.3 a formulation was given for acoustic propagation in a cylindrical duct with a uniform mean flow. It was shown that a modal solution could be obtained by using the method of separation of variables, so the wave equation was reduced to solving a Sturm Liouville equation. However, when the mean flow is not uniform the wave equation is not separable, and we must resort to other methods to find a solution. The most general approach is to use a numerical method to solve the appropriate differential equation with the correct boundary conditions as described by Astley and Eversman [1,2] and Golubev and Atassi [3]. However, some insight to the problem is obtained if a high-frequency approximation is used to obtain the solution to the wave equation in a hard-walled duct with nonuniform flow (Cooper and Peake [4]).

To show how the high-frequency approximation can be applied, it is assumed that the mean flow is only a function of the radial coordinate and is given by the sum of an axial component $U(R)$ and an azimuthal component $W(R) = \Omega(R)R$. We will assume no mean radial flow, which is an important simplification and ignores the interaction of the mean flow with the turbulent wakes of the fan blades. It will also be assumed that there are no vortical waves in the duct, so the acoustic field is completely described by the velocity potential that satisfies the convected wave equation given in Eq. (17.2.1). We can assume that the velocity potential has a harmonic time dependence and make use of the periodicity in the azimuthal direction to expand the potential in a Fourier series as given by Eq. (17.2.2). The resulting Fourier series coefficients satisfy the differential equation given by

$$\left( \kappa_m^2(R) - \frac{m^2}{R^2} \right) \hat{\phi}_m + 2i\kappa_m(R)M(R)\frac{\partial \hat{\phi}_m}{\partial x} + \left(1 - M^2(R)\right)\frac{\partial^2 \hat{\phi}_m}{\partial x^2} + \frac{1}{R}\frac{\partial}{\partial R}\left(R\frac{\partial \hat{\phi}_m}{\partial R}\right) = 0$$

$$(17.6.1)$$

where

$$\kappa_m(R) = (\omega + m\Omega(R))/c_\infty \qquad (17.6.2)$$

is the effective acoustic wavenumber in a swirling flow and will be a function of radius unless the mean flow is in solid-body rotation. The difference between Eq. (17.6.1) and Eq. (17.2.3) is that it is not separable because the coefficients are dependent on the radius. However, we can find a solution using the Wentzel–Kramers–Brillouin (WKB) method which gives an approximate solution in the high-frequency limit that $ka \gg 1$. To implement this approximation, we specify the potential as an exponential or phase function so that

$$\hat{\phi}_m = B_m e^{iq_m(R/a) + i\mu x} \qquad (17.6.3)$$

where $\mu$ and $B_m$ are constants and $q_m(R/a)$ is to be determined from the solution to Eq. (17.6.1). The mode shape is then given by $\exp(iq_m(R/a))$, and Eq. (17.6.1) becomes

$$\lambda_m^2(R) - \left(q_m'\right)^2 + \frac{ia}{R}q_m' + iq_m'' = 0 \qquad (17.6.4)$$

where the prime represents differentiation with respect to the argument of $q_m$ and

$$\lambda_m^2(R) = \kappa_m^2(R)a^2 - \left(1 - M^2(R)\right)\mu^2 a^2 - 2\kappa_m(R)M(R)\mu a^2 - \left(\frac{ma}{R}\right)^2 \qquad (17.6.5)$$

For the analysis of the sound field in the duct with uniform flow we expect $\mu/k$ and $m/ka$ to be of order one, and so it follows that $\lambda_m$ is of order $O(ka)$, which is a large parameter in the high-frequency limit $ka \gg 1$. We consider a solution that is of the form

$$q_m = q_m^{(o)} + ka q_m^{(1)}$$

where both $q_m^{(o)}$ and $q_m^{(1)}$ are the same order of magnitude. In terms of these variables Eq. (17.6.4) becomes

$$(ka)^2 \left( \frac{\lambda_m^2(R)}{(ka)^2} - \left( q_m'^{(1)} \right)^2 \right) + (ka) \left( \frac{ia}{R} q_m'^{(1)} + i q_m''^{(1)} - 2 q_m'^{(1)} q_m'^{(o)} \right)$$
$$+ \left( \frac{ia}{R} q_m'^{(o)} + i q_m''^{(o)} - \left( q_m'^{(o)} \right)^2 \right) = 0$$
(17.6.6)

each term in this equation is therefore defined in descending orders of $ka$. The principle of the WKB method is that when $ka \gg 1$ the first term is large compared to the second two terms, and so an approximate solution is obtained for $q_m^{(1)}$ as

$$q_m'^{(1)}(R) = \pm \frac{\lambda_m(R)}{(ka)}$$

To obtain a solution that is accurate to second order the second term in Eq. (17.6.6) can also be set to zero, giving

$$q_m'^{(o)} = \frac{ia}{2R} + \frac{i q_m''^{(1)}}{2 q_m'^{(1)}}$$

We have therefore reduced the problem to solving two first-order differential equations that have the solutions

$$q_m^{(1)}(R) = \pm \frac{1}{a} \int_b^R \frac{\lambda_m(R)}{(ka)} dR, \qquad q_m^{(o)}(R) = \frac{i}{2} \ln \left( \frac{|\lambda_m(R)| R}{ka^2} \right) + const$$
(17.6.7)

(where $b$ is the radius of the inner duct wall, and we have used the fact that $\left( \ln \left( |q_m'| \right) \right)' = q_m''/q_m'$). The approximate solution to the wave equation in the high-frequency limit is then given by

$$\hat{\phi}_m \approx \frac{B_m \exp \left( \pm i \int_b^R (\lambda_m(R)/a) dR + i\mu x \right)}{\sqrt{|\lambda_m(R)| R}}$$
(17.6.8)

This gives two alternative solutions for the acoustic field in a duct with a mean flow that is a function of radius, and the two solutions can be combined to match the boundary conditions at the duct walls. The radial dependence of the modes depends on the integral defining $q_m^{(1)}$ in Eq. (17.6.7) which takes the form

$$q_m^{(1)}(R)$$

$$= \pm \frac{1}{ka^2} \int_b^R \left( \kappa_m^2(R)a^2 - \left(1 - M^2(R)\right)\mu^2 a^2 - 2\kappa_m(R)M(R)\mu a^2 - \left(\frac{ma}{R}\right) \right)^{1/2} dR$$

The first point to note from this result is that at small radii the term $(m/R)^2$ may be large enough that the integrand is imaginary, but as $R$ increases a branch point is reached where the integrand is zero and at larger radii it becomes real valued. This causes the integrand to have a critical point defined by the value of $R = R_c$ where the integrand is zero.

To match the boundary conditions, we specify

$$\hat{\phi}_m \approx \frac{B_m e^{i\mu x}}{\sqrt{|\lambda_m(R)|R}} \left( \exp\left( i \int_b^R \frac{\sqrt{\lambda_m^2(R)}}{a} dR \right) + \exp\left( -i \int_b^R \frac{\sqrt{\lambda_m^2(R)}}{a} dR \right) \right)$$

$$(17.6.9)$$

where the square roots are evaluated so that their real parts are positive. Using this form of the solution the hard-walled boundary condition is met on the inner duct wall to first order in $1/ka$, and to the same order we must solve for $\mu$ in Eq. (17.6.5) so that $kaq^{(1)}(a) = n\pi$ to match the boundary condition on the outer wall, which can be done iteratively if $\lambda_m^2(R) > 0$ for $b < R < a$.

However, this solution does not apply close to the so-called critical radius where $\lambda_m$ is approximately zero, and so the high-frequency approximation is no longer strictly valid. A solution can be obtained by solving Eq. (17.6.1) in the vicinity of the critical point and the result is given by Cooper and Peake [4] in terms of Airy functions. This needs to be matched to the solution valid outside the critical region defined by Eq. (17.6.9). In the region where $\lambda_m^2$ is negative the mode shapes are given by a hyperbolic cosine, which needs to be matched to the solution in the critical region. The net effect is that the mode amplitude tends to zero when $\lambda_m^2(R) < 0$.

In the absence of swirl and for a uniform axial flow the approximate solution reduces to the form of the Bessel function solutions given in the previous section in Eq. (17.2.12) and shown in Fig. 17.4. The accuracy of the approximation is shown in Fig. 17.6A for $m = 2$ and $ka = 50$ for no axial flow. The advantage of the high-frequency analysis is that it allows the mean axial flow and swirl to be a function of radius. To illustrate this Fig. 17.6B shows the mode shape when the mean flow speed varies with radius as $M(R) = M_t R/a$ in comparison with the zero flow case.

The effect of solid-body swirl, in which $\Omega$ is constant, is to alter the effective frequency, so depending on the sign of $m$, the cut on frequency is shifted from $\omega_o$ without swirl to the effective frequency $|\omega_o \pm m\Omega|$, and this can have important implications for the number of modes that propagate.

It was also pointed out by Cooper and Peake [4] that some interesting possibilities occur when $\lambda_m(R)$ has more than one zero in the range $b < R < a$. In this case there is the possibility for two or more critical radii, and so the sound field can be trapped between these points, giving rise to loud zones and quiet zones radially in the duct. These zones exist if both $\lambda_{m}^{2}(a)$ and $\lambda_{m}^{2}(b)$ are less than zero and that $\lambda_{m}^{2}(R)$ is greater than zero at some location across the duct. If the swirl velocity is of the form $\Omega(R) = \Omega_{o} + \Gamma/R^{2}$, corresponding to the combination of solid-body rotation and a potential vortex, then $\omega + m\Omega(R)$ will be less at the outer wall than at the inner wall, and a mode that is cut on at a small radius could be cut off at the larger radius because the effective frequency is less.

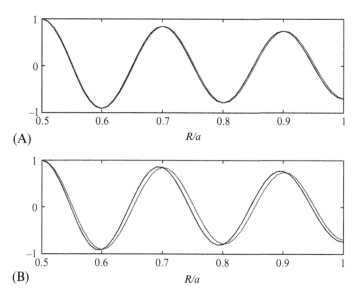

**Fig. 17.6** The duct modes obtained using the WKB approximation: (A) the duct mode for $m=2$, $ka=40$, $b=0.5a$ and no mean flow, *solid line* is WKB approximation, and *dashed line* is exact solution. (B) The same case with the mean flow given by $M=M_{t}R/a$ with $M_{t}=0.3$, *solid line* is WKB approximation, and *dashed line* is for no flow.

## 17.7 The radiation from duct inlets and exits

So far in this chapter we have discussed the propagation of acoustic modes in circular ducts of infinite length. In a turbofan engine (Fig. 17.1) the duct will vary in cross section and will be of finite length, and it is the radiation out of the duct inlets and exits that is of primary concern. There are a number of numerical approaches that can be used to address a real duct, including Finite Element Methods [1,2] and the Geometrical Theory of Diffraction [5,6]. Small variations in the cross section of the duct can also be considered using multiple scale analysis [7]. We can also gain insight into radiation from a duct inlet or exit using analytical methods, and these are discussed in this section.

Analytical models of the sound radiation from a duct are most readily obtained by considering a semi-infinite circular duct as illustrated in Fig. 17.7. In the case of the duct exit or jet pipe (Fig. 17.7A) the flow inside the duct is taken to be uniform and the flow speed outside the duct can be different so that a shear layer is formed between the two flows, and this can cause shear layer refraction effects and the possibility of instability waves at the interface. In the case of an inlet (Fig. 17.7B) the flow is taken to be uniform both inside and outside the duct.

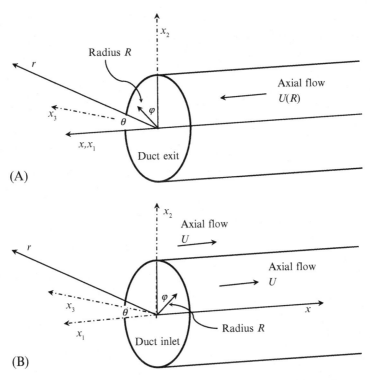

**Fig. 17.7** Modeling the duct exit (A) or inlet (B) as a semi-infinite circular duct.

The sound radiation from semi-infinite pipes with flow is described by Munt [8], who analyzed the problem using the Weiner Hopf method. However, Tyler and Sofrin [9] argued that a good approximation to the far-field sound from an inlet without flow is obtained by modeling the inlet by a circular duct that is mounted in a baffle, as shown in Fig. 17.8. Lansing [10] studied the equivalence between these two approximations and showed that there were only small differences in the radiated sound power between the two cases. The differences occurred at frequencies close to cutoff, as might be expected. However, the Tyler and Sofrin model cannot give the correct directivity at angles >90 degrees to the duct axis because this observer would be inside the baffle.

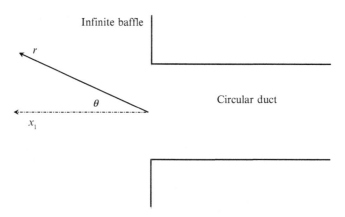

**Fig. 17.8** The Tyler-Sofrin model for sound radiation from a duct exit.

An alternative approximation was given by Cargill [11] who used a Ffowcs-Williams and Hawkings surface on the external surface of a semi-infinite pipe, and the pipe exit, for the configuration shown in Fig. 17.7A. A Green's function was developed for a source in the jet pipe flow with different flow speeds on either side of the shear layer. A key assumption was that the contribution from the external surface of the pipe was negligible compared to radiation from the duct exit. In spite of this apparently major simplification it was shown that the far-field sound was almost identical to Munt's exact solution at all but a few observer angles in the region upstream of the jet pipe exit.

To illustrate Cargill's method, we will consider the sound radiation from an inlet without flow, using the coordinates given in Fig. 17.7A, with $U=0$. The solution for the acoustic field outside the duct is given by Eq. (3.10.5) with the surface taken as the external surface of the duct and the duct exit. In Cargill's approximation only the duct exit is included, and the acoustic pressure and its normal derivative are defined by the waves propagating along the duct toward the duct exit. This implies that no waves are reflected back toward the source inside the duct, which is a reasonable approximation when $ka \gg 1$. The far-field sound is then given by Eq. (3.10.5) for a harmonic time dependence

$$\hat{p}(\mathbf{x}) = \int_S \left( \hat{p}(\mathbf{y}) \frac{\partial \tilde{G}_o(\mathbf{x}|\mathbf{y})}{\partial y_i} - \tilde{G}_o(\mathbf{x}|\mathbf{y}) \frac{\partial \hat{p}(\mathbf{y})}{\partial y_i} \right) n_i dS$$

If we limit consideration to a single duct mode and integrate over the duct exit, we obtain the pressure as

$$\hat{p}_{mn}(\mathbf{x}) = B_{mn} \int_0^{2\pi} \int_0^a \left( \frac{\partial \tilde{G}_o(\mathbf{x}|\mathbf{y})}{\partial y_1} - ik_{mn} \tilde{G}_o(\mathbf{x}|\mathbf{y}) \right) \psi_m(\alpha_{mn}R) e^{-im\varphi} R dR d\varphi$$

where $B_{mn}$ is the pressure amplitude of the duct mode. For a duct without a center body $\psi_m(\alpha_{mn}R) = J_m(\alpha_{mn}R)$. The free field Green's function can be approximated for observers in the acoustic far field of the duct by

$$\tilde{G}_o = \frac{e^{ikr - ikx_iy_i/r}}{4\pi r}$$

where $r$ is the distance from the center of the duct exit to the observer located at $x_1 = r\cos\theta$, $x_2 = r\sin\theta$, $x_3 = 0$ as shown in Fig. 17.7A. Using $y_1 = 0$, $y_2 = R\cos\varphi$, $y_3 = R\sin\varphi$, the acoustic far field is given by

$$\hat{p}_{mn}(x) = -iB_{mn}\frac{e^{ikr}}{4\pi r}\int_0^{2\pi}\int_0^a (k\cos\theta + k_{mn})J_m(\alpha_{mn}R)e^{-im\varphi - ikR\sin\theta\cos\varphi}R\,dR\,d\varphi$$

The integral over the azimuth gives

$$\int_0^{2\pi} e^{-im\varphi - ikR\sin\theta\cos\varphi}d\varphi = 2\pi J_m(kR\sin\theta)e^{-im\pi/2}$$

and the integral over the radius is a standard integral given by

$$\int_0^a J_m(\alpha_{mn}R)J_m(kR\sin\theta)R\,dR$$

$$= \frac{\alpha_{mn}aJ_{m+1}(\alpha_{mn}a)J_m(ka\sin\theta) - ka\sin\theta J_m(\alpha_{mn}a)J_{m+1}(ka\sin\theta)}{\alpha_{mn}^2 - k^2\sin^2\theta}$$

For a hard-walled duct $J_m'(\alpha_{mn}a) = 0$, and we can use the Bessel function recurrence relationships to simplify this result to

$$\frac{ka\sin\theta J_m(\alpha_{mn}a)J_m'(ka\sin\theta)}{\alpha_{mn}^2 - k^2\sin^2\theta}$$

The far-field sound is then given by

$$\hat{p}_{mn}(\mathbf{x}) = -2\pi iB_{mn}J_m(\alpha_{mn}a)D_{mn}(\theta)\frac{ae^{ikr}}{4\pi r} \tag{17.7.1}$$

where the directivity is

$$D_{mn}(\theta) = \frac{(k\cos\theta + k_{mn})(k\sin\theta)J_m'(ka\sin\theta)}{\alpha_{mn}^2 - k^2\sin^2\theta} \tag{17.7.2}$$

Fig. 17.9 shows the directivity of the far-field sound based on Eq. (17.7.2) and is compared to computations based on the full Weiner Hopf solution [12]. It is seen that the approximate solution is very accurate at angles where the directionality is a maximum but fails to give the correct result at large angles where the levels are about 10 dB less than the peak values. This is consistent with Cargill's [11] observation that the error was in the region of the far field where that particular mode did not significantly contribute to the overall level.

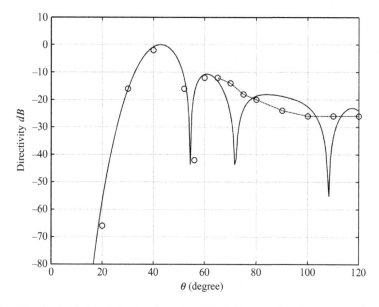

**Fig. 17.9** The far-field directivity based on Cargill's [11] approximation compared to the calculations given by Gabbard and Astley [12] for $ka = 30$, $m = 17$, and $n = 0$.

It is clear from Fig. 17.9 that the far-field sound peaks at an angle to the duct axis where $k\sin\theta = \alpha_{mn}$ where the denominator of Eq. (17.7.2) is zero. At this angle the directivity can be evaluated using L'Hôpital's rule because $J'_m(ka \sin \theta)$ is also zero when $k\sin\theta = \alpha_{mn}$. The directivity therefore has similar characteristics to a $\sin(x)/x$ function with a peak at the location where $x = 0$, and side lobes of a lower level on either side of the peak.

# References

[1] R.J. Astley, W. Eversman, A finite element formulation of the eigenvalue problem in lined ducts with flow, J. Sound Vib. 65 (1979) 61–74.
[2] R.J. Astley, W. Eversman, Acoustic transmission in non-uniform ducts with mean flow, part II, J. Sound Vib. 74 (1981) 103–121.

[3] V.V. Golubev, H.M. Atassi, Sound propagation in an annular duct with mean potential swirling flow, J. Sound Vib. 198 (1996) 601–616.

[4] A.J. Cooper, N. Peake, Upstream-radiated rotor–stator interaction noise in mean swirling flow, J. Fluid Mech. 523 (2005) 219–250.

[5] C.J. Chapman, Sound radiation from a cylindrical duct. Part 1. Ray structure of the duct modes and of the external field, J. Fluid Mech. 281 (1994) 293–311.

[6] G.M. Keith, N. Peake, High-wavenumber acoustic radiation from a thin-walled axisymmetric cylinder, J. Sound Vib. 255 (2002) 129–146.

[7] S.W. Rienstra, Sound transmission in slowly varying circular and annular ducts with mean flow, J. Fluid Mech. 380 (1999) 279–296.

[8] R.M. Munt, The interaction of sound with a subsonic jet issuing from a semi-infinite cylindrical pipe, J. Fluid Mech. 83 (1977) 609–640.

[9] J.M. Tyler, T.G. Sofrin, Axial flow compressor noise studies, SAE Trans. 70 (1962) 309–332.

[10] D.L. Lansing, Exact solution for radiation of sound from a semi infinite circular duct with application to fan and compressor noise, in: Analytical Methods in Aircraft Aerodynamics, 1970, pp. 323–332. NASA SP-228.

[11] A. Cargill, The radiation of high frequency sound out of a jet pipe, J. Sound Vib. 83 (1982) 313 337.

[12] G. Gabbard, R.J. Astley, Theoretical model for sound radiation from annular jet pipes: far- and near-field solutions, J. Fluid Mech. 549 (2006) 315–341.

# Fan noise

**18**

This chapter will discuss the application of aeroacoustic theories to the prediction of ducted fan noise. The focus will be on a fan in a circular duct with a center body as a model of a typical aero engine, but the methods are also applicable to other geometries. The most important difference between a ducted fan and an open rotor is that in the ducted fan the acoustic sources excite duct modes that are determined by the boundary conditions at the duct walls. This has a large impact on how the acoustic sources are coupled to the acoustic far field outside the duct and this chapter will show how these effects are modeled.

## 18.1    Sources of sound in ducted fans

In the early days of jet aircraft the far field noise was dominated by jet noise. It was shown by Lighthill [1] that the sound intensity was proportional to the eighth power of the jet exit velocity, and Morfey [2] showed that for heated jets the sound intensity scaled with the sixth power of the flow speed. In the 1970s high bypass ratio turbofan engines were introduced (see Fig. 17.1) which enabled a large increase in engine diameter so the same thrust could be achieved with a lower jet speed. The jet noise component of the sound was significantly reduced by the lower jet exit velocity and other sources such as fan noise became significant contributors to the overall noise level. Initially the far field sound, especially at low thrust conditions used when an aircraft approached an airport, was dominated by tone noise from the engine fan. This was attributed to the wakes of the fan blades impinging on the downstream stator vanes that excited spinning modes in the fan duct, as discussed in Chapter 17. However, these modes only propagate along the duct if their rate of rotation has a supersonic phase speed at the duct wall. As we will show in the next section a judicial choice for the number of blades and stator vanes in an engine enables the fan tone noise to be controlled and minimized. As the fan diameter is increased the broadband fan noise also increases and contributes more energy to the far field, so in very high bypass ratio turbofan engines the tone noise and broadband noise are of equal importance [3].

The noise sources in ducted fans have many of the same properties as the sources on an open rotor, but there are also some important differences. The most significant difference is the effect of the duct since sound from the fan or stator vanes must propagate along the duct before it radiates to the acoustic far field. This gives the opportunity to include sound absorbing materials in the duct wall so that sound can be attenuated before it reaches the duct exit and hence the acoustic far field. In addition, as we showed in Chapter 17, only certain modes will propagate in a duct at a given frequency. The duct therefore allows for noise control measures to be introduced that decouple the acoustic sources from the propagating duct modes. It is important therefore to treat ducted fan noise as a coupled system that includes the correct interaction

**Aeroacoustics of Low Mach Number Flows. http://dx.doi.org/10.1016/B978-0-12-809651-2.00018-7**

between the fan noise source and its local environment. The exception to this approach is when the duct is very short (less than an acoustic wavelength) in which case the nonpropagating modes can still reach the engine inlet or duct exit and couple with the acoustic far field. In this chapter we will focus on ducted fans and the duct mode amplitudes, referring to Chapter 17 for the effect of duct mode propagation and far field radiation.

In Chapter 16 we identified rotor noise sources as thickness noise, loading noise, unsteady loading noise, trailing edge noise, and quadrupole noise. As we will show in the next section, thickness noise only couples with modes that are spinning with the same speed as the fan. For fans with subsonic tip speeds these modes are typically cut off, and so thickness noise in ducted fans is not usually an issue. However, when the fan tip speed is supersonic then sources that rotate with the fan will propagate and cause additional radiation. In high speed fan applications that have supersonic tip speeds the quadrupole sources associated with shock waves on each blade propagate along the duct and couple with the acoustic far field. This is known as Buzz-saw noise and can often be important in aircraft cabins as well as to the observer on the ground. The term Buzz-saw noise was chosen because the sound resembles that of a chain saw cutting wood, and it typically has a spectrum that is rich in harmonics of the shaft rotation frequency. However, this source does not occur in low Mach number flows and will not be discussed in detail here.

The primary sources of fan noise are therefore unsteady loading noise and trailing edge noise. The unsteady loading noise from the fan is caused by disturbances in the inflow that result in a nonuniform flow entering the rotor. This flow can be significantly altered by the duct inlet shape and any obstructions upstream of the fan. A relatively small turbulent eddy in the atmosphere can be accelerated and stretched by the mean flow as it enters the fan inlet (see Chapter 6) so that it gets a thin elongated sausage shape that is repeatedly cut by the rotating blades. This results in fan tone noise that is referred to as haystacking, and is particularly important in engines that are operating on test stands close to the ground, where the atmospheric turbulence includes relatively small scale eddies. For this reason, aircraft engines on test stands are fitted with an inlet control device that consists of a cloth screen designed to break up the small scale eddies entering the engine. At flight conditions this effect is greatly reduced by the motion of the aircraft relative to the atmosphere. The acceleration of the outside flow into the inlet is reduced because the engine moves towards the turbulent eddy, rather than the eddy being sucked into the fan inlet. Consequently, haystacking tones are not as important in flight situations as they are in ground testing.

In flight, the dominant source of fan noise has been found to be the impingement of the turbulent wakes from the fan onto the downstream stator vanes [4] (see Fig. 17.1). This mechanism includes both the tone and broadband noise, and is complicated by the swirling flow in the duct downstream of the fan. The mean flow in the wakes of the fan blades is relatively well defined, and can be computed using Reynolds averaged Navier Stokes (RANS) methods (see Chapter 8) in a rotating frame. These disturbances can then be used with a suitable blade response function to calculate the unsteady loading on the vanes, and hence estimate the amplitude of the duct modes excited by this mechanism upstream and downstream of the stator vanes. However, sound radiated from the engine inlet will be attenuated by both the duct liners, as described in Chapter 17, and the propagation of sound through the fan.

In addition to tone noise caused by the interaction between the mean flow deficit in the blade wakes with the downstream stator vanes, there is broadband noise from rotor/ stator interactions. This is caused by random turbulent flow that is uncorrelated from wake to wake and has a time scale that is small compared to the shaft rotation period. This is a modulated turbulent flow and causes sound with a wide band spectrum that can be peaked around the blade passage frequency if the modulation is distinct. Many of the methods given in Chapter 16 can be used to describe this source, and, as will be shown below, the broadband noise level critically depends on the turbulent intensity and the turbulent length scale at the leading edge of the stator vanes.

The blade response to an incoming disturbance for a ducted fan or stator is distinctly different from that of an isolated blade. In rotor or propeller applications we were able to treat each blade section independently as if it were a section of a uniform blade of infinite span. The edge effects were accounted for by the limited spanwise correlation length scale of the incoming disturbance. For the fan or stator application the situation is different in two respects: first, the blade sets are of relatively high solidity which means that the blades are sufficiently close together to be overlapped and to interact aerodynamically. Second, the duct walls introduce spanwise boundary conditions that impact the blade response especially at frequencies close to the duct mode cut off frequency. It was shown in Chapter 17 that the Green's function for sources in a duct was infinite at the cut on frequency of each duct mode unless the source strength tended to zero. We will show in Section 18.3 that duct modes will always have this characteristic if the correct spanwise duct mode is used in the blade response function, but not if it is approximated using the strip theory approach used in Chapter 16 for open rotors. This has a large impact on broadband noise calculations at high frequencies for large turbo fan engines because, within any one-third octave band of frequencies, a large number of radial modes will cut on and the level averaged over frequency will include modes at their cut on frequency. If the incorrect blade response function is used in these calculations then each mode that is cutting on will have an infinite sound power and will dominate the predicted spectrum, giving an erroneous result. The tailoring of the blade response function to its environment is therefore an important aspect of fan noise prediction.

In addition to rotor/stator interaction noise there is also broadband fan self noise, which can be an important contributor to the far field noise spectrum at high frequencies. The flow speed over the fan blades is always much higher than the flow speed over the stator vanes so self noise from the stators is always relatively insignificant compared to the self noise from the fan. Fan self noise is caused by three distinctly different mechanisms: (1) the interaction of the fan blades with the duct wall boundary layer, (2) the tip flow between the duct wall and the fan blade, and (3) trailing edge noise from the fan. For large diameter fans the trailing edge noise is considered to be the most important of these three mechanisms [5], but for small or model scale fans the turbulence in the duct wall boundary layer can be a contributor [6]. In automotive fans, which have small diameter and have a low tip Mach number, tip effects and recirculation have been shown experimentally to dominate the noise generation [7].

The objective of this chapter is to outline procedures for the calculation of duct mode amplitudes and duct mode sound power levels for rotors and stators in an infinite duct. This gives the necessary inputs required for the calculation of the propagation effects described in Chapter 17. We will start with a general formulation applicable to all the source types described above, and then idealize the approach to correctly account for the effects of the duct walls on the source levels.

## 18.2    Duct mode amplitudes

To describe the sound field in a circular duct we need to specify the duct mode amplitudes in terms of the acoustic sources in the duct. The approach we will use initially is based on Lighthill's analogy in which the aeroacoustic sources are specified in a uniform mean flow in the duct. These concepts can be extended to nonuniform mean flows by using the linearized Euler equations, but first we will identify the most important sources of sound using Lighthill's approach.

In Section 5.4 the solution to Lighthill's equation in a uniform flow was specified in terms of a Green's function that applied to the local environment. For the case of a circular duct with a uniform axial flow the Green's function was derived in Section 17.4, and this allows the duct mode amplitudes to be directly related to Lighthill's source terms. In principle, the uniform flow assumption eliminates the cases where the mean flow includes swirl as a function of radius, but these issues will be addressed in Section 18.6. To apply Lighthill's theory it will be assumed that the flow in the duct is given by $U_i^{(\infty)} + w_i$ where $U_i^{(\infty)}$ represents the uniform axial flow (see Fig. 18.1) and $w_i$ is a small perturbation to the mean flow.

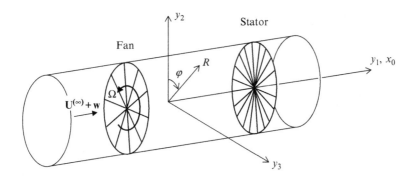

**Fig. 18.1** The coordinate system for a rotating fan in a circular duct.

The acoustic field is then given by Eq. (5.4.4) with a Green's function that satisfies Eq. (5.4.3), which is given by Eq. (17.4.3). Since the fan blades will be moving we take the boundary condition on the blade surface to be $\left(U_i^{(\infty)} + w_i\right)n_i = V_i n_i$ where $V_i$ is the blade speed. For stator vanes $V_i = 0$. Both the flow and the derivative of the Green's function normal to the duct walls are zero. Furthermore, we can ignore the quadrupole terms in Eq. (5.4.4) at low Mach numbers, and in the thickness noise term we can use

$$\rho w_j - \rho'\left(V_j - U_j^{(\infty)}\right)n_j = \rho_\infty\left(V_j - U_j^{(\infty)}\right)n_j$$

to give the acoustic field as

$$\rho'(\mathbf{x},t)c_\infty^2 = \int_{-T}^{T}\int_{S_o(\tau)}\frac{\partial G_e}{\partial y_i}p_{ij}n_j dS(\mathbf{y})d\tau$$
$$-\int_{-T}^{T}\int_{S_o(\tau)}\frac{D_\infty G_e}{D\tau}\left(\rho_\infty\left(V_j - U_j^{(\infty)}\right)n_j\right)dS(\mathbf{y})d\tau \qquad (18.2.1)$$

and from Eq. (17.4.2) the Green's function is

$$G_e(\mathbf{x},t|\mathbf{y},\tau) = \frac{1}{2\pi}\int_{-\infty}^{\infty}\sum_{m=-\infty}^{\infty}\sum_{n=0}^{\infty}\left(\frac{\psi_{mn}(R)\psi_{mn}(R_o)}{2i\Lambda_{mn}k_{mn}}\right)e^{im(\varphi-\varphi_o)-i\mu_{mn}^\pm(x-x_o)-i\omega(t-\tau)}d\omega$$

$$(18.2.2)$$

where the subscript $o$ refers to the source position in cylindrical coordinates.

In Section 17.2 the acoustic field in the duct was described by the modal expansion given in Eq. (17.2.10) for the velocity potential at a fixed frequency $\omega$. The amplitude of each duct mode was specified as $A_{mn}(\omega)$. Since the velocity potential in a uniform flow and the acoustic pressure are related by $p' = -\rho_\infty D_\infty\phi/Dt$ and $p' = \rho'c_\infty^2$ we can define the acoustic pressure in the duct in a similar modal expansion, given by

$$p'(\mathbf{x},t) = \int_{-\infty}^{\infty}\sum_{m=-\infty}^{\infty}\sum_{n=0}^{\infty}P_{mn}(\omega)\psi_{mn}(R)e^{im\varphi-i\mu_{mn}^\pm x-i\omega t}d\omega \qquad (18.2.3)$$

where

$$P_{mn}(\omega) = -i\rho_\infty\left(\omega + U\mu_{mn}^\pm\right)A_{mn}(\omega) \qquad (18.2.4)$$

(where $U$ is the axial flow speed along the duct as used in Chapter 17). The mode amplitudes can then be obtained by combining the results above, and as with rotor noise, may be split into terms representing thickness and loading noise, which can be considered separately.

## 18.2.1 Thickness noise for a ducted fan

The mode amplitudes for the thickness noise term in Eq. (18.2.1) are obtained by combining the second integral in Eq. (18.2.1) with the Green's function in Eq. (18.2.2) and separating out the modal content to give

$$P_{mn}^{(\text{thickness})} = -\int_{-T}^{T}\int_{S_o(\tau)}\left(\rho_\infty\left(V_j - U_j^{(\infty)}\right)n_j\right)\left(\frac{\left(\omega + \mu_{mn}^{\pm}U\right)\psi_{mn}(R_o)}{4\pi\Lambda_{mn}k_{mn}}\right)$$

$$e^{-im\varphi_o(\tau) + i\mu_{mn}^{\pm}x_o + i\omega\tau}dS(\mathbf{y})d\tau \tag{18.2.5}$$

where $\varphi_o(\tau)$ gives the azimuthal location of the blade surfaces at source time $\tau$. If the fan consists of $B$ identical blades that are equally spaced, then the duct mode amplitude will be the sum of the contributions from each blade. For a given axial and radial location of each blade element the azimuthal location is given by

$$\varphi_o(\tau) = \frac{2\pi(s-1)}{B} + \varphi_b^{(\pm)}(R_o, x_o) - \Omega\tau$$

where $s = 1, 2, 3, \ldots, B$ is the blade number, $\Omega$ is the blade angular velocity, and $\varphi_b^{(\pm)}$ defines the suction and pressure surfaces of the blade (see Fig. 18.2). It is important to note that in fan noise analysis we choose the $x_o$ coordinate to be pointing in the downstream direction, which is in contrast to the coordinates used in Chapter 16 that used a system in which the $x_1$ coordinate pointed upstream.

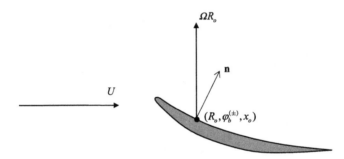

**Fig. 18.2** Coordinates for a fan blade in blade fixed coordinates. A point on the surface of the blade is given by the cylindrical coordinates $R_o, \varphi_b^{(\pm)}, x_o$.

The only other term in this equation that could vary with time is the blade normal velocity but since the blade is moving at constant angular speed in the azimuthal direction we can define $\left(V_j - U_j^{(\infty)}\right)n_j$ as independent of time and only a function of the location on the blade surface. (This is readily shown by expressing the velocities in cylindrical coordinates.)

The surface integral in Eq. (18.2.5) can then be carried out over each blade and the results added with the correct phase shift to give

$$P_{mn}^{(\text{thickness})} = -\left(\frac{\left(\omega + \mu_{mn}^{\pm}U\right)}{2\Lambda_{mn}k_{mn}}\right)\left\{\sum_{s=1}^{B}e^{-2\pi im(s-1)/B}\right\}$$

$$\times \int_{S_o(\tau)}\left(\rho_\infty\left(V_j - U_j^{(\infty)}\right)n_j\right)\psi_{mn}(R_o)e^{-im\varphi_b + i\mu_{mn}^{\pm}x_o}dS(\mathbf{y}) \tag{18.2.6}$$

$$\times \frac{1}{2\pi}\int_{-T}^{T}e^{i(\omega + m\Omega)\tau}d\tau$$

This result has allowed us to separate out each of the integrals so that they can be evaluated independently, and has important features. First we note that the summation given in { } is zero unless $m$ is an exact multiple of the blade number $B$ as discussed in Chapter 16. Alternatively, if $m = jB$ then each term in the summation is one and the sum is equal to $B$. This ensures that only modes that have azimuthal indices $m = jB$ will be excited. Second the integral over the source time $\tau$ is a Dirac delta function $\delta(\omega + jB\Omega)$. When this result is used in Eq. (18.2.4) and the integration is carried out over frequency we obtain

$$p'(\mathbf{x}, t) = \sum_{j=-\infty}^{\infty} \sum_{n=0}^{\infty} C_{jB,n} \psi_{jB,n}(R) e^{ijB\varphi - i\chi_{jBn}^{\pm} x + ijB\Omega t} \tag{18.2.7}$$

where the mode amplitudes are

$$C_{jB,n} = -\left[ \left( \frac{(\omega + \mu_{mn}^{\pm} U)}{4\pi\Lambda_{mn}k_{mn}} \right) \int_{S_o} \left( \rho_o \left( V_j - U_j^{(\infty)} \right) n_j \right) \psi_{mn}(R_o) e^{-im\varphi_b + i\mu_{mn}^{\pm}x_o} dS(\mathbf{y}) \right]_{\substack{\omega=-jB\Omega \\ m=jB}}$$

and the axial wavenumber is

$$\chi_{jB,n}^{\pm} = \left[ \mu_{jB,n}^{\pm} \right]_{\omega=-jB\Omega} \tag{18.2.8}$$

The result given in Eq. (18.2.7) shows that the sound field in the duct is the combination of harmonic waves at the blade passing frequencies $jB\Omega$ and that the duct mode amplitudes can be obtained from the surface integral of the thickness noise terms, as was done for an open rotor in Chapter 16. However, the axial wavenumber will determine if the modes propagate, and it was shown in Section 17.2 that for the modes to be cut on we require that $ka/\beta > |jB|$ where $a$ is the outer duct radius. Since in this case $k = |jB\Omega/c_\infty|$, we find that $ka/\beta = |jB(\Omega a/c_\infty)/\beta|$ and so the cut on condition will not be met unless $(\Omega a/c_\infty)/\beta > 1$ and this does not occur if the blade tip speed is subsonic and the axial flow speed is sufficiently subsonic that $(\Omega a/c_\infty) < \beta$. As a consequence we can state that if the fan tip speed or the axial flow speed is not in the transonic regime then the duct mode excited by thickness noise will be cut off and will not propagate along the duct. Hence it will not radiate to the acoustic far field unless the duct is very short. This is an important conclusion because it effectively eliminates thickness effects as a source of ducted fan noise unless there is transonic flow at the blade tips. However, at these conditions the blade will start to form leading edge shock waves which can propagate efficiently and cause Buzz-saw noise which is usually a much more significant source.

## 18.2.2   Blade loading noise

In the previous section we showed that thickness noise was not an important acoustic source for ducted fans, and quadrupole noise was only important when the blade tip's speed was transonic. Consequently, the most important source of sound in ducted fans is attributed to loading noise. As with thickness noise the steady loading sources will only couple with nonpropagating modes and as we will see below it is the unsteady loading on the blade surfaces that is of primary concern. The loading noise is specified by the first surface integral in Eq. (18.2.1) and, as in Section 16.2.1 we can represent loading noise by the surface pressure jump across the blade planform and represent the source term in Eq. (18.2.1) as

$$\left( \left[ p_{ij} n_j \right]_{\text{upper}} - \left[ p_{ij} n_j \right]_{\text{lower}} \right) dS = f_i^{(s)} d\Sigma$$

where $d\Sigma$ represents an element of the blade planform and $f_i^{(s)}(\mathbf{y}, \tau)$ is the unsteady loading per unit area on the planform surface as shown in Fig. 18.3 for blade number $s$.

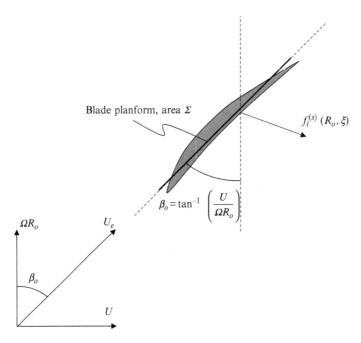

**Fig. 18.3** The surface loading per unit area $f_i$ as a function of position on the blade planform. The angle $\beta_o$ is the angle of the blade planform relative to the direction of rotation.

The mode amplitudes for loading noise are then obtained from Eqs. (18.2.1), (18.2.3) as

$$P_{mn}^{(\text{loading})} = \sum_{s=1}^{B} \int_{-T}^{T} \int_{\Sigma_s(\tau)} f_i^{(s)}(\mathbf{y}, \tau) \frac{\partial}{\partial y_i} \left( \frac{\psi_{mn}(R_o) e^{-im\varphi_s(\tau) + i\mu_{mn}^{\pm} x_o}}{4\pi i \Lambda_{mn} k_{mn}} \right) e^{i\omega\tau} d\Sigma(\mathbf{y}) d\tau$$

$$(18.2.9)$$

where we have defined

$$\varphi_s(\tau) = \frac{2\pi(s-1)}{B} + \varphi_c(R_o, \xi) - \Omega\tau \tag{18.2.10}$$

and $\varphi_c(R_o,\xi)$ is the azimuthal location of the blade planform relative to the blade leading edge for blade number $s=1$, and $x_o$ is its downstream location.

This result is further simplified by introducing the direction cosines of the loading vector in cylindrical coordinates. Since the loading is normal to the blade planform when viscous stresses are ignored (which will be reasonable for high Reynolds number flows) we find that

$$f_x^{(s)}(R_o, \xi, \tau) = \Delta p^{(s)}(R_o, \xi, \tau)\cos\beta_o \quad f_\varphi^{(s)}(R_o, \xi, \tau) = \Delta p^{(s)}(R_o, \xi, \tau)\sin\beta_o$$

where $\Delta p^{(s)}(R_o,\xi,\tau)$ is the magnitude of the pressure jump across the blade surface at the distance $\xi$ from the leading edge at radius $R_o$. This enables the evaluation of Eq. (18.2.9) as

$$P_{mn}^{(\text{loading})} = -\sum_{s=1}^{B} \int_{-T}^{T} \int_{\Sigma_s} \frac{\left(m\sin\beta_o/R_o - \mu_{mn}^{\pm}\cos\beta_o\right)\Psi_{mn}(R_o)\Delta p^{(s)}(\xi, R_o, \tau)}{4\pi\Lambda_{mn}k_{mn}}$$

$$e^{i\omega\tau - im\varphi_s(\tau) + i\mu_{mn}^{\pm}x_o}d\Sigma d\tau \tag{18.2.11}$$

This result applies to both fan tone noise and broadband noise and requires the evaluation of the unsteady loading on the blade planform.

This result is very similar to the result given by Eqs. (4.7.10)–(4.7.12) in which the far field radiation from unsteady loading on a flat plate was directly related to the wavenumber transform of the unsteady pressure jump across the plate. The difference here is that the blade planform is in a circular duct and includes curvature to allow for the camber of the blades.

### 18.2.3  Fan tone noise

Fan tone noise is the result of blade loadings that are periodic in time and so we can evaluate Eq. (18.2.11) by using a Fourier series expansion of the loading magnitude. Furthermore, if each blade is identical and passes through the same inflow disturbance then each blade loading will be the same, but shifted in time by the blade passing interval, so

$$\Delta p^{(s)}(R_o, \xi, \tau) = \Delta p^{(1)}(R_o, \xi, \tau - 2\pi(s-1)/B\Omega)$$

The Fourier series expansion of the blade loads then takes the form

$$\Delta p^{(s)}(R_o, \xi, \tau) = \sum_{p=-\infty}^{\infty} F_p(R_o, \xi)e^{ip\Omega\tau - 2\pi ip(s-1)/B} \tag{18.2.12}$$

Using this expression in Eq. (18.2.11) and substituting for $\varphi_s$ from Eq. (18.2.10) gives

$$P_{mn}^{(\text{loading})} = -\sum_{p=-\infty}^{\infty} \left(\frac{1}{2\Lambda_{mn}k_{mn}}\right) \left\{\sum_{s=1}^{B} e^{-2\pi i(m+p)(s-1)/B}\right\}$$

$$\times \int_{\Sigma_o(\tau)} \left(\frac{m\sin\beta_o}{R_o} - \mu_{mn}^{\pm}\cos\beta_o\right) F_p(R_o,\xi)\Psi_{mn}(R_o)e^{-im\varphi_c + i\mu_{mn}^{\pm}x_o} d\Sigma(y)$$

$$\times \frac{1}{2\pi}\int_{-T}^{T} e^{i(\omega+(m+p)\Omega)\tau} d\tau$$

As with the thickness noise case the summation over the blades is only nonzero if

$$m+p = jB$$

which restricts the mode order $m$ to specific values of $p$ and $j$. This is referred to as the Tyler Sofrin condition. Also we note that in this case the integral over source time gives a Dirac delta function $\delta(\omega+(m+p)\Omega)$ and when this is used in Eq. (18.2.3) we obtain a result similar to Eq. (18.2.7)

$$p'(\mathbf{x},t) = \sum_{p=-\infty}^{\infty}\sum_{j=-\infty}^{\infty}\sum_{n=0}^{\infty} B_{jB,p,n}\Psi_{jB-p,n}(R)e^{i(jB-p)\varphi - i\chi_{jB,p,n}^{\pm}x + ijB\Omega t} \qquad (18.2.13)$$

where

$$B_{jB,p,n}$$
$$= \left[\left(\frac{-1}{2\Lambda_{mn}k_{mn}}\right)\int_{\Sigma}\left(\frac{m\sin\beta_o}{R_o} - \mu_{mn}^{\pm}\cos\beta_o\right)F_p(R_o,\xi)\Psi_{mn}(R_o)e^{-im\varphi_c + i\mu_{mn}^{\pm}x_o}d\Sigma(y)\right]_{\substack{\omega=-jB\Omega \\ m=jB-p}}$$

and

$$\chi_{jB,p,n}^{\pm} = \left[\mu_{jB-p,n}^{\pm}\right]_{\omega=-jB\Omega} \qquad (18.2.14)$$

As with the thickness noise example we have an expression for fan tone noise that is the superposition of harmonic waves and causes fan tones at multiples of the blade passing frequency $B\Omega$. However, unlike thickness noise there are multiple azimuthal modes that exist at any blade passing frequency, one for each value of the integer $p$. However, not all of these modes will propagate and we can use the criteria developed in Chapter 17 to determine the propagating modes. This requires that $|ka/\beta| > |m|$ at each frequency where $m$ is the mode order, given in this case by $m = jB - p$. The frequency of the tone associated with this mode is $jB\Omega$ and so we require that

$$\left| jB \left( \frac{\Omega a}{\beta c_\infty} \right) \right| > |jB - p|$$

Consequently, for values of the integer $p$ that satisfy the criterion

$$\left| \frac{jB - p}{jB} \right| < \frac{\Omega a}{\beta c_\infty} \tag{18.2.15}$$

the duct modes will propagate. This is an important result because it shows the range of Fourier coefficients of the unsteady loading that will excite propagating modes in the duct. In the low Mach number limit only a few modes for which $|p|$ is approximately equal to $|jB|$ will propagate, but when $\Omega a / c_\infty \beta$ approaches one a much larger number of modes are cut on. Clearly if the loading harmonics can be controlled so that the amplitudes of the Fourier coefficients are small for propagating modes then significant noise reductions can be achieved. For example, if a fan is operating downstream of a set of fixed guide vanes that introduce a velocity disturbance in the flow that is periodic, then the unsteady part of the flow incident on the fan will have the same periodicity. The time period for the loading on each blade will repeat at intervals of $2\pi/\Omega V$ where $V$ is the number of upstream guide vanes and it follows that

$$\Delta p^{(1)}(R_o, \xi, \tau) = \sum_{k=-\infty}^{\infty} F_{kV}(R_o, \xi) e^{-ikV\Omega\tau}$$

which implies that the integer $p$ in Eq. (18.2.12) will only take on values that are integer multiples of the vane number $V$. We can then use Eq. (18.2.15) to find a suitable choice of $V$ and $B$ can eliminate many propagating modes.

The worst case occurs when $B = V$ because then there will always be an integer $k = j$ that meets the criterion for propagation given by Eq. (18.2.15). This is referred to as the plane wave mode because $m = jB - kV = 0$, and has no azimuthal variation. In contrast, if there are five guide vanes and three blades and the tip Mach number is 0.35 then for blade passing harmonics $j = 1, 2, 3$ we find that the criterion for propagation is not met for the first harmonic but is met for values of $k = 1$ for the second harmonic and $k = 2$ for the third harmonic of blade passing frequency. This example shows that by choosing the blade and vane numbers correctly the unsteady loading at the lower harmonics can be eliminated as a source of sound, which is clearly important in fan design.

### 18.2.4　In duct sound power

The duct mode amplitudes given by Eq. (18.2.13) for unsteady loading sources can be used in Eq. (17.5.3) to obtain the sound power in the duct. This is useful because, in the absence of reflections at the duct exit the modal sound power gives a direct measure of the far field sound. In Eq. (17.5.3) the sound power in each duct mode depends on the mode amplitude expressed as a velocity potential. The relationship to modes defined in terms of pressure amplitude is given by Eq. (18.2.4). However, some caution needs to be used in directly applying these results because the analysis given has not considered the blade response function, which is different for each of the duct modes, and this issue will be addressed in the next section.

## 18.3  The cascade blade response function

In the previous section we showed that the duct mode amplitudes in a cylindrical duct could be related to the distribution of unsteady loading, per unit area, on the planform of the fan blades or stator vanes. To predict the noise from ducted fans that may include a set of fixed stators, we need to calculate the unsteady loading, and this will depend on the response of the blades or stators to incoming perturbations to the mean flow. In ducted fans that have two stages (either the fan upstream of a set of stator vanes or a set of stator vanes upstream of a fan) the most significant unsteady flow will be caused by the wake of the upstream stage, and so we are most concerned with the unsteady loading on the downstream stage caused by the wake flow. The problem can be split into two parts: (i) the response of the downstream fan or stator to an arbitrary inflow disturbance and (ii) the specification of the unsteady inflow to the downstream stage caused by the upstream stage. In this section we will consider the response of a blade row to an arbitrary inflow disturbance and this will be followed in the subsequent sections by models of the unsteady inflow.

In Chapter 14 we discussed the details of the unsteady loading on a single blade in response to an arbitrary gust. For ducted blade rows the problem is significantly different in two respects. First, the duct will control the characteristics of the sound field, requiring it to match boundary conditions on the duct walls and to be periodic in the azimuthal direction. This leads to propagating and nonpropagating modes that do not occur in an open rotor situation, but are a central part of the problem for ducted fans. Second, the blade count on ducted fans is usually large and so the solidity of the fan or stator is high. This means that adjacent blades tend to overlap and that sound waves can be trapped in the gaps between the blades (see Fig. 18.4). This will always occur if the blade spacing is small enough that $d < c$, as shown in Fig. 18.4. The trapped waves significantly alter the response of a blade row to an incoming gust and its coupling with the acoustic waves that propagate up or downstream.

The complete solution for the response of a fan or stator to an incoming gust can be obtained using the linearized Euler equations (Eqs. 6.1.8, 6.1.12). However, analytical solutions to these equations in cylindrical coordinates, for a spinning fan and/or a swirling mean flow, have not been obtained. Numerical solutions to this problem are given by Atassi et al. [8] and Montgomery and Verdon [9]. Atassi and Golubev [10] also identify the importance of almost vortical waves and almost acoustic waves in a swirling mean flow and their numerical approaches allow for the inclusion of real blade geometries and transonic flow Mach numbers.

To obtain insight into the characteristics of the blade response an analytical approach is required. One approximation is to treat each radial blade section separately and calculate its response to the local unsteady inflow. However, this does

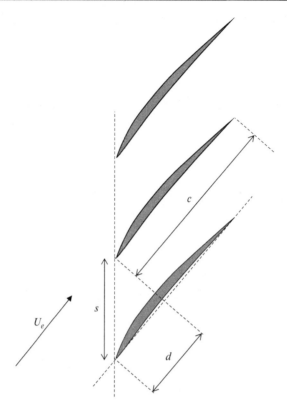

**Fig. 18.4** A set of overlapping blades in a typical fan or stator set. The blades are overlapped if $d < c$.

not allow the scattered field to satisfy the nonpenetration boundary conditions on the rest of the blade, which may include significant twist. To address this problem Cooper and Peake [11] developed an approach based on matched asymptotic expansions using a rectilinear model for the local blade response function and matching the output to the duct modes in the high frequency limit. The advantage of the rectilinear cascade model is that a complete solution can be obtained for an arbitrary gust, and the characteristics of the radiated sound field can be related to the duct and mean flow parameters. This provides some important insights into the problems that are not readily apparent in approximate solutions. Therefore in the following sections we will discuss the rectilinear cascade model in detail, and then, in Section 18.4, we will show how the rectilinear cascade matches the field in a cylindrical duct in the high frequency limit.

### 18.3.1  The rectilinear cascade model

The rectilinear cascade model is shown in Fig. 18.5. Each blade is assumed to be of zero thickness and infinite span, but may be restricted to a finite span between rigid plates for specific problems discussed later in this section. The blade chord is aligned with the mean flow, which is of speed $U_e$ in the $x$ direction. The effect of blade thickness and camber are assumed to be of second order, as discussed in earlier chapters where a flat plate was used to model an airfoil of finite thickness at an angle of attack to the flow.

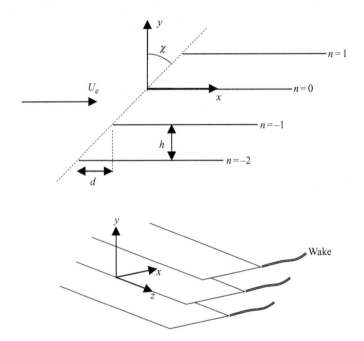

**Fig. 18.5** The rectilinear cascade in uniform flow.

The blades are numbered $n = 0, \pm1, \pm2, \pm3$ and the coordinate system has its origin at the leading edge of the blade $n = 0$. The blades are separated in the direction normal to the flow by the distance $h$ and the leading edge of the $n$th blade lies at $x = nd$. The blade chord is $c$ and for the blades to be overlapped we require that $d < c$. The distance between the leading edges is $s = (h^2 + d^2)^{1/2}$ and comparing Figs. 18.4 and 18.5 shows the equivalence of the rectilinear model to the blade row in cylindrical coordinates in the duct.

We will assume that the unsteady inflow to the blade causes a velocity perturbation $\mathbf{w}$ which may be of general form, and that we can expand it as a harmonic function of space and time such that the upwash normal to the blade surfaces is given by

$$\mathbf{w} \cdot \mathbf{n} = w_o e^{-i\omega t + ik_1 x + ik_2 y + ik_3 z} \tag{18.3.1}$$

To satisfy the Kutta condition a wake is shed from the trailing edge of each blade, as shown in Fig. 18.5, and we will assume that the wake cannot support a pressure jump.

The nonpenetration boundary condition and the Kutta condition define the boundary conditions of the problem and can be combined with the linearized Euler equations (see Section 13.3.1) to give the pressure and velocity perturbations throughout the fluid in terms of a velocity potential that is a solution to

$$\frac{1}{c_\infty^2}\frac{D_\infty^2\phi}{Dt^2} - \nabla^2\phi = 0 \qquad \frac{D_\infty\phi}{Dt} = \frac{\partial\phi}{\partial t} + U_e\frac{\partial\phi}{\partial x} \qquad (18.3.2)$$

The pressure perturbation in the flow is given by $p' = -\rho_o D_\infty\phi/Dt$ and the boundary condition on each blade surface requires that

$$\frac{\partial\phi}{\partial y} + \mathbf{w}\cdot\mathbf{n} = 0 \qquad (18.3.3)$$

so that for a harmonic gust given by Eq. (18.3.1) we have

$$\left[\frac{\partial\phi}{\partial y}\right]_{\substack{nd<x<nd+c \\ y=nh}} = -w_o\left[e^{-i\omega t + ik_1x + ink_2h + ik_3z}\right]_{nd<x<nd+c} \qquad (18.3.4)$$

This shows that the upwash on each blade has the same amplitude and is shifted in phase by $nk_1d + nk_2h = n\sigma$ where $\sigma = k_1d + k_2h$ is referred to as the interblade phase angle.

To find a solution to Eq. (18.3.2) we note that to match the nonpenetration boundary condition on the blade surfaces the velocity potential must be harmonic in time and, if the flow speed is independent of the span, then the potential is also harmonic in the spanwise direction. We can then seek a solution for the velocity potential in the form

$$\phi(\mathbf{x}, t) = \hat{\phi}(x, y)e^{-i\omega t + ik_3z} \qquad (18.3.5)$$

where $\hat{\phi}(x, y)$ is the two dimensional potential that satisfies the wave equation given by Eq. (18.3.2),

$$\beta^2\frac{\partial^2\hat{\phi}}{\partial x^2} + 2ikM\frac{\partial\hat{\phi}}{\partial x} + \frac{\partial^2\hat{\phi}}{\partial y^2} + (k^2 - k_3^2)\hat{\phi} = 0 \qquad (18.3.6)$$

where $\beta^2 = 1 - M^2$, $M = U_e/c_\infty$, and $k = \omega/c_\infty$.

In Section 13.2 we solved this equation for a single blade by noting that the potential was discontinuous across the blade surface on $y = 0$ and finding solutions for $y > 0$ and $y < 0$. In this case we have an infinite number of discontinuous sheets, each of which represents a blade surface and has its own acoustic field. The solution is therefore obtained by superimposing the solutions for each blade. Following the steps between Eqs. (13.3.4), (13.3.5) we obtain the solution to the wave equation for a blade row as

$$\hat{\phi}(x, y) = \sum_{n=-\infty}^{\infty} \text{sgn}(y - nh)\int_{-\infty}^{\infty} C_n(\alpha)e^{i\zeta|y-nh| - i\alpha x}d\alpha \qquad (18.3.7)$$

where $2C_n(\alpha)$ represents the Fourier transform of the potential jump across blade number $n$. In this section we have chosen to be consistent with Ref. [12] and use a different convention for the Fourier transform, changing the sign of $\alpha$ and defining the propagation of waves away from the plate in terms of a wavenumber

$$\zeta = \sqrt{(\omega + \alpha U_e)^2 / c_\infty^2 - \alpha^2 - k_3^2} \tag{18.3.8}$$

where the branch cut is chosen so that $\zeta$ has a positive imaginary part.

Since the amplitude of the upwash at each blade is identical and has a phase shift of $n\sigma$, and the leading edge of each blade is at $x = nd$, we can define the potential jump on each blade using a single function $D(\alpha, k_1, \sigma)$ for a specific gust as

$$C_n(\alpha) = \frac{1}{2} D(\alpha, k_1, \sigma) e^{in\sigma + in\alpha d}$$

(where we have introduced a factor of 2 so that $D(\alpha, k_1, \sigma)$ is the Fourier transform of the potential jump over each blade), so

$$\hat{\phi}(x, y) = \frac{1}{2} \sum_{n=-\infty}^{\infty} \mathrm{sgn}(y - nh) \int_{-\infty}^{\infty} D(\alpha, k_1, \sigma) e^{i\zeta|y - nh| - i\alpha(x - nd) + in\sigma} d\alpha \tag{18.3.9}$$

This shows that the solution is identical in each layer $nh < y < (n+1)h$ relative to the leading edge of blade number $n$, provided the interblade phase angle is correctly accounted for.

To solve Eq. (18.3.9) for the unknown function $D(\alpha, k_1, \sigma)$ we need to match it to the boundary conditions. Since there is only one variable we can use the boundary condition on one blade alone to obtain this solution. Using Eq. (18.3.4) for $y = 0^+$ and $0 < x < c$ we find that

$$w_o e^{ik_1 x} = \frac{-1}{2} \int_{-\infty}^{\infty} D(\alpha, k_1, \sigma) \left\{ \sum_{n=-\infty}^{\infty} , i, \zeta, e^{i\zeta|nh| + in(\alpha d + \sigma)} \right\} e^{-i\alpha x} d\alpha \quad 0 < x < c$$

We can then rearrange this equation so that it is in a form suitable for solution using the Weiner Hopf method, so

$$w_o e^{ik_1 x} = -2\pi \int_{-\infty}^{\infty} D(\alpha, k_1, \sigma) j(\alpha) e^{-i\alpha x} d\alpha \quad 0 < x < c \tag{18.3.10}$$

and

$$j(\alpha) = \frac{i\zeta}{4\pi} \sum_{n=-\infty}^{\infty} e^{i\zeta|nh| + in(\alpha d + \sigma)} \tag{18.3.11}$$

We have now defined a boundary value problem that is of the same form as the edge scattering problem discussed in Chapter 13. As before the boundary condition given by Eq. (18.3.10) is only defined over a limited range $0 < x < c$ and so the integral in this equation cannot be inverted to obtain the unknown function $D(\alpha, k_1, \sigma)$. However, we have defined $D(\alpha, k_1, \sigma)$ as the Fourier transform of the jump in potential across $y = 0$, and so the jump in potential across blade $n = 0$ is

$$\Delta \hat{\phi}(x) = \int_{-\infty}^{\infty} D(\alpha, k_1, \sigma) e^{-i\alpha x} d\alpha \tag{18.3.12}$$

which is zero when $x < 0$. This provides an additional boundary condition that can be used to complete the solution. As in Chapter 13 we can use the Weiner Hopf method to solve this problem, but only for semi-infinite surfaces. The application of this method to blades of finite chord requires a step by step solution. The first step is to find a solution for the case when the chord $c$ is infinite, and the second step is to add a correction that ensures there is no pressure jump across the wake downstream of the trailing edge so that the Kutta condition is satisfied. The first step is a well-defined Weiner Hopf problem with a known solution. The second step can be solved by using the same approach with the rigid surface extending upstream of the trailing edge to upstream infinity, $-\infty < x < c$, and requiring that the potential jump across the wake $c < x < \infty$ exactly cancels the pressure jump caused by the solution to the first step. This corrects the first solution so the Kutta condition is satisfied but also introduces a discontinuity of potential upstream of the leading edge where $x = 0$. The process therefore has to be repeated by adding additional solutions that cancel the jumps in potential upstream of the leading edge or the pressure jump across the wake. The full solution is given in Ref. [12] and the results are summarized in Appendix C.

### 18.3.2  The acoustic duct modes

To obtain the acoustic field from a stationary cascade the summation and integral in Eq. (18.3.9) need to be evaluated. The solution is given in Ref. [12] as

$$\hat{\phi}(x, y) = \frac{\pm w_o}{\beta s_e} \sum_{m=-\infty}^{\infty} \left\{ \frac{\pi \zeta_m^{\pm} D\left(\lambda_m^{\pm}, k_1, \sigma\right)}{\sqrt{\kappa_e^2 - f_m^2}} \right\} e^{i(\sigma - 2\pi m) y/h - i\lambda_m^{\pm}(x - yd/h)}$$

$$\kappa_e = \sqrt{k_o^2 - (k_3/\beta)^2} \quad f_m = (\sigma + k_o M d - 2\pi m)/s_e \tag{18.3.13}$$

$$s_e = \sqrt{d^2 + (h\beta)^2} \quad \lambda_m^{\pm} = k_o M - f_m \sin\chi_e \pm \cos\chi_e \sqrt{\kappa_e^2 - f_m^2}$$

and $M$ is the mean flow Mach number relative to the blade, $\beta = (1 - M^2)^{1/2}$, $k_o = \omega/c_\infty \beta^2$, $\chi_e = \tan^{-1}(h\beta/d)$, and $\zeta_m^{\pm} = \beta(\kappa_e^2 - (\lambda_m^{\pm} - k_o M)^2)^{1/2}$. The blade spacing and stagger angle is defined in Fig. 18.5.

In the result given by Eq. (18.3.13) the terms in { } represent the amplitude of each mode of propagation, while the phase terms give the spatial dependence of the modes.

Surfaces of constant $x - yd/h$ are parallel to the fan face shown in Fig. 18.4 and so $\lambda_m^\pm$ is the wavenumber of propagation up or downstream in duct-based coordinates. When $\lambda_m^\pm$ is imaginary the modes are cut off and this occurs when $f_m > \kappa_e$. From this expression it is also possible to obtain the upstream or downstream sound power for a harmonic gust by integrating over a surface where $x - yd/h$ is constant as described in Chapter 17:

$$W_\pm(\omega) = \left(\frac{w_o^2}{2}\right) \frac{\omega \rho_o B s b}{\beta s s_e} \sum_{m=-\infty}^{\infty} \mathrm{Re}\left[\frac{\left|\pi \zeta_m^\pm D\left(\lambda_m^\pm, k_1, \sigma\right)\right|^2}{\sqrt{\kappa_e^2 - f_m^2}}\right] \tag{18.3.14}$$

In this expression $(w_o^2/2)$ represents the mean square magnitude of the harmonic upwash component incident on the cascade.

In Eq. (18.3.14) each of the terms inside the summation sign represents the modal sound power, and this will only be nonzero if $\kappa_e > f_m$. Therefore, for modes to be cut on, the effective wavenumber $\kappa_e$ must exceed a certain value which is determined by the interblade phase angle, the Mach number, and the mode order. The modal sound power would appear to be infinite at cut on, but this is not the case because the blade response function tends to zero at cut on and so the radiated sound power remains finite. This is illustrated in Fig. 18.6A which shows the downstream sound power for the first three modes for a set of stator vanes. Note that no sound power is generated below the cut on frequency for each mode, and that the sound power remains finite at the cut on frequency in each case. The first mode also generates very little power at nondimensional frequencies between 13 and 14. This occurs when the direction of propagation of the mode is aligned with the blade chord and so the acoustic dipoles on the blade surface do not contribute to the acoustic field for that mode.

The effective wavenumber $\kappa_e$ that controls the cut on characteristics increases with frequency, but also depends on the spanwise wavenumber. Gusts with significant spanwise variations will cut on at higher frequencies than gusts with no spanwise variations. Fig. 18.6B shows the modal sound power output as a function of spanwise wavenumber for a fixed frequency. At this frequency the second mode has more sound power than the first mode at zero spanwise wavenumber, but not necessarily at all spanwise wavenumbers. Notice that as the first mode cuts off the sound power is transferred to the second mode at the cut off wavenumber. In a two dimensional model the spanwise variation of the gust is ignored, and it is assumed that $\kappa_e = \kappa$, so the cut on frequency of each mode is only a function of the mode order and interblade phase angle. Fig. 18.6B shows that significant errors could result from this approximation, especially in broadband noise modeling where the spanwise variations of the gusts are very significant.

### 18.3.3  The acoustic modes from an arbitrary gust

The theory described in Section 18.3.2 gives the sound radiation from a cascade of blades in response to an incident harmonic vortical gust which is convected with the flow. To apply this theory to broadband noise the gust must be defined in a more general form and is also required to satisfy various boundary conditions. The first modification we will impose is that the cascade is bounded by rigid end walls (see

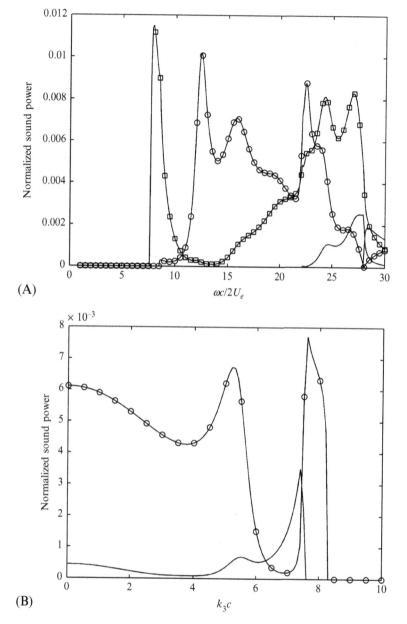

**Fig. 18.6** (A) Downstream sound power for the first three modes for a set of stator vanes. (B) Modal sound power output as a function of spanwise wavenumber for a fixed frequency ($M=0.3$, $\chi=40$ degrees, $s/c=0.6$, and $\sigma=3\pi/4$).

Reproduced with permission from S.A.L. Glegg, The response of a swept blade row to a three-dimensional gust, J. Sound Vib. 227 (1999) 29–64.

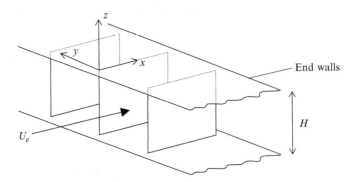

**Fig. 18.7** The rectilinear cascade between rigid end walls.

Fig. 18.7) and so both the acoustic field and the incoming gust must satisfy the non-penetration condition at these boundaries. Secondly the cascade flow represents an unwrapped version of a blade row in a cylindrical duct, and so must be periodic in the azimuthal direction. Finally, we will assume that the gust is convected with the mean flow at a uniform speed.

To develop a flow model which satisfies these conditions we will only consider a set of stator vanes and assume that the turbulent gust is convected by the mean flow with velocity $\mathbf{V}$. The flow speed relative to the blade is then obtained as $\mathbf{V}=(U_e,0,0)$ defined in $(x,y,z)$ coordinates (see Fig. 18.5). The gust must be periodic (repeating itself after $B$ blade passages), and is described as having an upwash velocity $\sum_{m=-\infty}^{\infty} w_o(\mathbf{x}-m\mathbf{d})$, where $\mathbf{d}=(Bd,Bh,0)$. Using this approach the gust $w_o(\mathbf{x})$ is zero unless $0<y<Bh$. The unsteady upwash relative to the blades is then

$$\mathbf{n}\cdot\mathbf{w}(\mathbf{x},t) = \sum_{m=-\infty}^{\infty} w_o(\mathbf{x}-m\mathbf{d}-\mathbf{V}t) \qquad (18.3.15)$$

For a vortical gust between two parallel end walls the flow components parallel to the walls can be described by a Fourier cosine series expansion, so in general we can write

$$w_o(x,y,z) = \iint \sum_{n=0}^{\infty} \tilde{w}_n(k_1,k_2)e^{ik_1x+ik_2y}\varepsilon_n\cos\left(\frac{n\pi z}{H}\right)dk_1dk_2 \qquad (18.3.16)$$

where $H$ is the height of the annulus and $\varepsilon_n=1$ for $n>0$ and $\varepsilon_o=1/2$. The integrand is defined as the wavenumber transform of $w(\mathbf{x})$ in the form

$$\tilde{w}_n(k_1,k_2) = \frac{2}{(2\pi)^2 H}\int_{-R}^{R}\int_{0}^{Bh}\int_{0}^{H} w(x,y,z)\,e^{-ik_1x-ik_2y}\cos(n\pi z/H)dxdydz \qquad (18.3.17)$$

The upwash is then obtained as

$$\mathbf{n} \cdot \mathbf{w}(\mathbf{x}, t) = \iint \sum_{n=0}^{\infty} \sum_{m=-\infty}^{\infty} \widetilde{\widetilde{w}}_n(k_1, k_2) e^{ik_1(x - U_e t - mBd) + ik_2(y - mBh)} \varepsilon_n \cos(n\pi z/H) dk_1 dk_2$$

$$(18.3.18)$$

Making use of the Poisson sum formula

$$\sum_{m=-\infty}^{\infty} e^{-ik_1 mBd - ik_2 mBh} = 2\pi \sum_{k=-\infty}^{\infty} \delta(k_1 Bd + k_2 Bh - 2\pi k) \qquad (18.3.19)$$

and integrating over $k_2$ gives

$$\mathbf{n} \cdot \mathbf{w}(\mathbf{x}, t)$$
$$= \frac{2\pi}{Bh} \int_{-\infty}^{\infty} \sum_{n=0}^{\infty} \sum_{k=-\infty}^{\infty} \widetilde{\widetilde{w}}_n\left(k_1, \frac{2\pi k}{Bh} - \frac{k_1 d}{h}\right) e^{ik_1(x - U_e t) + i\left(\frac{2\pi k}{Bh} - \frac{k_1 d}{h}\right)y} \varepsilon_n \cos(n\pi z/H) dk_1$$

$$(18.3.20)$$

If we define $\omega = k_1 U_e$ then we obtain the upwash velocity at the blade as

$$\mathbf{n} \cdot \mathbf{w}(\mathbf{x}, t) = \sum_{n=0}^{\infty} \sum_{k=-\infty}^{\infty} \int_{-\infty}^{\infty} w_{kn}(\omega) e^{-i\omega t + ik_1(x - yd/h) + 2\pi iky/Bh} \varepsilon_n \cos(n\pi z/H) d\omega$$

$$(18.3.21)$$

where

$$w_{kn}(\omega) = \frac{2\pi}{BhU_e} \widetilde{\widetilde{w}}_n\left(k_1, \frac{2\pi k}{Bh} - \frac{k_1 d}{h}\right) = \frac{1}{\pi BhHU_e} \int_{-R}^{R} \int_{0}^{Bh} \int_{0}^{H} w(\mathbf{x})$$

$$e^{-ik_1(x - yd/h) - 2\pi iky/Bh} \cos(n\pi z/H) dx dy dz \qquad (18.3.22)$$

Eq. (18.3.21) gives a Fourier series expansion of the incoming vortical gust that satisfies the periodicity condition and the end wall boundary conditions. Note that $x - yd/h$ is constant at the blade leading edges and so the interblade phase angle of the gust is $\sigma = 2\pi k/B$.

To obtain the acoustic field from this gust we can combine Eq. (18.3.13) with Eq. (18.3.22) to give

$$\phi(\mathbf{x}, t) = \frac{\pm 1}{\beta s_e} \sum_{m=-\infty}^{\infty} \sum_{k=-\infty}^{\infty} \sum_{n=0}^{\infty} \int_{-\infty}^{\infty} T_{mn}^{(k)}(\omega) w_{kn}(\omega)$$

$$e^{-i\omega t + i(\sigma - 2\pi m)y/h - i\lambda_m^{\pm}(x - yd/h)} \varepsilon_n \cos\left(\frac{n\pi z}{H}\right) d\omega$$

$$\text{where } T_{mn}^{(k)}(\omega) = \left(\frac{\pi \zeta_m^{\pm} D(\lambda_m^{\pm}, \omega/U_e, \sigma)}{\sqrt{\kappa_e^2 - f_m^2}}\right) \qquad (18.3.23)$$

as the amplitude of a mode of order $m$ with a spanwise mode order $n$ and an interblade phase angle $2\pi k/B$. The computation of the triple summation in this expression can be simplified by using the property that $T_{mn}^{(k)} = T_{on}^{(k+mB)}$. Then if we introduce the integer variable $p = mB - k$ it follows that $\sigma - 2\pi m = -2\pi p/B$. We can then write Eq. (18.3.23) as

$$
\phi = \frac{\pm 1}{\beta s_e} \sum_{p=-\infty}^{\infty} \sum_{n=0}^{\infty} \varepsilon_n \cos\left(n\pi z/H\right) \int_{-\infty}^{\infty} T_{0n}^{(p)}(\omega) \left\{ \sum_{m=-\infty}^{\infty} , w_{mB-p,n}, (\omega) \right\}
$$
$$
e^{-i\omega t + 2\pi ipy/h)/Bs - i\lambda_m^{\pm}(x - yd/h)} d\omega
$$
(18.3.24)

In an equivalent annular system, the azimuthal coordinate $\theta$ is related to $y$ by $2\pi y/Bh = \theta$ on a surface where $x - yd/h$ is constant, and so the integer $p$ defines the mode order in the cylindrical coordinate system. Also note that the summation over $m$ is only required over the gust coefficients and not over the blade response terms which simplifies the evaluation of Eq. (18.3.24).

### 18.3.4  The sound power spectrum

For a periodic gust the sound power for each mode is given by Eq. (18.3.14). However, for a stochastic input, such as broadband noise we need to calculate the sound power spectral density which is defined in a moving fluid as

$$
S_{WW}(\omega) = \frac{\pi}{T} Ex[W_{\pm}(\omega)]
$$
(18.3.25)

As in Eq. (18.3.14) the modes are orthogonal and contribute independently to the sound power giving

$$
S_{WW}(\omega) = \frac{\rho_o B s H U_e}{2} \sum_{m=-\infty}^{\infty} \sum_{k=-\infty}^{\infty} \sum_{n=0}^{\infty} H_{mn}^{(k)}(\omega) E_{kn}(\omega)
$$

$$
\text{where } H_{mn}^{(k)}(\omega) = \text{Re}\left( \frac{\omega \left| \pi \zeta_m^{\pm} D\left(\lambda_m^{\pm}, \omega/U_e, \sigma\right)\right|^2}{U_e \beta s s_e \sqrt{\kappa_e^2 - f_m^2}} \right)
$$
(18.3.26)

$$
\text{and } E_{kn}(\omega) = \varepsilon_n \frac{\pi}{T} Ex\left[ |w_{kn}(\omega)|^2 \right]
$$

Using $H_{mn}^{(k)} = H_{0n}^{(k+mB)}$ we find

$$
S_{WW}(\omega_o) = \frac{\rho_o B s b U_e}{2} \sum_{p,n} H_{0n}^{(p)}(\omega) \left\{ \sum_m E_{mB-p,n}(\omega) \right\}
$$
(18.3.27)

The evaluation of this expression is dependent on the definition of the disturbance spectrum function $E_{kn}(\omega)$ which is obtained from Eq. (18.3.22), which gives,

$$E_{kn}(\omega) = \frac{\varepsilon_n}{\pi(BhHU_e)^2 T} \int_0^{Bh} \int_0^H \int_0^{Bh} \int_0^H \int_{-R}^R \int_{-R}^R Ex[w_o(\mathbf{x})w_o(\mathbf{x}_1)]$$

$$e^{i\omega(x-x_1)/U_e + i(2\pi k + \omega B d/U_e)(y-y_1)/Bh}$$

$$\cos(n\pi z/H)\cos(n\pi z_1/H)dx dx_1 dz dy dz_1 dy_1 \qquad (18.3.28)$$

where the extent of the flow in the $x$ direction is $\pm R$ where $R$ tends to infinity. Defining $T = R/U_e$ and using the results of Chapter 9 shows that the integrals in Eq. (18.3.28) define the wavenumber spectrum of the turbulence, and so we find that

$$E_{kn}(\omega) = \frac{(2\pi)^2}{U_e BhH}\Phi_{ww}(-k_1, 2\pi k/Bh + k_1 d/h, n\pi/H) \qquad k_1 = \omega/U_e \qquad (18.3.29)$$

These results give a complete set of equations for the sound power from a linear cascade subject to an arbitrary inflow disturbance. They may be used for both broadband noise and tone noise (see Section 18.6) and will be discussed in more detail in the subsequent sections.

## 18.4 The rectilinear model of a rotor or stator in a cylindrical duct

### 18.4.1 Mode matching

In the previous section we solved the compressible flow boundary value problem for a blade row in a rectilinear duct subject to an unsteady velocity perturbation. In practice we are more often concerned with a fan or a set of stator vanes in a cylindrical duct, and the question arises as to whether the rectilinear model can be used to model the blade response in cylindrical coordinates. There are some significant differences between these two geometries which are highlighted by the wave equation in cylindrical coordinates for a steady mean axial flow with solid body rotation, which is

$$\frac{1}{c_\infty^2}\frac{D_\infty^2 \phi}{Dt^2} - \frac{\partial^2 \phi}{\partial x_o^2} - \frac{1}{R^2}\frac{\partial^2 \phi}{\partial \varphi^2} - \frac{1}{R}\frac{\partial \phi}{\partial R} - \frac{\partial^2 \phi}{\partial R^2} = 0 \qquad \frac{D_\infty \phi}{Dt^2} = \frac{\partial \phi}{\partial t} + U\frac{\partial \phi}{\partial x_o} + \Omega\frac{\partial \phi}{\partial \varphi}$$

where $(R,\varphi,x_o)$ are the cylindrical coordinates defined in Fig. 17.2, $U$ is the axial flow speed, and $\Omega$ is the angular velocity of the mean flow. The boundary condition on each blade surface requires that

$$\cos\beta_o\frac{\partial \phi}{\partial x} + \frac{\sin\beta_o}{R}\frac{\partial \phi}{\partial \varphi} + \mathbf{w}\cdot\mathbf{n} = 0 \qquad (18.4.1)$$

where $\mathbf{w}$ defines the incident unsteady disturbance and $\mathbf{n} = (0,\sin\beta_o,\cos\beta_o)$ is the normal to the blade planform as defined in Fig. 18.3.

If we consider a cylindrical duct with a center body as shown in Fig. 17.2B with an inner radius $b$ and an outer radius $a$ then we can define a set of coordinates

$$\xi_1 = x, \quad \xi_2 = R_c \varphi, \quad \xi_3 = R - b$$

where $R_c = (a+b)/2$ is the mean radius of the duct. In terms of these coordinates the wave equation becomes

$$\frac{1}{c_\infty^2} \frac{D_\infty^2 \phi}{Dt^2} - \frac{\partial^2 \phi}{\partial \xi_i^2} = \left( \frac{1}{R^2} - \frac{1}{R_c^2} \right) \frac{\partial^2 \phi}{\partial \varphi^2} + \frac{1}{R} \frac{\partial \phi}{\partial R} \tag{18.4.2}$$

where

$$\frac{D_\infty \phi}{Dt} = \frac{\partial \phi}{\partial t} + U \frac{\partial \phi}{\partial \xi_1} + \Omega R_c \frac{\partial \phi}{\partial \xi_2}$$

The terms on the left side of the wave equation given by Eq. (18.4.2) are identical to the wave equation in rectilinear coordinates $(\xi_1, \xi_2, \xi_3)$ aligned with the axial flow. For a duct mode with azimuthal order $m$ and radial wavenumber $\alpha_{mn}$ the order of magnitude of the terms on the right side of Eq. (18.4.2) will be

$$m^2 \left( \frac{1}{R^2} - \frac{1}{R_c^2} \right) \phi + \frac{\alpha_{mn} \phi}{R}$$

while terms on the left are of order $O(k^2 \phi)$. In general, the terms on the right will be small compared to the terms on the left provided that

$$(ka)^2 \gg \left( \frac{ma}{R} \right)^2 \left| 1 - \frac{R^2}{R_c^2} \right| + \frac{\alpha_{mn} a^2}{R}$$

The largest error occurs when $R = b$ and so this criterion will apply when

$$(kb)^2 \gg m^2 + \alpha_{mn} b \tag{18.4.3}$$

In a cylindrical duct, acoustic modes will only propagate if the frequency is above cut off which requires that $ka > \alpha_{mn} \beta a > m\beta$ so the limit set by Eq. (18.4.3) will be met for modes that are well above cut off of the lowest order radial mode provided that $b/a$ is not too small. This criterion can also be expressed in terms of the blade number and the blade tip Mach number $M_{tip} = \Omega a/c_\infty$ where $\Omega$ is the shaft rotation speed. At the blade passing frequency $ka = BM_{tip}$ and so the criterion (18.4.3) is met for fans with a large number of blades providing $BM_{tip} \gg ma/b$. In this limit the rectilinear approximation of the wave equation in a cylindrical duct appears to be correct to first order.

In addition to the approximation of the wave equation the boundary condition on the blade surface is not matched exactly by the rectilinear model. Expressing Eq. (18.4.1) in the rectilinear coordinate system leads to

$$\cos\chi\frac{\partial\phi}{\partial\xi_1} + \sin\chi\left(\frac{R_c}{R}\right)^2\frac{\partial\phi}{\partial\xi_2} = -\mathbf{w}\cdot\mathbf{n}\left(\frac{U_e(R)R_c}{U_e(R_c)R}\right) \tag{18.4.4}$$

where $\tan\beta_o = U/\Omega R$, $U_e^2 = U^2 + (\Omega R)^2$, and $\tan\chi = U/\Omega R_c$. We can rearrange this as

$$\cos\chi\frac{\partial\phi}{\partial\xi_1} + \sin\chi\frac{\partial\phi}{\partial\xi_2} \approx -\mathbf{w}\cdot\mathbf{n}\left(\frac{U_e(R)R_c}{U_e(R_c)R}\right) + \sin\chi\frac{\partial\phi}{\partial\xi_2}\left(1 - \frac{R^2}{R_c^2}\right)$$

and since $\partial\phi/\partial\xi_2 \ll \partial\phi/\partial\xi_1$ for well cut on modes for which $ka = BM_{tip} \gg ma/b$, the second term on the right can be dropped leading to a boundary condition that is independent of radius, provided that the gust amplitude is corrected as in Eq. (18.4.4).

It follows that the wave field in a rectilinear duct is representative of the waves in a cylindrical duct with a center body, at least in the high frequency limit or when blade count is large. The advantage of this approximation is that there is a complete solution to the boundary value problem for the blade response function in the rectilinear duct that is self consistent, but this is not the case for the cylindrical duct. Other approaches to solving this problem include using strip theory in which the blade loading at each spanwise location is modeled by a two dimensional blade response function. While this allows for the details of the gust and blade section to be included in the model it leads to a mismatch with the dispersion relationship that can cause large errors at the cut off frequency of the higher order modes. To overcome this issue Cooper and Peake [11] used a matching technique on a surface just upstream of the blade row. The rectilinear model described above was used to obtain the acoustic near field just upstream of the blade row, and then a least squares optimization was used to match the near field to the duct modes in a cylindrical duct. One of the features of Cooper and Peake's model was that the matching was only carried out for propagating acoustic modes and so numerical errors associated with the nonpropagating modes was eliminated. A similar justification can be used in the rectilinear model if only the sound power output of the modes is considered. Small errors in amplitude will tend to be smoothed out by averaging the acoustic intensity over the duct cross section and so it can be argued that the sound power estimate using the rectilinear model will match the sound power in the cylindrical duct more accurately than the acoustic mode amplitude, which is consistent with the scaling approach used by Cooper and Peake [11].

### 18.4.2 An axial dipole example

To illustrate this equivalence consider a stationary point dipole source with its axis parallel to the duct axis in the annular duct described above. The acoustic field is given by the derivative of the Green's function specified in cylindrical coordinates by Eq. (17.4.2), and gives, for a source at $\mathbf{y} = (R_o, 0, 0)$

$$\frac{\partial\widetilde{G}_e}{\partial x} = -\sum_{m=-\infty}^{\infty}\sum_{n=1}^{\infty}\left[\mu_{mn}^{\pm}\right]^{cyl}\frac{\psi_{mn}(R)\psi_{mn}(R_o)e^{im\varphi + i\left[\mu_{mn}^{\pm}\right]^{cyl}x}}{2\Lambda_{mn}\left[k_{mn}\right]^{cyl}} \tag{18.4.5}$$

where the wavenumbers for the annular duct have been specifically identified by the brackets $[]^{cyl}$.

For the annular duct it was shown in Section 17.2, Eq. (17.2.14), that

$$\psi_{mn}(R) \approx \left(\frac{R_c}{2R}\right)^{1/2} \cos\left(n\pi(R-b)/H\right) \tag{18.4.6}$$

where $H = a - b$ is the width of the annulus and the mode normalization factor is

$$[\Lambda_{mn}]^{cyl} = \int_b^a \int_0^{2\pi} |\psi_{mn}|^2 R dR d\varphi = \pi R_c H$$

and the wavenumber is

$$[\mu_{mn}^{\pm}]^{cyl} = \frac{-kM \pm [k_{mn}]^{cyl}}{\beta^2}, \qquad [k_{mn}]^{cyl} = k\sqrt{1 - (\beta n\pi/kH)^2} \tag{18.4.7}$$

If the same source is placed in a rectilinear duct with the same height $H$ and a cross axis dimension $L = 2\pi R_c$ then the sound field is given by the derivative of the Green's function for rectangular duct modes. For the source located at $(0,0,\xi_3^{(o)})$ we obtain

$$\frac{\partial \tilde{G}_e}{\partial x} = -\sum_{m=-\infty}^{\infty} \sum_{n=1}^{\infty} [\mu_{mn}^{\pm}]^{rec} \frac{[\psi_{mn}(\xi_3)]^{rec} \left[\psi_{mn}\left(\xi_3^{(o)}\right)\right]^{rec} e^{2\pi i m\xi_2/L + i[\mu_{mn}^{\pm}]^{rec}\xi_1}}{2[\Lambda_{mn}]^{rec}[k_{mn}]^{rec}} \tag{18.4.8}$$

where the mode shape for a rectangular duct is $[\psi_{mn}(\xi_3)]^{rec} = \cos(n\pi\xi_3/H)$ and the normalization is $[\Lambda_{mn}]^{rec} = HL/2$

To show the equivalence of this result with the field in the cylindrical duct substitute $\xi_1 = x$, $\xi_2 = \varphi R_c$, $\xi_3 = R - b$, and $L = 2\pi R_c$. The mode normalization factors and the phase terms match exactly in each case, but the mode amplitudes differ by a factor of

$$\frac{[-kM \pm k_{mn}]^{rec}[k_{mn}]^{cyl}}{[-kM \pm k_{mn}]^{cyl}[k_{mn}]^{rec}} \sqrt{\frac{R R_o}{R_c^2}} \tag{18.4.9}$$

For the rectilinear duct we have that

$$[k_{mn}]^{rec} = k\sqrt{1 - (\beta n\pi/kH)^2 - (\beta m\pi/kL)^2}$$

which when compared to Eq. (18.4.7) shows that the wavenumbers are almost identical in the limit that $\beta m\pi/kL \ll 1$ which occurs at frequencies well above cut on frequency of the radial mode. However, close to the cut on frequency the error may be large because the modes in the rectangular duct will cut on at a lower frequency than the modes in the cylindrical duct.

The amplitude ratio (18.4.9) shows that the modal amplitudes are similar in each case with an error that tends to zero for well cut on modes and a factor that depends on the source position in the duct. For broadband sources that are uniformly distributed across the duct the factor of $\sqrt{R_o/R_c}$ will average to a factor close to one so the main source of error is the mismatch in wavenumbers close to the cut on frequency of the radial modes. However, the modal sound power tends to zero at this frequency and so the largest error tends to occur where the sound power is zero anyway. Remarkably therefore the rectilinear model of the annular duct has a well defined and simple correction factor to give the mode amplitudes for a dipole source in an annular duct at frequencies not close to cut off, and can be used quite accurately.

## 18.5    Wake evolution in swirling flows

The dominant source of noise from high bypass ratio turbofan engines is the interaction of the wake from the fan with downstream stator vanes. The wake originates at the trailing edges of the fan blades and evolves as a function of distance downstream. To correctly calculate the interaction noise caused by the stator vanes the wake propagation must be correctly estimated. This is a nontrivial problem because there is also a mean swirling flow downstream of the fan associated with axial vorticity that is concentrated in the blade wakes, and this swirl can be an important contributor to the evolution of the wakes and thus the unsteady flow that impacts the stator vanes.

To illustrate how wakes propagate in a swirling mean flow consider the idealized situations illustrated in Fig. 18.8. Fig. 18.8A and B illustrates the evolution of a fan-blade in a nonrotating flow. In this case the wake remains in the same plane as it started and evolves in an approximately self-similar manner as discussed in Chapter 9. If a solid body rotation is added to the mean flow, or the fan rotates, then the wake follows a helical path as shown in Fig. 18.8C and D. At the downstream position shown in Fig. 18.8D the wake is still radial, but no longer in the plane in which it started. A more complex situation occurs when the swirling flow includes both solid body rotation and a flow component that is represented by an irrotational vortex, so the azimuthal velocity is $U_\varphi = \Omega_o R - \Gamma/2\pi R$. In this case the wake is distorted as shown in Fig. 18.8E and no longer lies along a radial line at the downstream position.

We can estimate the impact of the swirl on the velocity perturbation in the wake by considering its evolution using Cauchy's equation, which was derived in Chapter 6. It was shown that the vorticity at a downstream location $\boldsymbol{\omega}$ was related to its initial value $\boldsymbol{\omega}^{(\infty)}$ by the solution to the vorticity equation that may be written in the form

$$\boldsymbol{\omega}(\mathbf{x}, t) = \frac{\delta \mathbf{l}^{(i)}}{h_i} \omega_i^{(\infty)}(\mathbf{X} - \mathbf{i}U_\infty t) \tag{18.5.1}$$

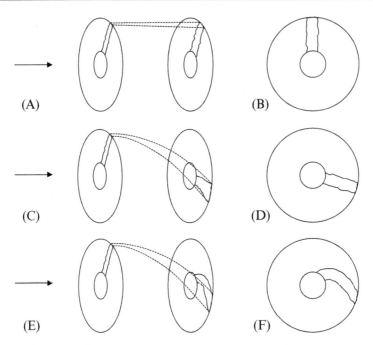

**Fig. 18.8** Wake evolution in a swirling flow downstream of a fan blade. (A) No rotation, 3D view and (B) downstream cross-sectional view. (C) Fan blade rotation at angular speed $\Omega$ in a swirling flow in solid body rotation with angular speed $\Omega_o$, 3D view and (D) downstream cross-sectional view. (E) Effect of swirling flow $U_\varphi = \Omega_o R - \Gamma/2\pi R$ on wake shape, 3D view and (F) downstream cross-sectional view.

where $\mathbf{X}$ represents drift coordinates and the vectors $\delta\mathbf{l}^{(i)}$ are the sides of the material volume as shown in Fig. 6.3. In the case of a wake that is initiated along a radial line the initial vorticity is primarily in the radial direction. There is also streamwise vorticity produced by the spanwise shedding of lift from the blades that contributes to the rotational component of the overall swirl. Near the spanwise ends of the blade this vorticity can become concentrated into hub and tip vortices that cause an unsteady flow near the duct walls.

Consider the fate of the vorticity under the action of this swirl. If the swirl approximates a solid body rotation, as shown in Fig. 18.8D, then the vortex lines remain radial or streamwise and are not stretched by the mean flow. However, if the angular velocity of the swirl varies significantly with radius then an initially radial material line in the wake will be rotated as shown in Fig. 18.9, to an angle $\beta_v$ to the radial direction. Consequently, the initially radial vorticity of the wake will be reoriented to have both radial and azimuthal components. Since there is no radial mean flow in this idealized view, the radial extent of the material lines will be unchanged by this distortion so that the radial component of the vector material volume length $\delta l^{(R)}$ remains constant. Consequently, the overall cross-sectional length of these material lines will be

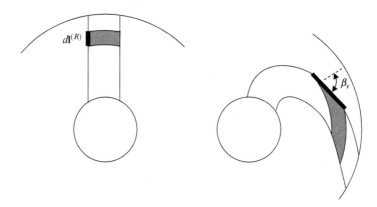

**Fig. 18.9** Distortion of a material volume by a swirling flow.

increased by a factor of $(1+\tan^2\beta_v)^{1/2} = \sec\beta_v$ and, by Eq. (18.5.1), the strength of the initially radial vorticity must be increased by the same factor. The swirl therefore amplifies the cross-plane vorticity in the wake as well as changing its structure. This amplification tends to increase the streamwise velocity deficit in the wake, countering its turbulent decay. Note that if the axial mean flow velocity downstream of the fan varies radially, the initially radial wake vorticity will suffer a similar distortion that will tend to reorient it into the axial direction.

It was shown in Chapter 9 that the wake downstream of an airfoil of large span was self similar and can be modeled as having a streamwise mean-velocity distribution,

$$u(\xi,x) = U_\infty - U_w(x)e^{-(\xi/L_w(x))^2} \tag{18.5.2}$$

where $x$ is the distance downstream of the trailing edge, $\xi$ is the distance from the centerline of the wake and $U_w(x)$ is the centerline velocity deficit. The lengthscale $L_w(x)$ is the wake half width which increases as $(x-x_c)^{1/2}$ where $x_c$ is a reference location near the trailing edge. The continuity of the mass flux requires that $U_w(x)L_w(x)$ is constant. This provides a simple model for the velocity deficit in the wake that can be used as an input to rotor stator interaction noise calculations.

In a swirling flow the streamwise decay of a fan-blade wake can impact the axial vorticity embedded within it. As the velocity deficit in the wake decays in the downstream direction the material volume of the particles in the wake must be stretched as illustrated in Fig. 18.10 (note this is the same effect of the stretching shown in Fig. 6.4 because the velocity on the centerline of the wake $U_\infty - U_w(x)$ increases with $x$). The amount of stretching that takes place is given by

$$\frac{\left|\delta l^{(1)}\right|}{h_1} = \frac{U_\infty - U_w(x)}{U_\infty - U_w(0)} > 1 \tag{18.5.3}$$

To ensure that the material volume remains constant, there must be a simultaneous contraction of the streamlines as the wake evolves, which is indicative of particles being entrained into the wake. The effect of the stretching is therefore to rotate the diagonal of the material volume is illustrated in Fig. 18.10. As a consequence the mean axial vorticity will also be rotated so as to have an azimuthal component that will be experienced as a disturbance when it encounters a downstream stator. In the far wake $U_w(x)$ tends to zero and so Eq. (18.5.3) shows that the stretching asymptotes to a constant value that depends on $U_\infty/(U_\infty - U_w(0))$ and is independent of downstream distance. Consequently, in a swirling mean flow, the axial vorticity field in the far wake will be distorted in a way that retains a characteristic of the initial wake deficit.

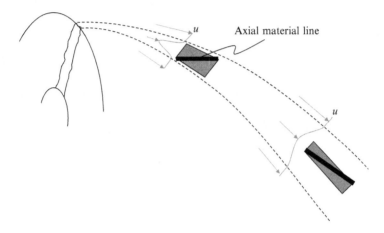

**Fig. 18.10** The evolution of the material volume on the centerline of a wake in a swirling flow shown along a helical line of constant radius.

The discussion above describes some of the issues with wake evolution and demonstrates how the swirl in the flow can affect the velocity deficit in the far wake. More detailed discussion of these effects are given by Cooper and Peake [11] who give a high frequency asymptotic method for calculating the evolution of the wake. To obtain greater accuracy a fully viscous RANS calculation of the mean flow can provide the details of the wakes downstream of the fan and this is the accepted approach for calculating the velocity deficits in the wakes as they enter a downstream set of stator vanes.

## 18.6   Fan tone noise

As described in Section 18.1 fan noise can be either tonal or broadband. Tone noise is caused by the blades or stators being subjected to a periodic disturbance that causes sound radiation at the blade passing frequencies. In this section we will use the rectilinear cascade model to calculate the amplitude of each mode in the duct that is caused by the interaction of the mean blade wakes that are incident on a set of down-stream stator vanes. It will be assumed that there are $B$ identical blades in the fan that are incident on a stator with $V$ vanes and that the mean-velocity deficit in the blade wakes is given by the self-similar wake profile specified by Eq. (18.5.2).

### 18.6.1   The upwash coefficients

The acoustic power in each mode was defined by Eq. (18.3.14) for a harmonic gust of the type

$$\mathbf{w} \cdot \mathbf{n} = w_o e^{-i\omega t + ik_1 x + ik_2 y + ik_3 z}$$

in the coordinate system defined in Fig. 18.5. To determine the mode amplitudes for tone noise caused by regular blade wakes we need to determine the wave amplitudes and wavenumbers that correspond to a suitable model of the blade wakes. In general, the wakes can be defined in the $(\xi_1', \xi_2', \xi_3')$ coordinates illustrated in Fig. 18.11.

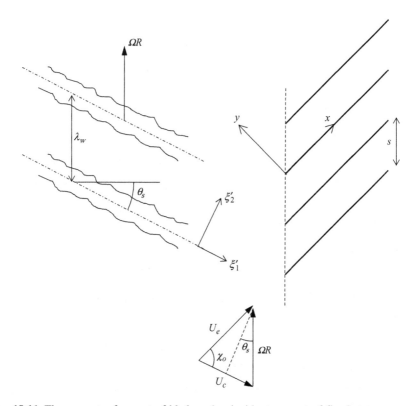

**Fig. 18.11** The geometry for a set of blade wakes incident on a set of fixed stator vanes.

This figure depicts a set of parallel rotor wakes that are rotating with angular speed $\Omega$ at a radius $R$. The wakes are aligned at an angle $\theta_s$ to the duct axis to account for swirl and the local convection velocity is $U_c$ along the wakes. It is assumed that the wakes are evolving slowly so that there is no change in perturbation velocity in the $\xi'_1$ direction. The wakes are identical so the perturbation velocity is periodic and the distance between the wakes normal to the axis of the duct is $\lambda_w$, and is equal to the blade spacing of the upstream fan. Consequently, the perturbation velocity $\mathbf{w}$ will be periodic in the $\xi'_2$ direction with a period given by $\lambda_w \cos\theta_s$.

In addition we will assume that the wakes are trapped by the end walls of an annular duct and can be described by a set of modes that satisfies the duct wall boundary conditions and, if we assume the wakes are incompressible, are divergence free. The velocity component will then have its own modal shape, and the components normal to the duct walls must be zero. The perturbation velocity can then be expanded in a Fourier series

$$\mathbf{w}\cdot\mathbf{n} = \sum_{j=-\infty}^{\infty}\sum_{n=0}^{\infty} w_{nj}\cos\left(\frac{n\pi\xi'_3}{H}\right)e^{2\pi i j\xi'_2/\lambda_w \cos\theta_s} \tag{18.6.1}$$

To calculate the upwash on the blades and the wavenumbers of the gust we need to express the $\xi'_2$ coordinate in terms of the $x$ and $y$ coordinate that are aligned with the vanes, as shown in Fig. 18.11. A rotation of the axes and a correction for the motion of the wakes relative to the stationary frame of reference gives

$$\xi'_2 = (x - U_e t)\sin\chi_o + y\cos\chi_o$$

The periodicity of the wakes is determined by the rotational speed of the fan $\Omega$, the swirl speed is $\Omega_o$, and the axial flow speed $U$ which convects the wakes downstream of the fan. If there was no swirl then the stator vanes would be aligned with the duct axis, but in the presence of swirl they have a finite pitch angle. Taking this into account Fig. 18.11 shows that the angle of the wake to the axial flow direction and the resolved flow speeds are given by

$$\theta_s = \tan^{-1}\left(\frac{(\Omega-\Omega_o)R}{U}\right) \qquad U_e = \sqrt{U^2 + \Omega_o^2 R^2} \qquad U_e\sin\chi_o = \Omega R\cos\theta_s$$

Substituting for $\sin\chi_o$ and noting that if there are $B$ blades on the fan then the periodicity of the wakes requires that $B\lambda_w = 2\pi R$, we obtain

$$\frac{\xi'_2}{\lambda_w\cos\theta_s} = \frac{B\Omega}{2\pi U_e}(x - U_e t + y\cot\chi_o)$$

Consequently, the upwash on the blade for which $y=0$ is given as

$$\mathbf{w}\cdot\mathbf{n} = -\sum_{j=-\infty}^{\infty}\sum_{n=0}^{\infty} w_{jn}\cos\left(\frac{n\pi z}{H}\right)e^{-ijB\Omega(t-x/U_e)} \tag{18.6.2}$$

The blade wakes therefore excite the stator vanes at frequencies $jB\Omega$ which are multiples of the blade passing frequency and the upwash is expanded in a harmonic series that is consistent with the inputs required in Section 18.3.

For each blade we can also calculate the interblade phase angle. Since the distance between the vane leading edges gives a displacement in $\xi'_2 = s\cos\theta_s$ we obtain the phase shift of the gust between each blade passage as

$$\sigma = 2\pi js / \lambda_w$$

This further simplifies if there are $B$ wakes so $\lambda_w = 2\pi R/B$ and $V$ stator vanes so $s = 2\pi R/V$ giving

$$\sigma = 2\pi jB/V \tag{18.6.3}$$

for each coefficient $j$.

We can then use these results in Eq. (18.3.14) to obtain the sound power in each mode for a harmonic gust so that the modal sound power is (in terms of the coefficients defined in Eq. (18.3.27)), is

$$W_{\pm} = \frac{\rho_o VsbU_e}{2} \sum_{p=-\infty}^{\infty} \sum_{n=0}^{\infty} \varepsilon_n H_{0n}^{(p)}(jB\Omega) \sum_{m=-\infty}^{\infty} |w_{mV-p}|^2 \tag{18.6.4}$$

## 18.6.2   Unskewed self-similar wakes

To estimate rotor stator interaction noise we need to specify the Fourier series coefficients $w_{jn}$. For wakes that are unskewed as shown in Fig. 18.8C and D, only the $n=0$ term is excited and we can obtain the Fourier series coefficients from a simple model of the wake deficit. In Chapter 9 we showed that wakes were in general self similar and well modeled by the function

$$u(\xi'_1, \xi'_2) = U_{\infty} - U_w(\xi'_1)e^{-\xi'^2_2/L^2_w(\xi_1)}$$

In general, the velocity deficit $U_w(\xi'_1)$ and the wake half width $L_w(\xi'_1)$ are slowly varying and so can be taken as constant in the evaluation of the Fourier series coefficients. Since there are multiple wakes and the axis of the wake makes an angle $\chi_o$ with the mean flow direction we can specify the upwash velocity on each blade as

$$\mathbf{w} \cdot \mathbf{n} = U_w \sin\chi_o \sum_{m=-\infty}^{\infty} e^{-\left(\xi'_2 + m\lambda_w/\cos\theta_s\right)^2/L^2_w}$$

The Fourier series coefficients in Eq. (18.6.1) are then determined as

$$w_{j0} = \frac{U_w \sin\chi_o}{\lambda_w \cos\theta_s} \sum_{m=-\infty}^{\infty} \int_{-(\lambda_w \cos\theta_s)/2}^{(\lambda_w \cos\theta_s)/2} e^{-\left(\xi_2' - m\lambda_w \cos\theta_s\right)^2/L_w^2 - 2\pi i j \xi_2'/\lambda_w \cos\theta_s} d\xi_2'$$

$$= \frac{U_w \sin\chi_o}{\lambda_w \cos\theta_s} \int_{-\infty}^{\infty} e^{-\xi_2'^2/L_w^2 - 2\pi i j \xi_2'/\lambda_w \cos\theta_s} d\xi_2'$$

$$= U_w \sin\chi_o \sqrt{\pi} e^{-(j\pi L_w/\lambda_w \cos\theta_s)^2}$$

This result highlights the relative importance of the wake width $L_w$ and the wake spacing $\lambda_w$. For narrow wakes that are widely spaced a large number of terms will contribute to the unsteady load. However, if the wake is dispersed so that $L_w$ approaches the value of $\lambda_w$ then the number of terms that have significant amplitudes are greatly reduced.

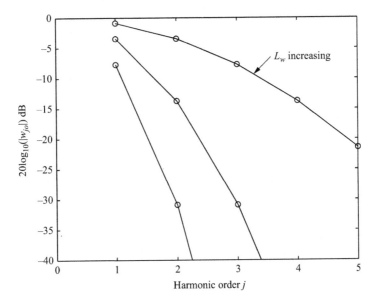

**Fig. 18.12** The upwash velocity coefficients for different wake widths in dB for different harmonics for $L_w/\lambda_w \cos\theta_s = 0.1$, 0.2, and 0.3.

To illustrate this point Fig. 18.12 shows the effect of increasing the wake width on the magnitude of the upwash harmonics plotted on a dB scale. It is seen that the levels of each harmonic decay rapidly as the wake width is increased, and so filling in the wakes can effectively reduce the tone noise from a fan dramatically. Conversely very narrow wakes can increase the levels significantly. The corresponding effect on the radiated sound power is shown in Fig. 18.13.

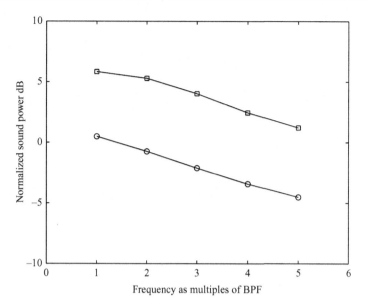

**Fig. 18.13** The normalized tone sound power levels for a set of blade wakes incident on a fan showing the effect of increase in the wake width. The blade wake spectrum is based on a wake width ratio of $L_w/\lambda_w \cos \theta_s = 0.4$ *(circles)* and $L_w/\lambda_w \cos \theta_s = 0.1$ *(squares)*.

## 18.7 Broadband fan noise

In Section 18.3 we showed that the power spectrum from a fan or set of stator vanes for an arbitrary gust was defined by Eq. (18.3.27) as

$$S_{WW}(\omega_o) = \frac{\rho_o B s b U}{2} \sum_{p,n} H_{0n}^{(p)}(\omega) \left\{ \sum_m , E_{p-mB,n,}(\omega) \right\} \qquad (18.3.27)$$

This represents a double summation of the product of two terms. Those terms which are embraced using { } represent the influence of the turbulence model on the result. The terms outside the braces represent the blade response to the turbulence and give the radiated sound power for each mode, where $p$ represents the azimuthal mode order and $n$ the radial mode order. The definition of the function $H_{0n}^{(p)}$ can be found in Eq. (18.3.27) and the definition of the function $E_{kn}$ for a turbulent flow is given by Eq. (18.3.29) as

$$E_{kn}(\omega) = \frac{(2\pi)^2 \delta}{UBhb^2} \Phi_{ww}(-\gamma_o, 2\pi k/Bh + \gamma_o d/h, 0) \qquad (18.3.29)$$

where $\Phi_{ww}$ is the wavenumber spectrum of the incoming turbulence. Various empirical models are available for the wavenumber spectra of the turbulent flow. The most commonly used is the isotropic von Kármán spectrum, Eq. (9.1.8), which is rewritten here as

$$\Phi_{ww}(k_x, k_y, k_z) = \left\{ \frac{\bar{w}^2 55\Gamma(5/6)}{k_e^3 9\sqrt{\pi}\Gamma(1/3)} \right\} \left( \frac{\left(1 - k_y^2/k_s^2\right)\left(k_s^2/k_e^2\right)}{4\pi\left(1 + k_s^2/k_e^2\right)^{17/6}} \right) \tag{18.7.1}$$

$$k_e = \frac{\sqrt{\pi}\,\Gamma(5/6)}{L_f\,\Gamma(1/3)} \qquad\qquad k_s^2 = k_x^2 + k_y^2 + k_z^2$$

The important issue here is that the wavenumber spectrum depends on only two parameters, $k_s$ and $\left(1 - k_y^2/k_s^2\right)$, which, by using the parameters required for the evaluation of Eq. (18.3.29) become, after some manipulation,

$$k_s = \left[ \left(\frac{\omega}{U_o}\right)^2 + \left(\frac{2\pi(p - mB)}{Bs}\right)^2 \right]^{1/2} \tag{18.7.2}$$

$$1 - \frac{k_y^2}{k_s^2} = \left(\frac{\omega}{k_s U}\right)^2$$

where $U_o = U_e \cos\chi$ is the flow velocity along the axis of the duct. The wavenumber spectrum is a maximum when $k_s$ is smallest and the result given in Eq. (18.7.2) shows that this will occur when $p = mB$.

The characteristics of the inflow turbulence are changed if the turbulence is anisotropic. This can be modeled by introducing different lengthscales for different directions. In general, if the turbulence lengthscale is stretched in a particular direction then, to ensure the flow remains divergenceless, the turbulence intensity in that direction must be increased by the same ratio. An expression for the wavenumber spectrum for anisotropic turbulence $\Phi_{ij}^{(A)}$ can then be obtained from the isotropic wavenumber spectrum $\Phi_{ij}$ by using the lengthscales $L_i$ in each direction and the wavenumber $k_i^{(A)} = (k_1 L_1/L_o, k_2 L_2/L_o, k_3 L_3/L_o)$ where $L_o$ is a reference lengthscale. The result is

$$\Phi_{ij}^{(A)}(\mathbf{k}) = \frac{L_1 L_2 L_3}{L_o^3} \left[ \frac{L_i L_j \Phi_{ij}\left(\mathbf{k}^{(A)}\right)}{L_o^2} \right]_{ij} \tag{18.7.3}$$

where no summation of repeated indices is implied for the term in square brackets given by $[\,]_{ij}$. In this case we assume that the stretching takes place in the direction of the axial flow in the duct which lies at an angle $\chi$ to the $x$ axis in Fig. 18.11. In this case, using the parameters required for (Eq. 18.7.3) we find $k_i^{(A)} = (-\omega L_1/U_o L_o, 2\pi(p - mB)L_2/BsL_o, 0)$ and so the anisotropic spectrum can be evaluated by multiplying $\Phi_{ww}$ by $(L_1 L_2 L_3/L_o^3)$ and using

$$k_s^{(A)} = \left[\left(\frac{\omega L_1}{U_o L_o}\right)^2 + \left(\frac{2\pi(p - mB)L_2}{BsL_o}\right)^2\right]^{1/2}$$

$$\left(1 - \frac{k_y^2}{k_s^2}\right)^{(A)} = \left(\frac{\omega L_1 L_2}{k_s^{(A)} UL_o^2}\right)^2$$

(18.7.4)

Increasing the lengthscale in the axial flow direction relative to the lengthscale in the cross flow direction will cause $L_1 > L_2$ and so $k_s^{(A)}$ will be influenced more strongly by the frequency dependent factor.

To illustrate the modal sound power distribution for broadband noise from a set of stator vanes downstream of a fan, Fig. 18.14 shows a modal plot of the sound power for each spinning mode, propagating in the downstream direction from the fan. The contribution from each radial mode has been summed to give the power in each spinning mode order. The vertical axis shows frequency and the horizontal axis shows the spinning mode order. The spectrum is given by summing the modal powers for each frequency. The plot shows that the power is concentrated in certain bands of spinning modes and the corotating modes that spin in the same direction as the fan tends to dominate for this configuration. The plot also shows how more and more spinning modes cut on as the frequency increases, and that the sound power levels in the modes that are just cut on (at the edges of the inverted triangle) can be dominant. This result emphasizes the importance of correctly modeling broadband noise predictions at frequencies close to cut off.

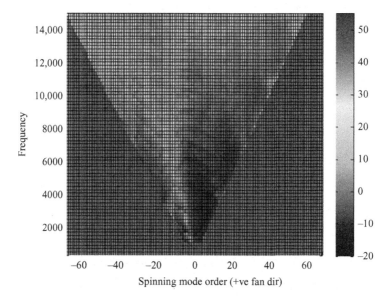

**Fig. 18.14** A spinning mode plot showing the sound power in each spinning mode for a high bypass ratio turbofan engine. The vertical scale shows the frequency and the horizontal scale shows the spinning mode order. The number of modes that are cut on and radiate nonzero power increases with the frequency.

# References

[1] M.J. Lighthill, On sound generated aerodynamically. I. General theory, Proc. R. Soc. Lond. A Math. Phys. Sci. 211 (1952) 564–587.

[2] C.L. Morfey, Amplification of aerodynamic noise by convected flow inhomogeneities, J. Sound Vib. 31 (1973) 391–397.

[3] P.R. Gliebe, Fan broadband noise—the floor to high bypass engine noise reduction, in: Noise Con 96, Seattle, WA, 1996.

[4] E. Envia, Fan noise reduction: an overview, NASA TM 2001-210699, 2001.

[5] S. Glegg, C. Jochault, Broadband self-noise from a ducted fan, AIAA J. 36 (1998) 1387–1395.

[6] U.W. Ganz, P.D. Joppa, T.J. Patten, D.F. Scharpf, Boeing 18 inch fan rig broadband noise test, NASA CR 1998-208704, 1998.

[7] R.E. Longhouse, Vortex shedding noise of low tip speed, axial flow fans, J. Sound Vib. 53 (1977) 25–46.

[8] H.M. Atassi, A.A. Ali, O.V. Atassi, I.V. Vinogradov, Scattering of incident disturbances by an annular cascade in a swirling flow, J. Fluid Mech. 499 (2004) 111–138.

[9] M.D. Montgomery, J.M. Verdon, A three dimensional linearized unsteady Euler analysis for turbomachinery blade rows, NASA CR 4770, 1997.

[10] V.V. Golubev, H.M. Atassi, Sound propagation in an annular duct with mean potential swirling flow, J. Sound Vib. 198 (1996) 601–616.

[11] A.J. Cooper, N. Peake, Upstream-radiated rotor-stator interaction noise in mean swirling flow, J. Fluid Mech. 523 (2005) 219–250.

[12] S.A.L. Glegg, The response of a swept blade row to a three-dimensional gust, J. Sound Vib. 227 (1999) 29–64.

# Appendix A: Nomenclature

## A.1 Symbol conventions, symbol modifiers, and Fourier transforms

*Vectors* are denoted using a bold character, such as $\mathbf{x}$ for position vector, or in terms of their Cartesian components, e.g., $x_1$, $x_2$, $x_3$, or $x_i$. *Tensors* are denoted in terms of their Cartesian components using double subscript notation, for example, $p_{ij}$ for the compressive stress tensor.

In general, the *mean value* of a variable is denoted using subscript "$o$" (as in the mean pressure, $p_o$), or using the expected value operator denoted as $E[\ ]$ or using an overbar (as in the mean square pressure fluctuation, $\overline{p'^2}$). The *fluctuating* part of a variable is indicated using a prime, as in the fluctuating density $\rho' = \rho - \rho_o$. For some common variables special symbols are defined for the mean and fluctuating parts that do not follow these conventions. For example, $U_i$ and $u_i$ for the mean and fluctuating velocity components, respectively. The *estimated value* of a variable or statistic (used in discussing measurements) is indicated by triangular brackets, as in the estimated value of the mean square velocity fluctuation $\left\langle \overline{u_1^2} \right\rangle$. The dot accent is used to indicate partial derivative with respect to time, as in the time rate of change of velocity potential, $\dot{\phi}$.

In certain situations, most notably Lighthill's analogy, variations in the thermodynamic variables are properly referenced to their ambient values indicated using an infinity subscript, for example $\rho_\infty$. In these circumstances, the prime is used to indicate *variation from the ambient value*, e.g., $\rho' = \rho - \rho_\infty$. The text has been written to make clear which meaning of the prime is intended whenever the distinction is significant.

The *complex amplitude* is indicated using a caret accent, as in the pressure amplitude of a harmonic wave $\hat{p}$ where $p' = \hat{p}e^{-i\omega t}$. The time Fourier transform is indicated using a tilde accent, and a wavenumber transform is indicated using a double tilde accent. We repeat here the Fourier transform definitions used in this book (also given in Chapters 1 and 3) for easy reference. Specifically, we define the *Fourier transform of a time history* as

$$\widetilde{p}(\omega) = \frac{1}{2\pi} \int_{-T}^{T} p'(t)e^{i\omega t} dt$$

where $T$ tends to infinity, and the inverse Fourier transform as

$$p'(t) = \int_{-\infty}^{\infty} \widetilde{p}(\omega)e^{-i\omega t}d\omega$$

where $\omega$ is angular frequency and we are using the symbol $i$ to represent the square root of $-1$. We define the one-dimensional *Fourier transform of a variation over distance* as

$$\widetilde{\widetilde{f}}(k_1) = \frac{1}{2\pi}\int_{-R_\infty}^{R_\infty} f(x_1)e^{-ik_1 x_1}dx_1$$

where $R_\infty$ tends to infinity, and the inverse transform as

$$f(x_1) = \int_{-\infty}^{\infty} \widetilde{\widetilde{f}}(k_1)e^{ik_1 x_1}dk_1$$

with two and three dimensional forms that are the result of repeated application of the above two expressions. Here $k_1$ is the wavenumber in the $x_1$ direction. Note that in the forward time transform the exponent is positive, whereas it is negative in the forward spatial transform. Thus the four-fold Fourier transform of a quantity $a(\ )$ varying in space and time would be calculated as,

$$\widetilde{\widetilde{a}}(\mathbf{k},\omega) = \frac{1}{(2\pi)^4}\int_{-R_\infty}^{R_\infty}\int_{-R_\infty}^{R_\infty}\int_{-R_\infty}^{R_\infty}\int_{-T}^{T} a(\mathbf{x},t)e^{i\omega t - i\mathbf{k}\cdot\mathbf{x}}dt dx_1 dx_2 dx_3$$

In other texts or fields of study the convention used for the fourfold Fourier transform is often different. Most importantly some more mathematically oriented texts, such as the book by Noble [1] on the Weiner Hopf Method, the exponent $+i\omega t + i\mathbf{k}\cdot\mathbf{x}$ is used, and the factors of $2\pi$ may be shifted to the inverse transform, or replaced by $\sqrt{2\pi}$ in both the transform and inverse transform. The final results of any derivation may of course be used to obtain the results in another convention by changing the sign of $\mathbf{k}$ (or $\omega$ or multiplying by factors of $2\pi$, etc.). However, some care needs to be exercised if the result includes a multivalued function for which a branch cut has been defined, such as in the results presented in Chapter 13.

## A.2   Symbols used

Symbols used are tabulated below in alphabetical order with Roman symbols listed before Greek symbols, and lower-case characters before upper case.

| Symbol | Definition |
|---|---|
| $a$ | Distance, representing; |
| | Airfoil semichord |
| | Radius of small sphere, Sections 3.4 and 3.5 |
| | Streamwise spacing between shed vortices, Section 7.4 |
| | Duct outer radius, Fig. 17.2B |
| | Amplitude of undisturbed gust, representing; |
| | Velocity, Eq. (6.3.4) |
| | Velocity potential, Eq. (13.4.1) |
| $b$ | Distance representing; |
| | Span |
| | Duct inner radius, Fig. 17.2B |
| $\mathbf{b}$, $b_j$ | Spectral densities of the output of a phased array at the $j$th focus point, Eqs. (12.2.7), (12.3.4) |
| $c$ | Airfoil chord length |
| | Local unsteady speed of sound |
| $c_m$ | Weighting function, Chapter 12 |
| $c_n$ | Fourier coefficients of sound from a rotor blade, Eq. (16.3.5) |
| $c_o$ | Speed of sound |
| $c_p$ | Specific heat at constant pressure |
| $c_v$ | Specific heat at constant volume |
| $c_\infty$ | Free stream or ambient sound speed |
| $d$ | Distance, representing; |
| | Semispan |
| | Distance between monopole sources |
| | Radial position of line vortex, Section 7.2 |
| | Cylinder diameter, Section 7.4 |
| | Pinhole diameter, Section 10.4 |
| | Source separation, Fig. 12.4 |
| | Chordwise blade displacement, Fig. 18.4 |
| $e$ | Specific internal energy |
| $\mathbf{e}$ | Viscous force per unit mass |
| $e_T$ | Specific total energy |
| $f$ | Frequency in Hz |
| $f(r)$ | Longitudinal correlation function of homogeneous turbulence, Section 9.1 |
| $f(\mathbf{x},t)$ | Scalar function defining a surface, Section 5.1 |
| $f_i$ | Rotor blade surface loading per unit area, Section 16.2.1 |
| $\bar{f}_i$ | Normalized frequency $2\pi\omega\delta_i^*/U$, Eq. (15.3.2) |
| $f_n$ | Third octave band mid-band frequency |
| $f_s$ | Sampling frequency |
| $g(r)$ | Lateral correlation function of homogeneous turbulence, Section 9.1 |
| $g^{(1)}$ | First order leading edge blade response function, Eq. (13.4.5) |
| $g^{(1+2)}$ | Second order leading edge blade response function, Eq. (13.4.6) |
| $g_{te}$ | Trailing edge blade response function, Eq. (13.3.9) |

| | |
|---|---|
| $h$ | Specific enthalpy |
| | Distance, representing |
| |    Off center position of shed vortex, Section 7.4 |
| |    Off chord position of incident vortex, Section 7.5 |
| |    Test section height, Section 10.1 |
| |    $x_2$ distance between source and shear layer, Section 10.2, Fig. 10.17 |
| |    Cavity depth, Section 10.4 |
| |    Perpendicular distance from source to array, Fig. 12.4 |
| |    Vortex-blade separation, Fig. 14.7 |
| |    Root-mean-square roughness height, Eq. (15.4.22) |
| |    Rotor blade thickness, Chapter 16 |
| |    Cascade blade spacing, Fig. 18.5 |
| $h_i$ | Initial length of material volume $i$ |
| $h_o$ | Mean specific enthalpy |
| $h_\infty$ | Free stream specific enthalpy |
| $i$ | Square root of $-1$ |
| $\mathbf{i}, \mathbf{j}, \mathbf{k}$ | Unit vectors in directions $x_1, x_2, x_3$ |
| $k$ | Acoustic wavenumber |
| | Turbulence wavenumber magnitude, Section 9.1 |
| $\mathbf{k}^{(o)}$ | Acoustic wavenumber vector in the direction of the observer, Eq. (4.7.8) |
| $\mathbf{k}^{(w)}$ | Wavenumber vector of sinusoidal ribs $(k_1^{(w)}, 0, k_3^{(w)})$, Eq. (15.4.28) |
| $\mathbf{k}, k_i$ | Wavenumber vector |
| $k_1^{(o)}$ | Streamwise acoustic wavenumber with Prandtl Glauert scaling, Eq. (6.5.6) |
| $k_{13}$ | $\sqrt{k_1^2 + k_3^2}$ |
| $k_3^{(o)}$ | Spanwise acoustic wavenumber with Prandtl Glauert scaling, Eq. (6.5.6) |
| $k_e$ | Wavenumber scale of the largest eddies, Eq. (9.1.9) |
| $K_n()$ | Modified Bessel function of the second kind of order $n$ |
| $k_o$ | Acoustic wavenumber with Prandtl Glauert scaling, Eq. (6.5.6) |
| $l_p(\omega)$ | Frequency-dependent spanwise pressure lengthscale, Eq. (15.2.12) |
| $m$ | Azimuthal mode order, Eq. (17.2.2) |
| $\mathbf{n}, n_i$ | Surface normal unit vector |
| | Unit vector normal to a streamline in two dimensions, Fig. 6.3 |
| $n$ | Radial mode order, Section 17.2.2 |
| $\mathbf{n}^{(o)}, n_j^{(o)}$ | Unit outward normal vector |
| $p$ | Pressure |
| $\mathbf{p}$ | Vector of Fourier transforms of measured microphone signals, Eq. (12.2.15) |
| $p'$ | Pressure fluctuation, pressure perturbation |
| $\hat{p}$ | Complex amplitude of the acoustic pressure |
| $p_{bl}$ | Boundary layer pressure fluctuation in the absence of the trailing edge, Section 15.2 |
| $p_c$ | Corrected sound pressure, Section 10.2 |
| $p_i$ | Incident acoustic pressure |
| $p_{ij}$ | Compressive stress tensor, $p\delta_{ij} - \sigma_{ij}$ |
| $p_m$ | Measured sound pressure, Section 10.2 |

| | |
|---|---|
| $p_o$ | Mean pressure |
| $p_{ref}$ | Reference pressure |
| $p_{rms}$ | Root mean square pressure |
| $p_s$ | Scattered acoustic pressure, Section 3.5 |
| $p_t$ | Sound pressure just after refraction, Section 10.2 |
| $p_\infty$ | Free stream or ambient pressure |
| $\mathbf{q}, q_n$ | Time Fourier transform of effective source strengths, Eq. (12.2.2) and Section 12.2.7, diagonal elements of $\mathbf{B}$ |
| $q_m$ | Radial phase function, Eq. (17.6.3) |
| $r$ | Radial coordinate, radial distance |
| $r_c$ | Corrected propagation distance of measured sound, Fig. 10.17 |
| $r_e$ | Observer radius with Prandtl Glauert scaling, Eq. (6.5.5) |
| $r_g$ | Source to observer distance with Prandtl Glauert scaling, Eq. (6.5.3) |
| $r_m$ | Observer distance from source, Fig. 10.17 |
| $r_o$ | Distance to flow origin in distortion example, Section 6.3 |
| | Distance from source to array center, Chapter 12 |
| $r_r$ | Distance from the retarded source position to the observer, Figs. 15.3 and 15.6 |
| $r_y$ | Distance of the source point from the trailing edge, Fig. 15.1 |
| $s$ | Specific entropy |
| | Distance traveled by shed vortex, Section 7.4 |
| | Laplace transform frequency, Chapter 13 |
| | Blade index number, Chapter 18 |
| | Blade spacing, Fig. 18.4 |
| $\mathbf{s}$ | Unit vector along a streamline in two-dimensions, Fig. 6.3 |
| $t$ | Time, observer time |
| $\mathbf{t}$ | Unit vector out of plane of flow of Fig.6.3, $\mathbf{s} \times \mathbf{n}$ |
| $t_g$ | Observer time with Prandtl Glauert scaling, Eq. (6.5.1) |
| $u$ | Scale of velocity fluctuation due to largest eddies, Section 8.1 |
| $u(\ )$ | Time varying convection speed of shed vortex during acceleration, Section 7.4 |
| $\mathbf{u}^{(\infty)}, u_i^{(\infty)}$ | Undisturbed gust velocity at the inflow boundary, Eq. (6.2.6) |
| $\mathbf{u}^{(g)}$ | Goldstein's velocity perturbation, Eq. (6.1.9) |
| $\mathbf{u}^{(h)}$ | Goldstein's composite velocity perturbation, Eq. (6.1.12) |
| $\mathbf{u}, u_i$ | Velocity fluctuation |
| $u^+$ | Mean velocity in boundary layer inner variables, Eq. (9.2.15) |
| $u_2$ | Upwash velocity of gust, Eq. (14.1.7) |
| $u_n$ | Surface normal velocity fluctuation |
| | Velocity component in direction of separation distance, Section 9.1 |
| $u_o$ | Surface velocity of sphere, Eq. (3.4.1) |
| $u_r$ | Velocity fluctuation in the direction perpendicular to the trailing edge, Fig. 15.1 |
| $u_s$ | Velocity component normal to direction of separation distance, Section 9.1 |
| $u_t$ | Velocity component normal to direction of separation distance and $u_s$, Section 9.1 |
| $u_\eta$ | Kolmogorov velocity scale, Section 8.1 |
| $u_\tau$ | Friction velocity, Section 9.2.2 |

| $\hat{\mathbf{v}}$ | Complex amplitude of the acoustic velocity |
|---|---|
| $\mathbf{v}, v_i$ | Velocity vector |
| $v_o$ | Amplitude of sphere oscillations, Section 3.4 |
| $v_r, v_\theta$ | Polar velocity components aligned with the trailing edge, Fig. 15.1 |
| $w$ | Complex potential, Eq. (2.7.10) |
| $w'$ | Complex velocity, Eq. (2.7.11) |
| $w(\ )$ | Window function |
| $\mathbf{w}, w_i$ | Deviation of the velocity from $U_i^{(\infty)}$ |
| $\tilde{\tilde{w}}_2$ | Planar wavenumber transform of $u_2(y_1, 0, y_3)$, Eq. (14.1.10) |
| $w_c'$ | Convection velocity in the mapped domain, Section 2.7 |
| $w_m^{(j)}$ | Array steering vector for microphone $m$ and focus point $j$, Eq. (12.2.4) |
| $w_o$ | Amplitude of step gust |
| $\mathbf{x}, x_i$ | Position, far-field position of observer |
| $x$ | Axial duct coordinate, Fig. 17.2A |
| $\mathbf{x}'$ | Observer position in frame moving with uniform flow, Eq. (5.4.1) |
| $x, x'$ | Cascade chordwise position, Fig. 18.5 and Fig. 18.7 |
| $x_2^+$ | Distance from the wall in inner variables, Eq. (9.2.15) |
| $x_o$ | Downstream-pointing axial location, Fig. 18.1 |
| $\mathbf{y}, y_i$ | Position, near-field position of source |
| $\mathbf{y}'$ | Observer position in frame moving with uniform flow, Eq. (5.4.1) |
| $y, y'$ | Cascade blade-normal position, Fig. 18.5 and Fig. 18.7 |
| $\mathbf{y}^{(c)}$ | Centroid of noise generating surface |
| $\mathbf{y}^{(v)}$ | Line vortex coordinate, Section 7.2 |
| $\mathbf{z}, z_i$ | Position, moving coordinates, rotor blade based coordinates |
| $z$ | Complex coordinate, $x_1 + ix_2$, $y_1 + iy_2$ |
| | Liner acoustic impedance, Section 17.3 |
| $\mathbf{z}$ | Unit vector aligned with line vortex, Fig. 14.7 |
| $A$ | Acoustic wave amplitude, Eq. (3.3.1) |
| | Pinhole area, Section 10.4 |
| | Wavenumber multiplier, Eq. (13.3.9) |
| $A(\ )$ | Fourier transform of the scattered pressure, Eq. (13.2.7) |
| $A_-(\ ), A_+(\ )$ | Half-range Fourier transforms of the scattered pressure, Eqs. (13.2.8), (13.2.9) |
| $A_{mn}$ | Duct mode amplitude, Chapter 17 |
| $B$ | Integer factor used in frequency averaging, Section 11.6.1 |
| | Parameter equal to $U_c/(c_\infty(1 - M^2))$, Eq. (15.2.11) |
| | Number of blades, Chapters 16 and 18 |
| $\mathbf{B}$ | Matrix of estimated source auto and cross spectra from which the source image is extracted, Eq. (12.2.18) |
| $B(\ )$ | Component function of the wavenumber transform of the scattered field, Section 13.2.2 |
| $\mathbf{C}, C_{mn}$ | Cross-spectral matrix of microphone signals, Eqs. (12.2.8), (12.2.16) |
| $C_-(\ ), C_+(\ )$ | Half-range Fourier transform of the gradient of the scattered pressure, Chapter 13 |
| $C_{ab}$ | Cospectrum between $a$ and $b$, Section 8.4 |
| $C_d$ | Drag coefficient of 2D body |
| $C_f$ | Friction coefficient $C_f \equiv \tau_w / \frac{1}{2}\rho U_e^2$ |

| $C_L$ | Coefficient of fundamental of cylinder lift fluctuations |
|---|---|
| $C_p$ | Nondimensional acoustic pressure due to loading noise, Eq. (16.2.5) |
| $C_q$ | Nondimensional acoustic pressure due to thickness noise, Eq. (16.2.19) |
| $D$ | Cavity diameter, Section 10.4 |
| $D(\ )$ | Fourier transform of the potential jump across a blade, Eq. (18.3.9) |
| $D/Dt$ | Substantial derivative |
| $D_o/Dt$ | Substantial derivative for convection with the mean flow |
| $D_\infty/Dt$ | Substantial derivative relative to uniform motion at $U_\infty$ or $U_i^{(\infty)}$ |
| $E$ | Energy spectrum function of homogeneous turbulence |
| $E[\ ]$ | Expected value |
| $E_2(\ )$ | Modified Fresnel integral function, Eq. (13.2.3) |
| $\mathbf{F}, F_i$ | Force on the fluid (imposed by an aerodynamic body for example) |
| $F_2$ | Negative of the lift force on an airfoil |
| $F_-(\ )$ | Laplace transform of the scattered field in the limit as the $x_1$ axis is approached from the positive side for $x_1 < 0$, Eq. (13.2.14) |
| $F()$ | Array sensitivity function, Chapter 12 |
| $F_+(\ )$ | Laplace transform, with respect to $x_1$ of the scattered field in the limit as the $x_1$ axis is approached from the positive side for $x_1 > 0$, Eq. (13.2.13) |
| $F_A, F_B, F_{K1}, F_{K2},$ $F_{\Delta K}$ | Component parts of model trailing edge noise spectral forms $F_i$, Section 15.3 |
| $F_i$ | Model spectral forms for trailing edge noise spectra SPL$_i$, Section 15.3 |
| $F_D$ | Rotor blade drag force due to an element of its span $\Delta R$ |
| $F_{jn}$ | Point spread function at focus point $\mathbf{y}_j$ due to a source at $\mathbf{y}_n$, Eq. (12.2.11) |
| $F_L$ | Rotor blade thrust force due to an element of its span $\Delta R$ |
| $G$ | Green's function, $G(\mathbf{x},t|\mathbf{y},\tau)$ |
| $\mathbf{G}$ | Matrix of source Green's functions, Eq. (12.2.15) |
| $G_-(s), G_+(s)$ | Positive and negative range Laplace transforms of the $x_2$ gradient of the scattered field, in the limit as the $x_1$ axis is approached for $x_1 < 0$, Section 13.4.1 |
| $G_{aa}$ | Single sided time autospectrum of quantity $a$, Section 8.4 |
| $G_e$ | Green's function in the fixed frame with a free stream, Eq. (5.4.2) |
| $G_g$ | Green's function with Prandtl Glauert scaling, Eq. (6.5.1) |
| $G_o$ | Free field Green's function, Eq. (3.9.17) |
| $G_o^{\#}$ | Free field Green's function for image sources in the wall, Eq. (4.5.3) |
| $G_T$ | Tailored Green's function, Section 4.5 |
| $H$ | Stagnation enthalpy |
| | Distance in $x_2$ between source and observer, Fig. 10.17 |
| $H'$ | Stagnation enthalpy fluctuation |
| $H(\ ), H_s(\ )$ | Heaviside step function |
| $H_n^{(1)}(\ )$ | Hankel function of the first kind of order $n$ |
| $H_o$ | Mean stagnation enthalpy |
| $\mathbf{I}$ | Acoustic intensity vector $E[(\rho\mathbf{v})'H']$, Eq. (2.6.16). |
| $I$ | Integrated source level, Section 12.3.5 |
| $I_r$ | Radial component of the acoustic intensity vector |
| $J_n(\ )$ | Bessel function of the first kind of order $n$ |
| $J_+(s), J_-(s)$ | Factorizations of $\gamma$, Eq. (13.2.16) |

| $K_1$ | $\omega/U_c$ |
|---|---|
| $L$ | Flow scale, representing |
| |    Size of the eddies |
| |    Lengthscale of the turbulence |
| |    Scale of the mean flow distortion |
| |    Vortex length |
| | Pinhole depth (Section 10.4), microphone array length (Chapter 12) |
| $\mathcal{L}$ | Amiet's generalized lift function, Eq. (15.2.11) |
| $L_{eff}$ | Effective pinhole depth, Section 10.4 |
| $L_f$ | Longitudinal integral scale, Eq. (9.1.4) |
| $L_g$ | Lateral integral scale, Eq. (9.1.4) |
| $L_{ij}$ | Integral scale of $u_i$ in direction $x_j$, Eq. (8.4.29) |
| $L_w$ | Wake half width, Section 9.2.1 |
| | Streamwise gust scale, Eq. (14.1.2) |
| $M$ | Mach number |
| | Mass of fluid oscillating in pinhole |
| | Total number of array microphones, Chapter 12 |
| $M_r$ | Mach number in direction of observer |
| $N_{rec}$ | Number of records used in spectral analysis, Chapter 11 |
| $P_o$ | Amplitude of pressure perturbation |
| $Q$ | Acoustic monopole strength |
| $\mathbf{Q}, Q_i$ | Heat flux vector, Section 2.6.1 |
| $\mathbf{Q}, Q_{mn}$ | Cross-spectral matrix of source strengths, Eq. (12.2.17) |
| $\hat{Q}$ | Complex amplitude of the potential disturbance, Section 6.3 |
| $Q_{ab}$ | Quadrature spectrum between $a$ and $b$, Section 8.4 |
| $Q_{m,n}$ | Fourier series components of the rotor noise source term, Eq. (16.3.15) |
| $R$ | Gas constant |
| | Distance, representing |
| |    Radius of circle, Section 2.7 |
| |    Distance from rotor axis, Fig. 16.11 |
| |    Radial distance from the duct axis, Fig. 17.2A |
| | Shear layer reflection coefficient, reflected over incident pressure |
| |    amplitude |
| $R_\infty$ | Distance interval of Fourier wavenumber transform chosen such that |
| |    $-R_\infty$ to $R_\infty$ encompasses the entire spatial variation |
| $R_{aa}$ | Auto correlation function of quantity $a$, Eq. (8.4.3) |
| $R_{ab}$ | Cross correlation function between $a$ and $b$, Eq. (8.4.18) |
| Re | Reynolds number, see Section 2.3.2 |
| $\text{Re}_d$ | Cylinder diameter Reynolds number, Section 7.4 |
| $\text{Re}_\theta$ | Boundary layer momentum thickness Reynolds number, Section 9.2.2 |
| $R_{ij}$ | Velocity correlation tensor, Section 8.4.3, Eq. (9.1.3) |
| $R_n$ | Radius segment in Amiet's approximation, Eq. (16.4.4) |
| $R_{tip}$ | Rotor tip radius |
| $S$ | Surface, area |
| $S(\ )$ | Sears function |
| $S(\omega)$ | Normalized spectral shape function, Chapter 15 |
| $S^{(1)}$ | Unsteady lift per unit span as a function of frequency, Eq. (14.1.4) |
| $S_{aa}$ | Double sided time autospectrum of quantity $a$, Eq. (8.4.2) |

| | |
|---|---|
| $S_{ab}$ | Cross spectral density between $a$ and $b$, Eq. (8.4.20) |
| $S_{FF}$ | Wavenumber frequency spectrum of the unsteady blade loading, Eq. (14.3.1) |
| $S_o$ | Closed surface of integration in Ffowcs-Williams Hawkings equation, Section 5.1 |
| SPL | Sound pressure level |
| $SPL_i$ | One-third octave band spectra due to the suction ($i = s$) and pressure ($i = p$) side boundary layers, and angle of attack ($i = \alpha$), Section 15.3 |
| $SPL_n$ | $n$th band of one-third octave sound pressure level, Eq. (8.4.9) |
| $S_{pp}$ | Far-field sound frequency spectrum |
| $St$ | Strouhal number |
| $S_\infty$ | Exterior surface of infinite volume, Section 5.1 |
| $T$ | Time period of Fourier frequency transform chosen such that $-T$ to $T$ encompasses the entire time history |
| | Time period of Green's function integration ($-T$ to $T$), Eq. (3.9.8) |
| | Shear layer transmission coefficient, transmitted over incident pressure amplitude |
| $\mathcal{T}$ | Integral timescale, Eq. (8.4.5) |
| $T_e$ | Temperature |
| $T_{ij}$ | Lighthill stress tensor, Eq. (4.1.4) |
| $T_o$ | Total sampling time |
| $T_p$ | Rotor rotation period |
| $T_v$ | Thrust disturbance timescale, Eq. (16.2.7) |
| $U$ | Reference flow velocity |
| $\mathbf{U}, U_i$ | Mean velocity |
| $U_\infty$ | Nominal wind tunnel free stream velocity, Section 10.1 |
| $\mathbf{U}^{(\infty)}, U_i^{(\infty)}$ | Constant velocity vector of uniformly moving medium, Eq. (4.2.3) |
| $U_c$ | Convection speed |
| $U_e$ | Boundary layer edge velocity |
| $U_o$ | Axial forward velocity of rotor, Fig. 16.11 |
| $U_r$ | Source velocity in direction of observer |
| $\mathbf{U}_s$ | Translational velocity of surface, Eq. (5.2.11) |
| | Velocity of shear layer surface wave, Eqs. (10.2.3), (10.2.4) |
| $U_w$ | Wake centerline velocity deficit, Section 9.2.1 |
| $U_\infty$ | Free stream velocity |
| $V$ | Volume |
| | Number of stator vanes, Section 18.6 |
| $\mathbf{V}, V_i$ | Velocity of moving surface, Eq. (5.1.12) |
| $\mathbf{V}_b$ | Blade velocity, Section 18.3.5 |
| $V_o$ | Volume exterior to $S_o$ in Ffowcs-Williams Hawkings equation, Section 5.1 |
| $V_\infty$ | Infinite volume, Section 5.1 |
| $W$ | Complex potential in the unmapped domain, Section 2.7 |
| $W'$ | Complex velocity in the unmapped domain, Section 2.7 |
| $W_a$ | Expected acoustic sound power output, Eq. (2.6.14) |
| $W_c'$ | Convection velocity in the unmapped domain, Section 2.7 |
| $W_i$ | Uniform axial flow, Chapter 18 |
| $W_m$ | Array weighting for $m$th microphone, Chapter 12 |

| | |
|---|---|
| $W_s$ | Expected power generated due to steady contributions, Eq. (2.6.13) |
| $W_T$ | Total power generated by a system, Eq. (2.6.11) |
| $X_1, X_2, X_3$ | Drift coordinates, Eq. (6.2.1) |
| $X_m$ | Axial dependency of $\hat{\phi}_m$, Eq. (17.2.4) |
| $Y_i$ | Kirchoff coordinates (Eq. 7.3.3) |
| $Y_m(\ )$ | Bessel function of the second kind of order $m$ |
| $\alpha$ | Free-stream angle, angle of attack, angle of surface |
| | Wavenumber of the scattered pressure field in the $x_1$ direction, Eq. (13.2.4) |
| | Wavenumber parameter for duct acoustics, Eq. (17.2.8) |
| $\alpha'$ | Wind tunnel geometric angle of attack, Section 10.1 |
| $\alpha_w$ | Orientation of ribs in the $y_1 y_3$ plane, Fig. 15.13 |
| $\beta$ | $\sqrt{1 - M^2}$ |
| | Negative of the zero lift angle of attack for a Joukowski foil, Section 2.7 |
| $\beta_o$ | Location on the real axis of the inverse Laplace transform, Eq. (13.2.15) |
| | Blade pitch angle, Fig. 16.22 |
| $\beta_a$ | Nondimensional liner admittance, Eq. (17.3.5) |
| $\chi_{mn}$ | Cut off ratio, Eq. (17.2.17) |
| $\chi_o$ | Angle between wake and stator-relative flow direction, Fig. 18.11 |
| $\theta_s$ | Angle between wake and duct axis, Fig. 18.11 |
| $\delta$ | Boundary layer thickness |
| $\delta(\ )$ | Dirac delta function, Eq. (3.9.3) |
| $\delta(\mathbf{x})$ | Dirac delta function (3D), Eq. (3.9.4) |
| $\delta^*$ | Boundary layer displacement thickness |
| $\delta[\ ]$ | Uncertainty interval |
| $\delta H$ | Far-field sound (in terms of stagnation enthalpy) from segment of vortex pair, Section 7.2 |
| $\delta_i^*$ | Trailing edge boundary layer displacement thickness for the suction $(i = s)$ and pressure $(i = p)$ sides, Section 15.3 |
| $\delta_{ij}$ | Kronecker delta |
| $\delta \mathbf{l}^{(i)}$ | Displacement coordinate giving the edge length of a material volume |
| $\delta q$ | Heat added per unit mass, Eq. (2.4.2) |
| $\delta w$ | Work done, per unit mass, Eq. (2.4.2) |
| $\varepsilon$ | Small parameter, number tending to zero |
| | Rate of viscous dissipation per unit mass, Section 8.1 |
| $\phi$ | Velocity potential |
| | Phase angle, Eq. (3.3.1) |
| | Azimuthal rotor angle in the observer frame, Fig. 16.11 |
| $\phi_o$ | Azimuthal angle of rotor far-field observer, Eq. (16.3.10) |
| $\phi_1$ | Azimuthal rotor angle in the blade frame, Fig. 16.11 |
| $\phi_{22}$ | Planar wavenumber upwash frequency spectrum |
| $\phi_{ij}$ | Planar wavenumber spectrum of $u_i u_j$, e.g., Eq. (8.4.32) |
| $\phi_{LE}, \phi_{TE}$ | Azimuthal angle of rotor blade leading and trailing edges, in the blade-fixed frame |
| $\hat{\phi}_m$ | Complex Fourier coefficients of the velocity potential of the in-duct sound field, Eq. (17.2.2) |
| $\phi_m$ | Phase shift needed to steer array to direction $\theta_s$, Chapter 12 |

| | |
|---|---|
| $\phi_{qq}$ | Spanwise wavenumber transform of airfoil surface pressure jump, Section 15.2 |
| $\phi_r$ | Out of plane directivity angle measured from the retarded source position, Fig. 15.6 |
| $\phi_v$ | Angle between vortex and blade span, Fig. 14.7 |
| $\phi_x$ | Directivity angle measured from the trailing edge, Fig. 15.1 |
| $\gamma$ | Ratio of specific heats |
| | Combination of $k$ and $\alpha$, Eq. (13.2.5) |
| $\gamma_{ab}^2$ | Coherence spectrum between $a$ and $b$, Eq. (8.4.23) |
| $\eta$ | Kolmogorov lengthscale, Section 8.1 |
| | Wake similarity coordinate $x_2/L_w$, Section 9.2.1 |
| $\varphi$ | Angle given by the gust wavenumbers scaled using Mach number, Section 14.1.3 |
| | Azimuthal angle in duct, Fig. 17.2A |
| $\varphi_e$ | Angle of propagation of acoustic wave produced by gust, Section 14.1.3 |
| $\varphi_{ij}$ | Wavenumber spectrum, e.g., Eq. (9.1.2) |
| $\kappa$ | von Karman constant, Eq. (9.2.18) |
| | Magnitude of the wavenumber vector component in the $x_1$, $x_2$ plane scaled on $\beta^2$, Eq. (13.3.7) |
| | Wavenumber parameter representing constant terms in Goldstein's equation for duct acoustics, see Eq. (17.2.6) |
| $\boldsymbol{\kappa}$ | Product of the wavenumber vector and the drift gradient, Eq. (6.3.6) |
| | Modified wavenumber, Eq. (15.4.17) |
| $\kappa_e$ | Turbulence kinetic energy, Eq. (8.3.3) |
| $\lambda$ | Acoustic wavelength |
| $\mu$ | Dynamic viscosity |
| | Angle of the observer to the path of the source, Eq. (5.3.6) |
| | Scaled frequency, $k_o(1-M)c$, Eq. (14.1.5) |
| $\mu$, $\mu_{mn}^{\pm}$ | Axial wavenumber of the in-duct sound field, Eqs. (17.2.7), (17.2.16) |
| $\mu_t$ | Boussinesq eddy viscosity, Eq. (8.3.3) |
| $\nu$ | Kinematic viscosity |
| $\theta$ | Angle measured from the $x_1$ axis, directivity angle |
| | Polar angle in the complex plane, $\arctan(x_2/x_1)$, Section 2.7 |
| | Momentum thickness of a wake (Section 9.2.1) or boundary layer (Eq. 9.2.9) |
| | Angle subtended by source to array normal, Chapter 12 |
| $\theta_o$ | Polar angle of rotor far-field observer, Eq. (16.3.10) |
| $\theta_{ab}$ | Phase spectrum between $a$ and $b$, Eq. (8.4.24) |
| $\theta_c$ | Corrected directivity angle of measured sound, Fig. 10.17 |
| $\theta_e$ | Directivity angle, Eq. (14.2.7) |
| $\theta_t$ | Incident wave polar angle, Fig. 10.16 |
| $\theta_m$ | Observer angle from source, Fig. 10.17 |
| $\theta_r$ | Reflected wave polar angle, Fig. 10.16 |
| | Directivity angle from the flow direction measured from the retarded source position, Figs. 15.3 and 15.6 |
| $\theta_s$ | Direction in which array is steered, measured relative to array normal, Chapter 12 |

| | |
|---|---|
| $\theta_t$ | Transmitted wave polar angle, Fig. 10.16 |
| $\theta_x$ | Directivity angle in a plane perpendicular to the trailing edge, Fig. 15.1 |
| $\theta_y$ | Angle of the source point in a plane perpendicular to the trailing edge, Fig. 15.1 |
| $\rho$ | Density |
| $\rho'$ | Density fluctuation, density perturbation |
| $\rho_\infty$ | Free stream or ambient density |
| $\rho_o$ | Mean density |
| $\rho_{aa}$ | Correlation coefficient function of quantity $a$, Eq. (8.4.5) |
| $\rho_{ab}$ | Cross correlation coefficient function between $a$ and $b$, Eq. (8.4.19) |
| $\sigma$ | Reduced frequency $\omega a/U_\infty$, nondimensional frequency |
| | Interblade phase angle, Section 18.3 |
| | Distance along a streamline, Section 6.2 |
| $\sigma_{ij}$ | Viscous stress tensor |
| $\tau$ | Time, source time |
| $\tau^*$ | Retarded time, Eq. (3.9.19) |
| $\tau_c$ | Corrected time of measured sound, Section 10.2 |
| $\tau_g$ | Source time with Prandtl Glauert scaling, Eq. (6.5.1) |
| $\tau_\eta$ | Kolmogorov timescale, Section 8.1 |
| $\tau_m$ | Time of measured sound, Section 10.2 |
| | Time segment in Amiet's approximation for broadband rotor noise, Eq. (16.4.4) |
| $\tau_w$ | Viscous shear stress at a wall |
| $\upsilon$ | Specific volume, Section 2.6 |
| $\omega$ | Angular frequency, radians per second |
| $\boldsymbol{\omega}, \boldsymbol{\omega}_t$ | Vorticity, disturbance vorticity |
| $\boldsymbol{\omega}^{(\infty)}$ | Disturbance vorticity at the inflow boundary |
| $\boldsymbol{\omega}_o$ | Vorticity of the mean flow |
| $\omega_o$ | Angular frequency of unsteady force, Section 5.3 |
| $\omega_n$ | Natural frequency of microphone cavity, Section 10.4 |
| $\xi$ | Shear layer displacement normal to the flow, Chapter 10 |
| | Rough surface height in the $y_2$ direction, Fig. 15.10 |
| | Displacement at a liner surface, Eq. (17.3.1) |
| $\xi_1, \xi_2$ | Position in the unmapped domain, Section 2.7 |
| $\xi_1$ | Chordwise distance from the rotor blade leading edge |
| $\xi_1, \xi_2, \xi_3$ | Vortex aligned coordinates, Fig. 14.7 |
| $\xi'_1, \xi'_2, \xi'_3$ | Wake aligned coordinates, Fig. 18.11 |
| $\boldsymbol{\xi}$ | Observer position with Prandtl Glauert scaling, Eq. (6.5.1) |
| $\psi_1, \psi_2, \psi$ | Stream function, Eq. (2.7.1) |
| $\psi_i$ | Incident wave azimuthal angle, Fig. 10.16 |
| $\psi_m$ | Radial dependency of $\hat{\phi}_m$, Eq. (17.2.4) |
| $\psi_t$ | Transmitted wave azimuthal angle, Fig. 10.16 |
| $\zeta$ | Complex coordinate in the unmapped domain, $\xi_1 + i\xi_2$, Section 2.7 |
| | Source position with Prandtl Glauert scaling, Eq. (6.5.1) |
| | Mach number parameter, Eq. (10.2.28) |
| $\Delta p$ | Pressure jump across the airfoil chord (upper minus lower surface) |
| $\Delta t$ | Sampling period |
| $\Delta \phi$ | Potential jump across the airfoil chord |

| | |
|---|---|
| $\Delta\omega$ | Frequency resolution of numerical Fourier transform, Section 11.5 |
| $\Phi_{ij}$ | Wavenumber frequency spectrum of $u_i u_j$, e.g., Eq. (8.4.33) |
| $\Phi_{pp}$ | Wavenumber frequency spectrum of pressure, e.g., Eq. (8.4.35) |
| $\Gamma$ | Circulation, Eq. (2.7.3), line vortex strength |
| | Wavenumber spectrum of roughness height normalized on $h^2$, Eq. (15.4.23) |
| $\Lambda$ | Nondimensional blade response function for acoustic far field, Eq. (14.2.4) |
| $\Lambda_o$ | Sweep angle of the trailing edge, Fig. 15.1 |
| $\Lambda_{mn}$ | Normalization parameter for $\psi_m$, Eqs. (17.2.11) |
| $\Sigma$ | Area of rough surface projected onto the $y_1 y_3$ plane, Eq. (15.4.22) |
| | Area measured on the rotor blade planform, Fig. 16.10 |
| $\Sigma_o$ | Total planform area of rotor, Fig. 16.10 |
| $\Omega$ | Angular velocity of; |
| | Surface about origin of $\mathbf{z}$, Eq. (5.2.11) |
| | Rotor, Fig. 16.11 |
| | Vortex pair system, Section 7.2 |
| | Angular frequency of fundamental of vortex shedding, Section 7.4 |
| $\Omega_o$ | Sampling frequency in radians per second |

# Reference

[1] B. Noble, Methods Based on the Wiener-Hopf Technique for the Solution of Partial Differential Equations, Chelsea Publishing Company, New York, NY, 1958.

# Appendix B: Branch cuts

We have frequently referred to choosing the correct branch cut to fully define a multi-valued function. For example, in Chapter 13 we have required that the real part of $\gamma = (\alpha^2 - k^2)^{1/2} = (-(s^2 + k^2))^{1/2}$ and the imaginary part of $\kappa = \left(k_o^2 - \nu^2\right)^{1/2}$ be positive. These are multivalued functions and so they can only be fully defined for restricted values of the complex variables $s$, $\alpha$, $\nu$, or $k_o$. The purpose of this appendix is to fully explain what is implied by these conditions.

First we note that the square root of a complex number $z$ is given by

$$z^{1/2} = r^{1/2} e^{i\theta/2}$$

and is only completely defined if $\theta$ is restricted to the range

$$\vartheta < \theta < \vartheta + 2\pi$$

where $\vartheta$ defines the angle of the branch cut relative to the branch point at $z = 0$. The value of the complex number is discontinuous across the branch cut and its value jumps from $r^{1/2} \exp(\vartheta/2 + \pi)$ to $r^{1/2} \exp(\vartheta/2)$. Consequently, the magnitude of $z^{1/2}$ stays the same but its phase jumps by a factor of $\pi$.

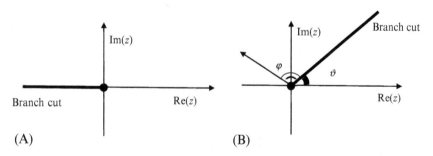

(A)  (B)

**Fig. B.1** The definition of branch cuts and branch points: (A) the branch cut for $[z]^{1/2}$ and (B) the branch cut for an arbitrary branch.

Computer languages such as Matlab or Fortran define the square root of a complex number to have a positive real part, and so we will use the notation that a square root in rectangular brackets represents this case, so

$$[z]^{1/2} = \mathrm{R} e^{i\phi} \quad -\pi/2 < \phi < \pi/2$$

This branch cut lies along the negative real axis as shown in Fig. B.1A and is defined by $\vartheta = -\pi$. If we want to evaluate a square root with a different branch cut at a different angle then we evaluate

$$\left[ze^{-i(\vartheta+\pi)}\right]^{1/2} e^{i(\vartheta+\pi)/2} = Re^{i\phi} \quad \vartheta/2 < \phi < \vartheta/2 + \pi$$

as shown in Fig. B.1B. It is sometimes convenient to define $z = r\exp(i(\varphi+\vartheta))$ so

$$\phi = \varphi/2 + \vartheta/2 \quad 0 < \varphi < 2\pi$$

An important example is given by the evaluation of $[z^2]^{1/2}$ which has a positive real part, and so is not equal to $z$ which can have both positive and negative real parts. In this case the branch cut prevents the correct evaluation of the function $(z^2)^{1/2}$ for negative real values of $z$. To overcome this problem we need to evaluate

$$\left(z^2\right)^{1/2} = [z]^{1/2}[z]^{1/2}$$

each term on the right has the same branch cut and so the jump in phase across the branch cut is now $2\pi$ and the function is correctly evaluated.

An important extension of this occurs when we evaluate $(z^2 - a^2)^{1/2}$, where $a$ is real, because if it is evaluated directly as $[z^2 - a^2]^{1/2}$ then the result will only have a positive real part. In some applications that is a requirement, but in others, such as in potential flow, we need to evaluate this function for all values. To eliminate the ambiguity we specify the function as

$$\left(z^2 - a^2\right)^{1/2} = [z - a]^{1/2}[z + a]^{1/2}$$

This function has two branch points, as shown in Fig. B.2, and the branch cuts from each branch point extend to $-\infty$ on the real axis. However, when $z < a$ on the branch cut both $[z - a]^{1/2}$ and $[z + a]^{1/2}$ have a phase shift of $\pi$ and so the net phase shift is $2\pi$, and the function is continuous. The remaining part of the branch cut forms a slit in the complex plane between $z = -a$ and $z = a$.

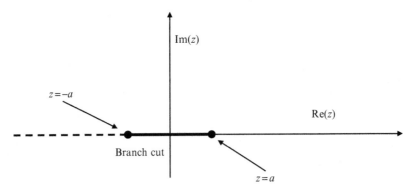

**Fig. B.2** The branch cut for $(z^2 - a^2)^{1/2}$ that forms a slit in the complex plane.

We can extend this concept and define branch cuts at different angles for each point, and allow the branch points to be at some arbitrary location $z_o$ so

$$\left(z^2 - z_o^2\right)^{1/2} = \left[(z - z_o)e^{-i(\vartheta_1 + \pi)}\right]^{1/2}\left[(z + z_o)e^{-i(\vartheta_2 + \pi)}\right]^{1/2} e^{i(\vartheta_1 + \vartheta_2)/2 + i\pi} = Re^{i\phi}$$

where the phase lies in the range

$$\frac{\vartheta_1 + \vartheta_2}{2} < \phi < \frac{\vartheta_1 + \vartheta_2}{2} + 2\pi$$

An example is shown in Fig. B.3 in which the branch cuts have angles $\vartheta_1 = 0$ and $\vartheta_2 = -\pi$. We can also specify the phase $\phi$ using the angles to the branch cut defined in Fig. B.1B. These give

$$\phi = (\varphi_1 + \varphi_2)/2 + (\vartheta_1 + \vartheta_2)/2$$

For example, in Fig. B.3 we see that on the imaginary axis at $z = i\infty$ the value of $(\varphi_1 + \varphi_2)$ is $2\pi$, but in the vicinity of the origin it is reduced to a minimum but is never less than $\pi$. At $z = -i\infty$ the value has increased again to $2\pi$. On the imaginary axis it follows that the phase $\phi$ must lie in the range

$$0 < \phi < \pi/2$$

and so this choice of branch cut ensures that the function $\left(z^2 - z_o^2\right)^{1/2}$ has a positive real and imaginary part on the imaginary axis.

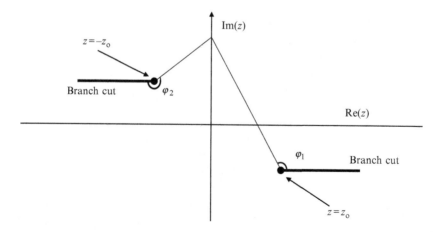

**Fig. B.3** The branch cuts for $\left(z^2 - z_o^2\right)^{1/2}$ that ensure it is real valued on the imaginary axis.

# Appendix C: The cascade blade response function

This appendix describes the solution to the problem defined in Section 18.3. The solution for the jump in potential across a blade that is part of a rectilinear cascade, as shown in Fig. 18.5, in response to a harmonic gust defined as $w_o\exp(-i\omega t + ik_1 x + ik_2 y + ik_3 z)$ is given by

$$\Delta\hat{\phi}(x) = \int_{-\infty}^{\infty} D(\alpha, k_1, \sigma)e^{-i\alpha x}d\alpha$$

where it is shown in Chapter 18, Ref. [12] that

$$D(\alpha, k_1, \sigma) = \frac{-iw_o}{(2\pi)^2(\alpha+k_1)J_+(\alpha)J_-(-k_1)} - \left\{\sum_n \frac{(A_n+C_n)e^{i(\alpha-\delta_n)c}}{i(\omega+\alpha U)(\alpha-\delta_n)}\left[\frac{J_-(\delta_n)}{J_-(\alpha)}\right]\right\}$$
$$- \left\{\sum_m \frac{B_m}{(\alpha-\varepsilon_m)}\left[\frac{J_+(\varepsilon_m)}{J_+(\alpha)}\right]\right\}$$

where

$$A_o = \frac{w_o(\omega-k_1 U)}{(2\pi)^2 j(-k_1)} \qquad \delta_o = -k_1$$

$$A_n = \frac{w_o(\omega+\delta_n U)}{(2\pi)^2(\delta_n+k_1)J'_+(\delta_n)J_-(-k_1)} \qquad \delta_n = \kappa M + \theta_{n-1} \quad n>0$$

with

$$J'_+(\delta_n) = \frac{j'(\delta_n)}{J_-(\delta_n)}$$

$$j'(\delta_n) = \frac{(\kappa M-\delta_n)h\beta^2}{4\pi(1-\cos(\delta_n d+\sigma)\cos((n-1)\pi))} \begin{cases} 2 & n=1 \\ 1 & n>1 \end{cases}$$

and the coefficients $B_m$ are obtained as the solutions to the equations

$$\{B_m\} = [[\mathbf{1}] - [\mathbf{F}_{mn}][\mathbf{L}_{mn}]]^{-1}\{[\mathbf{F}_{mn}]\{A_n\} + A_o G_m\}$$

where

$$F_{mn} = -\frac{e^{i(\varepsilon_m - \delta_n)c}}{i(\omega + \varepsilon_m U)(\varepsilon_m - \delta_n)} \left[\frac{J_-(\delta_n)}{J'_-(\varepsilon_m)}\right]$$

$$G_m = -\frac{e^{i(\varepsilon_m - \delta_o)c}}{i(\omega + \varepsilon_m U)(\varepsilon_m - \delta_o)} \left[\frac{J_-(\delta_o)}{J'_-(\varepsilon_m)}\right]$$

$$L_{mn} = \frac{i(\omega + \delta_m U)}{(\varepsilon_n - \delta_m)} \left[\frac{J_+(\varepsilon_n)}{J'_+(\delta_m)}\right]$$

The coefficients $C_n$ are obtained from

$$C_o = 0$$

$$C_n = \sum_m \frac{i(\omega + \delta_n U)}{(\varepsilon_m - \delta_n)} \left[\frac{J_+(\varepsilon_m)}{J'_+(\delta_n)}\right] B_m \quad n > 0$$

The split functions are defined as

$$J_+(\alpha) = \frac{\kappa_e \beta \sin(\kappa_e h \beta)}{4\pi(\cos(\kappa_e h \beta) - \cos(\rho))} \frac{\prod_{m=0}^{\infty}(1 - (\alpha - k_o M)/\theta_m)}{\prod_{m=-\infty}^{\infty}(1 - (\alpha - k_o M)/\eta_m^-)} e^{\Phi}$$

$$J_-(\alpha) = \frac{\prod_{m=0}^{\infty}(1 - (\alpha - k_o M)/\vartheta_m)}{\prod_{m=-\infty}^{\infty}(1 - (\alpha - k_o M)/\eta_m^+)} e^{-\Phi}$$

The function $\Phi$ must be chosen so that both $J_+$ and $J_-$ have algebraic growth as $\alpha$ tends to infinity and is given by

$$\Phi = \frac{-i(\alpha - k_o M)}{\pi} \{h\beta \log(2\cos\chi_e) + \chi_e d\}$$

The singularities of these functions are given by

$$\theta_m = -\sqrt{\kappa_e^2 - \left(\frac{m\pi}{\beta h}\right)^2}$$

$$\vartheta_m = \sqrt{\kappa_e^2 - \left(\frac{m\pi}{\beta h}\right)^2}$$

$$\eta_m^\pm = -f_m \sin \chi_e \pm \cos \chi_e \sqrt{\kappa_e^2 - f_m^2}$$

$$f_m = \frac{\sigma + k_o M d - 2\pi m}{\sqrt{d^2 + (h\beta)^2}}$$

where $\tan \chi_e = d/h\beta$. Finally, we have used the variables

$$M = U_e/c_o$$

$$\beta^2 = 1 - M^2$$

$$k_o = \omega/c_o \beta^2$$

$$\kappa_e^2 = k_o^2 - (k_3/\beta)^2$$

$$\rho = \sigma + k_o M d$$

so

$$j(\alpha) = \frac{\zeta}{4\pi} \left\{ \frac{\sin(\zeta h)}{\cos(\zeta h) - \cos((\alpha - k_o M)d + \rho)} \right\}$$

$$\zeta = \beta \sqrt{\kappa_e^2 - (\alpha - k_o M)^2}$$

# Index

Note: Page numbers followed by *f* indicate figures, and *t* indicate tables.

Printed in the United States
By Bookmasters